Biological Investigations

Seventh Edition

Form, Function, Diversity, and Process

Warren D. Dolphin

Iowa State University

 **Higher Education**

Boston   Burr Ridge, IL   Dubuque, IA   Madison, WI   New York   San Francisco   St. Louis
Bangkok   Bogotá   Caracas   Kuala Lumpur   Lisbon   London   Madrid   Mexico City
Milan   Montreal   New Delhi   Santiago   Seoul   Singapore   Sydney   Taipei   Toronto

Higher Education

BIOLOGICAL INVESTIGATIONS: FORM, FUNCTION, DIVERSITY, AND PROCESS
SEVENTH EDITION

Some ancillaries, including electronic and print components, may not be available to customers outside the United States.

This book is printed on recycled, acid-free paper containing 10% postconsumer waste.

1 2 3 4 5 6 7 8 9 0 QPD/QPD 0 9 8 7 6 5 4

ISBN 0–07–255285–9

Editorial director: *Kent A. Peterson*
Senior sponsoring editor: *Patrick E. Reidy*
Developmental editor: *Margaret B. Horn*
Marketing manager: *Tami Petsche*
Senior project manager: *Jayne Klein*
Senior production supervisor: *Sherry L. Kane*
Designer: *Rick D. Noel*
Cover designer: *Lindsey Huber*
Cover image: *©Minden Pictures, Flip Nicklin, 087614*
Lead photo research coordinator: *Carrie K. Burger*
Photo research: *David Tietz*
Compositor: *Shepherd, Inc.*
Typeface: *10/12 Times Roman*
Printer: *Quebecor World Dubuque, IA*

The credits section for this book begins on page 480 and is considered an extension of the copyright page.

www.mhhe.com

BRIEF CONTENTS

iii

CONTENTS

PREFACE

This lab manual is dedicated to the many students and colleagues who have been my patient teachers over the years. I hope that it returns some of what has been taught so that a new generation of biologists may soon add to our wonder of nature's ways while advancing our understanding of life's diverse forms and processes.

As reflected in the subtitle, this lab manual addresses fundamental biological principles based on intertwining themes in evolution: form reflects function; unity despite diversity; and the adaptive processes of life. The manual was written for use in a two-semester introductory biology course serving life science majors. I have emphasized investigatory, quantitative, and comparative approaches to studying the life sciences, integrating physical science principles and statistics where appropriate. In choosing topics for inclusion, I sought to achieve a balance between experimental, observational, and comparative activities. Throughout the manual, the concept of hypothesis testing as the basic method of inquiry has been emphasized. Starting with Lab Topic 1 on the scientific method, and reiterated in topics throughout the manual, students are asked to make hypotheses and to test them through their lab work. At the end of a lab topic students are asked to reach a conclusion regarding accepting or rejecting their hypotheses and their reasons for doing so. The activities included in each lab topic have been tested in multi-section lab courses and are known to work well in the hands of students.

Nature of the Revisions

There is not a section of the lab manual that was not revised in some way. Some changes were small while others were substantial. All were made to add clarity, give more emphasis, or cut redundancy. A significant number of illustrations have been upgraded, and many new ones have been added to support the new lab topics. The result is a lab manual that works. Major changes are discussed below.

Two new lab topics have been added to this manual. Lab Topic 6 is a new type of activity for general biology lab manuals: an exercise on molecular structures using computer visualization techniques. We have used this for a few years in our labs and find that it promotes a familiarity with molecular structures that cannot be achieved with static textbook figures alone. Lab Topic 24 is a new lab on plant cells and tissues. Part of this lab was in-

cluded in the topic on roots in the last edition, but its significance was lost. Here the importance of the development of plant tissues is given the prominence that it deserves in understanding plant anatomy. The root section was combined with the stem topic in this lab manual to keep the number of topics on plants the same.

The fruit fly genetics lab (Lab Topic 11) was recast into a focused format. Students are still asked to determine the genotype of an unknown fly, but the number of traits has been reduced from five to two, making the lab easy to prepare while illustrating the same principles of dominance and recessiveness, segregation, independent assortment, and autosomal and sex linkage. A new interchapter on sterile technique was also added to support the microbiological techniques used in Lab Topics 12 through 14, which investigate transformation, mutation, and bacterial diversity. Major rewrites of the protist lab (Lab Topic 15) and animal diversity labs (Lab Topics 20 through 23) now introduce the cladistic approach to phylogenetic relationships found in major textbooks. This was a particularly important change in the protostomes where two new clades, Lophotrochozoa and Ecdysozoa, have been described. A new phylum, the Bryozoans, was added to the animal diversity labs to illustrate one of the morphological characteristics of the new clades, the lophophore. The approach in the protist lab is now more focused on diversity rather than individual specimens as examples. In the diversity labs, new summary tables have been added to encourage phylogenetic comparisons.

At the request of reviewers, the use of an oil immersion objective was added to the microscopy lab (Lab Topic 2). Interpreting of electron micrographs was added to Lab Topic 3. The plant diversity section (Lab Topics 16 and 17) now starts with a look at the multicellular algae *Chara,* a stonewort, and the evolutionary appearance of the gametangium. In the fungal lab (Lab Topic 18), a discussion of *Allomyces* has replaced the discussion of *Chytridium* as representative of the Chytridiomycota.

A new design element has been introduced at the request of reviewers. Numbered arrow icons in the margins make it easier to direct students to a section. This should aid instructors who do parts of a lab topic, but not the entire topic. For example, an instructor can now say we will be doing parts 1 through 7 but not the others.

Organization of Lab Topics

The lab topics have a standard format. All start with lists of equipment, organisms and other materials, and solutions to be used during the lab. This is followed by the *Student Prelab Preparation* section that indicates the essential vocabulary and understandings needed to benefit fully from the lab, and the *Objectives* section, which gives the goals of the lab. A brief *Background* introduction summarizes the essential concepts and relationships on which the lab is based. These introductions are not meant to replace a textbook. The *Lab Instructions* are detailed and allow students to proceed at their own pace through either experimental or observational lab work. Dangers are noted and explained. Data tables help students organize their lab observations. Questions, indicated by the question mark icon, are interspersed to avoid a cookbook approach to science, and spaces are provided for answers and sketches. New terms are in boldface the first time used and are followed by a definition. Appendices on *Significant Numbers, Graphing, Statistics,* and *Report Writing* can be used to support each exercise. At the end of each lab topic, several alternative suggestions are given for summarizing the lab work. A *Learning Biology by Writing* section usually describes a writing assignment or lab report. *Lab Summary Questions* organize the reporting of lab activities in a more stepwise approach. *Critical Thinking Questions* emphasize applications. An *Internet Sources* section points the students toward information sources on the www.

Laboratory Preparation Guide

The Laboratory Preparation Guide is onlined at *http://www.mhhe.com/dolphin*. This preparator's manual gives recipes of chemical solutions and sources of supplies for each of the exercises. Linda Westgate, who prepares labs at Iowa State University, revised the prep manual so that it supports the changes made in this edition. The site is password protected, but users of the manual can contact a McGraw-Hill sales representative for access. I maintain a site for my students at *http://www.biology.iastate.edu/Courses/Courses1.html* (201L or 202L). It has many photographs of the materials used in the lab.

Acknowledgments

I would especially like to thank Linda Westgate, my lab coordinator, who brought numerous inconsistencies to my attention and often suggested corrections. Also, I would like to recognize James Colbert, Associate professor of Botany at Iowa State University, for his patience in explaining plant biology. In addition, I wish to thank the critical reviewers who made constructive suggestions throughout the writing of this manual: William Barstow, University of Georgia; Daryl Sweeney, University of Illinois; Gerald Gates, University of Redlands; Marvin Druger, Syracuse University; Thomas Mertens, Ball State University; Cynthia M. Handler, University of Delaware; San Eisen, Christian Brothers College; Paul Biebel, Dickinson College; Stephen G. Saupe, St. Johns University (Minnesota); Sidney S. Herman, Lehigh University; Margaret Krawiec, Lehigh University; Charles Lycan, Tarrant County Junior College; Olukemi Adewusi, Ferris State University; Karel Rogers, Adams State College; Peter A. Lauzetta, Kingsborough Community College (CUNY); Maria Begonia, Jackson State Univesity; Thomas Clark Bowman, Citadel Military College; Gary A. Smith, Tarrant County Junior College; Timothy A Stabler, Indiana University Northwest; William J. Zimmerman, University of Michigan-Dearborn; and Nancy Segsworth, Capilano College (British Columbia).

The following reviewers of the sixth edition provided excellent comments and suggestions:

Linda L. Allen, *Lon Morris College*
Gordon Atkins, *Andrews University*
Brenda Blackwelder, *Central Piedmont Community College*
Natalie Bronstein, *Mercy College*
Christian Chauret, *Indiana University—Kokomo*
Mary Anne Clark, *Texas Wesleyan University*
Naomi D'Alessio, *Nova Southeastern University*
Renata Dusenbury, *St. Augustina's College*
Professor Becky Green-Marroquin, *Los Angeles Valley College*
Ida Greidanus, *Passaic County Community College*
Dana Brown Haine, *Central Piedmont Community College*
Peter King, *Francis Marion University*
Raymond Lewis, *Wheaton College*
Lewis Lutton, *Mercyhurst College*
Lee Anne Martinez, *University of Southern Colorado*
Susan Peckham Petro, *Ramapo College of New Jersey*
Mary Schmall, *McDaniel College*
Mindy Skarda, *Southwestern Community College*
Gary Smith, *Tarrant County College N.E.*
Stacy Smith, *Lexington Community College*
Conrad Toepfer, *Millikin University*
Linda R. Van Thiel, *Wayne State University*
Miryam Wahrman, *William Paterson University of New Jersey*
Linda M. Westgate, *Iowa State University*
Jan Whitson, *Concordia University*
Lynne Zeman, *Kirkwood Community College*
Lise Wilson, *Siena College*

My sincere thanks go to Margaret Horn, developmental editor at The McGraw-Hill Companies, who patiently organized reviews and kept me on track during the revision preparation and Jayne Klein, senior project manager, who was an accommodating master at coordinating all of the elements that go into making a book. Special thanks go to my friend and illustrator, Dean Biechler, who operates Chichaqua Bend Studios, and to students of the Biological/Pre-Medical Illustration Program at Iowa State University. They prepared many of the illustrations for this and earlier editions of the lab manual. By working directly with them, I have clarified many of my understandings of biology and have truly developed an appreciation of how form reflects function. Last, but certainly not least, I thank my family—Judy, Jenny, Garth, Shannon, and Lara as well as a new generation Aidan, Brendan, and Hunter—for their support and inspiration throughout the preparation of this and earlier editions.

If you have questions or comments, please contact me by E-mail (wdolphin@iastate.edu.).

Lab Topic	Audesirk et al. Biology, 6/e	Campbell & Reece Biology, 6/e	Freeman Biological Science	Lewis et al. Life, 5/e	Mader Biology, 8/e	Purves, Sadava, Orianes & Heller Life, 7/e	Raven & Johnson Biology, 6/e	Starr & Taggert Biology: Unity and Diversity, 10/e
1. Science: A Way of Gathering Knowledge	1	1	1	1	1	1	1	1
2. Techniques in Microscopy	5	7	5	3	4	4	5	4
3. Cellular Structure Reflects Function	5	7	5	3	4	4	5	4
4. Determining How Materials Enter Cells	4	8	4	4	5	5	6	5
5. Using Quantitative Techniques and Statistics	NA	NA	NA	NA	NA	NA	NA	NA
6. Modeling Biological Molecules	2, 3	3, 4, 5	2, 3	2	3	2, 3	2, 3	2, 3
7. Determining the Properties of an Enzyme	6	6	6	5	6	6	8	6, 7
8. Measuring Cellular Respiration	8	9	6	7	8	7	9	8
9. Determining Chromosome Number in Mitotic Cells	11	12	8	8	9	9	11	9
10. Observing Meiosis and Determining Crossover Frequency	11	13	9	9	10	9	13	10
11. Determining Genotypes of Fruit Files	12	14, 15	10	10, 11	11, 12	10	13	11
12. Isolating DNA and Working with Plasmids	9, 13	16, 20	12, 14, 17	12, 13	13, 16	11, 16	14, 19	13, 16
13. Testing Assumptions in Microevolution and Inducing Mutations	15	23	22	15	18	23	20, 21	17
14. Working with Diverse Bacteria	19	27	25	20	21	27	34	21
15. Diversity Among Protists	19	28	27	21	22	28	35	22
16. Investigating Plant Phylogeny: Seedless Plants	21	29	28	22	24	29	37	23
17. Investigating Plant Phylogeny: Seed Plants	21	30	28	22	24	30	37	23
18. Observing Fungal Diversity and Symbiotic Relationships	20	31	29	23	23	31	36	24
19. Investigating Early Events in Animal Development	36	32, 47	18, 19	40	44	20	60	43
20. Animal Phylogeny: Investigating Animal Body Plans	22	33	30	24	29	32	44	25

Lab Topic	Audesirk et al. Biology, 6/e	Campbell & Reece Biology, 6/e	Freeman Biological Science	Lewis et al. Life, 5/e	Mader Biology, 8/e	Purves, Sadava, Orianes & Heller Life, 7/e	Raven & Johnson Biology, 6/e	Starr & Taggert Biology: Unity and Diversity, 10/e
21. Protostomes I: Lophotrochozoans and Development of Complexity	22	33	30	24	30	32	45	25
22. Protostomes II: Ecdysozoa and Great Diversity	22	33	30	24	30	32	46	25
23. Deuterostomes and the Origins of the Vertebrates	22	33, 34	30	25	31	34	47, 48	26
24. Investigating Plant Cells, Tissues, and Primary Growth	23	35	31	26	25	35	38, 39	29
25. Investigating Primary and Secondary Growth in Roots and Stems	23	36	32	27	26	36	39	30
26. Investigating Leaf Structure and Photosynthesis	7	10	7	6	7	8	10	7
27. Angiosperm Reproduction, Germination, and Development	24	38	36	28	28	39	40, 42, 43	31, 32
28. Investigating Digestive and Gas Exchange Systems	28, 29	41, 42	40, 41	36, 37	36, 37	48, 50	51, 53	40, 41
29. Investigating Circulatory Systems	27	42	41	35	34	49	52	38
30. Investigating the Urogenital System	30, 35	44, 46	39, 45	38	38, 43	43, 51	58, 59	42, 43
31. Investigating the Properties of Muscle and Skeletal Systems	34	49	43	34	41	47	50	37
32. Investigating Nervous and Sensory Systems	33	48, 49	42, 43	31, 32	39, 40	44, 45, 46	54, 55	34, 35
33. Statistically Analyzing Simple Behaviors	37	51	47	41	45	52	27	46
34. Estimating Population Size and Growth	38	52	48	42	46	54	24	45

LAB TOPIC 1

Science: A Way of Gathering Knowledge

Supplies

Preparator's guide available at
  http://www.mhhe.com/dolphin

Materials

Meter sticks
Photo copies of newspaper, magazine, and journal
  articles about biology (AIDS, rainforests, or cloning
  would be good examples, especially if articles
  were coordinated so students see same material
  intended for different audiences.)

Student Prelab Preparation

Before doing this lab, you should read the introduction
and sections of the lab topic that have been scheduled
by the instructor.
  You should find definitions for the following terms:

Dependent variable
Hypothesis
Independent variable
Scientific literature

  You should be able to describe in your own words
the following concepts:

Critical reading
Experimental design
Reaction time
Scientific method

  As a result of this review, you most likely have
questions about terms, concepts, or how you will do the
experiments. Write these questions in the space below
or in the margins of the pages of this lab topic. The lab
experiments should help you answer these questions, or
you can ask your instructor during the lab.

Objectives

1. To understand the central role of hypothesis testing
   in the modern scientific method
2. To design and conduct an experiment using the
   scientific method
3. To summarize sample data as charts and graphs
4. To learn to draw conclusions from data
5. To evaluate writing samples for science content
   and style

Background

Many dictionaries define science as a body of knowledge
dealing with facts or truths concerning nature. The emphasis is on facts, and there is an implication that absolute truth
is involved. Ask scientists whether this is a reasonable definition and few will agree. To them, science is a process. It
involves gathering information in a certain way to increase
humankind's understanding of the facts, relationships, and
laws of nature. At the same time, they would add that this
understanding is always considered tentative and subject to
revision in light of new discoveries.

  Science is based on three fundamental principles:

The *principle of unification* indicates that any explanation
  of complex observations should invoke a simplicity of
  causes such that the simplest explanation with the least
  modifying statements is considered the best; also known
  as the law of parsimony.

The second principle is that *causality is universal;* when
  experimental conditions are replicated, identical results
  will be obtained regardless of when or where the work is
  repeated. This principle allows science to be self-
  analytical and self-correcting, but it requires a standard of
  measurement and calibration to make results comparable.

The third principle is that of the *uniformity of nature;* it
  states that the future will resemble the past so that what
  we learned yesterday applies tomorrow.

  For many, science is just a refined way of using common sense in finding answers to questions. During our
everyday lives, we try to determine cause and effect relationships and presume that what happened in the past has a
high probability of happening in the future. We look for relationships in what we experience and observe. We ask our-
selves questions about these daily experiences and often
propose tentative explanations that we seek to confirm
through additional observations. We interpret new information in light of old and always make decisions about

whether our hunches are right or wrong. In this way, we build experience from the past and apply it to the future. The process of science is similar.

The origin of today's scientific method can be found in the logical methods of Aristotle. He advocated that three principles should be applied to any study of nature:

1. One should carefully collect observations about the natural phenomenon.

2. These observations should be studied to determine the similarities and differences; *i.e.,* a compare and contrast approach should be used to summarize the observations.

3. A summarizing principle should be developed.

While scientists do not always follow the strict order of steps to be outlined, the modern scientific method starts, as did Aristotle, with careful observations of nature or with a reading of the works of others who have reported their observations of nature. A scientist then asks questions based on this preliminary information-gathering phase. The questions may deal with how something is similar to or different from something else or how two or more observations relate to each other. The quality of the questions relates to the quality of the preliminary observations because it is difficult to ask good questions without first having an understanding of the subject.

After spending some time in considering the questions, a scientist will state a research **hypothesis,** a tentative answer to a key question. This process consists of studying events until one feels safe in predicting future events. In forming a hypothesis, the assumptions are stated and a tentative explanation proposed that links possible cause and effect. A key aspect of a hypothesis, and indeed of the modern scientific method, is that the hypothesis must be falsifiable; *i.e.,* if a critical experiment were performed and yielded certain information, the hypothesis would be declared false and would be discarded, because it was not useful. If a hypothesis cannot be proven false by additional experiments, it is considered to be tentatively true and useful, but it is not considered absolute truth. Possibly another experiment could prove it false, even though scientists cannot think of one at the moment. Thus, recognize that science does not deal with absolute truths but with a sequence of probabilistic explanations that when added together give a tentative understanding of nature. Science advances as a result of the rejection of false ideas expressed as hypotheses and tested through experiments. Hypotheses that over the years are not falsified and which are useful in predicting natural phenomena are called theories or principles—for example, the principles of Mendelian genetics.

In designing experiments to test a hypothesis, predictions are made. Experiments or reviews of previously conducted experiments provide the data and are therefore the means for testing hypotheses. If the hypothesis is accurate, predictions based on it should be true. In converting a research hypothesis into a prediction, a deductive reasoning approach is employed using if-then statements: if the hypothesis is true, then this will happen when an experimental variable is changed. The experiment is then conducted and as certain variables are changed, the response is observed. If the response corresponds to the prediction, the hypothesis is supported and accepted; if not, the hypothesis is falsified and rejected.

The design of experiments to test hypotheses requires considerable thought! The variables must be identified, appropriate measures developed, and extraneous influences must be controlled. The **independent variable** is that which will be varied during the experiment; it is the cause. The **dependent variable** is the effect; it should change as a result of varying the independent variable. **Control variables** are also identified and are kept constant throughout the experiment. Their influence on the dependent variable is not known, but it is reasoned that if kept constant they cannot cause changes in the dependent variable and confuse the interpretation of the experiment.

Once the variables are defined, decisions must be made regarding how to measure the effect of the variables. Measures may be quantitative (numerical) or qualitative (categorical) and imply the use of a standard. The metric system has been adopted as the international standard for quantitative science. If the independent variables are to be varied, a decision must be made concerning the scale or level of the treatments. For example, if something is to be warmed, what will be the range of temperatures used? Most (but not all) biological material stops functioning (dies) at temperatures above $40°$ C and it would not be productive to test at temperatures every $10°$ C throughout the range $0°$ to $100°$ C. Another aspect of experimental design is the idea of replication: how many times should the experiment be repeated in order to have confidence in the results and to develop an appreciation of the variability in the response.

Once collected, experimental data are reviewed and summarized to answer the question: does the data falsify or support the hypothesis? The research conclusions then state the decision regarding the acceptability of the hypothesis and discuss the implications of the decision.

If the hypothesis is in a popular area of research, others may independently devise experiments to test the same hypothesis. A hypothesis that cannot be falsified, despite repeated attempts, will gradually be accepted by others as a description that is probably true and worthy of being considered as suitable background material when making new hypotheses. If, on the other hand, the data do not conform to the prediction based on the hypothesis, it is rejected.

Modern science is a collaborative activity with people working together in a number of ways. When a scientist reviews the work of others in journals or when scientists work in lab teams, they help one another with interpretation of data and in the design of experiments. When a hypothesis has been tested in a lab and the results are judged to be significant, she or he then prepares to share this information with others. This is done by making a presentation at a scientific meeting or writing an article for a journal. In both forms of communication, the author shares the preliminary observations that led to the forming of the hypothesis, the data from the experiments that tested the hypothesis, and the conclusions based on the data. Thus, the information becomes public and is carefully scrutinized by peers who may find a flaw in the logic or who may accept it as a valuable contribution to the field. Thus, the scientific discussion fostered by presentation and publication creates an evaluation function that makes science self-correcting. Only robust hypotheses survive this careful scrutiny and become the common knowledge of science.

LAB INSTRUCTIONS

You will create a research hypothesis, design an experiment to test it, conduct the experiment, summarize the data, and come to a conclusion about the acceptability of the hypothesis. You will also practice evaluating scientific information from various published sources.

Using the Scientific Method

Description of the Problem

Working in groups of three, you are to develop a scientific hypothesis and test it. The topic will be neuromuscular reaction time. This can be easily determined by measuring how quickly a person can grasp a falling meter stick and will be recorded as mm free fall. The person whose reaction time is being measured sits at a table with her or his forearm on the top and the hand extended over the edge, palm to the side and the thumb and forefinger partially extended. A second person holds a meter stick just above the extended fingers and drops it. The subject tries to catch it. The distance the meter stick drops is a measure of reaction time.

Your assignment is to create a scientifically answerable question regarding reaction time in individuals with different characteristics and to express this as a testable hypothesis. You will then design an experiment to test the hypotheses, collect the data, analyze, and come to a decision to reject or accept your hypothesis. For example, you might investigate the differences between those who play musical instruments and those who do not or try a more complex design that investigates gender differences in reaction time for students who are in some type of athletic training versus those who are not. You could test whether there are differences between the right and left hands and whether this correlates with handedness. The design will depend on the hypotheses that you decide to test as a group in your lab section. Continuing the example, you might propose a hypothesis that there will be no significant differences in reaction time between musicians and nonmusicians. An alternative hypothesis would be that there is a significant difference in the reaction times between the two types.

Summarizing Observations

1 ▷ Start your discussion of this assignment by summarizing the collective knowledge of your group about neuromuscular response time. Are these responses the same for all people or might they vary by athletic history, gender, body size, age, hobbies requiring manual dexterity, left versus right hand, or other factors? Be sure to consider these factors in both a qualitative and quantitative light. You might expect differences in the physiological responses of those who exercise. What other factors might influence the response time? As your group discussion proceeds, make notes below that summarize the group's knowledge and observations about what characteristics influence reaction time.

Asking Questions

2 ▷ Research starts by asking questions which are then refined into hypotheses. Review the group observations that you listed and write down scientifically answerable questions that your group has about reaction time in people with different characteristics. Be prepared to present your group's best questions to the class and to record the best questions from the class on a piece of paper.

Forming Hypotheses

3▷ With your group, review the questions posed in the class discussion. Examine the questions for their answerability. Do some lack focus? Are they too broad? Are others too simple, with obvious answers? By what criteria would you judge a good question?

Describe how your hypothesis is testable.

4▷ As a group take what you think is the best question and state it as a prediction. For example, because piano players constantly train their neuromuscular units you might expect that they would have short reaction times. Use this prediction as a basis for forming a testable hypothesis. Continuing with the example, you might propose for a hypothesis that there would be no significant difference between piano players and nonmusicians in reaction time. The alternate hypothesis would be that there is a significant difference. Remember that hypotheses must be testable through experimentation or further data gathering. State your hypotheses below.

Designing an Experiment

To test the hypothesis, a controlled experiment must be devised. It should be designed to collect evidence that would prove the hypothesis false. Discuss what the experiment should be. Your discussion should address the variables in the experiment.

5▷ Which of the variables is (are) the independent variable(s)?

6▷ Which of the variables is (are) the dependent variable(s)?

7▷ What variables will be controlled and how will they be controlled?

TABLE 1.1   Sample Data Table: Reaction times measured as millimeters free fall.

subject	gender	musician (yes, no)	right/left handed	athlete (yes, no)	reaction times (mm free fall)			
					trial 1	trial 2	trial 3	average
John	M	yes	right	no				
Maria	F	no	right	no				
Jianzhao	M	yes	right	yes				
etc.								

8 ▶ Having decided which variables fit into these categories, you must now decide on a level of treatment and how it will be administered. How will you standardize measurements across groups in the lab so that the results are comparable?

9 ▶ Recognizing that the subject may anticipate the dropping of the meter stick or be momentarily distracted when it is dropped, how many times will you repeat the experiment to have confidence in your results?

Procedure

10 ▶ After answering these questions as a group, write a set of standard operating procedures about how the experiment should be performed. All groups will follow these procedures. Your group should then perform the experiment. One person should be the subject, chosen according to the procedures. The others should each take different jobs. One can be the person who drops the stick, and another can record the data after each try. Each person in the group can rotate through these positions so data are gathered about everyone in the lab.

Data Recording

11 ▶ On a separate piece of paper, design a data recording table similar to that illustrated in table 1.1 but column headings should be customized to the factors being investigated. If anonymity is desired, assign each person in the lab a number rather than using names. There should be as many rows as there are people in the lab. After all measurements are completed, data should be shared among groups.

Data Summarization

12 ▶ Different groups should now share their data. If a computer is available in the lab, the data may be entered in a spreadsheet set up similar to table 1.1. The file may be printed or shared electronically with all class members. This data can be analyzed in a number of different ways. What is the average reaction time in the lab section? _____ What is the average reaction time for females? _____ Males? _____ For right hand? _____ Left hand? _____ Musicians? _____ Nonmusicians? _____ Other factors investigated?

Data Interpretation

13 ▶ Write a few sentences that summarize the trends that you see in the data and the differences between groups.

Conclusion

14▶ Return to the hypotheses that you made at the beginning of the experiment. Compare them to the experimental results. Must you accept or reject the hypothesis? Why? Cite the data used in making the decision. If you determine that there is a difference in reaction time between categories of people, how can you decide if it is a significant difference?

Discussion

15▶ Discuss with your partners how the experiments added to the class knowledge base which was outlined before the experiment began. Do you see any significance to the knowledge gained? Explain.

16▶ As you conducted this experiment and analyzed the results, additional questions probably came into your mind. As a result of this thinking and the results of this experiment, what do you think would be a significant hypothesis to test if another experiment were to be done?

17▶ Evaluate the design of your experiment. Be as critical as you can. Were any variables not controlled that should have been? Is there any source of error that you now see but did not before?

Scientific Method Assignment

Your instructor may ask you to write up this experiment as a scientific report and to hand it in at the next lab meeting. Refer to appendix D for instructions on how to write such a report.

Evaluating Published Information

(adapted from notes prepared by Chuck Kugler at Radford University and Chris Minor when at Iowa State University)

Daily, we are exposed to scientific information in newspapers, magazines, over the World Wide Web, and through scholarly reports in journals and books. How do you evaluate such information? Is a newspaper best because it is available daily or is the WWW better because no editors have changed words to fit a story in the column space? In classes throughout your undergraduate years, in your future jobs, and in everyday life, you will be asked to evaluate what you read and make decisions about the quality of information.

In this section you will learn how to evaluate a written report. Your instructor will pass out photocopies of a newspaper, magazine, and journal article reporting on the same scientific discovery. Read the articles quickly so that you have a rough idea of what is in them. When you are finished with the articles, read the following material in the lab manual. Refer back to the photocopies as you read and try to find examples of the writing styles mentioned in the lab manual.

Evaluate Format

First, be suspicious of any scientific report that is not written in a style that parallels the scientific method where the hypotheses are clearly identified, data are presented, and the reasoning leading to the conclusions is explained. The formal elements of a scientific paper are discussed in appendix D. If a report lacks these elements, it is not a scientific report. On the other hand, reading about a discovery in the newspaper can alert you to locate the actual report in a journal that the reporter read before writing the story.

Evaluate the Source

Several thousand journals publish information of interest to biologists. The journals range from magazines such as *National Geographic* and *Scientific American* to scholarly journals, published by professional associations, such as the *American Journal of Botany, Journal of Cell Biology, Genetics, Ecology, Science, etc.* Magazine articles are usually written by science journalists and not by scientists who did the research. They can be quite helpful in developing a general appreciation for a topic, but they are not ultimate sources of scientific information. Scholarly journals are considered the most reliable sources and even these will vary in the quality of the work that is published.

What makes these journals so reliable is the use of a peer review system. Articles are written by scientists and sent to the journal editor, who is usually a scientist. When he/she receives the article, it is sent to three other scientists who are working on similar problems and they are asked to make comments about the work. Often these reviews can be harsh and may criticize writing style and content. The reviewers' comments are returned to the author who then revises the paper before it is published. It is this peer review system that maintains the quality of the information appearing in journals. Popular magazines such as *Time,* or television shows (even those on the Discovery channel), or movies have been created for entertainment purposes and are not sources of evidence.

Evaluate Writing Style

Good scientific writing is factual and concise. It is not overly argumentative, nor should it be an appeal to the emotions. As you read any scientific report, watch for the following:

1. Repetition: some authors believe that the more they say something, the more likely you are to believe it;

2. Dichotomous simplification: expressing a complex situation as if there were only two alternatives;

3. Exaggeration: often identifiable by the use of the words "all" or "never";

4. Emotionally charged words or forceful statements: the author is attempting to get you to agree based on "feelings," not reason.

Evaluate the Arguments

Analyze how the author seeks to convince you that what is reported is true, significant, and applicable to science. Be on guard for the following types of rhetorical arguments:

1. Appeals to authority: citing a well-known person or organization to make a point, *e.g.,* "the American Dental Association recommends. . . ." Recognize that

authorities can be biased, be experts in fields other than the one under consideration, and be wrong.

2. Appeals to the democratic process: using the phrase "most people" believe, use, or do. Remember, only 200 years ago, most people erroneously believed in the spontaneous generation of life.

3. Use of personal incredulity: implying that you could not possibly believe something, *e.g.,* "how could something as complex as the human just evolve, didn't it need a designer?"

4. Use of irrelevant arguments: statements that might be true but which are not relevant, *e.g.,* "suggesting that complex animals could have resulted by chance is like saying that a clock could result from putting gears in a box and shaking it."

5. Using straw arguments: presenting information incorrectly and then criticizing the information because it is wrong, *e.g.,* "the evolution of a wing requires 20 simultaneous mutations—an impossibility." There is no basis for saying that the evolution of a wing requires 20 mutations; it could be fewer but most likely many more.

6. Arguing by analogy: using an analogy to suggest that an idea is correct or incorrect, *e.g.,* "intricate watches are made by careful designers, so complex organisms must have had a designer."

Evaluate the Evidence

Before getting too involved in interpreting trends in the data, spend a few moments thinking about the type of evidence that is presented. Was the evidence collected using the scientific method and is a hypothesis being tested? Be especially skeptical of reports that have the following flaws in their evidence:

1. Distinguish between evidence and speculation: evidence includes data, whereas speculation is simply a statement based on an educated guess.

2. Use of anecdotal evidence: stories usually involving single events that are not the results of designed experiments, *e.g.,* "bee stings are lethal; my uncle died when he was stung."

3. Correlation used to imply cause and effect: correlation is a probability of two events occurring together. While it is interesting to speculate that one might cause the other, this is not necessarily so; *e.g.,* at the instant before a major earthquake has struck a major city, there is a high probability that someone was slamming a car door. Did the slam cause the earthquake?

4. Sample size and selection: in statistical studies, a large number of situations should be examined and the procedures used to select the situations should be free of bias. You do not choose to report only the experiments that support your beliefs.

A 60-pound bushel of soybeans contains
about 48 pounds of meal and 11 pounds of oil.

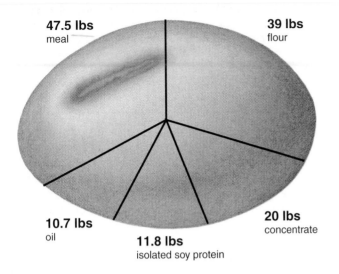

47.5 lbs
meal

39 lbs
flour

10.7 lbs
oil

11.8 lbs
isolated soy protein

20 lbs
concentrate

What does this chart tell you?

Do the numbers add up?

Would you use the information in this chart to make a decision?

Do you trust this data?

SOURCE: Iowa Soybean Review, 1995/1996 Soya Bluebook

5. Misrepresentation of source: source material can be quoted out of context or badly paraphrased; *e.g.*, "Moderate drinking of alcohol may benefit the consumer" could be misrepresented as "Drinking is good for you."

Check the Data

When data are presented, get in the habit of doing routine checks. If percentages are involved, do you know the sample size? It is an impressive statement to say that 75% of the people surveyed preferred brand X, but it is less impressive to find out that this calculation is based on a sample size of four rather than 400 or more. When percentages are reported, be sure to check that they add up to 100. If on the eve of the election 42% of the voters are for Gore and 41% are for Bush, it would seem that Gore has won, except that 17% of the voters are unaccounted for and could swing the election one way or the other.

Continue the habit of doing simple arithmetic checks when examining data in tables. If totals are given for columns of numbers, do some quick math to see if things check out. If they do not, you might not want to base major decisions on the report. Besides you do not know what other kinds of errors went undetected!

With the advent of computer graphics, it is now rather easy to use computer programs to produce interesting looking and appealing graphics. However, one should not accept data based on its beauty of presentation. To illustrate this point, see figures 1.1 and 1.2 for some interesting graphics that appeared in newspapers or trade publications.

Evaluate the Conclusions

In a scientific paper, the conclusions should come near the end of the article. Conclusions are not a summary of the data. Conclusions deal with the decision that is to be made about the hypothesis that was being tested. You should ask, "Are the data thoroughly reviewed to test the hypothesis?" Ask yourself whether there is another explanation to what the author is telling you. Once a decision is made to accept or reject the hypothesis, the implications of the decision are discussed. In some cases, the implications are then extrapolated to new situations, but overextrapolation can result in problems. For example, raising a frog's body temperature from 10° to 20° C may increase the frog's metabolic rate twofold, but this does not mean by extrapolation that raising it to 100° C will increase metabolic rate tenfold. In fact, the frog will die when its body temperature approaches 40° C.

As you look back through the newspaper, magazine, and journal articles, determine which one of these forms of publication used more of the nonscientific forms of writing and indicate what you found below.

Vital statistics

These numbers may help you decide where you stand on the issues.

Married mothers in the workplace

Percentage of married working women with children under the age of six.

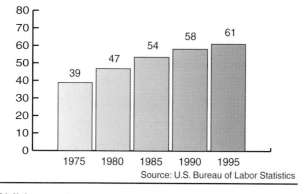

Source: U.S. Bureau of Labor Statistics

Child poverty

Percentage of children living in poverty.

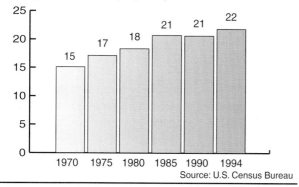

Source: U.S. Census Bureau

Juvenile violent crime

Violent crime arrests per 100,000 juveniles.

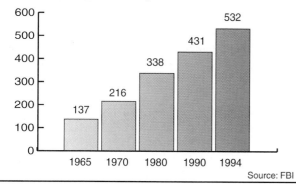

Source: FBI

Most non-custodial fathers don't pay child support

Percentage of non-custodial fathers who paid child support 1989

63% - Paid nothing
26% - Paid full amount
12% - Paid partial amount

Source: U.S. Census Bureau

Figure 1.2 This graphic appeared in a major newspaper. Focus on the trends in the data. Has the number of working mothers increased significantly in the past five years? Do significantly more children live in poverty in 1994 compared to 1985? What is the sample size? Were the same populations of people compared? What is the message of this collection of graphics when considered as a whole? Why is this not an acceptable scientific report? What is the difference between correlation and cause and effect when considering two or more trends?

Evaluating Scientific Literature Assignment

Go to the library and choose a science-related article from a periodical of your choice, such as a newspaper, popular magazine, or a science journal. Your instructor or a librarian can suggest some journals to skim through to locate an article that is of interest to you. Photocopy the article because you will be writing on it. Once photocopied, at the top of the first page write the name of the journal from which it was copied, the volume, and the number (or month) of the issue. Your assignment is to analyze the article using the information given on the Journal Analysis Form. You will be marking all over this article as you analyze it. Next week, you should turn in the marked article along with the following form.

Journal Analysis Form

Evaluate the Source

Where does (do) the author(s) work?

Could the author(s) have vested interests?

What type of source is this?

Are articles peer reviewed in this source?

What Is the Hypothesis?

State the hypothesis tested in the work reported. If none, so indicate.

Examine the Writing Style

Use a light-colored marker to highlight on the photocopy any passages that seem to deviate from a factual and concise style. According to the following key, write a number next to the highlighted area indicating the type of writing style used.

1. Use of repetition
2. Dichotomous simplification
3. Exaggeration
4. Use of emotionally charged words or forceful statements

Examine the Arguments

Use a light-colored marker to highlight on the photocopy any arguments used in the article. According to the following key, write a number next to the highlighted area indicating the type of argument that is used.

1. Appeals to authority
2. Appeals to the democratic process
3. Uses personal incredulity
4. Uses irrelevant arguments
5. Uses straw arguments
6. Argues by analogy

Analyze the Evidence

Underline the sections of the photocopied article that present the arguments of the author. According to the following key, write a letter next to the arguments.

A. Speculation
B. Evidence collected using the scientific method
C. Anecdotal evidence
D. Correlation, not cause and effect
E. Description of sample size and selection method
F. Possible place to check for misrepresentation of source

Check the Data

Do all percentages given add up to 100? If not, circle in the text where the omission is located.

Do all numbers in columns or charts add up to the indicated totals or are there math mistakes?

Circle the mistakes.

Are flashy graphics used to catch your attention?

Write comments next to the questionable graphics.

Examine the Conclusions

Are the conclusions easy to find and clearly stated?

Are the conclusions based on a review of the data and a test of the hypotheses presented in the introduction?

Are the conclusions supported by evidence collected using the scientific method?

Has the author extrapolated beyond the range of data collected?

Attach this evaluation form to your photocopied article and turn it in for grading.

LAB TOPIC 2

Techniques in Microscopy

Supplies

Preparator's guide available at
http://www.mhhe.com/dolphin

Equipment

Compound microscope
Dissecting microscope

Materials

Ocular micrometer
Stage micrometer
Small colored letters from printed page
Slides and coverslips
Dropper bottles with water
Dissecting needles and scissors
Prepared slide
Mammalian blood smear, Wright's stained
Electron micrographs from old textbooks or the
WWW mounted on poster board

Student Prelab Preparation

Before doing this lab, you should read the introduction
and sections of the lab topic that have been scheduled
by the instructor.

You should find definitions for the following terms:

Brightness
Calibration
Contrast
Magnification
Resolution

You should be able to describe in your own words
the following concepts:

Light path through parts of a microscope
How to make wet-mount slide
How to calculate an ocular micrometer

As a result of this review, you most likely have
questions about terms, concepts, or how you will do
this lab. Write these questions in the space below or in
the margins of the pages of this lab topic. The lab
experiments should help you answer these questions,
or you can ask your instructor during the lab.

Objectives

1. To learn the parts of a microscope and their functions
2. To investigate the optical properties of a microscope, including image orientation, plane of focus, and measuring objects
3. To understand the importance of magnification, resolution, and contrast in microscopy

Background

Since an unaided eye cannot detect anything smaller than 0.1mm (10^{-4} meters) in diameter, cells, tissues, and many small organisms are beyond our visual capability. A light microscope extends our vision a thousand times, so that objects as small as 0.2 micrometers (2×10^{-7} meters) in diameter can be seen. The electron microscope further extends our viewing capability down to 1 nanometer (10^{-9} meters). At this level, it is possible to see the outlines of individual protein or nucleic acid molecules. Needless to say, microscopy has greatly improved our understanding of the normal and pathological functions of organisms.

Although 300 years have passed since its invention, the standard light microscope of today is based on the same principles of optics as microscopes of the past. However, manufacturing technology has developed to a point that quality instruments for classroom use are now mass produced. Your microscope is as good as those used by Schleiden, Schwann, and Virchow, the biologists who founded cell theory in the mid-nineteenth century, and is far superior to the one used by Robert Hooke, the first person to use the word "cell" in describing biological materials.

Microscope quality depends upon the capacity to resolve, not magnify, objects. **Resolution** is the capacity of an optical instrument to distinguish two points that are close together. High resolution allows us to see separate points (detail) while low resolution blurs close-together points into one image rather than showing them as separate ones. The distinction between microscopic resolution and magnification can best be illustrated by an analogy. If a photograph of a newspaper is taken from across a room, the newspaper image would be small, too small for us to read the words. If the photograph were enlarged, or magnified, the image would be larger, but the print would still be unreadable. Regardless of the magnification used, the photograph would never make a fine enough distinction between the points on the printed page. Therefore, without **resolving power,** or the ability to distinguish detail, magnification is worthless.

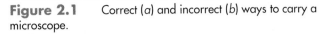

Figure 2.1 Correct (*a*) and incorrect (*b*) ways to carry a microscope.

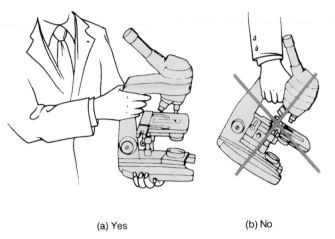

(a) Yes   (b) No

Modern microscopes increase both magnification and resolution by matching the properties of the light source and precision lens components. Today's light microscopes are limited to practical magnifications in the range of 1000× to 2000× and to resolving powers of 0.2 micrometers. Most student microscopes have magnification powers to 450×, or possibly to 980×, and resolving properties of about 0.5 micrometers. These limits are imposed by the expense of higher power objectives and the accurate alignment of the lens elements and light sources.

The theoretical limit for the resolving power of a microscope depends on the **wavelength** of light (the color) and a value called the **numerical aperture** of the lens system, times a constant (0.61). The numerical aperture is derived from a mathematical expression that relates the light delivered to the specimen by the condenser to the light entering the objective lens. If all other factors are equal, resolving power is increased by reducing the wavelength of light used. Microscopes are often equipped with blue filters because blue light has the shortest wavelength in the visible spectrum. Therefore,

minimum distance that can be resolved

$$= \frac{\text{wavelength}}{\text{numeric aperature}} \times 0.61$$

For example, if green light with a wavelength of 500 nanometers is used and the numerical aperture is 2, the theoretical resolving power is 153 nanometers, or 0.153 micrometers.

Even with sufficient magnification and resolution, a specimen can be seen on a microscope slide only if there is sufficient **contrast.** Contrast is based on the differential absorption of light by parts of the specimen. Often a specimen will have opaque parts or will contain natural pigments, such as chlorophyll, but how is it possible to view the majority of biological materials that consist of highly translucent structures?

Microscopists improve contrast by using stains that bind to cellular structures and absorb light to provide con-

trast. Some stains are specific for certain chemicals and bind only to structures composed of those chemicals. Others are nonspecific and stain all structures.

To summarize, good microscopy involves three factors: resolution, magnification, and contrast. A beginning biologist must learn to manipulate a microscope with these factors in mind to gain access to the world that exists beyond what can be seen with the unaided eye.

LAB INSTRUCTIONS

AVOIDING HAZARDS IN MICROSCOPY

Use care in handling your microscope. The following list contains common problems, their causes, and how they can be avoided.

1. Microscope dropped or ocular falls out
   a. Carry microscope in upright position using both hands, as shown in figure 2.1.
   b. When placing the microscope on a table or in a cabinet, hold it close to the body; do not swing it at arm's length or set it down roughly.
   c. Position electric cords so that the microscope cannot be pulled off the table.

2. Image blurred
   a. High-power objective was pushed through the coverslip (see number 3) and lens is scratched.
   b. Slide was removed when high-powered objective was in place, scratching lens. Remove slide only when low-power objective is in place.
   c. Use of paper towels, facial tissue, or handkerchiefs to clean objectives or oculars scratched the glass and ruined the lens. Use only *lens tissue* folded over at least twice to prevent skin oils from getting on the lens. Use distilled water to remove stubborn dirt.
   d. Clean microscope lenses before and after use. Oils from eyelashes adhere to oculars, and wet-mount slides often encrust the objectives or substage condenser lens with salts.

3. Objective lens smashes coverslip and slide
   a. Always examine a slide first with the low- or medium-power objective.
   b. Never use the high-power objective to view thick specimens.
   c. Never focus downward with the coarse adjustment when using high-power objective.

4. Mechanical failure of focus mechanism
   a. Never force an adjustment knob; this may strip gears.
   b. Never try to take a microscope apart; you need a repair manual and proper tools.

Figure 2.2 A binocular compound microscope.

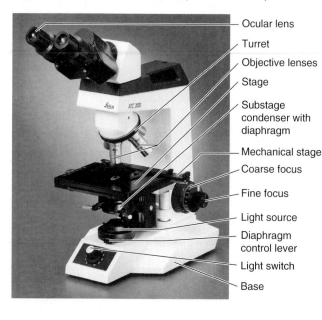

- Ocular lens
- Turret
- Objective lenses
- Stage
- Substage condenser with diaphragm
- Mechanical stage
- Coarse focus
- Fine focus
- Light source
- Diaphragm control lever
- Light switch
- Base

The Compound Microscope

1▶ Get your microscope from its storage place, using the precautions just mentioned. Depending on its age, manufacturer, and cost, your compound microscope may have only some of the features discussed in this section. Look over your microscope and find the parts described, referring to figure 2.2.

Parts of a Microscope

Ocular Lens

The ocular lens is the lens you look through. If your microscope has one ocular, it is a **monocular** microscope. If it has two, it is **binocular.** In binocular microscopes, one ocular is adjustable to compensate for the differences between your eyes. Ocular lenses can be made with different magnifications. What magnification is stamped on your ocular lens housing?

The ocular lens is actually a system of several lenses that may include a pointer and a measuring scale called an ocular micrometer. Never attempt to take an ocular lens apart.

On a binocular microscope, note that the distance between the oculars, called the interpupillary distance, can be adjusted by either pulling the oculars outward to the sides or squeezing them together.

Body Tube

The body tube is the hollow housing through which light travels to the ocular. If the microscope has inclined oculars, as in figure 2.2, the body tube contains a prism to bend the light rays around the corner.

Objective Lenses

The **objective lenses** are a set of three to four lenses mounted on a rotating **turret** at the bottom of the body tube. Rotate the turret and note the click as each objective comes into position. The objective gathers light from the specimen and projects it into the body tube. Magnification ability is stamped on each lens. What are the magnification abilities of your objectives?

Scanning (small) Lens   _____

Low-power (medium) Lens   _____

High-power (large) Lens   _____

Oil Immersion (largest) Lens   _____ (optional)

Stage

The horizontal surface on which the slide is placed is called the **stage.** It may be equipped with simple clips for holding the slide in place or with a **mechanical stage,** a geared device for precisely moving the slide. Two knobs, either on top of or under the stage, move the mechanical stage. If your microscope has these knobs, twist them to see how the stage moves.

Substage Condenser Lens

The substage **condenser** lens system, located immediately under the stage, focuses light on the specimen. An older microscope may have a mirror instead.

Diaphragm Control

The **diaphragm** is an adjustable light barrier built into the condenser. It may be either an **annular** or an **iris** type. With an annular control, a plate under the stage is rotated, placing open circles of different diameters in the light path to regulate the amount of light that passes to the specimen. With the iris control, a lever projecting from one side of the condenser opens and closes the diaphragm. Which type of diaphragm does your microscope have?

Use the smallest opening that does not interfere with the field of view. The condenser and diaphragm assembly may be adjusted vertically with a knob projecting to one side. Proper adjustment often yields a greatly improved view of the specimen.

Light Source

The light source has an off/on switch and may have adjustable lamp intensities and color filters. To prolong lamp life, use medium to low voltages whenever possible. A second diaphragm may be found in the light source. If present, experiment with it to get the best image.

Figure 2.3 Procedure for making a wet-mount slide. (*a*) Place a drop of water on a clean slide. (*b*) Place specimen in water. (*c*) Place edge of coverslip against the water drop and lower coverslip onto slide.

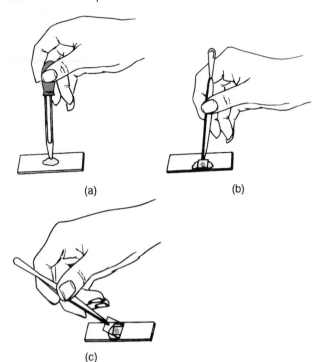

(a) (b)

(c)

Base and Body Arm
The base and body arm are the heavy cast metal parts.

Coarse Focus Adjustment
Depending on the type of microscope, the **coarse adjustment** device either raises and lowers the body tube or the stage to focus the optics on the specimen. Use the coarse adjustment only with the scanning (4×) and low-power (10×) objectives. Never use it with the high-power (40×) objective. (The reasons for this will be explained later.)

The Focus Adjustment
The **fine adjustment** changes the specimen-to-objective distance very slightly with each turn of the knob and is used for all focusing of the 40× objective. It has no noticeable effect on the focus of the scanning objective (4×), and little effect when using the 10× objective.

Making Slides and Using a Microscope

2 ▶ Figure 2.3 shows how to prepare a wet mount microscope slide. Take a magazine or printed page and cut out a colored lowercase letter *e* or *a* or the number *3, 4,* or *5*. Clean a microscope slide with a tissue, add a drop of water to the center, and place the letter in the drop. Add a coverslip and place the slide in its normal orientation on the microscope stage with the scanning objective in place. To view the slide, follow the steps listed in the box.

1. Check that the ocular and all objective lenses as well as the slide are clean.

2. Turn the illuminator on and open the diaphragm. Center specimen over stage opening.

3. Starting with the scanning objective as close to the slide as possible while looking through the oculars, back off with the coarse adjustment knob until the specimen is in focus.

4. Readjust the light intensity and center the specimen in the field of view by moving the slide. Close down the iris diaphragm and, if possible, adjust the substage condenser height until the edges of the diaphragm are in focus.

5. Switch from scanning to the low-power (10×) objective. The turret should click when the objective is in place. Sharpen the focus, adjust the specimen centering, and readjust the condenser height and diaphragm opening.

6. Switch to the high-power (40×) objective. Adjust the focus with the *fine focus adjustment only*. If you use the coarse adjustment, you may hit the slide and damage the high-power objective.

7. If you have a binocular microscope, adjust the ocular lenses for the differences between your eyes. Adjust the interpupillary distance by pulling the oculars apart or squeezing them together. Then determine which ocular is adjustable. Close the eye over that lens and bring the specimen into sharp focus for the open eye. Open the other eye, and close the first. If the specimen still is not in sharp focus, turn the adjustable ocular until the specimen is in focus. You need not repeat this procedure when you look at other specimens, but should do it each time you get the microscope from the cabinet because other students may also be using your microscope and adjusting it for their eyes.

8. Students wonder if they should remove their glasses when using a microscope. If you are nearsighted or farsighted, you need not. The focus adjustments will compensate. If you have an astigmatism, wear your glasses.

9. If your microscope is monocular, you will tend to close one eye. Eyestrain will develop if this is continued. Learn to keep both eyes open and ignore what you see with the other eye. This will be hard at first. Remove all light-colored papers from your field of view or try covering your eye with your hand.

The first seven steps listed are the usual procedures for using the microscope. Always start with a clean scanning objective and proceed in sequence to high power, making minor adjustments to the focus and light source. Using a microscope is similar to changing the channels on a television set and adjusting the picture at each new setting. Your skill in using and tuning your microscope will determine what you will see on microscope slides throughout this course.

The following activities are designed to familiarize you with your microscope. Use the wet-mount slide you just made to carry out these activities.

The Compound Microscope Image

A compound microscope image has several properties, including image orientation, magnification, field of view, brightness, focal plane, and contrast.

Image Orientation

3 With the scanning objective in place, observe the letter on the slide through the microscope and then with the naked eye. Is there a difference in the orientation of the images? While looking through the microscope, try to move the slide so that the image moves to the left. Which way did you have to move the slide? Try to move the image down. Which way did you have to move the slide?

When showing someone an interesting specimen, you can describe the location of the specimen by referring to the field of view as a clock face. (Thus, the point of interest might be described as being at one o'clock or seven o'clock.) Some microscopes have pointers built into the ocular. In such cases, the structure of interest can simply be moved to the end of the pointer.

Magnification

Compound microscopes consist of two lens systems: the objective lens, which magnifies and projects a "virtual image" into the body tube, and the ocular lens, which magnifies that image further and projects the enlarged image into the eye.

Figure 2.4 Comparison of the relative diameters of fields of view, light intensities, and working distances at three different objective magnifications.

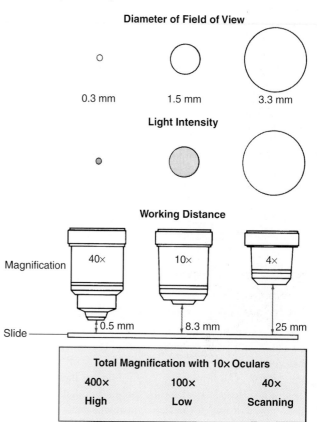

The ocular lens only increases the magnification of the image and does not enhance the resolution. The objective lens magnifies and resolves. The total magnification of a microscope is the product of the magnification of the objective and the ocular. If the objective lens has a magnification of 5× and the ocular 12×, then the image produced by these two lenses is 60 times larger than the specimen.

What magnifications are possible with your microscope?

Scanning power =   _____
Low power   =   _____
High power   =   _____
Oil immersion =   _____ (if present)

Field of View and Brightness

4 Observe your microscope slide with the scanning, low-, and high-power objectives. Note that as magnification increases, the diameter of the field of view decreases and the brightness of the field is reduced. Note also that the **working distance,** the distance between the slide and the objective, decreases as the magnification is increased. (This is the reason you never focus on thick specimens with a high-power objective.) These relationships are summarized in figure 2.4.

Techniques in Microscopy

Figure 2.5 (a) Sequentially focusing at depths (1), (2), and (3) yields (b) three different images that can be used to reconstruct the original three-dimensional structure.

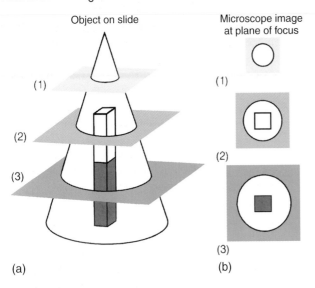

Object on slide

Microscope image at plane of focus

(1)

(2)

(3)

(a)

(1)

(2)

(3)

(b)

Focal Plane and Optical Sectioning

The concept of the **focal plane** is important in microscopy. Like the eye, a microscope lens has a limited depth of focus; therefore, only part of a thick specimen is in focus at any one setting. The higher the magnification, the thinner the focal plane. In practical terms, this means that you should make constant use of the fine adjustment knob when viewing a slide with the high-power objective. If you turn the knob a quarter turn back and forth as you view a specimen, you will get an idea of the specimen's three-dimensional form. It would be possible, for example, to reconstruct the three-dimensional structure in figure 2.5 from sections (1), (2), and (3).

Image Contrast

The contrast of the image can be changed by closing the diaphragm, although this usually results in poorer resolution. Light rays are deflected from the edges of the diaphragm and enter the slide at oblique angles. Scattered light makes materials appear darker because some rays of light take longer to reach the eye than others. This can be an advantage when looking at unstained specimens. Thus, the benefits of greater contrast sometimes outweigh the loss of resolving power. Contrast is also improved by reducing light intensity or brightness.

Measuring with a Microscope (Optional)

Measuring microscopic structures requires a standardized **ocular micrometer.** It is a small glass disc on which are etched uniformly spaced lines in arbitrary units. The disc is inserted into an ocular of the microscope, and the etched scale is superimposed on the image of the specimen when you look through a microscope. Does your microscope have an ocular micrometer? _____ The spacing between the lines on the disc must be calibrated with a very accurate standard ruler called a **stage micrometer.**

The ocular micrometer must be calibrated for each objective. Any object can then be measured by superimposing the ocular scale on it and measuring its size in ocular units. These units can then be multiplied by the calibration factor to obtain the actual size of the object.

To calibrate an ocular micrometer, obtain a stage micrometer from the supply area. Look at it with the scanning objective. What are the units? What is the smallest space equal to in these units? Follow the steps given in figure 2.6.

Determine how many spaces on the stage micrometer are equal to 100 spaces on the ocular micrometer at the following powers and record in the table below. Divide the number of stage units in millimeters by 100 to determine the calibration for one ocular unit when using the scanning objective. Record below. Repeat for each objective. *Be careful not to push the high-power objective through the stage micrometer. (They are expensive!)*

	Stage Units (mm)	Ocular Units	mm per Ocular Unit	Converted to µm
Scanning	____	____	____	____
Low	____	____	____	____
High	____	____	____	____

Now use your microscope to measure the sizes of some cells. Obtain a prepared slide of a mammalian blood smear. Look at it first with the scanning objective and then with the 10×. Switch to high power and measure the diameter of a red blood cell. What is the diameter expressed in micrometers?

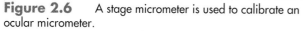

Figure 2.6   A stage micrometer is used to calibrate an ocular micrometer.

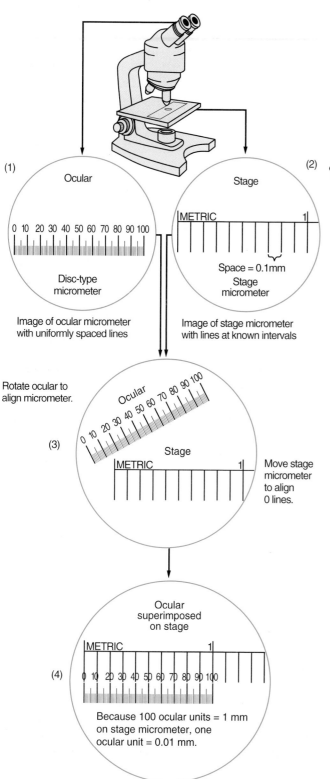

(1) Ocular

0 10 20 30 40 50 60 70 80 90 100

Disc-type micrometer

Image of ocular micrometer with uniformly spaced lines

(2) Stage

METRIC   1

Space = 0.1mm
Stage micrometer

Image of stage micrometer with lines at known intervals

Rotate ocular to align micrometer.

(3)

Ocular
0 10 20 30 40 50 60 70 80 90 100

Stage
METRIC   1

Move stage micrometer to align 0 lines.

(4)

Ocular superimposed on stage

METRIC   1

0 10 20 30 40 50 60 70 80 90 100

Because 100 ocular units = 1 mm on stage micrometer, one ocular unit = 0.01 mm.

Using an Oil Immersion Lens (optional)

C A U T I O N

Oil immersion lenses allow you to appproach 1000× magnifications with an increase in resolution. However, the distance from the lens surface to the slide is very small, and it is quite easy to push the lens through the slide, possibly breaking the slide and ruining the lens.

1. Focus first on the object on the slide by proceeding from the scanning to high-power objectives as you have done before. Now you are ready to try the oil immersion lens.

2. Do NOT touch the focus knobs or the stage knobs. Turn the turret to swing the high-power objective out of the way. Place a single drop of immersion oil on the slide right over where the light is coming through the stage, and rotate the **oil immersion lens** into place. The lens will actually contact the oil drop, making a column of oil from the slide surface to the lens surface. This column of oil prevents light scattering and improves resolution.

3. Now look through the oculars and open the substage diaphragm to increase the light. The object on the slide should still be in the field of vision but will probably be out of focus. Use the fine-adjustment knob to focus clearly. Never use the coarse-adjustment knob.

4. Once you have an oil immersion lens in place, do NOT swing the 40× objective back into place. Because the objective focuses close to the slide, the 40× objective will get oil on it; is difficult to clean the oil from the lens surface. The 10× can be used because that lens focuses far away from the slide so that oil contamination is not likely.

5. When you have finished using the oil immersion objective, you must clean the oil from its surface and from the slide using lens paper. Because oil immersion lenses require extra cleanup and the danger of breaking slides is great, they are not used for routine examination of slides in beginning laboratories.

Stereoscopic Dissecting Microscopes

The stereoscopic microscope (fig. 2.7), usually called a **dissecting microscope,** differs from the compound microscope in that it has two (rather than one) objective lenses for each magnification. This type of microscope always has two oculars. The stereoscopic microscope is essentially two microscopes in one. The great advantage of this instrument is that objects can be observed in three dimensions. Because the alignment of the two microscopes is critical, the resolution

Figure 2.7 Stereoscopic microscope.

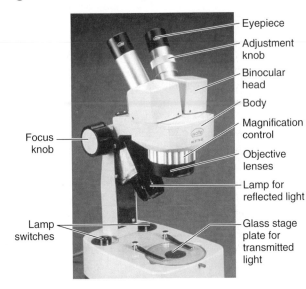

- Eyepiece
- Adjustment knob
- Binocular head
- Body
- Magnification control
- Objective lenses
- Lamp for reflected light
- Glass stage plate for transmitted light

Focus knob

Lamp switches

Figure 2.8 A transmission electron microscope.

and magnification capabilities of a stereoscopic microscope are less than in a compound microscope. Magnifications on this type of microscope usually range from 4× to 50×. The oculars can be adjusted for individual eye spacing and for focus, as in the compound binocular microscope.

Stereoscopic microscopes are often used for the microscopic dissection of specimens. The light source may come from above the specimen and be reflected back into the microscope, or it may come from underneath and be transmitted through the specimen into the objectives. The stage may be clear glass or an opaque plate, white on one side and black on the other. The choice of illumination source depends on the task to be performed and on whether the specimen is opaque or translucent.

10 Set up your dissecting microscope with reflected light. Place your hand on the stage and observe the nail on your index finger. Move your hand so the image travels to the right and down. How does image movement correspond to actual movement?

Change the illumination to transmitted light. Place the previously prepared slide of a printed letter on the stage and focus on it using the highest magnification. Determine which ocular is adjustable. Close the eye over the adjustable ocular and focus the microscope sharply on the edge of the letter. Now close the other eye and open the first. Is the edge still in sharp focus? If not, turn the adjustable ocular until it is. To avoid eyestrain this procedure should be followed whenever a stereoscopic microscope is used for long periods.

Your instructor may have a supply of flowers, seeds, or dead insects to examine with the stereoscopic microscope.

Electron Microscopes

A fundamental problem in microscopy is the limit to the resolving power of light microscopes. Nothing smaller or closer together than 0.2 μm can be seen very clearly through a light microscope. Most structures inside cells are smaller.

About 50 years ago, biologists realized that a microscope could be built to use electrons rather than light as an illumination source. The theoretical resolution limit for such an instrument would be 100,000× greater than that of the light microscope because electrons have a wavelength of 0.005 nm, a hundred thousand times smaller than the wavelength of visible light. The development of the electron microscope has virtually revolutionized science's understanding of the cell in the last 50 years.

Transmission Electron Microscope

A transmission electron microscope (TEM) is shown in figure 2.8. The central column is the microscope proper, and all the rest is electronic equipment, vacuum pumps, and plumbing.

The TEM is essentially a vertical television tube with the electron gun at the top and the fluorescent screen at the bottom. Electrons leave a hot filament at the top, are accelerated by high voltage, and pass down the tube to the screen. The tube must be in a vacuum so that the electrons can pass without interference.

As the electrons strike the screen, a glow appears. If a biological specimen is treated and placed in the column, it deflects electrons away from the screen, and an image of the electron-opaque and electron-transmitting sections of the specimen appear on the screen as shadows and light areas, respectively. When this image is recorded on photographic film, a picture like that in figure 2.9 is obtained. Clearly, this technique reveals a great deal of detail inside a cell.

However, a specimen must be carefully treated and prepared before it is viewed in the TEM. This treatment includes **fixing** the tissue to preserve structure. After the cells are fixed and the structure stabilized it is infiltrated and embedded in hard epoxy plastic. The plastic then may be sliced thinly, cutting the embedded cells, much as a butcher slices a loaf of salami. These sections are stained with heavy metal salts and finally are placed in the microscope for study. Tissue preparation typically takes about 24 hours.

Keep in mind that transmission electron micrographs are pictures of thinly sectioned material. If a structure does not appear in a particular picture, it does not mean that the struc-ture does not occur in that cell. When the section was cut, the structure simply may not have been included in that section, in much the same way that a peppercorn does not occur in every slice from a salami loaf. This is especially important to remember when you are trying to count structures in cells.

Interpreting TEMs

Figures 2.9 and 2.10 are transmission electron micrographs of animal and plant cells. Your instructor may have additional photos taken through a transmission electron microscope. Your task is to work with a partner to interpret these images. Eukaryotic cellular structure is based on membranes that form compartments within the cells. These compartments are called **organelles** and are areas where enzymes for special functions are found.

You should be able to identify the following organelles and structures in the photographs. Place a check in table 2.1 when you have identified them. Use your textbook to add a brief description of their functions.

Figure 2.9   Transmission electron microscope image of a pituitary gland cell.

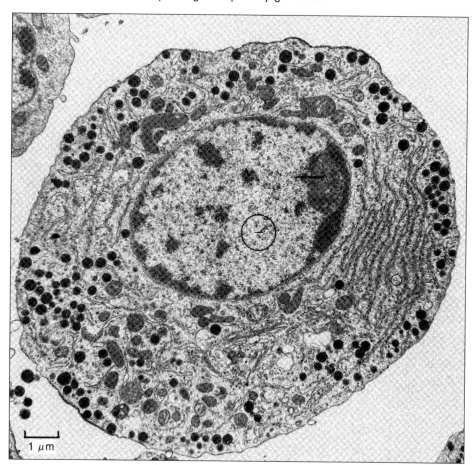

1 μm

Organelle	Animal Cell	Plant Cell	Function
Cell wall			
Plasma membrane			
Cytoplasm			
Nucleus			
Chromatin			
Nucleolus			
Endoplasmic reticulum			
Golgi apparatus			
Ribosomes			
Lysosomes			
Mitochondria			
Chloroplasts or plastids			

What structures are found in plant cells that are absent from animal cells?

Do you expect that bacterial cells would have a similar or a different structural basis? Why?

If plant and animal cells have some of the same structures, does this mean that the cells can perform many of the same functions?

Figure 2.10   Transmission electron micrograph of a plant parenchyma cell from a coleus leaf, magnification 10,000× as printed.

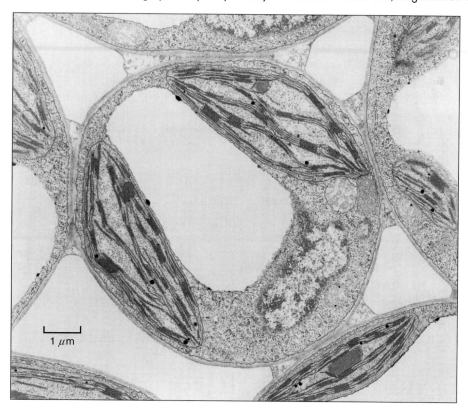

1 μm

The magnifications for the photographs in the lab manual are given in the figure legends. If you have a millimeter ruler, you can determine the actual size of the organelle in a cell, by using the following formula:

$$\text{Cell size} = \frac{\text{Measured size of cell}}{\text{Magnification of photo}}$$

What is the diameter of the animal cell nucleus? The plant cell nucleus?

Another way to determine size is to use the size bar at the lower right in each of the figures. The length of this line represents the size of 1 μm when it is magnified the same amount as the structures in the photo. To use the size bar, measure its length in millimeters and, using the following formula, calculate how may micrometers equal a millimeter at this magnification:

$$\mu\text{m/mm} = \frac{1\mu\text{m}}{\text{bar length in mm}}$$

What value did you obtain? _____

12▶   Now measure in mm some of the structures in the figures and multiply by this conversion factor to determine the actual size of each organelle. Enter the sizes in the work table next to the checks you placed in the columns. Be sure to include the units. What are the units?

Learning Biology by Writing

Write a short essay (about 150 words) describing how magnification, resolution, and contrast are important considerations in microscopy. Indicate how microscopists can increase contrast in viewing specimens.

As an alternative assignment, your instructor may ask you to complete some or all of the lab summary and critical thinking questions.

Lab Summary Questions

1. Define magnification and resolution. How do these properties of a microscope differ?
2. In the table below, use the words *least, intermediate, or greatest* to describe how the indicated image properties change as you change objectives on your compound microscope.
3. When you calibrated your microscope, what were the sizes of one ocular unit at:

40× _____
100× _____
400× _____

Critical Thinking Questions

1. When looking through the oculars of a binocular compound light microscope, you see two circles of light instead of one. How would you correct this problem? If you saw no light at all, just a dark field, what correction would you make? Now, you finally have an interesting structure in view using your 10× objective lens, but, when you switch to the 40× objective lens, the structure is not in the field of view. What happened? How would you correct this?
2. What type of microscope would you use to observe the tube feet of a sea star? What type of microscope would you use to determine the sex of a live fruit fly? If you wanted to look at the chromosomes of the fruit fly, what type of microscope would you use?

Image Properties	Objectives		
	Scanning	*Low*	*High*
Magnification			
Field of view			
Brightness of field			
Resolving power			

LAB TOPIC 3

Cellular Structure Reflects Function

Supplies

Preparator's guide available at
http://www.mhhe.com/dolphin

Equipment

Compound microscopes with ocular micrometers and
oil immersion objectives, if available
Optional: microtome and wax-embedded specimens
for sectioning demonstration

Materials

Living organisms
Mix cultures of *Anabaena* and *Gloeocapsa*
together
Culture of mixed protozoa and green algae
Fungal hyphae (any culture)
Onion(red)
Elodea
Yogurt or fresh sauerkraut from natural food store
Prepared slides of
Gram-stained mixed cocci, bacilli, and spirilla
Human skin
Frog skin
Columnar epithelium from intestine (demo)
Ciliated epithelium from trachea (demo)
Areolar connective tissue
Neurons from cow's spinal-cord smear
Pine secondary xylem macerate
Coverslips and slides
Razor blades and forceps

Solutions

Methyl cellulose or *Protoslo*
Neutral red stain
India ink

Define in lab report

Student Prelab Preparation

Before doing this lab, you should read the introduction
and sections of the lab topic that have been scheduled
by the instructor.
You should use your textbook to review the
definitions of the following terms:

Bacteria   Epithelium
Cyanobacteria   Eukaryotic
Epidermis

Plant vascular tissues   Prokaryotic
(xylem and phloem)   Protists

You should be able to describe in your own words
the following concepts:

Cell theory
Structure reflects function
Cell compared to tissue

As a result of this review, you most likely have
questions about terms, concepts, or how you will do
this lab. Write these questions in the space below or
in the margins of the pages of this lab topic. The lab
experiments should help you answer these questions,
or you can ask your instructor during the lab.

Objectives

1. To learn the differences between prokaryotic and
   eukaryotic cell types
2. To observe living cells
3. To introduce students to staining methods
4. To observe representative cell types
5. To identify an unknown tissue
6. To collect evidence that cellular structure reflects
   function

Background

In 1665, Robert Hooke first used the word **cell** to refer to
the basic units of life. One hundred and seventy-three years
later, after other scientists had observed cells in many different organisms, two German biologists, M. Schleiden and
T. Schwann, published what is called the **cell theory.** This
theory states that the cell is the basic unit of life and that all
living organisms are composed of one or more cells or the
products of cells. The cell theory is not the result of one
person's work but is based on the observations of many microscopists. Today the cell theory is accepted as fact. All
living organisms are constructed of cells and the products
of cells. Only viruses defy inclusion in this generalization.

If cells are the basic units of life, then the study of basic
life processes is the study of cells. Today cell biologists strive

to understand how cells function by using such as microscopes and biochemical analyses. This quest for knowledge is driven by a logical relationship; if normal organismal function is dependent on cell function, then disease and abnormal functioning can also be understood at the cellular level.

Biologists recognize two organizational plans for cells. **Prokaryotic cells** lack a nuclear envelope, chromosomal proteins, and membranous cytoplasmic organelles. Bacteria and blue-green algae (called Cyanobacteria) are prokaryotic cells. **Eukaryotic cells** have the structural features that prokaryotes lack. The cells found in protists, fungi, plants, and animals are eukaryotic.

Although these two types of cells are distinctly different, they share many characteristics. A plasma membrane always surrounds a cell and regulates the passage of materials into and out of the cell. Both types of cells have similar types of enzymes, depend on DNA as the hereditary material, and have ribosomes that function in protein synthesis. The eukaryotic types evolved after prokaryotic cells and are more complex.

Many organisms are only single cells and are said to be **unicellular.** Others are **colonial**, composed of many single cells that cluster together, but all of the cells are alike, having the same cellular structure and functions. **Multicellular** organisms are also composed of many cells, but the cells are specialized for different functions. Consequently, their shapes, contents, and sizes as well as their biochemistry may be quite different. For examples in humans, red blood cells tend to be small, disk-shaped cells filled with the oxygen-binding protein hemoglobin, while muscle cells are large, tubular cells filled with the contractile proteins, actin, and myosin. As a result, the blood cells pass easily through the blood vessels, carrying oxygen with them while the muscle cells are capable of contraction, but one cannot do the other's function. In this lab you will have a chance to see some cells whose form indicates that they have a particular function.

LAB INSTRUCTIONS

You will observe the differences between prokaryotic and eukaryotic cells, as well as variations within these groups. This will introduce you to the paradox that biologists constantly face: the unity and the diversity of living forms. Moreover, you should come to appreciate a maxim in biology: form reflects function.

Prokaryotic Cells

Prokaryotic cells are found in all members of the Kingdoms Archaebacteria and Eubacteria.

Bacteria

In 1884, the Danish bacteriologist Christian Gram developed a diagnostic staining technique, which is used to separate bacteria into two groups: Gram positive and Gram negative.

Dead Gram-positive bacteria retain crystal violet dye while being washed in alcohol, but Gram-negative bacteria do not.

Modern microscopists know this is due to chemical differences in the composition of the bacterial cell walls. Gram-negative bacteria have more lipid material in the cell wall. When washed with alcohol, the lipids are extracted and the crystal violet stain no longer binds to the cell. Gram-negative cells are colorless after the Gram staining, but then a second stain (safranin) makes them visible as pink cells. The identification of thousands of different types of bacteria is based on this diagnostic test in combination with other traits. This is an important test because Gram-positive bacteria can be killed by certain antibiotics that do not affect Gram-negative bacteria. This test tells doctors what to use in treating illnesses.

1▷ Obtain a prepared slide of mixed types of bacteria. Observe first with the 10× objective and then with the 40× objective or with an oil immersion objective, if available. (Directions for using an oil immersion objective are given in Lab Topic 2, page 17.) The slide should contain both Gram-positive and Gram-negative bacteria and three shapes of bacterial cells. **Cocci** are sphere-shaped bacteria, **bacilli** are rod-shaped bacteria, and **spirilla** are comma- or corkscrew-shaped bacteria (fig. 3.1).

Indicate whether both Gram-positive (violet) and Gram-negative (pink) forms are found for each shape. If your microscope has a calibrated ocular micrometer, measure the bacterial cell sizes and record below:

Sizes of Bacteria

Cocci _____
Bacilli _____
Spirilla _____

Because of their small size, it is impossible to see detail inside bacterial cells with the light microscope. Figure 3.2 is a transmission electron micrograph of a section of a bacterial cell. Note the **cell wall**, **cell membrane**, **cytoplasm**, **ribosomes**, and **nucleoid region** containing DNA. Notably absent is any evidence of a nuclear envelope, chromosomes, or any internal organelles.

Yogurt is made by adding the bacteria *Lactobacillus* sp. and *Streptococcus thermophilus* to milk and allowing the bacteria to anaerobically metabolize milk sugar. Lactic acid is produced and excreted by the bacteria. It curdles the milk, producing the semisolid yogurt. Sauerkraut is made by similar fermentation process, starting with cabbage leaves.

2▷ Take a very small amount of dilution made from yogurt or sauerkraut and mix it on a slide with a drop of water. Add a coverslip and observe the slide through your microscope. What are the shapes of the bacteria in the sample? What are their approximate sizes? Record your observations below.

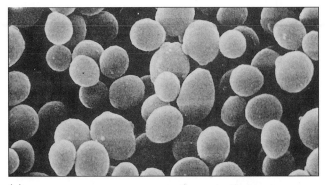

(a)

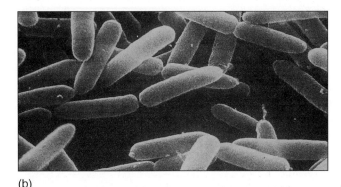

(b)

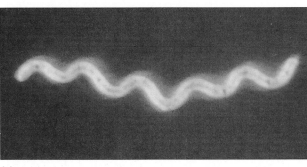

(c)

Cyanobacteria (Blue-Green Algae)

The common name "blue-green algae" characterizes the predominant feature of about half the organisms found in this group: they are blue because of the presence of a pigment called phycocyanin and green because they contain chlorophyll. However, some species may be brown or olive because of other pigments. All cyanobacteria are prokaryotes and most are surrounded by a gelatinous matrix. They live in soils, on moist surfaces, and in water.

3 ▷ Two cyanobacteria are available for study in this lab—*Anabaena* (see fig. 14.3) and *Gloeocapsa*. They have been mixed together in a single culture. Make a wet-mount slide by placing a small drop of the culture on a slide. Take a dissecting needle and dip it in India ink and touch the wet needle to the drop of blue-green algae culture. Some of the ink

Figure 3.2 Transmission electron micrograph of bacterium *Pseudomonas aeroginosa*. Magnification, ×67,200.

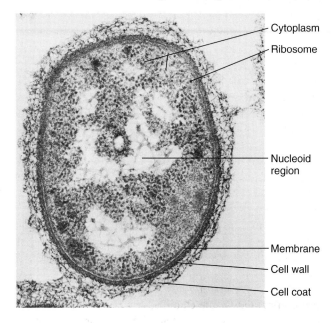

will transfer and improve the viewing. Press a coverslip down and blot away excess liquid.

Can you see structures inside the individual cells? Below make a sketch of each organism. Which species is surrounded by an extensive gelatinous matrix?

Anabaena cells are filaments that contain three cell types: small spherical **vegetative cells;** elongate spores called **akinetes;** and large spherical **heterocyst** cells, which function in nitrogen fixation. Label these cells in your drawing above.

Eukaryotic Cells

Eukaryotic cells are found in protists, fungi, plants, and animals.

Figure 3.3   Common types of protozoa.

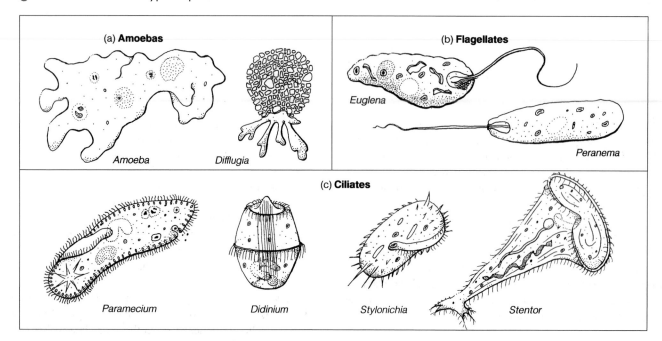

- **(a) Amoebas** — *Amoeba*, *Difflugia*
- **(b) Flagellates** — *Euglena*, *Peranema*
- **(c) Ciliates** — *Paramecium*, *Didinium*, *Stylonichia*, *Stentor*

Protists

Protozoa and algae are single-celled and colonial eukaryotic organisms that some scientists include in a single kingdom, Protista, while others separate these organisms into several kingdoms too numerous to discuss here. The term "protozoa" means first animals and at one time was used as a phylum name by zoologists. Now it is a term of convenience including organisms from several kingdoms, as does the term "algae." What cellular structure not found in bacteria should you be able to see in protozoa and algae?

4 ▷   The answer you just gave to this question is a hypothesis that you can test by observation. A culture of mixed protozoa and green algae is available in the lab. Make a wet-mount slide with some of the culture debris from the bottom of the container. Look at it first with the scanning objective and then with high power to identify the three protozoan types (fig. 3.3) and algae. To observe internal cellular structure, you may have to add methyl cellulose, a thickening agent, or neutral red stain.

As you find species that look like the protozoa in figure 3.3, place a check on the image. If you see anything that is not in figure 3.3, draw it in the circle in the next column for future reference. Also note how these organisms move. Do you see any differences? What allows them to move and how does their form reflect this function?

Green algae are also mixed in the culture with the protozoans. What do you suppose gives them their green color? Is this another example of form reflecting function?

? Do your observations support or contradict your hypothesis? What evidence do you now have that protists are eukaryotes?

Fungal Cells

The cells found in fungi are also of the eukaryotic cell type. Most fungi have stages in their life cycle. The fruiting body stage is a reproductive stage illustrated by the structures like the familiar mushrooms or brackets seen on dead wood. The so called vegetative stage, when the fungus is not sexually mature, is microscopic and infrequently seen.

5 ▷ In the lab there is a fungal culture. It consists of cells that are joined end-to-end to form what are called hyphae. Take a few hyphae and mount them in a drop of water on a slide and observe. You may add some neutral red stain by "wicking" to improve viewing. What evidence do you see that they are eukaryotic cells?

Hyphae have a tremendous surface area per volume of cellular material; *i.e.,* every cell is exposed to its environment. Can you think of a reason why this form might promote a particular function? (Hint: fungi secrete enzymes to digest materials outside of the cells. What must happen if the fungus is to get an energy from this food?)

Plant Cells

The cells of plants differ from those of animals in several characteristics. Plant cells are always surrounded by a **cell wall** composed of cellulose and other materials. The living part of a cell surrounded by the cell wall is called the **protoplast.** In the cytoplasm of some protoplasts, unique organelles called **chloroplasts** are found. They carry out the complex chemical reactions of photosynthesis. Protoplasts also usually have a large central **vacuole** filled with water and dissolved materials.

As in animals, the cells of plants are organized into **tissues,** cells that are similar in structure and function. In this part of the lab exercise, you will look at three types of plant cells.

Epidermal Cells

6 ▷ Epidermal cells are found on the surfaces of plants and function as a protective barrier. What shape would you hy-

pothesize best suits the function of epidermal cells? Do you think they would be plate-like, cuboidal, or tubular? Why?

Figure 3.4 demonstrates how to prepare a wet-mount slide of onion epidermis. After making your slide, observe it under a 10× objective and test the hypothesis that you made concerning the shape of epidermal cells. Once the specimen is in focus, adjust the light intensity and condenser so that the cells are clearly visible.

Note the individual cells outlined by thin cell walls composed of cellulose. The **plasma membrane** lies just inside the cell wall but cannot be seen because its thickness is less than the resolution of the light microscope. A large, fluid-filled vacuole is in the center of some cells. Another membrane, the **tonoplast,** surrounds the vacuole and regulates what passes in and out.

Switch to the high-power objective and locate the **nucleus** in the periphery of the cell. Sketch a typical cell and label it, including dimensions, in the space below. Measure the length and width of one cell. Focus up and down through a single cell. Do you think the cell is thin or thick? Do your observations support or disprove your hypothesis about cell shape?

? How does the three-dimensional shape of an epidermal cell relate to its function?

Figure 3.4 Method for preparing and staining a wet-mount slide of onion epidermis.

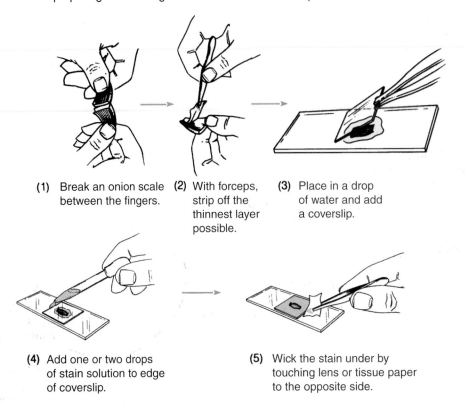

(1) Break an onion scale between the fingers.

(2) With forceps, strip off the thinnest layer possible.

(3) Place in a drop of water and add a coverslip.

(4) Add one or two drops of stain solution to edge of coverslip.

(5) Wick the stain under by touching lens or tissue paper to the opposite side.

Parenchyma Cells

7▷ *Elodea* sp. is an aquatic plant whose leaves are only two cells thick. Break off a leaf and mount it in a drop of water on a slide. Add a coverslip and observe.

◆ Besides the nucleus, what two other structures characteristic of plant cells can you see?

◆ What evidence have you observed that plants have eukaryotic cells?

Often when one is observing *Elodea* leaves, an interesting movement called **cyclosis** (also called cytoplasmic streaming) will be seen in the cytoplasm. Can you see it on your slide? Describe it below.

Figure 3.5 Two types of cells found in xylem.

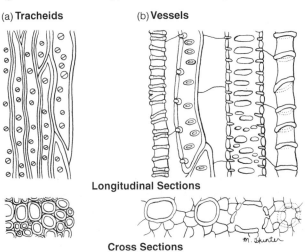

(a) **Tracheids**   (b) **Vessels**

Longitudinal Sections

Cross Sections

m. Hunter

Plant Vascular Tissues

Plants have vascular tissues that transport water, minerals, and other materials from the roots to the leaves, and also transport the products of photosynthesis from the leaves to other regions of the plant. Two basic tissues make up the plant vascular system: **xylem** transports water and minerals (fig. 3.5) and **phloem** transports photosynthetic products. These cells also provide structural support for a plant. Their elongate, tubular form with thick cell walls reflects their function. You will look at xylem.

8 ▷ *Xylem:* Obtain a slide of macerated pine wood from the supply area. The trunks of pine trees, except for a narrow ring just beneath the bark, are composed primarily of **tracheids.**

Examine your slide first under scanning and then in detail with the high-power objective. The tracheids are elongated cells with long, tapering end walls. The side walls of the cell are perforated by pits. Water passes to adjacent tracheids through the pits so that water moving from the roots to needles follows a zigzag pathway.

Note two important aspects of the tracheids: (1) the cell walls are thickened so that the tracheids not only transport but also structurally support the plant, and (2) no **protoplast** (living cell) is visible in the tracheid. The protoplast of a tracheid functions only to make the cell wall and then dies, leaving the tubelike wall to function in water transport and support.

Sketch a few tracheids below. Be sure to show the pit structure in the side walls. Use your ocular micrometer to measure a tracheid and add dimensions to your illustration.

In some plants, a second type of xylem cell is found. Called **vessels,** these cylindrical cells are not tapered at the ends and have cell walls that are reinforced by ringed and spiral thickenings.

◆ If form reflects function, do your observations of xylem cells support the idea that they conduct fluids and support the plant? Explain.

◆ A redwood tree is 100 m tall. Use your measurements of tracheid cell length to calculate how many tracheid cells must be stacked on end to form a conduit from the ground to the top of the tree.

Animal Cells

To prepare animal tissues (and many plant tissues) for microscopic observation, it is necessary to instantly kill and fix the cells with a chemical. The cells are then frozen or infiltrated with plastic or wax to make the tissue rigid. Thin sections then can be cut from this rigid block using a special cutting machine called a **microtome.** If a microtome is available in the lab, your instructor will demonstrate its use.

Tangential, longitudinal, or cross-sectional cuts may be made on embedded tissue. As figure 3.6 indicates, the same basic structure may look different, depending on the plane of the section. After they are cut, the sections are attached to slides and stained to illustrate various structures. (Staining reactions depend on the chemical composition of the structure to be illustrated.) Because the same tissue can be stained by different dyes, it is a good practice not to "memorize" tissues by color. For example, skin could be stained blue on one slide and red on another.

Permanent slides are made by mounting coverslips over the tissue sections using a resin, such as balsam, instead of water. This procedure takes several hours and is beyond the scope of a simple laboratory. Throughout the rest of the course, you will use slides that have been commercially prepared in this manner.

In most animals, cells are specialized to carry out particular functions. Four basic tissue types are found: epithelial, connective, nerve, and muscle. In this section of the exercise, you will look at examples of the first three on stained, prepared slides. At the end of this portion of the exercise, you should be able to look at an unknown tissue and identify it as one of these basic tissue types.

Epithelial Tissue

Tissues found on body surfaces, lining cavities, or forming glands are **epithelial** tissues (fig. 3.7). They are characterized by (1) having one surface not in contact with other cells, (2) having another surface in contact with a basement membrane, and (3) having no materials between adjacent cells.

Cellular Structure Reflects Function　　**29**

Figure 3.6 How plane of sectioning affects shape seen on slide. (a) Sections through a bent tube. (b) The effects of slicing through an egg in different directions. Clearly, preparation determines what is seen.

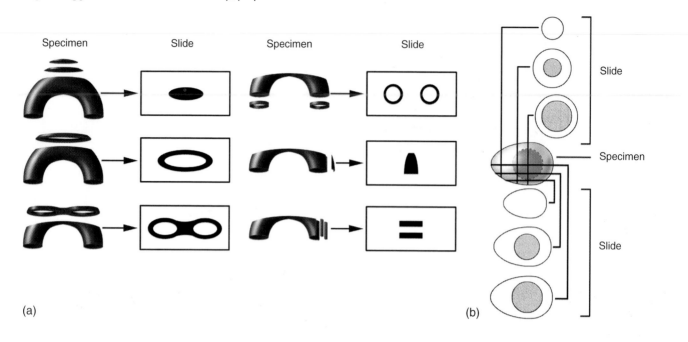

(a)

(b)

Figure 3.7 Epithelial tissue: (a) simple epithelium; (b) artist's interpretation.

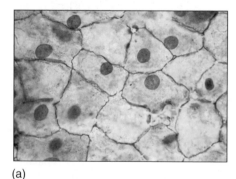

(a)

Nucleus

Basement membrane

Connective tissue

(b)

Human skin is a more effective barrier because it is stratified, squamous epithelium, meaning that it consists of many layers of flattened cells, while the frog's squamous epithelium is simply a single cell layer. The skin acts as a barrier to prevent water loss in dry environments. Which organism, frog or human, is more fully adapted to a terrestrial environment? Why do you say so?

Below, make two comparison sketches of the outer layers of frog and human skin.

9 Obtain two slides: one a cross section of a frog's skin and the other a cross section of human skin. Look at each under the low power of your compound microscope and determine which surface was outermost in the animal. As you study both, consider how the skin of both animals acts as an outer barrier that prevents water loss and invasion by microorganisms. (See figure 3.8.)

Figure 3.8   Cross section of human skin showing dried, compressed cells of outer stratified layers, the epidermal layer that divides to produce outer cells, pigmented cells that give color to skin, and connective tissue of layer beneath the skin (dermis).

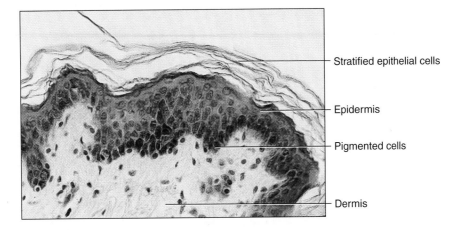

- Stratified epithelial cells
- Epidermis
- Pigmented cells
- Dermis

Epithelial cells lining the digestive system have a columnar shape rather than being flattened, and those lining the trachea and bronchi of the respiratory system are ciliated. Demonstration slides of these tissues are available in the lab and should be studied. Does each of these tissues demonstrate the three characteristics of epithelial tissues listed earlier? What are they?

Figure 3.9   Connective tissue: (a) loose, fibrous, connective tissue (areolar); (b) artist's interpretation.

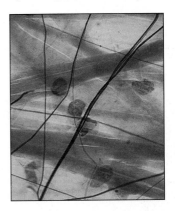

(a)

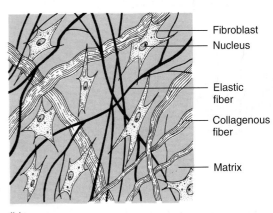

- Fibroblast
- Nucleus
- Elastic fiber
- Collagenous fiber
- Matrix

(b)

Connective Tissue

This tissue is characterized by a nonliving, extracellular **matrix,** which is secreted by the basic connective tissue cell type, the **fibroblast** (fig. 3.9). The matrix contains mucopolysaccharide gel material and many protein fibers. A common fiber in connective tissue is made of the protein collagen. Cartilage and bone are examples of supportive connective tissues. In bone, the matrix contains calcium salts as well as collagen fibers and the cells are confined to small chambers. (See lab topic 31.)

10 ▶ Obtain a slide of areolar connective tissue (fig. 3.9) and observe it under low power with your compound microscope. Note the many fibers and scattered fibroblasts. This tissue attaches the skin to the body and strengthens the walls of blood vessels and organs in the body. Below, sketch a small section of the slide and label the diagnostic features.

11 ▶ Use your ocular micrometer to measure the size of ten fibroblasts. Calculate the average size from your measurements.

Cell #	Longest Dimension in μm
1	____
2	____
3	____
4	____
5	____
6	____
7	____
8	____
9	____
10	____
Average length =	____

How is the extracellular matrix related to the function of connective tissues?

Nerve Tissue

Nerve cells are specialized to transmit messages from one part of the body to another as nerve impulses. In mammals, most nerve cell bodies reside in the spinal cord or brain, and cytoplasmic extensions pass out to muscles or to sensory receptors (fig. 3.10). If you have sensory receptors in your toes and cell bodies at the base of your spine, how long must the cytoplasmic extensions of the neurons be? _____

Figure 3.10 Nerve cell.

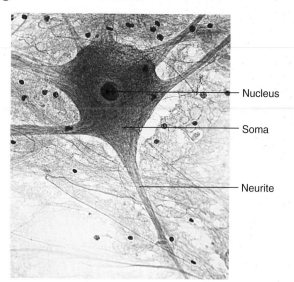

- Nucleus
- Soma
- Neurite

12 ▶ Obtain a slide of a neuron prepared by smearing a section of a cow's spinal cord on the slide. Observe it under low power with your compound microscope. Note the cell body (**soma**) containing the nucleus. Extending from the soma are cytoplasmic extensions called **neurites.** Those that conduct impulses away from the cell body are called axons and those that conduct impulses toward the cell body are called dendrites. Sketch a neuron below.

Use your ocular micrometer to measure the size of the soma of a neuron. Add the dimensions to your drawing.

Is a neuron larger, smaller, or the same size as the fibroblast that you measured when you looked at connective tissue?

? How does the shape of a neuron differ from that of other cells you have studied?

? How does the form of a neuron reflect its function?

Unknowns

13▶On a table in the laboratory, your instructor may have set up five microscopes. Each has a slide of an unknown prokaryote or eukaryote cell type. Identify the type of cell at the end of the pointer. In the following table list your reasons for naming each type.

Unknown Identification

Cell/Specimen Type	Reason
1.	
2.	
3.	
4.	
5.	

Learning Biology by Writing

Based on your observations during this exercise, write a short essay (about 150 words) on the theme "form reflects function" at the cellular level in plants and animals. Cite four examples that you looked at during this lab session.

As an alternative assignment, your instructor may ask you to complete the following summary and critical thinking questions.

Lab Summary Questions

1. List the structural differences found between prokaryotic and eukaryotic cells.
2. How do plant cells differ from animal cells?
3. Epidermal cells in both plants and animals are often flattened, that is, they are thin and wide. How is this an example of form reflecting function?
4. Using your observations of connective and nerve tissues, explain how form reflects function at the cellular level.
5. Explain how the shapes of xylem tracheids are related to their functions in plants.
6. Based on your observations, indicate the relative size of the following cells in relation to each other (smallest . . . largest). If you made measurements, indicate the actual size in micrometers.

Bacillus _____
Onion epidermal cell _____
Xylem tracheid _____
Fibroblast _____
Soma of neuron _____

Critical Thinking Questions

1. During your observation of cells, what similarities did you notice between prokaryotic cells and organelles of eukaryotic cells?
2. Give two examples of both plant and animal cells where form reflects function. Explain how this principle applies to your examples.
3. Describe what a chair would look like when viewed from the top at three different depths of field. Describe what it would look like from three different sides. Could you describe the whole structure if you saw only one cross section and one longitudinal section?
4. Why is it logical to assume that the first living cells on earth had a prokaryotic cell plan and that the eukaryotic cell plan was a later development?

see p. 16

LAB TOPIC 4

Determining How Materials Enter Cells

Supplies

Preparator's guide available at
    http://www.mhhe.com/dolphin

Equipment

Compound microscopes
Balance (0.1 g sensitivity)

Materials

Slides and coverslips
Small glass rods
Dropper bottles
Markers
Osmometer
Dialysis tubing precut to 15 cm
1 ml pipettes
Test tubes and rack
Albustix (drugstore)
Millimeter rulers, mm
Glass diffusion tube
Cork borer
250 ml beaker
Paper towels
Petri plates containing 5 mm of 2% plain agar
Live organisms
    Elodea
    Paramecium caudatum

Solutions

3% NaCl
$2 + 10^{-3}$M sodium azide (optional)
1% albumin solution in 3% NaCl
0.25% soluble starch in 1% Na_2SO_4
I_2Kl solution (5 g I_2: 10 g KI: 100 ml H_2O)
1 M $AgNO_3$
2% $BaCl_2$
0.1 M Na_2SO_4
Concentrated HCl
Concentrated NH_4OH
Light corn syrup dyed with neutral red
35% sucrose

Student Prelab Preparation

Before doing this lab, you should read the introduction and sections of the lab topic that have been scheduled by the instructor.

You should use your textbook to review the definitions of the following terms:

Diffusion   Osmoregulation
Hypertonic   Osmosis
Hypotonic   Plasmolysis
Isotonic   Turgor
Kinetic energy   Water expulsion vesicle

You should be able to describe in your own words the following concepts:

Facilitated diffusion
Osmotic pressure
Differentially permeable membrane

As a result of this review, you most likely have questions about terms, concepts, or how you will do the experiments. Write these questions in the space below or in the margins of the pages of this lab topic. The lab experiments should help you answer these questions, or you can ask your instructor for help during the lab.

Objectives

1. To observe the result of diffusion in liquid and gaseous systems
2. To determine if diffusion and osmosis both occur through differentially permeable membranes

Figure 4.1 Fluid mosaic model of an animal cell membrane.

Animal cell

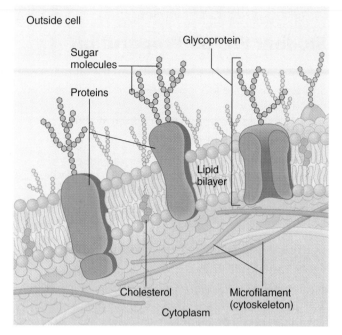

Figure 4.2 Model of osmosis through a differentially permeable membrane. Small water molecules can pass through pores in membrane but larger protein molecules cannot.

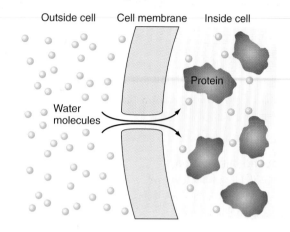

3. To observe the effects of turgor and plasmolysis in plant cells
4. To test experimental hypotheses about the effects of hypotonic, isotonic, and hypertonic solutions on the water-pumping rates of *Paramecium*

Background

The maintenance of a constant internal environment in a cell or organism is called **homeostasis.** In a constant environment, enzymes and other cellular systems are able to function at optimum efficiency. A cell's homeostatic mechanisms include the ability to exchange materials with the environment. Ions and organic compounds, such as sugars, amino acids, and nucleotides, must enter a cell, whereas waste products must leave. Regardless of the direction of movement, the common interface for these processes is the plasma membrane. The cell walls of plants and bacteria offer little, if any, resistance to the exchange of molecules.

The plasma membrane is a mobile mosaic of lipids and proteins (fig. 4.1). Materials cross this outer cell boundary by several processes. Large particles are engulfed in membrane, forming a vesicle or vacuole that can pass into or out of the cell. Some small molecules diffuse through the spaces between lipid molecules in the membrane. Others bind with proteins in the membrane and are transported into or out of the cell.

To understand cellular transport, you should recognize that atoms, ions, and molecules in solution are in constant

motion, continuously colliding with one another because of their **kinetic energy.** As the temperature of any phase is raised, the speed of movement increases so that molecules collide more frequently with greater force.

Diffusion results from the kinetic energy of molecules. For example, when a few crystals of a soluble substance are added to water, molecules break away from the crystal surface and enter solution, some traveling to the most remote regions of the solution. This process continues until the substance is equally distributed throughout the solvent. To generalize this example, in any localized region of high concentration, the movement of molecules is, on the average, away from the region of highest concentration and toward the region of lowest concentration. The gradual difference in concentration over the distance between high and low regions is called the **concentration gradient.** The steeper the concentration gradient, the more rapid the rate of diffusion. The rate of diffusion is also directly proportional to temperature and inversely proportional to the molecular weight of the substance involved. (All molecules move rapidly at high temperatures, but larger molecules move more slowly than small molecules at the same temperature.)

Substances diffuse into and out of cells by passing through the spaces between membrane molecules or dissolving in the lipid or protein portions of the membrane. However, due to size or charge, some substances cannot pass through membranes. Membranes that block or otherwise slow passage of certain substances are described as being **differentially permeable.** Differential permeability accounts for the phenomenon of **osmosis,** or the diffusion of water through a membrane under special conditions.

The conditions for osmosis are shown in figure 4.2, where a porous membrane is pictured with water on one side and protein in water on the other. The special condition is that the small water molecules can pass through the pores of the membrane, but the large proteins cannot. Hence, water at a greater concentration outside (because it is not diluted by protein) will tend to diffuse into the cell. If the cell were encased in a rigid box, the increasing water pressure would cause the water to flow back out of the cell to the low-pressure area. Eventually, an equilibrium would be reached when the flow of water into the cell, due to concentration differences, balances the flow out of the cell, caused by pressure differences. The pressure at equilibrium is called the **osmotic pressure** of the solution.

Since all cells contain molecules in solution that cannot pass through the membrane, osmosis always occurs when cells are placed in dilute aqueous solutions. In bacteria and plants, the cell wall prevents the cell from bursting by providing a rigid casing that helps regulate osmotic pressure within the cell. In animals, an osmoregulatory organ is found, such as the kidney, which adjusts the concentration of substances in the body fluids that bathe the cells.

Many ions and organic molecules important to cell metabolism are taken into cells by specific transport proteins found in the cell membranes. **Facilitated diffusion** occurs when such a protein simply serves as a binding and entry port for the substrate. In essence, the protein is a pipeline for a specific substance. The direction of flow is always from a region of high concentration to one of low concentration, but gradients are maintained because many molecules, upon entering the cell, are metabolically converted to other types of molecules.

For many other materials, favorable diffusion gradients do not exist. For example, sodium ions are found at higher concentrations outside mammalian cells, yet the net movement of sodium is from inside to outside the cell. For such materials, cellular energy must be expended to transport the molecules across the cell membrane. **Active transport** occurs when proteins in the cell membrane bind with the substrate and with a source of energy to drive the "pumping" of a material into or out of a cell.

LAB INSTRUCTIONS

You will observe diffusion and the properties of differentially permeable structures. Start the experiment "Simultaneous Osmosis and Diffusion" first so that sufficient time can elapse to see results. You will formulate and test experimental hypotheses regarding the activity of the water-expulsion vesicle in *Paramecium*.

Simultaneous Osmosis and Diffusion

Dialysis tubing is an artificial membrane material with pore sizes that allow small molecules (usually with molecular weights less than 1000) to pass through it but not large molecules. Water, NaCl, and Na_2SO_4 have molecular weights of 18, 58.5, and 142 respectively. Starch and proteins have molecular weights greater than 30,000. If dialysis tubing is a differentially permeable membrane, which

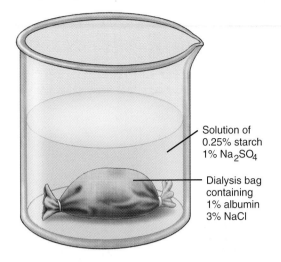

Solution of 0.25% starch 1% Na_2SO_4

Dialysis bag containing 1% albumin 3% NaCl

molecules would you hypothesize can pass through the membrane that has a cutoff between 12,000 and 14,000?

1 ▷ Obtain a 15 cm section of dialysis tubing that has been soaked in distilled water. Tie or fold and clip one end of the tubing to form a leakproof bag. Half fill the bag with a solution of 1% protein (albumin) dissolved in 3% NaCl. Also add a 1 ml sample of the same solution into each of four test tubes labeled **"Inside Initially."**

Now tie the bag closed with a leakproof seal. Wash the bag with distilled water, blot it on a paper towel, weigh it to the nearest 0.1 g, and record the weight in table 4.1, in the Inside Bag Initial column.

Place the bag in a 250 ml beaker containing a solution of 0.25% soluble starch dissolved in 1% Na_2SO_4. Place 1 ml samples of the fluid from the beaker in each of four test tubes labeled **"Outside Initially."** The starting conditions are summarized in figure 4.3. This experiment will run for approximately two hours. Go on to the other experiments while this experiment runs in the background.

2 ▷ At 15-minute intervals, swirl the beaker containing the bag or place the beaker on a slowly turning magnetic stirrer.

Analysis

3 ▷ After two hours or longer, take four 1 ml samples from the beaker and place them in four test tubes labeled **"Outside End."** Now remove the bag, rinse it with distilled water, blot

TABLE 4.1 Results of osmosis/diffusion experiment with dialysis tubing

	Outside Bag		Inside Bag	
	Initial	End	Initial	End
NaCl	−		+	
Na$_2$SO$_4$	+		−	
Protein	−		+	
Starch	+		−	
Weight	not needed for outside	− − −		

it on a paper towel, and weigh it to the nearest 0.1 g. Record the weight in table 4.1 in the Inside Bag End column.

Empty the contents into a beaker, take four 1 ml samples, and place them in four test tubes labeled **"Inside End."**

4▷ Now assay the inside and outside samples from the start and end for the presence of the compounds added at the beginning of the experiment. Record the results of your analysis in table 4.1, using plus and minus symbols to indicate the presence or absence of material both before and after incubation. The following are specific, easy-to-perform indicator tests:

Test for Chloride Ion

5▷ Add a few drops of 1 M AgNO$_3$ to one inside and one outside tube for the initial samples. Repeat for an inside and outside end sample. A milky white precipitate of AgCl indicates the presence of Cl$^-$. **Caution:** AgNO$_3$ stains skin and clothing.

Test for Sulfate Ion

6▷ Add a few drops of 2% BaCl$_2$ solution to one inside and one outside initial samples. Repeat for an inside and outside end sample. If SO$_4^-$ is present, a white precipitate of BaSO$_4$ will form.

Test for Protein

7▷ Dip Albustix reagent strips (usually used in urinalysis) into initial and end samples for both inside and outside solutions. The paper will turn green to blue-green if albumin is present.

Test for Starch

8▷ Add a few drops of freshly made I$_2$KI to each remaining tube. If a blue color appears before mixing, it indicates the presence of starch.

◈ Which set of test tubes served as a control in this experiment?

◈ Describe which substances were able to move through the dialysis membrane. Which direction did they move in relation to their concentration gradient? What are the molecular weights of these substances?

◈ Did starch and protein move through the dialysis membrane? What are their typical molecular weights?

◈ What evidence do you have that water moved through the dialysis membrane?

◈ Do the experimental results support or falsify your hypothesis? How?

Figure 4.4 Osmometer demonstrates passage of water across a differentially permeable membrane.

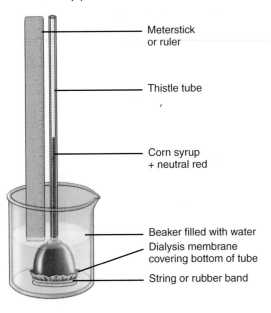

- Meterstick or ruler
- Thistle tube
- Corn syrup + neutral red
- Beaker filled with water
- Dialysis membrane covering bottom of tube
- String or rubber band

TABLE 4.2 Fluid height in osmometer

Time	Elapsed time	Height

Assuming you had access to a wide range of chemicals with different molecular weights, how would you determine the pore size for the dialysis tubing used in this experiment?

9 Early in the lab period, observe the height of the column of fluid in the thistle tube. At approximately 20-minute intervals during the lab, repeat. Record the time and height in table 4.2.

Describe what is happening to both sugar and water molecules in the osmometer.

Over time do you expect that the rate of water movement will increase, decrease, or remain the same? Why?

Osmosis

The rate of water movement in osmosis can be observed in an osmometer (fig. 4.4). A sugar solution in a thistle tube is separated from a beaker containing water by a dialysis membrane that allows water to pass through but sugar molecules cannot. What do you expect will happen over time in a setup such as in figure 4.4?

Diffusion in Gases (Optional Demonstration)

CAUTION

The HCl is a concentrated acid and can harm you and your clothes. The NH_4OH is very irritating to the eyes and nose. For these reasons, your instructor may choose to do this part of the exercise as a demonstration.

Figure 4.5 Setup for diffusion of gases. Use glass tube 1 inch in diameter and 16 inches long with cotton-tipped wooden applicator sticks inserted in rubber stoppers at ends.

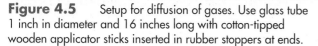
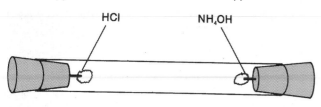

10▸ Figure 4.5 shows a simple apparatus for determining the effect of molecular weight on diffusion in gases. Your instructor may do this as a demonstration. If concentrated HCl is introduced at one end and NH_4OH at the other, the gases, HCl and NH_3, diffuse toward the center. HCl has a molecular weight of 36 and NH_3 of 17. Hypothesize which gas should diffuse faster.

When HCl and NH_3 meet, they form NH_4Cl, a white salt, not a gas. It precipitates in the tube.

$$HCl \text{ (gas)} + NH_3 \text{ (gas)} \rightarrow NH_4Cl \text{ (solid)}$$

To do the experiment, remove stoppers and simultaneously dip one in concentrated HCl and the other in concentrated NH_4OH. Simultaneously reinsert and observe for formation of white ring of NH_4Cl. Measure the distance each gas traveled in the tube.

Record the results of your experiment below. What is the ratio of these distances? What is the ratio of the molecular weights? What is the ratio of the square roots of the molecular weights? Is diffusion rate directly or inversely proportional to molecular weight or the square root of molecular weight? Refer back to your hypothesis regarding the rate of diffusion. Do you accept or reject this hypothesis? How would you now modify it to be more quantitative? Propose an experiment to test this hypothesis.

yes

Diffusion in Gels

11▸ Diffusion in highly viscous solutions can be demonstrated using agar as a gelling agent. Obtain a petri plate containing 1% agar poured to a depth of 5 mm. Use a 5 mm cork borer to make two wells about 1 cm apart. Add a solution of sodium sulfate to one and barium chloride to the other. Observe at intervals during the lab by placing the plate on a black paper. When these two soluble compounds diffuse through the water in the agar and meet, the white precipitate barium sulfate will be formed.

$$Na_2 SO_4 + BaCl_2 \rightarrow BaSO_4 \text{ (ppt)} + 2 NaCl$$

Record your results as a time diary below.

ye

Osmosis in Living Cells
Elodea Plasmolysis

Plant cells are surrounded by rigid cell walls, and under normal conditions, the cytoplasm of the cell is closely pressed against these cell walls. Cells in this condition are said to be **turgid.**

12▸ Remove a leaf from an *Elodea* plant, cut a very small piece from the tip, and mount it in water on a microscope slide. Using your compound microscope, identify individual cells and note the position of the cytoplasm in relation to the cell walls. Sketch a few cells. While looking at these cells, you may see cytoplasmic streaming (cyclosis). If you think in terms of molecular mixing, can you propose how this might benefit the cells in *Elodea?*

13▸ Now treat the cells with a concentrated solution of 35% sucrose by placing a drop or two of the solution next to the coverslip and wicking it under the coverslip. (See fig. 3.4.) Observe the cells looking at the position of the cytoplasm relative to the cell wall. When the cytoplasm pulls away from the cell wall due to water loss from the central vacuole, the cell is said to **plasmolysed.** Draw a few plasmolysed cells below.

⬥ What do you think would happen if you now placed these cells in plain water? Formulate this prediction as a hypothesis.

14▶ Test the hypothesis by wicking distilled water under the coverslip. Record your results below. Based on this observation must you accept or reject your hypothesis?

⬥ Use the concept of osmosis to explain why the cytoplasm pulled away from the cell wall and then expanded again under the different conditions tested.

Osmoregulation in Paramecium

Aquatic animals and plants can be exposed to environmental media that differ significantly in their osmotic concentrations. An **iso-osmotic** medium has the same concentration of osmotically active particles and, therefore, the same osmotic pressure as a cell does. This is sometimes also referred to as an isotonic medium. A cell in an iso-osmotic medium will neither gain nor lose water since there is no net "pulling" force on the water in either the internal or external environments.

If a cell is placed in a **hyperosmotic** (hypertonic) medium, or a medium that has a higher concentration of osmotically active particles and thus a higher osmotic pressure than the cell, the cell tends to lose water to the medium. A **hypo-osmotic** (hypotonic) medium contains a lower number of osmotically active particles than a cell and thus has a lower osmotic pressure. In a hypo-osmotic medium, a cell gains water by osmosis. You just finished experimenting with the effects of hypo-osmotic and hyperosmotic solutions on *Elodea* leaf cells.

Figure 4.6 Photomicrograph of *Paramecium caudatum* showing one of its two water expulsion vesicles (WEV). In life, these empty and fill in a regular cycle.

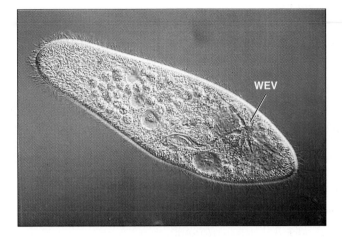

Paramecium is a freshwater, single-celled organism. Its habitat is a hypoosmotic environment and *Paramecium* continually gains water. A mechanism is required for the *Paramecium* to get rid of this excess water. The **water-expulsion vesicle** (or contractile vacuole) functions in this capacity (fig. 4.6). The osmotic regulation function of the water-expulsion vacuole can be observed on a microscope slide. The preparation is difficult and requires patience and careful observation.

Baseline Observations

15▶ Obtain a drop of culture fluid containing paramecia that have been kept in a boiled wheat grain culture. Pick up some culture debris to aid in trapping the organisms. Place the sample on a clean glass slide and add a coverslip. Using medium power, look for the water-expulsion vacuole. (*Note:* It will be necessary to slow the animal down. Draw some of the fluid off with a paper towel. Try to trap protozoans in culture debris under the coverslip. If this does not work, shred a small piece of lens paper or cotton and put it on a new slide. Add a drop of culture and a coverslip. Look for paramecia caught in the fiber matrix.)

⬥ Observe a trapped *Paramecium* with high power, looking for one of the two water expulsion vesicles. Are the water-expulsion vacuoles temporary or permanent? Do the anterior and posterior vesicle contract at the same time? At the same rate? The normal rates of pulsation vary somewhat with culture conditions. You may expect a rate of six to ten pulsations per minute in the culture fluid. Draw a *Paramecium* below and record your observations as notes. Record the number of pulsations per minute on the first row of table 4.3.

TABLE 4.3   Water expulsion vesicle pumping rates in *Paramecium*

	Pulsations/Min	Vacuolar Volume (μ^3)	Volume Pumped (μ^3/Min)
Culture fluid (control)			
Distilled water			
3% NaCl			
2 mM sodium azide (optional)			

Experimental Observations

Using your baseline observations for comparison, you should now determine the pumping rate of pulsation (pulsations/minute) of the vesicle when the organisms are in the following solutions:

1. distilled water

2. 3% NaCl

3. *Optional:* culture fluid + 2×10^{-3}M sodium azide

In each case, mix one drop of the culture fluid containing paramecia with one drop of the desired solution. Counts can be made at the magnification provided by a 10× objective and 10× ocular once the vesicles and their pulsations have been observed.

Would you predict that the pulsations per minute in distilled water would be higher or lower than the number when in culture medium? Formulate this prediction as a testable hypothesis and record below:

16 Now count the number of pulsations per minute in distilled water and record in table 4.3.

Review your hypothesis and make a decision to accept or reject it. Record your decision in the space above.

Formulate a new testable hypothesis that relates expected vacuole pulsation rate in 3% NaCl to that seen in normal culture media. State the hypothesis below:

17 Add 3% NaCl to your slide and "wick" it under the coverslip. Count the number of pulsations and record in table 4.3.

Review the hypotheses you made for this experiment. Based on your observations must you accept or reject your hypotheses?

C A U T I O N

Sodium azide is a deadly poison that blocks cellular respiration by binding to the cytochromes. Do not ingest.

Optional: Sodium azide is a metabolic poison that stops the ability of a cell to make ATP during aerobic respiration. If energy is required for the water-expulsion vesicle to operate, it should stop functioning in cells treated with sodium azide.

Formulate a testable hypothesis that relates azide treatment to normal vesicle pulsation rate. State the hypothesis below:

no

18 Make a fresh slide of *Paramecium* and wick a solution of sodium azide under the coverslip. Let sit for a few minutes and then measure the rate of vacuole pulsation. Record in table 4.3.

◆ Must you accept or reject your hypothesis? Why?

Analysis

If the volume of the water-expulsion vesicle at maximum expansion were known, it would be possible to estimate the amount of water expelled per minute. To calculate the vesicular volume, it is necessary to know the vesicle's diameter. This can be measured with an ocular micrometer. Since the use of the ocular micrometer is somewhat involved (especially in measuring the diameter of the water-expulsion vesicle in a *Paramecium*), you should use the following measurement: water expulsion vesicle diameter = 6.5 micrometers (10^{-6} m). However, keep in mind that vesicle diameters vary from individual to individual and that the diameter of a vesicle within an individual may change in response to the osmotic pressure of the environmental medium.

no

19 Using the following formula, calculate the vesicle volume. This vesicular radius should be in micrometers (10^{-6} meters). Remember to use the concept of significant figures in your calculations and answer (appendix A).

$$\text{volume (for a sphere)} = \frac{4\pi(\text{radius})^3}{3}$$
$$= 4.19r^3$$

Since you now know the volume of the water-expulsion vesicle and the pumping rates per minute of the vesicle, you can calculate the amount of water pumped per minute by a *Paramecium*. Enter the values in table 4.3. The units will be in cubic micrometers.

◆ How many cubic micrometers are in a cubic centimeter? (*Note:* a cubic centimeter equals a milliliter.) How many milliliters is the *Paramecium* pumping per minute? Per day?

◆ Why do you think the pumping rate was greatest in distilled water and least in the 3% NaCl solution? What does the sodium azide treatment demonstrate?

Learning Biology by Writing

Prepare a brief lab report about the water-expulsion vesicle experiments in this exercise. In the introduction to the report, state the hypotheses that were tested by your experiments. Next, describe the methods used to test each hypothesis. Finally, report the results of the tests, including the tables you completed. Discuss what these experiments demonstrated about the osmotic challenges that freshwater organisms face. Do marine organisms face the same osmotic challenges?

As an alternative assignment your instructor may ask you to complete the following lab summary and critical thinking questions.

Internet Sources

Reverse osmosis is a recently developed process to purify seawater for drinking water or to clean up polluted waters. Current information on this process can be found on the World Wide Web. Use an Internet browser to locate this information.

When in your browser, connect to the search engine GOOGLE at http://www.google.com. When you get the dialog box, enter the words "*reverse osmosis*" and submit the search.

Look over the results from your search. Answer the following questions. What is reverse osmosis? Describe the process. List three applications of reverse osmosis. List two URLs that would be good resources for learning about this process.

Lab Summary Questions

1. What data do you have to support the conclusion that diffusion is inversely related to molecular weight?
2. What data do you have to support the conclusions that (1) dialysis membranes are differentially permeable and that (2) osmosis and diffusion can occur at the same time through a dialysis membrane?
3. Explain how osmosis can be considered a special case of diffusion.
4. Explain why *Paramecium* continually gains water from its environment. How does it get rid of excess water?

Critical Thinking Questions

1. If a person's blood volume drops due to injury or severe dehydration, why do doctors administer isotonic saline intravenously instead of pure water?

2. How would an increase in temperature affect the rate of diffusion of gases? How might it affect osmosis? Explain.
3. How would increasing the concentration of albumin in your simultaneous osmosis and diffusion experiment affect the results?
4. When preparing sauerkraut, shredded cabbage is layered with coarse salt crystals. What effect would this have on the cabbage?
5. What osmotic regulatory challenges would a fish living in freshwater have versus a fish living in salt water?
6. If fruits are crushed, cooked, and then stored at room temperature, they spoil. If high concentrations of sugar are added as in making jams, their shelf life is greatly extended. Explain this in terms of the osmotic challenges facing invading bacteria and fungi.

LAB TOPIC 5

Using Quantitative Techniques and Statistics

Supplies

Preparator's guide available at
   http://www.mhhe.com/dolphin

Equipment

   Spectrophotometer and tubes
   Balances (sensitivity minimum 0.1 g)

Materials

   5 ml pipettes or micropipetters
   Suction devices for pipettes
   Test tubes and racks
   50 ml beakers
   Spectrophotometer cuvettes

Solutions

   Distilled water
   Bromophenol blue standard solution (0.02 mg/ml)
   Bromophenol blue solutions for unknowns

Student Prelab Preparation

Before doing this lab, you should read over the entire lab topic.
   You should use your textbook to review the definitions of the following terms:

   Absorbance
   Balance
   Gram
   Micropipet
   Milliliter
   Nanometer
   Spectrophotometer
   Wavelength

   You should be able to describe in your own words the following concepts:

   Beer's law
   Absorption spectrum
   Standard curve
   Review appendices A,B, and C
   Mean and Standard Deviation
   Histogram

   As a result of this review, you most likely have questions about terms, concepts, or how you will do the experiments. Write these questions in the following space or in the margins of the pages of this lab topic. The lab experiments should help you answer these questions, or you can ask your instructor during the lab.

Objectives

1. To learn to use pipettes and balances
2. To use a spectrophotometer to measure light absorbance by colored solutions
3. To construct a standard curve to determine the concentration of dye in an unknown
4. To analyze data using simple statistical techniques

Background

This lab introduces you to skills that you will need later in order to perform experiments. The problems to be solved are not biological; some would say they are chemical. However, weighing, pipetting, and using spectrophotometers are performed daily in modern biological research labs and are needed skills. In addition, this lab introduces data and statistical analyses which are commonly used in experimental biology.

When using quantitative techniques, one should ask two fundamental questions: What precision is necessary and appropriate? Are the measurements accurate? Precision and accuracy are very different concepts, although they are often confused.

Precision has to do with exactness. While some might think that as we always want to be as precise as possible, it is often not necessary to be exact: close is good enough. When asked how much they weigh, very few people feel compelled to answer in pounds, ounces, and fractions of ounces: pounds is enough. Most bathroom scales are calibrated only in pounds and do not have ounce readings. Why? Less precise instruments are cheaper to produce and easier to use. Only in medical research labs, where greater precision might be required, do you find more precise devices for determining body weight.

Figure 5.1   One type of glass pipette used in laboratories. The serological pipette must drain dry to dispense the calibrated volume. Pipettes should have the letters TC or TD printed on them. TC means "To Contain" and such pipettes should have the last remaining drop blown out of them. TD means "To Deliver" and such pipettes should be allowed to drain only, leaving a small amount of fluid in the tip at the end when the full contents of the pipette are delivered.

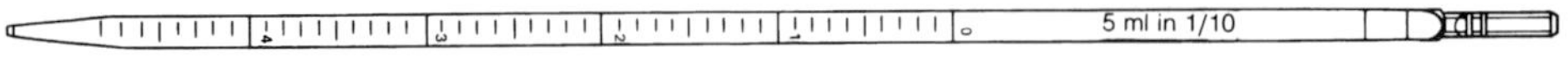

Serological pipette — 5 ml in 1/10

Accuracy has nothing to do with precision. An inaccurate bathroom scale may consistently report a person's weight as being under what it actually is. It could do so with great precision. It could report a weight of 130 pounds and 12.5 ounces, but if the person's real weight, the true value, was 135 pounds and 2.0 ounces, the scale, while being precise, was not accurate. The only way we can know whether an instrument is accurate is to standardize it. This is usually done by measuring a standard and calibrating the instrument. Standard weights, made of substances that do not change weight, can be weighed and the scale adjusted if it is "off."

In modern labs where many instruments have digital read-out devices, you will often see a number expressed to three or more decimal places. One should be skeptical before accepting such numbers and should first ask a few questions. Some examples might be: (1) Was this instrument calibrated with a standard that had an accuracy to three decimal places? (2) Did all instruments used in preparing the sample yielding this data have the same precision as the instrument I am reading? (3) Is it necessary to have great precision in the measurement or will a value with fewer decimal places be sufficient?

The implications of precision in measurement and laboratory calculations are discussed further in appendix A on page 467. Read this so that you know how to do rounding and arithmetic with values that have different levels of precision. You will have to apply these rules in the calculations that are called for in this lab.

LAB INSTRUCTIONS

In this lab you will learn to use pipettes, balances, and spectrophotometers. You will learn to calculate simple statistics.

How to Pipette

Your instructor will demonstrate the use of common volumetric glassware, including graduated cylinders, volumetric flasks, and beakers with approximate markings. In chemical tests, it is frequently necessary to measure a small volume of liquid (between 0 and 10 milliliters) quickly and accurately. Various calibrated automatic dispensers or syringes may be used to do so. For years, however, the standard measuring device used in research laboratories

Figure 5.2   A micropipetter has a dial that allows one to set the volume to be measured, and a disposable tip that is discarded after use. Depressing a plunger button on the top allows you to fill or empty the fluid in the tip.

was the glass **pipette** (fig. 5.1) because of handling ease and low cost. Recently, pipetting devices called micropipetters (fig.5.2) that have disposable plastic tips have replaced the standard glass pipette in most laboratories.

A glass pipette is filled by using a syringe or valved rubber bulb attached to the end of the pipette (fig. 5.3). To use a pipette, immerse just the tip in the appropriate fluid and draw it up beyond the zero mark using a suction device. Look at the surface of the fluid in the pipette and note the **meniscus,** the concave upper surface caused by surface tension. Hold the pipette vertically and allow the fluid to escape until the bottom of the meniscus touches the zero line. Any drops hanging from the tip of the pipette should be removed by touching the tip to the inside of the beaker from which the solution was drawn.

CAUTION

If you use glass pipettes, never draw chemical or biohazardous solutions into a pipette by mouth suction, as you would do with a straw, because you may accidentally ingest a poison or pathogen. Use a suction device (fig 5.3.)

Figure 5.3 Filler devices used with glass pipettes: (a) rubber bulb with valves, (b) syringe with rubber connector. Other suction devices may be used and, if available, will be demonstrated by your instructor.

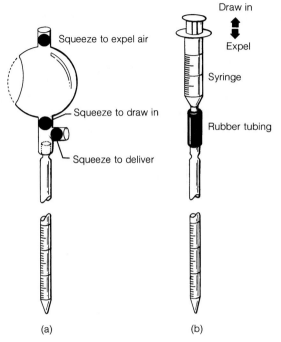

Using an automatic micropipetter is a bit more involved, and proper technique will be demonstrated by your instructor. Usually the following steps are involved.

1. Select a pipetter that dispenses volumes in the range you want.

2. Note the units and dial in the volume you wish to dispense.

3. Put the correct color tip on the pipette with a twisting motion.

4. Depress the plunger to the first stop and, while holding it in that position, submerge it in the fluid to be drawn up. Slowly release the plunger to draw the fluid up.

5. Without blotting, put the tip into the tube to receive the fluid. Press the button to dispense the fluid, continuing to press it past the first step to the second, thus blowing out the fluid. Touch the tip to the side wall to remove any drops clinging to the tip.

6. If a new solution is to be pipetted, press the tip ejection lever and put on a new tip. The same tip can be used over again when dispensing the same fluid.

Practice Pipetting

1▷ Practice your pipetting technique by obtaining a 5 ml pipette and a 50 ml beaker.

TABLE 5.1   Beaker and sample weights

Weight of beaker containing water	_____ g
Minus weight of empty beaker	_____ g
= Weight of water	_____ g

2▷ Weigh the empty beaker to the nearest tenth of a gram. Your laboratory may be equipped with electronic or triple-beam balances. Your instructor will give you directions on how to use the available equipment. Be sure that you know how to zero the balance and then to weigh a specimen. Record the beaker's weight on the second line of table 5.1.

Now use the pipette to add the following volumes of water in milliliters to the beaker:

5.0	4.0	3.0
4.5	3.5	2.5
4.3	3.3	1.5
4.2	3.2	1.0

Theoretically a total of 40 ml should have been added.

Verifying Techniques Using a Balance

To check the accuracy of your pipetting, you will weigh the amount of water added to the beaker and then use the density of water to convert this weight to a volume. The density of a substance is its weight per unit volume. The density of water at room temperature (23° C) is 0.998 g/ml.

3▷ Weigh the beaker containing the 40 ml of water and record the weight on the first line of table 5.1. Subtract the weight of the empty beaker from the filled beaker weight to obtain the actual weight of water in the beaker. This operation—subtracting the weight of the container from the weight of the material plus the weight of the container—is called **taring.**

 Divide the weight of water in the beaker by the density of water to calculate the volume of water in the beaker. Remember to apply the concept of significant figures discussed in appendix A. How much water, in milliliters, is in your beaker?

Calculate the experimental error as a percentage (called percentage error) using the following equation:

$$\% \text{ error} = \frac{\text{actual} - \text{theoretical}}{\text{theoretical value}} \times 100$$

$$= \frac{\text{actual volume of } H_2O - 40}{40} \times 100$$

$$\% \text{ error} = \underline{\hspace{2in}}$$

If your percentage error is not zero, think about how you did this experiment. What does a positive percent error mean? A negative? Honestly evaluate your work. Did you leave out a pipetting? Did you repeat one? Did you correctly read the pipette and meniscus? Did you properly use the balance? Is the balance accurate? Check it with a standard, known weight. What would be the effect on your data if the balance was improperly calibrated and consistently gave readings that were heavier than they were supposed to be? Would your % error be positive or negative?

List below your explanations for the error in your measurement.

If you did a similar experiment with larger pipettes in which you added two 20 ml portions, would you expect your error to change? Why?

Calculating Simple Statistics

Different students probably obtained slightly different values for their water volume measurements at the beginning of the exercise. Even if gross errors (such as wrong calcula-

tions, failure to follow directions, or incorrect readings) are ruled out, there would still be some variation due to minor experimental and chance errors. Therefore, the measurements you have made are not true values but are simply estimations of a true value. Some estimations obtained by the class are lower than the true value, while others are higher. When it is important for scientists to obtain a true measurement, they repeat the measurement several times and calculate a **mean** (average) value.

Chance errors in data sets cancel each other out when means are calculated—that is, a value that is too high due to chance error is balanced by a value that is low for the same reason. The **range** of observations gives some sense of the variability in measurement. However, range is not a very good estimator of variability because it can be artificially inflated by one or two outlying values. Consider the following two sets of hypothetical data:

	Set A	Set B
	30	30
	29	40
	31	20
	28	32
	32	31
	30	30
	29	31
	31	26
Σ (= sum)	240	240
N (= number of observations)	8	8
Mean	30	30
Range	28 – 32(±2)	20 – 40(±10)

One pair of values in set B created an extremely broad range even though the means were the same. Because of this problem and the need to convey information about the amount of variability in a set of measurements, scientists use prescribed calculations to obtain variability estimators called **variance** and **standard deviation.** (Appendix C contains a more thorough discussion of statistics.) These estimators are obtained by expressing all values in a data set as plus or minus variations from the mean, that is, as *measured value – mean value.* Variance and standard deviation are calculated as follows:

variance =

$$\frac{\Sigma \, (\text{measured value for each sample} - \text{mean})^2}{(N - 1) \text{ one less than number of observations}}$$

$$\text{standard deviation} = \pm\sqrt{\text{variance}}$$

Use the following work table to calculate the standard deviation of the hypothetical set of measured values given in the first column.

Measured value	Measured minus mean	(Measured minus mean)^2
15		
20		
25		
16		
20		
24		
10		
20		
30		

Σ _____   Σ _____

N _____   N – 1 _____

Mean _____   Variance _____

                    SD _____

? Do your answers agree with those of other students? Check your arithmetic if they do not.

Statistical Analysis of Class Data

Having practiced how to calculate a standard deviation, you will now apply this concept to the class data on the volume of water pipetted at the beginning of this exercise.

5 To create a class data set, all students should go to the blackboard and write the values they calculated from the density data for the volume of water pipetted into the beaker at the beginning of this exercise. Copy the data from the blackboard into the first two columns of table 5.2.

Scan the values that you entered in table 5.2 and answer the following questions:

? 1. Are some values in the class data grossly different from all the others? If so, check for errors in calculation or procedure that will allow you to objectively eliminate the data. Do any of the values have too many significant figures, given the precision of measurement? If so, round off. If any data are rejected, indicate why.

TABLE 5.2		Class data work table

Calculated Water Volume (ml)	(Measured Vol – Mean Vol)	(Measured Vol – Mean Vol)^2

Sum (Σ) _____   Σ

Range _____   Σ/(N – 1) _____

Number of observations (N) _____

Mean (Σ/N) _____ $\sqrt{Σ/(N-1)}$ _____

? 2. Now, calculate the mean value for the water volume and enter the range of values. Is it possible to determine if variation is due to measurement of weight or of water volume? Compare the mean values for water volume to the theoretical value of 40 ml. Is the mean closer to the real value than are many individual measurements? Why? Is the mean volume for water volume skewed significantly away from 40 ml? If so, check to see if the skewing is due to a few values or whether all values show bias. In the latter case, check the balance. Balances that weigh consistently heavy or light would cause such a bias.

3. Do the range values give you an estimate of the variability in the measurements? If good lab technique is used, should the range be large or small?

4. Calculate the standard deviation for the volumes of water from the class data. Use the last two columns in table 5.2 to organize your calculations. Record the mean and standard deviation below.

Mean =

SD = ±

Read the first part of appendix C and explain, in your own words, what it signifies when a scientist writes that the mean for a set of measurements was 40 plus or minus a standard deviation of 1.5. Write your explanation below.

Students who have experience with the use of spreadsheets on microcomputers should realize that such software often contains functions for simple statistical analyses. You need only enter the data set into a spreadsheet and call up the functions AVERAGE and STANDARD DEVIATION. In this lab manual, you will often collect data that consist of repeated measurements. You should learn how to use a spreadsheet program such as *Excel* to calculate a mean and standard deviation. Your instructor may have handouts to help you use these programs.

Using Histograms to Summarize

The class data on the volume of water can also be used to make a histogram, sometimes called a bar graph (read appendix B). Histograms are useful in that they visually convey to the reader the amount of variability in a given set of measurements. Many spreadsheet programs have built-in graph functions and can also be used to create histograms or line graphs directly from data entered into the program.

To create a histogram for the class data, look at table 5.2 and determine the range of water volumes. Follow the directions in appendix B to create a histogram of the class data for water volumes. Choose intervals to plot that are easy to manage, perhaps intervals such as 39.1 to 39.3, 39.4 to 39.5, *etc.* Label all axes.

Add an arrow above the histogram to indicate where the mean of the data set lies and add a cross bar perpendicular to the arrow that spans plus or minus one standard deviation.

Many spreadsheet programs have built-in graph functions and can also be used to create histograms directly from data entered into the program. If you are to use such a program, your instructor will give you directions.

Spectrophotometry

Many organic molecules absorb radiant energy because of the nature of their chemical bonds. (Light-absorbing organic molecules have a system of single and double bonds between adjacent carbons or carbon and nitrogen.) For example, proteins and nucleic acids absorb ultraviolet light in the wavelength interval 240 to 300 nanometers (nm), pigments and dyes absorb visible light (about 400 to 770 nm), and other organic compounds absorb infrared energy (above 770 nm). Our perception of color is related to the ability of pigment molecules in our eyes' cone cells to absorb light energy. If an object appears to be red, it contains molecules whose chemical bonds absorb blue or green **photons** of light while reflecting red light back to the red sensitive cones of our eyes.

A **spectrophotometer** is an instrument designed to detect the amount of radiant energy absorbed by molecules in a solution. Spectrophotometers have five basic components: a **light source,** a **diffraction grating,** an **aperture** or **slit,** a detector (a **photoelectric tube**), and a **readout** to display the output of the phototube. The arrangement of these parts is shown in figure 5.4.

Figure 5.4 Schematic drawing of the path of light through a spectrophotometer.

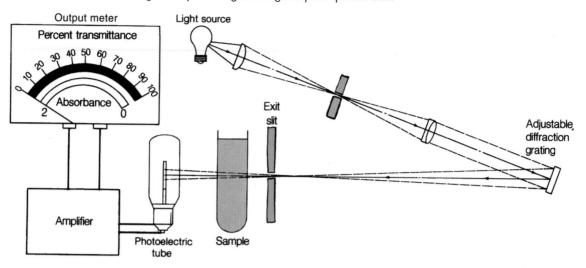

When light passes through the diffraction grating, it is split into its component colors or wavelengths, which then diverge. Sections of the projecting spectrum can be either blocked or allowed to pass through the slit, so that only one color will pass to the other sections of the spectrophotometer.

Light that passes through the slit travels to the phototube, where it creates an electric current proportional to the number of photons striking the phototube. If a current meter is attached to the phototube, the electric current output—which represents the quantity of light striking the phototube—can be measured and displayed on a meter or a digital readout. The meter scale is usually calibrated in two ways: **percent transmittance,** which runs on a scale from 0 to 100, and **absorbance,** which runs from 0 to 2 in most practical applications.

Before the light-absorbing properties of a solution can be measured, three adjustments on the spectrophotometer are necessary. First, the diffraction grating must be adjusted so that the desired color of light passes through the slit. Second, the output of the phototube must be adjusted or calibrated to correct for drift in the electronic circuit. Third, a compensation must be made for dirt or contaminating colored material in the light path between the source and the detector.

To calibrate the electrical circuits, an adjustment knob is turned or a button is pushed when there is no sample in the instrument to set the readout to 0% transmission. To adjust for materials in the light path, a clean sample tube is filled with the same solvent (usually water) to be used in dissolving the dye. Place the tube containing the solvent in the sample compartment of the spectrophotometer (see fig. 5.5). This tube, called a **blank,** serves as a control. The readout should indicate 100% transmittance, or 0 absorbance, depending on which scale you are using. If it does not, an adjustment must be made. This standardizes the spectrophotometer.

If a colored solution is put in the tube in place of the pure solvent, some of the light coming from the slit will be absorbed by the dye molecules, and some will be transmitted to the phototube. The amount absorbed will be proportional to the number of the dye molecules per unit volume (or **concentration**) of the solution.

The readout conveys the phototube current output, which will be less than that seen in the standardizing procedure when there were no dye molecules. If the transmittance scale is used, the amount of light transmitted by the solution is measured as a percentage by the spectrophotometer. This measurement is described by the following equation:

$$\text{Percent transmittance (T)}$$
$$= \frac{\text{intensity of light through sample}}{\text{intensity of light through blank}} \times 100$$
$$= \frac{I_s \times 100}{I_b}$$

If the **absorbance** scale is used, the measurement equation is somewhat more involved. Instead of measuring the amount of light transmitted by the solution, the spectrophotometer reads the amount of light absorbed and converts this measurement into absorbance (A) units described by the following equation:

$$A(\text{absorbance}) = \log_{10}\left(\frac{1}{T}\right)$$

An example may help to clarify the relationship between transmittance and absorbance. If a dye solution is placed in the spectrophotometer and is found to transmit 10% of the light, its absorbance can be calculated thus:

$$\text{Since } T = 0.10$$
$$\text{And } A = \log_{10}\left(\frac{1}{T}\right) = \log_{10}\left(\frac{1}{0.1}\right)$$
$$A = \log_{10}(10) = 1$$

Using Quantitative Techniques and Statistics

Figure 5.5　Steps for using a spectrophotometer. Two models are shown and the following steps apply to either one. (1) Turn on the instrument and wait 10 minutes for it to stabilize. (2) Adjust the wavelength to 425 nanometers with the appropriate control knob or button. (3) Adjust the meter reading to zero transmittance by either rotating the zero control knob or using the zero control button. (4) Insert a tube containing solvent into the sample compartment. Keep the index line of the tube aligned with the index line on the sample holder. (5) Close the cover and adjust the meter to read 100% transmittance, which is the same as zero absorbance, by either rotating the 100% transmittance control knob or pushing the appropriate button. (This will adjust the amount of light passing through the spectrophotometer slit, thus controlling the amount of light that reaches the sample and the phototube.) (6) Remove the tube and close the lid. If the meter readout does not return to zero, readjust accordingly. (7) Reinsert the solvent tube to see if the instrument still registers 100% transmittance; if not, readjust. If the spectrophotometer is used for any length of time, recheck these readings now and then. Once the meter is calibrated, measurements can be made on samples at that wavelength only. You must recalibrate at each new wavelength. (8) Insert sample to be measured and record results.

Zero transmittance control　　100% transmittance control　　Wavelength control

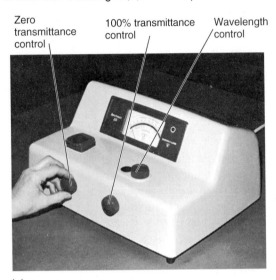

(a)

(b)

Sample compartment

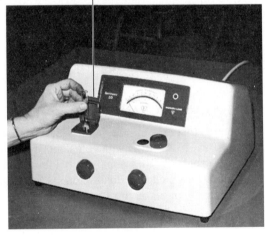

(c)

(d)

Scientists prefer to work with absorbance units because they are directly and linearly related to the concentration of the dye in solution (fig. 5.6). This relationship is described by the **Lambert-Beer Law,** which states that for a given concentration range and a sample tube of constant diameter, the absorbance is directly proportional to the concentration of solute molecules.

This relationship will be true regardless of the dye used, providing: (1) monochromatic (one-color) light is used, and (2) this color is the wavelength of light best absorbed by the dye. The wavelength of maximum absorption will always be given in the directions for any experiment or can be experimentally determined.

Figure 5.6   Standard curve of the absorbances of known dye concentrations. First, the absorbances of four known concentrations of dye are measured and plotted to make a graph. Then the absorbance of an unknown concentration is measured. The graph is used to find the concentration of dye that would give the measured absorbance.

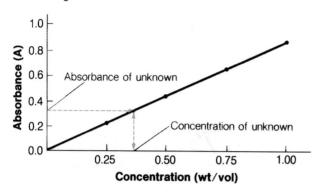

TABLE 5.3   Absorbance readings
for bromophenol blue

Wavelength (nm)	Absorbance Units
425	_____
450	_____
475	_____
500	_____
525	_____
550	_____
575	_____
600	_____
625	_____

Creating an Absorbance Curve

To learn how to use the spectrophotometer, you will explore the Lambert-Beer relationship using bromophenol blue dye. Your instructor will supply you with a 0.02mg/ml solution of bromophenol blue in water.

6▷   First, determine the wavelength of maximum absorption by following these steps:

1. Turn on the spectrophotometer and familiarize yourself with the position of the control knobs. Your lab instructor will discuss the features of the spectrophotometers in your lab.

2. Adjust the meter to 0% *transmittance,* using the zero control knob/button.

3. Obtain two matched sample tubes and clean them, if necessary, with tissue. Fill one tube with distilled water (solvent) and use it as a blank. Fill the other with bromophenol blue solution. This will be your experimental tube.

4. Set the wavelength control at 425 nm, insert the blank tube containing water, and adjust the reference knob, so that the meter reads 0 *absorbance* (= 100% transmittance).

5. Remove the blank tube and recheck the 0% *transmittance.* If you must readjust, repeat step 4. Insert the tube containing 0.02 mg/ml bromophenol blue.

6. Look at the readout and record the *absorbance* in table 5.3. Be sure to read the correct scale. Do not read the % transmittance scale.

7. Change the wavelength to 450 nm and repeat the adjustments in steps 4, 5, and 6. Continue these measurements at 25 nm intervals up to 625 nm. Remember to always recalibrate (steps 4 and 5) at each new wavelength.

8. Note the wavelength at which bromophenol blue absorbs the most light.

To be more precise in determining the absorption maximum, measure the absorbance by changing the wavelength of light to plus and minus 10 nm on either side of the absorption maximum. Enter the values in the space that follows table 5.3.

    Plot the measured *absorbance* as a function of wavelength on the graph paper at the end of this exercise. Wavelength, the independent variable, should be on the x-axis and absorbance on the y-axis. Instructions for drawing graphs are in appendix B. Read them!

Constructing a Standard Curve

In this section you will construct a *standard curve,* which demonstrates the linear relationship between absorbance and concentration. You will then use this standard curve to determine the concentration of dye in an unknown.

7▷   Obtain a stock solution of bromophenol blue containing 0.02 mg/ml. Prepare a series of dilutions in eight test tubes, using the proportions of dye and water listed in table 5.4. After adding the solutions, mix by holding the test tube in the left hand between your thumb and forefinger. Gently strike the bottom of the tube several times with the forefinger of your right hand to create a swirling motion of the fluid in the tube. Do not strike the tube so hard that the dye splashes out.

8▷   Set the spectrophotometer at the maximum absorption wavelength for bromophenol blue. Record the wavelength in the heading of table 5.4. After calibrating the spectrophotometer with water as a blank, read the *absorbance* for all eight tubes in the dilution series at this wavelength. Because you are not changing wavelength, you do not need

TABLE 5.4 — Concentration and absorbance for eight dye dilutions at _____ nanometers

Tube	ml of Dye	ml of H₂O	Concentration	A
1	0	4.00	0.00 mg/ml	____
2	0.50	3.50	____	____
3	1.00	3.00	____	____
4	1.50	2.50	____	____
5	2.00	2.00	____	____
6	2.50	1.50	____	____
7	3.00	1.00	____	____
8	3.50	0.50	____	____
9	4.00	0	0.02 mg/ml	____
10	unknown	0	____	____

to blank the spectrophotometer between readings. Record your results in the last column of table 5.4.

To calculate the actual concentration of bromophenol blue in each tube in units of mg/ml, do the following:

1. Determine the total mg of dye added to each tube by multiplying the number of ml of added dye by the dye concentration, which was 0.02 mg/ml.

2. Then divide the value by 4 ml, the total volume of fluid present after adding water. Record the concentrations in the fourth column of table 5.4.

These calculations are summarized in the work table that follows:

Tube	ml dye	total mg dye	Dye concentration
1	____	____	____
2	____	____	____
3	____	____	____
4	____	____	____
5	____	____	____
6	____	____	____
7	____	____	____
8	____	____	____

9 ▶ Use the graph paper at the end of this exercise to plot your data with absorbance as a function of dye concentration. Instructions for drawing graphs are given in appendix B. Remember that the x-axis, or abscissa, is always the independent variable and the y-axis, or ordinate, is always the dependent variable. In this experiment, what is the independent variable?

Label both axes. After plotting your data points, draw a straight line that, on the average, best fits all points of the data. Do not connect the points and create bumpy curves. The best straight line technique compensates for some of the random variability in the data.

The plot of absorbance as a function of dye concentration is called a **standard curve.** By reading the graph, you can determine the absorbance of any dye concentration within the range of concentrations tested. The line on the graph may be extrapolated to predict the absorbance of concentrations beyond the highest tested, but there is always a danger that the Lambert-Beer Law does not apply at very high concentrations.

10 ▶ Your instructor will now give you a solution of bromophenol blue that contains an unknown amount of dye.

How can you determine the dye concentration using the spectrophotometer and the standard curve you just constructed? Briefly outline the steps below. Perform the procedure and record the result below.

Unknown concentration = _____ mg/ml

Your instructor will now tell you the actual concentration of dye in the unknown. Calculate your percentage error.

Learning Biology by Writing

Prepare a lab report in which you present your data on determining the dye concentration in the unknown (include your graph). Be sure you state the problem and then discuss how you solved it. In your discussion, describe in general terms some sources of experimental error in lab work and the value of working with repetitive measurements.

As an alternative assignment, your instructor may ask you to complete the following summary and critical thinking questions.

Lab Summary Questions

1. Explain how you would use a standard curve to determine the amount of orange dye that had been added to a can of orange soda.
2. Describe the differences between mean, range, and standard deviation.
3. Hand in the histogram for the class data on volume of water. Show the mean and standard deviation for the class data.

4. Turn in your standard curve of absorbance versus dye concentration. Indicate the absorbance of your "unknown" on the graph and the dye concentration that corresponds.

Critical Thinking Questions

1. Explain why the sample cuvettes (sample tubes) used with the spectrophotometer are "matched." Why were the outsides cleaned each time before placing them in the spectrophotometer? What effect does the sample volume have on readings? What effect does tube placement have on readings?
2. Describe a "normal distribution" curve. What percentage of the data is encompassed by one standard deviation?
3. What is meant by "grading on the curve" or "curving" the grades?
4. A class has just finished doing the standard curve part of this exercise and determined the dye concentrations in unknown samples. All of the determinations are 25% lower than expected. Describe at least three sources of error that would give these results.

Using Quantitative Techniques and Statistics

LAB TOPIC 6

Modeling Biological Molecules

Supplies

Preparator's guide available at
http://www.mhhe.com/dolphin

Equipment and Software

Computers with RasMol installed. Program can be
downloaded from http://www.bernstein-
plus-sons.com/software/rasmol/
Download the structures for the following molecules by
using the GOOGLE search engine and typing the
molecule's name followed by a space and the
letters pdb: (1) water; (2) glucose; (3) sucrose;
(4) amylose; (5) stearic acid; (6) oleic acid;
(7) glycerol; (8) triglyceride; (9) phosphoglyceride;
(10) membrane lipids; (11) alanine; (12) serine;
(13) leucine; (14) phenylalanine; (15) glutathione
or tripeptide; (16) peroxidase; (17) collagen;
(18) deoxyribose; (19) thymine; (20) ATP; (21) DNA.

also Tryptophan (Trp)

Materials

Beakers
Spatulas, small
Magnetic stirrer and bars
Distilled water

Chemicals

Glucose
Phenylalanine
Serine
Leucine
Stearic acid

Student Prelab Preparation

Before doing this lab, you should read the introduction
and sections of the lab topic that have been scheduled
by the instructor.
　　You should use your textbook to review the
definitions of the following terms:

Acid	Ester linkage
Amino group	Functional group
Atomic mass	Glycosidic linkage
Base	Hydrogen bond
Carbonyl group	Hydrophilic
Carboxyl group	Hydrophobic
Covalent bond	Hydroxyl group

Molecular weight (mass)	Phosphodiester linkage
Methyl group	Polymer
Monomer	Ring structure
Phosphate group	Sulfhydryl group

You should be able to describe in your own words
the following concepts:

Role of covalent and hydrogen bonds in three-
dimensional structure of molecules
Carbon backbones in organic molecules
Role of functional groups in biochemical reactions
Condensation or dehydration synthesis reactions in
polymerization
How molecular shape can reflect molecular function

As a result of this review, you most likely have questions
about terms, concepts, or how you will do the
experiments included in this lab. Write these in the
space below or in the margins of the pages of this lab
topic. The lab activities should help you answer these
questions, or you can ask your instructor during the lab.

Objectives

1. To recognize carbon backbones and functional
groups of biological molecules
2. To learn to distinguish between covalent and
hydrogen bonds
3. To predict hydrophilic and hydrophobic
interactions based on functional group and carbon
backbone analysis
4. To recognize that the shapes and sizes of
biological molecules relate to their function
5. To recognize the role of condensation in polymer
formation

Background

Although virtually every element can be found in living organisms, the most common elements in biological molecules are carbon, hydrogen, oxygen, nitrogen, phosphorous,

Figure 6.1 Examples of carbon backbones in organic molecules. Each line represents a covalent chemical bond. Molecules differ in length of backbone, bonding pattern, and elements present.

Figure 6.2 Common functional groups in biological molecules.

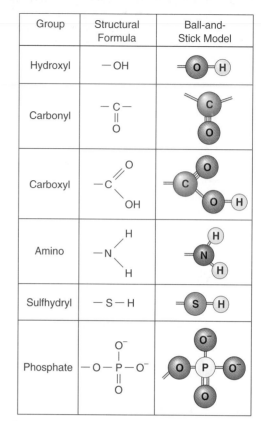

Group	Structural Formula	Ball-and-Stick Model
Hydroxyl	—OH	
Carbonyl		
Carboxyl		
Amino		
Sulfhydryl	—S—H	
Phosphate		

and sulfur. The acronym CHONPS is a convenient way to remember the elements. Carbon is by far the most common element. A carbon atom has the ability to chemically combine with other carbons or with H, O, N, P, S, as well as with other elements. Carbon-containing molecules are called organic molecules, and those common in living organisms are called biochemicals.

Carbon forms four covalent bonds by sharing electrons with up to four adjacent atoms. For example, a single hydrogen can form one covalent bond, so a carbon could share electrons with four hydrogen atoms (CH_4), or it could combine with two atoms of oxygen (CO_2) where two covalent bonds link each oxygen to the carbon. Carbon can share electrons with adjacent carbons to build chains of carbons called **carbon backbones** (fig. 6.1). Such backbones may vary from two carbons to over 18. Other elements can share electrons with the carbons in the backbone to form projecting "limbs" that give the organic molecule chemical and physical properties that the backbone alone does not have. Sometimes these limbs can be complex and consist of several atoms. The limbs, called **functional groups,** give the molecule functional properties and convey a unique shape to the molecule. Most biochemical reactions involve only the functional groups of a molecule as you will see when we discuss biological polymers.

Six types of functional groups are most often found in biological molecules although there certainly are others that are important. These are shown in figure 6.2. All biological molecules do not have the same functional groups. Some contain only one type of group while others have several. The single dashed line before the functional group indicates that the group will share one pair of electrons with a carbon. Theoretically, a single carbon could combine with up to four different functional groups.

The molecular weight (mass) of an organic molecule is a measure of its size. For example, glucose ($C_6H_{12}O_6$) has a molecular weight of 180. This is calculated by multiplying the number of each type of atom present in the molecule by the atomic mass and then summing the masses due to all atoms for a total molecular mass (weight). Molecular weights are measured in units called **daltons.** One dalton equals one atomic mass unit. Biochemical molecules can have molecular weights over a wide range from approximately 100 for many amino acids to over 1,000,000 daltons for proteins and nucleic acids.

Biochemists talk about classes of organic molecules: carbohydrates/polysaccharides, lipids, amino acids/proteins, and nucleotides/nucleic acids. All of these molecules have carbon backbones but differ from each other in the length of their backbones and in the types of functional groups that are present. There are certainly additional classes of molecules, but knowledge of the four basic classes allows one to appreciate many of life's molecular processes. This lab provides an opportunity to study some representative molecules of each of the four classes.

The medium of life is water. Almost all biochemical reactions involve molecules that are dissolved in the water found in the cell. The simple structure of water (H_2O) obscures its great importance and students often do not give it any thought. What is it about water that makes it worth thinking about? Water is a **polar molecule,** a dipole with one region that has a tendency to be positive

Figure 6.3 Covalent and hydrogen bonds in water. Two hydrogens share electrons with oxygen through covalent bonds. Oxygen attracts the electrons more strongly than do the hydrogens. Consequently, the oxygen bears a partial negative charge and the hydrogens bear a positive charge, creating a dipole molecule. The resulting dipoles are attracted to one another, forming hydrogen bonds.

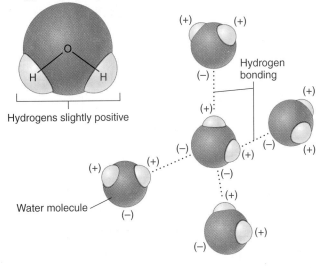

Figure 6.4 Ionized substances such as salt dissolve in water because the water dipoles are attracted to the ions, forming hydration shells around them. The hydration shells prevent the ions from binding back into the crystal.

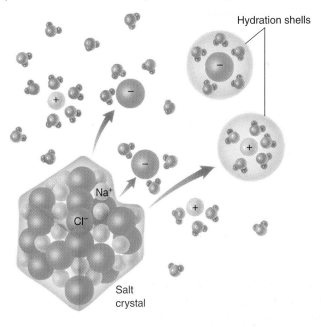

and two regions that tend to be negative (fig. 6.3). Consequently, water molecules interact with each other by forming weak opposite-charge interactions called **hydrogen bonds.** Hydrogen bonds are much weaker than covalent bonds, but nonetheless are very important in biological systems. They determine what dissolves in cells and are the basis for the three-dimensional structure of proteins and nucleic acids.

Hydrogen bonding explains why some molecules dissolve in water and others do not. Figure 6.4 shows why table salt (NaCl) readily dissolves. Water, being a polar solvent, will interact with any other molecule that is polar or that has a charge on it. Na^+ and Cl^- ions in a crystal are in constant thermal motion. As an ion starts to escape a crystal, it is usually pulled back by ions of the opposite charge, except in aqueous solutions. In water, the escaping ion is immediately surrounded by oriented water molecules that form a **hydration shell** around it. The hydration shell prevents the escaped ion from packing back into the crystal; consequently, it remains in solution.

A molecule that dissolves in water is said to be **hydrophilic,** or water loving. If a molecule is relatively neutral with no polar tendency, it will not dissolve in water and is said to be **hydrophobic,** or water hating. Among biochemicals, those with the functional groups -OH, -COOH, $-H_2PO_4$, and $-NH_2$, are hydrophilic, whereas those with -H and $-CH_3$ are hydrophobic. In larger molecules, it is the relative number and location of these groups that determines solubilities. In this lab, you will look at the structures of molecules and predict whether they are hydrophilic or hy-

drophobic. You will then test your predictions by trying to dissolve the substance in water.

Many biological molecules are **polymers,** very large molecules that consist of covalently linked units generally referred to as **monomers.** For example, a protein can be made from a few to several hundred amino acids whereas starch and cellulose are polymers made from thousands of sugar molecules. Biological polymers are made in **condensation** or dehydration synthesis reactions (fig. 6.5). In these reactions, the -OH group of a carboxyl group combines with a hydrogen from a hydroxyl or amino group on another molecule to join the two together through a new covalent bond. Because many monomers have more than one of these groups, they can combine with several other monomers to build up chains often thousands of monomers in length. Such synthetic reactions usually require the expenditure of cellular energy to make the bonds. These bonds are variously called glycosidic, ester, peptide, and phosphodiester linkages as you will soon see. These bonds can also be broken through hydrolysis reactions. Digestion of starches, lipids, proteins, and nucleic acids are examples of hydrolysis reactions.

In large, long molecules some regions may be hydrophilic and others hydrophobic depending on the functional groups and where they are located in the molecule. If one end has several functional groups containing oxygen or nitrogen, that end will be hydrophilic because these are polar functional groups. If the other end has no polar groups, it will be neutral. What this means is that one end of the molecule will interact (dissolve) in water but the other will not. This can cause the molecule to fold

Modeling Biological Molecules **59**

Figure 6.5 (a) Condensation reactions between functional groups form biological polymers. A hydrogen from one monomer is removed along with a hydroxyl from the other to form water and a larger organic molecule. (b) The bond can be broken in a hydrolysis reaction. This is what occurs in digestion.

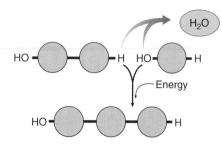

(a) Condensation or dehydration synthesis

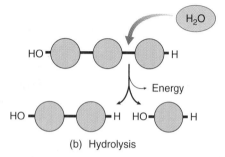

(b) Hydrolysis

into unique shapes, and it is shape that gives biomolecules particular functions. These functions can include structural support, recognition mechanisms, cellular signaling, information storage, and catalytic activities. In this lab, you will look at the structure of several biological polymers.

LAB INSTRUCTIONS

This lab uses computer modeling to investigate the composition of common types of biological molecules. You will make predictions about solubilities based on analysis of structures and then test the predictions by trying to dissolve the actual chemicals. You will look at how polymers form from monomers and at the unique shapes of several molecules.

Getting Started

Work with a partner so that you can discuss what you are seeing. On your computer, there is a folder called Molecular Modeling. In it is a program called RasMol, which is a program for visualizing molecules. The program works from molecular data files called pdb (protein data bank)

files that have three-dimensional coordinates for the atoms found in a particular type of molecule. These files should be numbered in the order that you will use them during this lab.

The easiest way to open the files is to click on them and drag them on top of the file called "drag molecules here." When it highlights, release the mouse and the file will open.

1 ▷ **Water** Click on the file entitled **01.Water.pdb** to open it.

When the file opens, you should see a wire frame representation of the molecule on a black background. If a second box called "RasMol Command Line" also opens, close it. At the top of the screen, click on DISPLAY, and note there are several ways to display the data. Try a few different ones so that you know the program's capabilities. When finished experimenting, click on *ball and stick*. This is the display mode you should use in most of the lab.

If you put the cursor on top of the water molecule and click and hold while moving the mouse, you will be able to rotate the molecular structure to view it from different angles. You can grab the lower corner of the display screen with your cursor and click and drag to increase its size. On a Macintosh computer the molecules can be positioned on the screen by holding down the COMMAND key on the keyboard and then moving the mouse. On a PC, hold down the right click key while moving the mouse.

In table 6.1 on the next page the atomic weights for the common elements in biochemicals are given. What is the molecular weight of water? _18_ Keep this number in mind as you study other molecules.

Note the general shape of a water molecule. The bond angle between the hydrogens is about 109° and the lines from the Hs to the O represent electrons shared through covalent bonds. What the model does not show is that the electrons tend to be found more frequently around the O nucleus than around the H nuclei. Because electrons carry a negative charge, the O has a tendency to be negative. Clustering of the negatively charged electrons around the oxygen leaves the positively charged hydrogen nuclei exposed. They have a tendency to be positive. Thus a **dipole** is established. When biochemicals containing polar functional groups are added to water, the dipoles orient around the groups in hydration shells. Sometimes these are sufficient to keep the molecule suspended in solution. When the biological molecules are very large with hydrophilic and hydrophobic regions, the hydration shells cause hydrophilic regions to extend and hydrophobic regions to fold. This gives unique three-dimensional shapes to molecules.

Close the data file for this molecule by going to FILE at the top of the screen and clicking on *CLOSE*.

During this lab you will look at 20 other molecules and answer questions about their structure, shape, and solubility. This is a compare and contrast exercise to help you recognize different classes of molecules, functional groups, monomers, polymers, molecular shapes, and hydrophobic and hydrophilic characteristics.

Carbohydrates

Commonly known as sugars or saccharides and by their polymer names such as starch and cellulose, carbohydrates are found in every living cell. They serve as sources of energy when broken down in metabolism, serve as a source of chemical building units for other types of molecules, as cell to cell recognition molecules, and for structural support in cell walls of plants, fungi, and bacteria.

You will start your investigation of carbohydrates by looking at simple sugars or monosaccharides. All sugars have a structural formula based on the ratio of one C to two Hs to one O. The number of carbons in the backbone of the molecule can vary between 3 and 7. Thus when there are five carbons, the formula is $C_5H_{10}O_5$. This would be called a pentose because it is a sugar with five carbons. How many carbons are found in a hexose? ___6___

All sugars have multiple hydroxyl (-OH) functional groups (fig 6.6) and some will have phosphate and amine groups as well. Glucose is a hexose. Its structure can be represented in a number of ways as is shown in figure 6.6. Crystals of glucose exhibit the straight chain structure, but when dissolved in water convert to the ring structure. The ring structures can be portrayed in a simplified or short hand notation, where the carbons located at each angle of the ring are not drawn. As a reader of chemical structures, you have to learn to mentally fill in the carbons. Look at one of the ring structures in figure 6.6, count the number of carbons and hydrogens in the molecule, and compare your count to the chemical formula for glucose, $C_6H_{12}O_6$.

2 ▶ *Glucose*   You will start your investigation by looking at the structure of glucose (fig. 6.6). Glucose is a common sugar in organisms. It is metabolized as an energy source and is used to make storage and structural polysaccharides.

Open the *02.glucose.pdb* data file. You should see a wire frame representation of the molecule on the screen. At the top of the screen, under DISPLAY, click on the *ball and stick* representations and the display will change. If you put the cursor on top of the molecular structure and click and hold while moving the mouse, you will be able to rotate the molecular structure to view it from different angles. You can grab the lower corner of the display screen with your cursor and click and drag to increase its size. Answer the following questions and rotate the molecule for better views.

What chemical elements are found in glucose?

carbon (gray)
oxygen (red)
hydrogen (white)

Figure 6.6   Several ways of representing the structure of glucose. (*a*) linear form found in dry crystals; (*b*) folded linear form; (*c*) combination linear-ring; (*d*) ring form found in aqueous solution in cells; (*e*) space-filling model; (*f*) folded boat form. The darker lines represent 3–D aspect of structure.

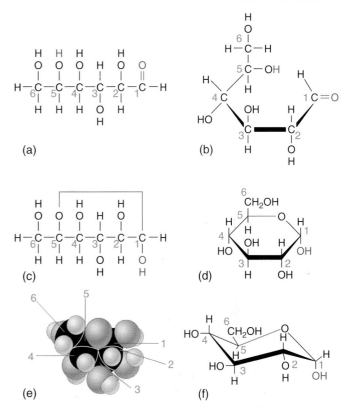

TABLE 6.1	Atomic masses	
Element	**Atomic mass**	**Color in model**
Hydrogen	1	*white*
Carbon	12	*gray*
Nitrogen	14	*blue*
Oxygen	16	*red*
Phosphorous	31	*orange*
Sulfur	32	*yellow*

In this and the other molecules you will study, look at the colors used to display the atoms of different elements. Record them in table 6.1 and add to the table as you work through the exercise and encounter other elements.

Modeling Biological Molecules

Figure 6.7 Condensation reaction between glucose and fructose forms sucrose when a glycosidic linkage forms between the monomers. The linkage is broken in hydrolysis reactions. Note that -OH groups are left off carbons in rings for clarity.

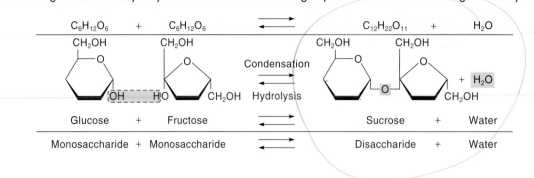

? Calculate the molecular weight (mass) of glucose by counting the number of each type of atom and multiply that number by the atomic mass. Sum these numbers to get the mass of the molecule.

$6 C \times 12 = 72$
$12 H \times 1 = 12$
$6 O \times 16 = 96$
_____
$180 d$

? From your earlier calculation, you know that water has a molecular weight of 18. How many times heavier (and larger) than a water molecule is a glucose molecule? __10X__

? What functional group is common in carbohydrates? __hydroxyl (–OH)__

Hydroxyl groups are polar functional groups. The oxygen tends to be negatively charged because of electron clustering, and the hydrogens tend to be positive. Draw a straight chain glucose molecule and then indicate how water molecules would form a hydration shell around it. Recognize that the shell would be several water molecules thick as the molecules in the first layer oriented to those in the second, and so on.

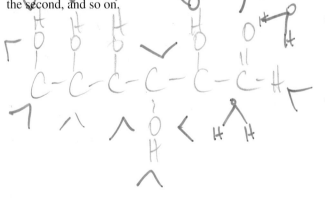

? Predict whether glucose is hydrophilic or hydrophobic? Why do you say so?

Hydrophilic – because the hydroxyl groups have partial charges

? Test your hypothesis by adding some glucose to water and stirring. Describe the experimental results. Must you accept or reject your hypothesis?

Fail to reject (the glucose dissolves easily)

Close the data file by going to FILE at the top of the screen and clicking on *CLOSE*.

3 ▶ *Sucrose.* Now open the data file *03.sucrose.pdb.* When the file opens, use DISPLAY at the top to change the view to *ball and stick.* Sucrose, also known as table sugar, is used by plants to transport photosynthetic products from leaves to roots as well as other parts of the plant.

? Can you see two ring structures in sucrose? How many rings were found in glucose?

One in glucose

Sucrose is a disaccharide, meaning that it is made from two monosaccharides. Both glucose and fructose are found in sucrose. They have the same composition, $C_6H_{12}O_6$, but differ in how the atoms are arranged in space. Look carefully at the two components of sucrose and describe the differences.

In glucose, 5 carbons are in ring, while fructose has 4 carbons in ring.

Condensation reactions occur when a hydroxyl group of one sugar reacts with a hydroxyl group of another (fig 6.7) to form a covalent bond called a **glycosidic linkage** between the two sugars with water as the by-product. This bond can be broken in a hydrolysis reaction, yielding the two starting sugars. When sucrose is made in plants, the condensation reaction occurs and when we metabolize sucrose, enzymes in our bodies perform the hydrolysis reactions. Locate the glycosidic linkage between the two sugars. Draw the atoms and bonds found in the linkage below.

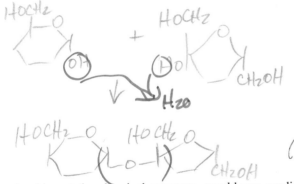

Looking at the chemical structure, would you predict that sucrose is hydrophilic or hydrophobic? Why? Is this prediction consistent with your observations when you add sugar to your tea or coffee?

Hydrophilic — lots of -OH w/ partial charges. Sugar dissolves easily in coffee

If you gently heat a sugar solution made up in distilled water, the water evaporates as hydrogen bonds are broken by the heat and the water molecules escape as a vapor. Eventually, the sugar will recrystalize and be the same as the sugar you added. What does this experiment tell you about the relative strengths of hydrogen and covalent bonds?

Covalent bonds are stronger than hydrogen bonds

Close the data file for this molecule by going to FILE at the top of the screen and clicking on *CLOSE*.

Polysaccharides

Polysaccharides are large molecules composed of several sugar units, called residues, in glycosidic linkage. There are many different polysaccharides found in living organisms. Cellulose is found in plant cell walls and chitin is found in the cell walls of fungi and in the exoskeletons in many invertebrates. Starch is one of many different storage polysaccharides that are synthesized when excess sugars are available, but is later broken down when sugars are needed. Mucus is another interesting polysaccharide that is very important in the animal kingdom and in plant roots as they push their way through the soil. The hydroxyl groups projecting from mucus polymers bind water molecules so that layers of water build up around each polymer. Mucus feels slippery because when such hydrated molecules are placed between two surfaces that are moving across each other, the hydrogen bonds are sheared and the water molecules slide by each other. You will look at the starch amylose as an example of a polysaccharide (fig. 6.8).

4 ▶ **Starch.** Open the file for *04.amylose.pdb*. Use DISPLAY to change the view to *ball and stick*. The starch molecule you see on your screen is but a small section of a complete amylose molecule. Amylose will typically contain over 1,000 glucose residues, the name given to the building units in a polymer. Amylose is one form of plant starch and was chosen because its structure is a straight chain. Other starches have branches and are more complex.

Find the glucose residues in the starch molecule. How many are shown in this sample?

Two

Figure 6.8 Structure of starch granules found in potato cells. Glycosidic linkages formed by condensation reactions join glucose residues into long chains. Some starches, called amylopectins, are branched; others, called amyloses, are not.

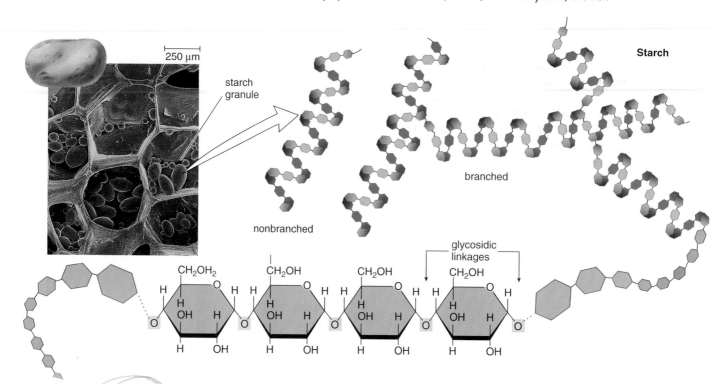

❖ Find the glycosidic linkages between adjacent glucose residues. How many are there?

one less than glucose residues

❖ Is this molecule hydrophilic or hydrophobic?

hydrophilic

❖ You might think that starch would dissolve in water. It does so only partially because the molecular weight is so great. If a thousand glucose units condense to form a starch molecule, what is the approximate molecular weight of the starch? *1.8×10⁵ d* Large molecules such as this often do not enter into true solution but are suspended as a colloidal mixture when added to water. Colloids consist of aggregations of molecules that remain suspended in a liquid.

Close the data file for this molecule by going to FILE at the top of the screen and clicking on *CLOSE.*

At the end of this lab topic, find the summary table (6.2) for characteristics of biological molecules. Fill in the information for carbohydrates.

Lipids

Lipids as a class of compounds share one characteristic: they are not soluble in water and are said to be hydrophobic. Common names for different kinds of lipids include fats,

glycerides, oils, phosphoglycerides, steroids, and waxes. Lipids vary widely in their functions in cells. Fats and oils store large amounts of energy. When an organism is taking in excess food, fats are synthesized and stored. Later when food intake is low, the fats may be broken down so that the energy can be used. Other fats are extremely important in forming cellular structures. Phospholipids are lipids containing phosphate groups. They are the basic building units of cellular membranes, defining the cell boundary as well as forming many structures (organelles) within eukaryotic cells. Steroids are also found in membranes, and in many animals some function as homones. Waxes are waterproofing agents that prevent desiccation in terrestrial environments. In this section, you will investigate the structures of fatty acids, glycerides, phospholipids, and membranes.

Lipids do not polymerize to form large molecules. They are assembled from smaller units and have molecular weights that are only several hundred rather than several thousand. The basic units in glycerides are glycerol and fatty acids (fig. 6.9). Glycerol is a polyalcohol. It has a three-carbon backbone, each carbon with a hydroxyl group. Fatty acids are long chain molecules with up to 22 carbons in their backbones. Except for a terminal carboxyl group, the only other functional group is hydrogen. Because of their composition, fatty acids are very hydrophobic and do not dissolve in water. Some fatty acids will have double or triple bonds between some of the carbons in the backbone. They are said to be **unsaturated fatty acids.** In contrast,

saturated fatty acids have no double bonds between adjacent carbons. Fatty acids undergo condensation reactions with alcohols to form esters. Glycerol can combine with up to three fatty acids to form a triglyceride.

The physical state of glycerides or phospholipids, *i.e.*, whether they are oils or solid fats, depends on their molecular weights and state of saturation. The general rules are the longer the backbone of the fatty acid and the greater the degree of saturation, the more likely it will be solid at room temperature. Plant lipids are generally oils because they contain many unsaturated fatty acids. Manufacturers of margarine have long known this and adjust the fatty acid content of their products so that they will spread on fragile toast when more or less taken directly from the refrigerator. Waxes are lipids made from very long chain, saturated fatty acids so that they are hard solids at room temperature. When you polish an apple, friction melts the natural wax and forms a smooth surface.

5 ▸ *Stearic acid.* Open the data file for *05.stearic acid.pdb.* Switch to the *ball and stick* view and use your mouse to rotate it for the best views.

◇ Trace the carbon backbone on your computer screen. What chemical elements are found in stearic acid? Calculate the molecular weight for stearic acid. Is this less than or greater than that of glucose?

Mass greater than glucose

C, H, O (mostly C & H)

◇ How would you describe the shape of stearic acid? How does the shape differ from that of glucose?

linear w/ long hydrocarbon tail

(glucose was a ring)

◇ What functional group is present in stearic acid?

— carboxyl group

◇ Do you see any regions of the molecule that could be polar? *Yes* How do you think hydration shells would form around this molecule? Draw a cartoon below showing them.

6 ▸ *Oleic acid.* Move the stearic acid molecule to the top of your screen by depressing the COMMAND key and moving the mouse or using the right click button on your mouse. If you cannot move the molecule, coordinate with the students using the computer next to you so one computer displays stearic acid and the other oleic acid. Go to FILE at the top of the screen and open the *06.oleic acid.pdb* data file. Display it as a *ball and stick* model.

How does oleic acid differ from stearic acid? How is it similar?

Differs: It bends

Similar — Carboxyl group
— long hydrocarbon tail

Stearic acid, which you looked at previously, is a saturated fatty acid. All of the carbons share one covalent bond with adjacent carbons and have at least two hydrogens except for the end carbons. Oleic acid is an unsaturated fatty acid. It has a two covalent bonds between two adjacent carbons and those carbons have only one hydrogen each. The double bond causes the backbone to have a kink in it. The kinks prevent tight packing when in pure state and this has implications about what its physical state will be, *i.e.*, a liquid versus a solid. Unsaturated fatty acids tend to remain fluid at temperatures that cause saturated fatty acids to solidify.

Figure 6.9   Structures of glycerol, fatty acid, and a triglyceride. (*a*) Glycerides are formed when the carboxyl of a fatty acid undergoes a condensation reaction with a hydroxyl of glycerol. (*b*) Up to three fatty acids, which may be the same or different, can condense with one glycerol to form a triglyceride.

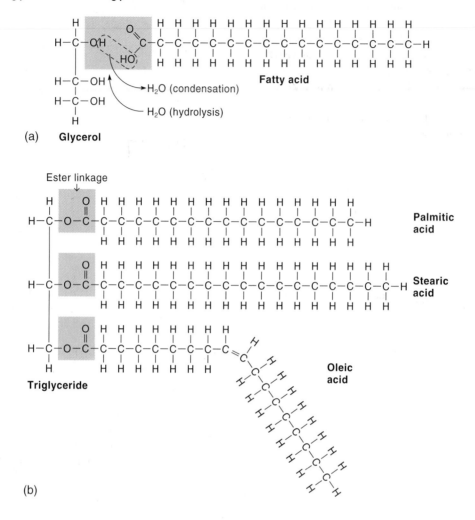

(a)   **Glycerol**

(b)

Triglyceride

Would you predict that stearic and oleic acid are hydrophilic or hydrophobic? Why?

hydrophobic — most of molecule is nonpolar hydrocarbon

Ask your lab instructor to demonstrate the solubility of fatty acids in water. Does this confirm or contradict your expectation based on molecular modeling?

confirms

Close the data files for these molecules by going to FILE at the top of the screen and clicking on *CLOSE* twice to close both data files.

7 ▸ *Glycerol.*   Open the structure for *07.glycerol.pdb.* Change the display to the *ball and stick* mode. Rotate the molecule and note its resemblance to a carbohydrate; *i.e.*, it is a polyhydroxyl molecule.

What elements are found in glycerol? What is it molecular weight?

C, H, O
3C × 12 = 36
8H × 8 = 8
3O × 16 = 48
92

What functional groups are present?

—OH hydroxyl

Close the data file for this molecule by going to FILE at the top of the screen and clicking on *CLOSE*.

Triglyceride. Carboxyl groups of fatty acids can undergo a condensation reaction with hydroxyl groups on other molecules to form an **ester linkage** (fig. 6.9). Because glycerol contains three hydroxyl groups, up to three fatty acids can form ester linkages with a single glycerol, forming a mono-, di-, or triglyceride. Different fatty acids can link with each hydroxyl group to form a complex molecule. Sometimes fatty acids will link to two of the hydroxyls and then a phosphate group will link to the third to form a phospholipid.

8 ▷ Open file *08.triglyceride.pdb.* Change to a *ball and stick* display. Locate the glycerol and fatty acid components of the molecule.

◆? Where are the ester linkages located?

Where fatty acids carboxyl attaches to glycerols hydroxyl

◆? Do you think that a triglyceride would be hydrophilic or hydrophobic? Why do you say so?

Hydrophobic — majority in nonpolar

Triglycerides and phospholipids are **amphipathic molecules,** large molecules that have both hydrophilic and hydrophobic regions.

Note the types of fatty acids linked to each position of the glycerol in this model. In cells, the three fatty acids could be the same or different. Sometimes they will have different backbone lengths or they will have different degrees of saturation. Which of the following do you think would be an oil at room temperature—a triglyceride containing saturated fatty acids that all had backbones that were 18 to 20 carbons long or one where the fatty acids were unsaturated with backbones of 14 to 16 carbons? Figure 6.10 shows how saturation affects the ability of lipids to pack closely together, leading to some being solids at room temperatures whereas others are liquids.

unsaturated — cant pack together

Move the triglyceride model to the bottom of the screen by either holding down the control button or right click and

Figure 6.10 (*a*) Hard fats such as those found in animals (*e.g.,* bacon grease) are composed of saturated triglycerides that can pack closely together. (*b*) Oils such as those found in plants (*e.g.,* cooking oil) are composed of unsaturated triglycerides that cannot pack closely together.

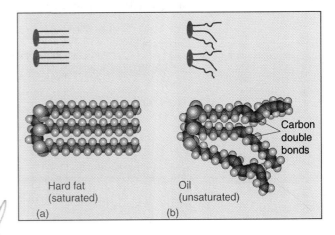

Hard fat (saturated)
Oil (unsaturated)
Carbon double bonds
(a) (b)

dragging the molecule. If this cannot be done, coordinate with another student group so one computer displays the triglyceride and the other displays the next molecule.

Phospholipid. Phospholipids are the types of lipids that are found in cell membranes. They are essentially a diglyceride with a phosphate group on the third hydroxyl group of the glycerol.

9 ▷ Open the file *09.phospholipid.pdb* and display it as a *ball and stick* model on the top half of the screen.

◆? Compare this molecule to the triglyceride. What do you see that is the same and what is different about the two molecules?

Both have fatty acids attached to glycerol and both are amphipathic. But triglyceride has 3 fatty acids, while phospholipid has 2 fatty acids plus a polar group

◆? How does a phospholipid differ from a fatty acid?

It has a larger polar region "head" attached to one end of glycerol

Figure 6.11   Phospholipid interactions with water.
(a) If phospholipids are added to an oil-water interface, they orient with hydrophobic tails in oil and hydrophilic heads in water. (b) If added only to water, they form bilayers with hydrophobic tails inward and hydrophilic heads outward.

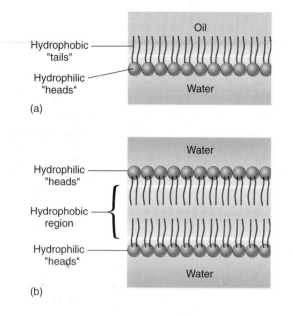

(a)

(b)

Find the glycerol in the phospholipid and lightly trace over it with the eraser end of a pencil.

Would you predict that this molecule is hydrophilic, hydrophobic, or that it has a little of both these properties, *i.e.*, it is amphipathic?

A little of both

Close the data files for these molecules by going to FILE at the top of the screen and clicking on *CLOSE* twice.

Membranes.   Phospholipids are amphipathic molecules. When added to water, phospholipids float with the charged "heads" down into the water and the hydrophilic tails up toward the air. In cells, phospholipids are the basic building blocks of membranes. When two or more phospholipids are synthesized, they tend to form a bilayer with the hydrophobic tails oriented toward each other, and the hydrophilic heads facing toward the water found within and around cells (fig. 6.11).

Open the file *10.membrane.pdb.* Leave this as a wireframe model so as not to clutter the view. Rotate the membrane so that you can see it from different angles. Note the lipid packing and how the kinks from the unsaturated fatty acids prevent close packing. Organisms with body temperatures that follow those of the environment can seasonally adjust the fluidity of their membranes by adjusting the types of fatty acids in their membranes. In hot months, the

membranes will have saturated long chain fatty acids so that they do not melt and lose their integrity. In cool winter months, flexibility is maintained by inserting more unsaturated fatty acids with shorter chain lengths. Cholesterol, which is normally found in eukaryotic membranes, is not shown in this diagram. It influences fluidity as well.

Close the data file for this structure by going to FILE at the top of the screen and clicking on *CLOSE*.

At the end of this lab topic, find the summary table (6.2) for the characteristics of biological molecules. Fill in the information for lipids.

Amino Acids

Approximately 20 types of amino acids are found in all living organisms, although theoretically several other amino acids are possible. The name "amino acid" recognizes two common functional groups that all of these molecules have: an amino group and an acid (carboxyl) group. At neutral pHs, the carboxyl group ionizes by losing a hydrogen ion and the amino group is protonated by gaining a hydrogen ion. Consequently, carboxyl groups are written as -COO⁻ and the amino groups as $-NH_3^+$

Amino acids are very important because they are the building units of proteins. Tens, sometimes hundreds, can chemically combine to form large molecules that are important structural, communication, transport, and catalytic molecules in cells. The chemical formulas for several amino acids are shown in figure 6.12. Note that each one has a single carbon, called the alpha carbon. It shares electrons with an amino group, a carboxyl group, a hydrogen, and a variable side group. Among these side groups, some are polar, others are charged, and most are non-polar. Consequently, some amino acids readily dissolve in water and are hydrophilic while others are hydrophobic. You will look at the three-dimensional structure of four amino acids: alanine, serine, leucine, and phenylalanine.

Open the data file for the amino acid *11.alanine.pdb.* Use your cursor and the command key or right click to move it to the upper left of the screen and then open the file *12. serine.pdb.* Position the molecule at the upper right of the screen. Now open the file *13.leucine.pdb* and position it on the lower left. Last, open the file for *14.phenylalanine.pdb* and position it at the lower right. Double click on each molecule model to select it and change the display to a *ball and stick* model. You may want to experiment with other DISPLAY modes to help you understand these structures.

What chemical elements are found in these molecules? Calculate the molecular weights for alanine and phenylalanine to give you an estimation of the range of molecular weights for amino acids.

C, H, O, and N

alanine 3C × 12 = 36
7H × 1 = 7
2O × 16 = 36
1N × 14 = 14
—————
93 d

phenyl-
alanine 9C × 12 = 108
11H × 1 = 11
2O × 16 = 36
1N × 14 = 14
—————
169 d

Figure 6.12 Some of the 20 common amino acids. Each has an alpha carbon that is bonded to both an amino group (H₂N—) and a carboxyl group (-COOH). At pH 7, the prevailing form of the amino is H₃N⁺ and of the carboxyl is COO-. They are categorized according to the main properties of their side chains. Nonpolar amino acids are hydrophobic. Polar and electrically charged amino acids are hydrophilic.

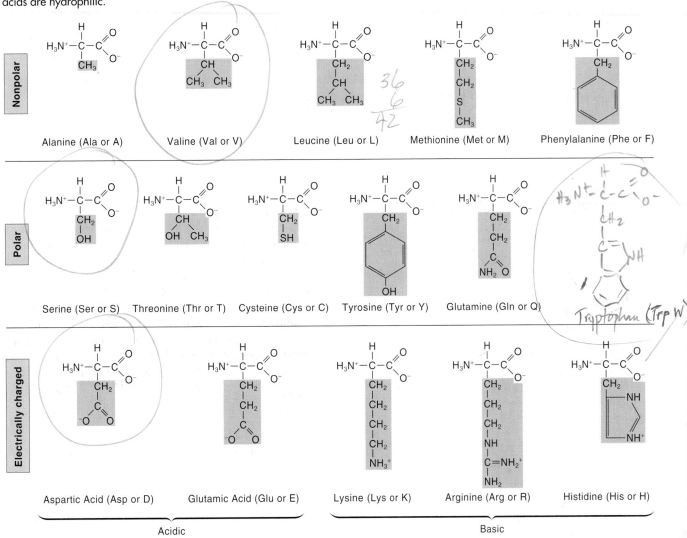

Nonpolar: Alanine (Ala or A), Valine (Val or V), Leucine (Leu or L), Methionine (Met or M), Phenylalanine (Phe or F)

(handwritten near Leucine: 36, 6, 42)

Polar: Serine (Ser or S), Threonine (Thr or T), Cysteine (Cys or C), Tyrosine (Tyr or Y), Glutamine (Gln or Q), Tryptophan (Trp W)

Electrically charged: Aspartic Acid (Asp or D), Glutamic Acid (Glu or E), Lysine (Lys or K), Arginine (Arg or R), Histidine (His or H)

Acidic — (Aspartic Acid, Glutamic Acid)
Basic — (Lysine, Arginine, Histidine)

How would you describe the shape of each of these amino acids? Are they similar or different?

(handwritten answer) Different — Phenylalanine is largest w/ a ring sidechain.
Size: Phe > Leu > Ser > Ala

Identify the α carbon in each of the molecules. Find the four different groups attached to it: the carboxyl group, the amino group, a hydrogen, and then a group that is different for all four amino acids. In textbooks this is often referred to as the R group. In the table below, describe what you find in each molecule.

Name	Molecular weight	Amino present?	Carboxyl present?	R group description
Phe	169 d	yes	yes	large ring (phenyl group)
Leu	135 d	↓	↓	branched hydrocarbon
Ser	109 d			hydroxyl + methyl
Ala	93 d			methyl

(handwritten bottom: 93, 42, 135 and 93, 16, 109)

R groups affect the solubility of the amino acids. Large R groups that contain mostly carbon and hydrogen will make an amino acid relatively hydrophobic while small R groups or those containing nitrogen or oxygen will make it hydrophilic (fig. 6.12).

Which of these four amino acids would you predict is hydrophilic? Why? Which is hydrophobic?

hydrophilic: alanine, serine, aspartic acid (high enough ratio of N+O to C) - maybe valine

hydrophobic: valine? leucine? phenylalanine, tryptophan - (too much C+H for O's/N to pull into solution)

Sketch below how you think hydration shells would form around the most hydrophilic of the four amino acids on your screen.

serine

Figure 6.13   Condensation reactions between amino acids form peptide bonds. Reaction is between the amino group of one amino acid and the carboxyl group of another. Free amino or carboxyl groups on the new dipeptide can condense with other amino acids leading to elongation of the peptide.

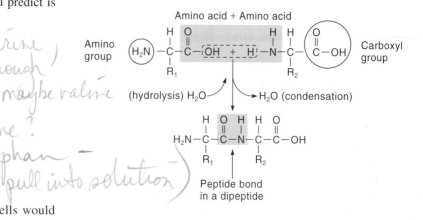

Amino acid + Amino acid

Amino group

Carboxyl group

(hydrolysis) H_2O ⟶ H_2O (condensation)

Peptide bond in a dipeptide

Do you accept or reject your hypothesis? Why?

Mostly accept, because solubility: Ser > Asp ≅ Val > Trp! Asp was less soluble than predicted (large size reduces solubility)

Close the data files for the amino acids by going to FILE at the top of the screen and clicking on *CLOSE* four times.

Polypeptides and Proteins

Amino acids will chemically combine with each other by forming **peptide bonds** to make polymers (fig. 6.13). This covalent bond is made when the amino group of one amino acid reacts with the carboxyl group of another in a condensation reaction. When two amino acids join through a peptide linkage, the resulting molecule, a dipeptide, has an amino group at one end and a carboxyl at the other. It can react with additional amino acids to make a tripeptide and so on, up to molecules called polypeptides, containing many amino acids. Each peptide has a unique shape, depending on which amino acids are in the sequence.

Testing a hypothesis.   Your instructor will discuss with the class which amino acid is predicted to be most hydrophobic and which the most hydrophilic. This hypothesis will be tested by adding crystals of these amino acids to water and measuring the time required for them to dissolve.

Record your hypothesis and the experimental results below.

Hypothesis: solubility: Serine ≥ Asp Acid > Valine > tryptophan

amino acid	time to dissolve
Ser	~30 seconds
Asp	most in 2-3 min.
Val	most in 2-3 min
Trp	didn't dissolve clumped together

15 ▶ **Tripeptide.**   Open the file *15.tripeptide.pdb* and display it as a *ball and stick* model. As you study this molecule, you may want to toggle back to the wire frame view. Find the peptide bonds between the amino acids and note how the bond involves the carboxyl group of one amino acid and the amino group of another. How many peptide bonds are in this molecule? *2*

Identify the backbone of the polymer as it passes along the length of the tripeptide. Note how the R groups project away from the backbone. Some of these are hydrophilic and will extend into the surrounding water in a cell while others are hydrophobic and will tend to cluster. Try to identify any potentially hydrophilic or hydrophobic R groups. If you start from the carboxyl terminus (end), are the hydrophobic amino acids in the first, second, third, *etc.* positions?

[handwritten: H₃N⁺ — philic — philic — phobic — COO⁻]
[handwritten drawings of R groups]

Close the data file for this molecule by going to FILE at the top of the screen and clicking on *CLOSE*.

Proteins. Proteins are large polymers of amino acids. Small proteins like the hormone insulin are made from 52 amino acids, whereas large ones like the blood pigment hemoglobin contain over 600. Cells contain thousands of different proteins, each of which is synthesized from information stored in the gene for that protein. This means that the types of proteins we have in our cells are inherited characteristics depending on our individual genetic legacies and mutation processes.

When describing proteins, four levels of structure are used (fig. 6.14). The primary structure is the sequence of amino acids. If you had two different proteins each consisting of 100 amino acids, both would contain the same 20 types, but one might have more of one type of amino acid and less of another so that the primary sequences would be different. The English language is based on an alphabet of 26 letters and an almost infinite variety of words is possible when these are combined in different sequences. So it is with proteins, where the alphabet consists of 20 amino acids and the words are very much longer than those in English.

Beyond the primary level of structure, there are three additional levels which describe the protein's shape. The secondary level of structure in proteins describes the shape of the protein's backbone. It can have regions that are helical or regions that are folded in a zig-zag pattern called a pleated sheet. The tertiary level of structure describes how the chain of amino acids with its helices or pleated sheets is folded to give the protein a unique shape. Proteins that have regions consisting of polar or charged amino acids will extend outward into the surrounding water. Those consisting of hydrophobic amino acids will fold inward so that the hydrophobic amino acids pack together. This folding pattern gives each protein a unique shape that is related to its function. Large proteins are often made of subunits. Each subunit is a polypeptide and the polypeptides associate to form the complete protein. The combining of subunits is called the quaternary level of structure.

All higher levels of structure are determined by the primary level. The sequence of amino acids with their hydrophilic or hydrophobic side groups determines how the protein will fold and associate with other structures. Therefore, in specifying a sequence of amino acids a gene is also specifying the folding pattern and shape of a protein. You will look at two proteins with distinct shapes. One is a globular protein and the other is a fibrous one.

Open the data file for *16.peroxidase.pdb.* Depress the COMMAND key or use the right click button and your mouse to move the molecule to the bottom of the screen. Now open file *17.collagen.pdb* and move it to the top of the screen. Later you may want to experiment with DISPLAY modes to better understand the structures you are viewing.

Peroxidase is an enzyme, and in fact one that you will study in the next lab topic. Its shape is globular, as are most enzymes. Collagen is a structural protein and like most such proteins has an elongated shape. These are called fibrous proteins. You are not looking at a whole collagen molecule. Normally it would be about ten times longer; only a small section is shown here. Collagen protein accounts for about one quarter of your body weight. It is found in tendons and ligaments. It also forms tough connecting sheets that support your organs, is a component of bone, and toughens the base of your skin.

Contrast the shapes of the two proteins.

[handwritten: Collagen: narrow & long (fibrous) like a rope]
[handwritten: Peroxidase: oval-shaped globular protein]

Locate the peptide bonds between the carboxyl and amino groups in these proteins. Can you trace the backbones of the molecules? *[handwritten: yes]*

Find where the side groups are located along the chain. Are there any free amino or carboxyl groups?

[handwritten: one free amino, one free carboxyl } per polypeptide]

How do proteins differ from amino acids? From a tripeptide?

[handwritten: Proteins are larger containing many more amino acids. Proteins fold into more complex shapes. Proteins are polymers, while the tripeptide has only 3 amino acids]

Modeling Biological Molecules

Figure 6.14 Levels of protein structure. The amino acid sequence of a protein, called its (*a*) primary structure, encourages folding and hydrogen bonds to form between nearby amino acids. This produces coils (*b*) called alpha helices and fold-backs called beta-pleated sheets; these coils and fold-backs constitute the protein's secondary structure. The protein folds up on itself further to assume a three-dimensional (*c*) tertiary structure. Many proteins are formed as aggregations of polypeptide chains in clusters; this clustering is called the (*d*) quaternary structure of the protein.

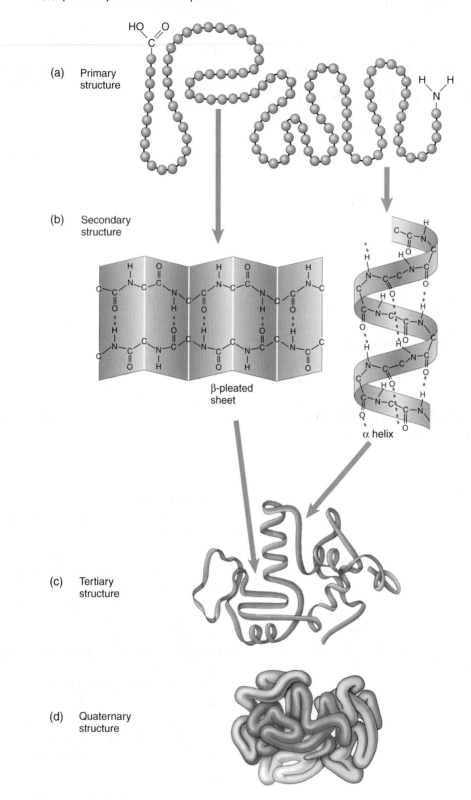

❓ How does the shape of peroxidase relate to its function as a catalyst?

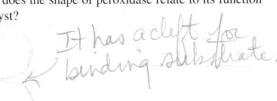

It has a cleft for binding substrate.

❓ How does the shape of collagen relate to its function?

It is long and rope-like, with 3 polypeptides contributing to its tensile strength

Close the files for peroxidase and collagen by going to FILE at the top of the screen and clicking on *CLOSE* twice.

Advanced Analysis of Peroxidase Structure

18 ▶ Reopen the file *16.peroxidase.pdb.* The RasMol program you are using has the capability of doing more advanced analyses than those you have done so far. In this section you will look at the location of specific amino acids in peroxidase and reveal the chemistry of its active site.

Polar amino acids are hydrophilic. To view the location of the polar amino acids in peroxidase, do the following.

1. Have the peroxidase file open and displaying in its window as a wireframe with a black background.

2. Go to the top of the screen and click on WINDOW to open the RasMol Command Line screen (a white background screen) so that both windows are open.

3. Click on the Command Line box and type the following after the prompt: *select polar.* Press the enter (return) key.

4. After the prompt type *color green.*

5. Go to the top of the screen and under DISPLAY choose *Ball and Stick.* All of the polar amino acids in peroxidase should now be displayed in green as ball and stick models.

❓ Rotate the molecule and decide whether there are more polar amino acids to the outside of the molecule than there are on the inside. What do you think? How does this relate to the shape of the protein in aqueous solutions?

Confirm this observation by now visualizing where several nonpolar amino acids are located. Review the amino acids shown in figure 6.12 and identify those that are nonpolar. Note the standard abbreviations for these amino acids. Enter the standard abbreviations into RasMol by typing the following:

1. *Select ala, val, leu, met, phe* (return)

2. *Color orange* (return)

3. Go to the top of the screen and under DISPLAY choose Ball and Stick. All of the nonpolar amino acids should now be shown in orange.

❓ Rotate the molecule so that you have good viewing angles. Are most of the nonpolar amino acids located on the outside of the protein or toward the inside? Why?

As you look inside the molecule, look for a polygonal structure. This is a heme group that marks the active site. It is a non-amino acid structure that gives the enzyme its ability to break down hydrogen peroxide to water and oxygen. To visualize the heme group, open the COMMAND WINDOW. Type the following:

1. *Select hem* (enter)

2. *Color red* (enter)

3. Go to DISPLAY at the top of screen and click on *Space Fill.*

The active site should now be clearly visible. Rotate peroxidase and find the channel that allows hydrogen peroxide to bind with the porphyrin ring.

19 ▶ Go to FILE at the top of the screen and close the file. Reopen the peroxidase file so that you have a fresh display of the original molecule. This time you will investigate where specific amino acids are located in the protein.

There are several ways to do this. The first, although easy to perform, yields a complex picture. Go to OPTIONS at the top of the screen and click on *labels.* The names of all of the amino acids in the molecule will appear followed by a number that represents the site of the amino acid in the primary sequence from the amino terminal end. (Remember that every protein, regardless of how large, has a carboxyl group on one end and an amino group on the other. You can enlarge the view of the molecule by going to the Command Line screen and typing *zoom 200* (or some other number greater than 100). This will help you see the labels. What is the largest number that you can see? _____ If you have found the amino acid that is on the carboxyl terminus, you have identified the number of amino acids in the protein. Go back to OPTIONS and click on labels again to turn them off.

Now you will experiment at finding where in peroxidase different amino acids are located. Be sure that you are starting with a wireframe display of peroxidase with the CPK color checked in the COLOURS drop-down menu. Click on the Command Line screen and enter the following:

1. *Select ala* (or the standard abbreviation of any other amino acids that interest you).

2. Go to the OPTIONS menu at the top of the screen and click on labels.

3. Go to the DISPLAY menu at the top of the screen and click on Ball and Stick.

You can now see where in the molecule alanine is located and the positions of this amino acid in the primary structure. How many alanine residues are found in peroxidase? _____

Repeat this analysis for another amino acid.

Close the data file for peroxidase by going to FILE at the top of the screen and clicking on *CLOSE*.

At the end of this lab topic, find the summary table (6.2) for characteristics of biological molecules. Fill in the information for amino acids and proteins.

Nucleotides/Nucleic Acids

Just as proteins are polymers of amino acids, nucleic acids are polymers of monomers called **nucleotides.** Nucleic acids range in size from a hundred or so nucleotides to well over several hundred million. A nucleotide is a complex monomer. It is made of three components: a sugar, a nitrogenous base, and a phosphate group (fig. 6.15). There are two types of sugars: ribose and deoxyribose. Both have five carbon sugars but deoxyribose has one less oxygen than does ribose. Nucelotides containing ribose function as temporary energy transfer compounds in cell metabolism and as components of different types of RNA which are part of a nucleic acid involved in protein synthesis. Nucleotides containing deoxyribose are found only in DNA where they make up genes. Forming a covalent bond with the sugar is a phosphate group. It is the acid part of a nucleic acid. Also in covalent linkage with the sugar is a nitrogenous base. There are five common bases: adenine, cytosine, guanine, thymine, and uracil. Thymine is never found in ribose-containing nucleotides (RNA) and uracil is never found in deoxyribose-containing nucleotides (DNA).

20 ▶ *Sugars in nucleotides.* Open the file *18.deoxyribose.pdb* and change to the *ball and stick* display.

How is this molecule similar to glucose? How is it different?

Both have ring structures with lots of -OH groups.

But Ribose has 5 carbons, while glucose has 6 carbons

Close the data file for this molecule by going to FILE at the top of the screen and clicking on *CLOSE*.

21 ▶ *Nitrogenous bases in nucleotides.* Open file *19.thymine. pdb.* Switch to the *ball and stick* view and use your cursor to rotate it for the best views.

What chemical elements are found in thymine?

C
H
N
O

How would you describe the shape of this molecule?

6-membered ring

Close the data file for this molecule by going to FILE at the top of the screen and clicking on *CLOSE*.

22 ▶ *A complete nucleotide.* Open the data file *20.ATP.pdb.* and change to a *ball and stick* display. Adenosine triphosphate is an important molecule in cells. It acts as an energy transfer molecule in most cellular reactions and a derivative of it is found in RNA.

How would you describe its shape?

several rings
kind of planar

Look carefully at the nucleotide and identify the nitrogenous base, sugar, and phosphate group in the molecule. Look again at the sugar. Is it ribose or deoxyribose? 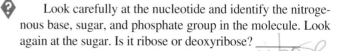 *ribose*

Figure 6.15 Nucleotide structure is shown in center. One of two types of 5 carbon sugars is found in a nucleotide. Nucleotides found in DNA have the sugar deoxyribose and those in RNA have ribose. The nitrogenous base portion can be one of five types of bases. However, thymine is never found in combination with ribose and uracil never in combination with deoxyribose.

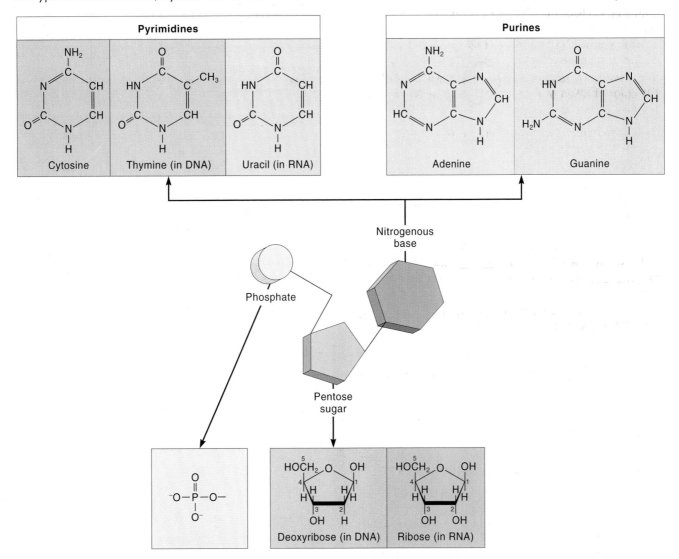

Is this molecule hydrophilic or hydrophobic? Why do you say so?

Hydrophilic – lots of O and N atoms around whole molecule

Close the data file for this molecule by going to FILE at the top of the screen and clicking on *CLOSE*.

23▶DNA (Deoxyribose Nucleic Acid). Open file *21.DNA. pdb.* Switch to the *ball and stick* view and use your cursor to rotate it for the best views. This is just a short sequence of a DNA double helix. DNA isolated from a human chromosome would be thousands of times longer. Nucleotides can chemically join together in condensation reactions to form **phosphodiester** linkages between the nucleotides.

Look carefully at the DNA double helix and identify where the nitrogenous base portion of each nucleotide is located. Where are the sugar and phosphate portions of each molecule located?

sugars (●)
phosphates
These form the ribbons (backbone) outsides of helix

Find the phosphodiester bond that holds them together. What functional groups are involved in the phosphodiester bond?

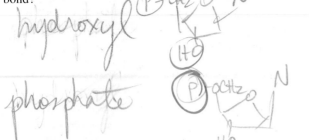

hydroxyl

phosphate

Using figure 6.15 as an aid, identify the organic bases in this sample of DNA. List those found in one strand as a sequence and then those found as the complementary strand across from those identified first. You can enlarge the image of the DNA double helix by typing in the Command Line window the following: *zoom 150*.

5' CGCAAATTTCGC 3'
3' GCGTTTAAAGCG 5'

The bases projecting to the inside of the double helix form hydrogen bonds with each other. It is these hydrogen bonds that hold together each of the single helices in a double helix. Go to WINDOW at the top of the screen and click on COMMAND LINE. A new box will appear. After the prompt, type *hbonds on*. The program will now draw the hydrogen bonds in as dashed lines. Rotate the molecule so that you can see them.

Close the data file for this molecule by going to FILE at the top of the screen and clicking on *CLOSE*.

At the end of this lab topic, find the summary table (6.2) for characteristics of biological molecules. Fill in the information for nucleotides and nucleic acids.

Learning Biology by Writing

Write a short essay describing the monomer/polymer concept as it applies to biology. Integrate into your discussion what the effects of hydrophilic and hydrophobic monomers are on polymer shape. Give examples from your lab work.

Your lab instructor may also ask you to answer the following Lab Summary and Critical Thinking Questions.

Internet Sources

The investigation of the three-dimensional structures of biological molecules is a very current area of research. Several WWW sites are repositories of data files for visualizing complex molecules such as proteins and nucleic acids. Check some of them out at http://www.rcsb.org/index.html or http://www.ncbi.nlm.nih.gov/. After looking these sites over, how would you describe the type of work that seems to be going on? What type of a background would you need to do such work?

Lab Summary Questions

1. What are polar molecules? Explain how functional groups can be polar.
2. Explain the difference between covalent and hydrogen bonds.
3. How many times larger than water is glucose? The amino acid serine?
4. What determines whether a molecule is hydrophilic or hydrophobic?
5. Explain how polar substances are able to dissolve in a polar solvent such as water. Describe the events at the molecular level.
6. Write a critique of this statement explaining why it is wrong: "Hydrophobic substances do not dissolve in polar solvents like water. Therefore, hydrophobic substances are not important in cells."
7. If a peptide bond forms between one amino acid and another between the amino and carboxyl functional groups, how is it possible for more than two amino acids to join to each other?
8. The primary structure of a protein is the sequence of amino acids. Describe how this sequence determines the folding and shape of the protein. What is the significance of the shape of proteins?
9. What are biological polymers? Give examples of several types of polymers investigated using RasMol.
10. How do the chemical properties of lipids relate to their functions in cells?
11. How do the shapes of different proteins relate to their functions in cells?

Critical Thinking Questions

1. Wood, which is composed mostly of cellulose, never completely dries and loses all of the water it contains. The wood in the walls of houses may contain 8% to 16% moisture, meaning that the indicated percentage of its weight is due to water. However, if you looked at the wood and felt it, you would say that it was dry. Explain where this water could be located and what type of chemical bonding is holding it in the wood.
2. A biochemist studying the fatty-acid composition of cell membrane phospholipids in normal and "hibernating" frogs found that the chemical composition of the membrane changed with season. In the summer there were more saturated fatty acids present and in the winter more unsaturated. Explain how this might be adaptive and allow the frog's survival.
3. The amino acid sequence of a protein is analyzed, and one end of the amino acid chain is composed almost exclusively of hydrophobic amino acids while the opposite end has hydrophilic amino acids. Others studying this protein found that it is an antibody found on lymphocyte cell surfaces. How do these two independent observations complement one another?
4. The molecular weight of the haploid human genome is about 1.83×10^{12}. The average molecular weight of a base pair in DNA is about 610, and each base pair has a length of 0.34 nm. How many base pairs are in the human haploid genome?
5. Using the data given in question 4, how long is the DNA found in a haploid human sperm or egg? Each of the 23 chromosomes in a human sperm or egg contains a single DNA molecule. What is the average length in nanometers of DNA found in a human chromosome? Convert this to centimeters.
6. Hydrogen bonds have been called the velcro® of biological systems. By analogy, covalent bonds are the thread stitches. Explain why both types of bonds are important in the functioning of cells.

Modeling Biological Molecules

TABLE 6.2   Summary of characteristics of biological molecules

Molecule name	Elements present	Estimated molec wt	Functional groups present	Hydro-philic or -phobic?	Polymer or monomer?	Type of bond in polymer	Describe shape
Glucose	C,H,O	180d	—OH (and/or =O)	philic	monomer	—	ring or linear
Sucrose	C,H,O	360d	"		dimer	glycosidic	2 rings
Amylose	O	lots	"		polymer	glycosidic	branched
Stearic acid	C,H,O	much larger than glycerol	carboxyl	-phobic	one fatty acyl	—	straight
Oleic acid	C,H,O		carboxyl	-phobic	"	—	bend
Glycerol	CHO	92d	hydroxyl	-philic	one glycerol	—	straight
Triglyceride	CHO	~3X fatty acid	ester	-phobic	1 glycerol 3 fatty acids	ester linkage	
Phospholipid	CHOPN	~2/4 X fatty acid	ester phosphate	-philic -phobic	1 glycerol + 2 fatty acids	ester + O ring "	
Alanine	CHON	93 d	methyl -CH3	-philic	mono-	—	globular small
Serine	CHON	106 d	hydroxyl	-philic	"	—	"
Leucine	CHON	135 d	methyl	amphipathic	"	—	medium
Phenylalanine	CHON	169 d	phenyl	-phobic	"	—	longer
Tripeptide	CHON	~300 d		-philic (overall) tri-	peptide	longer
Peroxidase	CHON	huge mass	many	-phobic -philic	"	peptide	linear globular
Collagen	CHON	huge mass	many	-philic	"	peptide	fibrous
Deoxyribose	C,O,H	~C5H10O5 150 d	-OH hydroxyl	-philic	sugar monomer	—	ring
Thymine	CO HN	~1500 d	methyl carbonyl amines	-philic	base monomer	—	planar ring
ATP	COHNP	~500 d	amino hydroxyl phosphate		monomer		planar
DNA	COHNP	huge	hydroxyl amino phosphate, methyl		polymer	phospho diester	double helix

LAB TOPIC 7

Determining the Properties of an Enzyme

Supplies

Preparator's guide available at
http://www.mhhe.com/dolphin

Equipment

Constant temperature water baths
Spectrophotometers at 500 nm

Materials

Blender or mortar and pestle
Fresh white turnip, horseradish root, or potato
Tissues and markers
Spectrophotometer tubes
10 ml test tubes and rack
50 ml beakers
5 ml pipettes graduated in 0.1 ml units with suction
devices or automatic pipetters
Thermometer (alcohol)

Solutions

10 mM H_2O_2
25 mM guaiacol
2% Hydroxylamine
Citrate-phosphate buffers at pHs 3, 5, 7, and 9
Peroxidase I (horseradish), 6.4 mg/L

Student Prelab Preparation

Before doing this lab, you should read the introduction
and sections of the lab topic that have been scheduled
by the instructor.

You should use your textbook to review the
definitions of the following terms:

Enzyme
Inhibitor
pH
Peroxidase
Product
Spectrophotometer
Substrate

You should be able to describe in your own words
the following concepts:

Structure of an enzyme
Effect of pH on enzyme structure

Effect of temperature on enzyme structure
Effect of competitive inhibitors on enzymes
Review appendix B

As a result of this review, you most likely have
questions about terms, concepts, or how you will do the
experiments. Write these questions in the space below
or in the margins of the pages of this lab topic. The lab
experiments should help you answer these questions, or
you can ask your instructor during the lab.

Objectives

1. To perform a quantitative assay of the activity of
   an enzyme in a tissue extract using a
   spectrophotometer
2. To organize the data as concise tables and graphs
   for inclusion in a lab report describing the
   properties of the enzyme peroxidase
3. To test the following hypotheses:
   a. The amount of enzyme does not influence the
      rate of reaction.
   b. The temperature of the solution does not
      influence the activity of an enzyme.
   c. The pH of the solution does not influence the
      activity of an enzyme.
   d. Boiling an enzyme before a reaction does not
      influence its activity.
   e. Other molecules with shapes similar to an
      enzyme's substrate have no effect on the activity
      of the enzyme.

Background

The thousands of chemical reactions occurring in a cell
each minute are not random events but are highly controlled by biological catalysts called **enzymes.** Like all catalysts, enzymes lower the **activation energy** of a reaction,
the amount of energy necessary to trigger a reaction.

Figure 7.1   Enzymes are large proteins that bind substrates at their active sites. The substrate is converted into products, which are released, allowing the enzyme to bind to another substrate molecule. In this example, sucrase converts sucrose into glucose and fructose. Note that enzyme and substrate sizes are not to the same scale. Enzymes are often thousands of times larger than substrates.

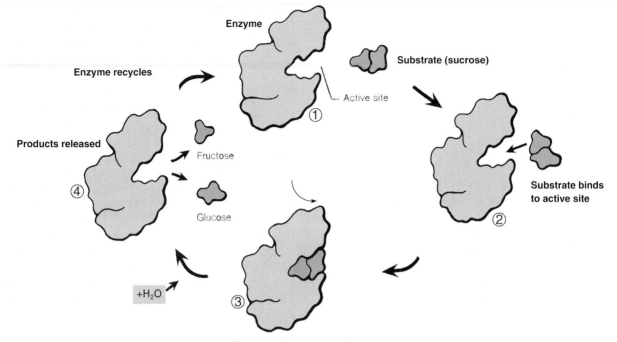

Most enzymes are proteins with individual shapes determined by their unique amino acid sequences. Since these sequences are spelled out by specific genes, the chemical activities that occur in a cell are under genetic control. The shape of an enzyme, especially in its **active site,** determines its catalytic effects (fig. 7.1). The active site of each type of enzyme will bind only with certain kinds of molecules—for example, some enzymes bind with glucose but not with ribose because the former is a six-carbon sugar while the latter has only five carbons.

A molecule that binds with an enzyme and undergoes chemical modification is called the **substrate** of that enzyme. Often metallic ions, such as Fe^{+++}, Mg^{++}, Ca^{++}, or Mn^{++}, aid in the binding process, as do vitamins or other small molecules called **co-factors** or **coenzymes.**

The binding between enzyme and substrate consists of weak, noncovalent chemical bonds, forming an **enzyme-substrate complex** that exists for only a few milliseconds. During this instant, the covalent bonds of the substrate either come under stress or are oriented in such a manner that they can be attacked by other molecules, for example, by water in a hydrolysis reaction.

The result is a chemical change in the substrate that converts it to a new type of molecule called the **product** of the reaction. The product leaves the enzyme's active site and is used by the cell. The enzyme is unchanged by the reaction and will enter the catalytic cycle again, provided other substrate molecules are available.

Individual enzyme molecules may enter the catalytic cycle several thousand times per second; thus, a small amount of enzyme can convert large quantities of substrate to product. Eventually enzymes wear out; they break apart and lose their catalytic capacity. Cellular proteinases degrade inactive enzymes to amino acids, which are recycled by the cell to make other structural and functional proteins.

The amount of a particular enzyme found in a cell is determined by the *balance* between the processes that *degrade* the enzyme and those that *synthesize* it. When no enzyme is present, the chemical reaction catalyzed by the enzyme does not occur at an appreciable rate. Conversely, if enzyme concentration increases, the rate of the catalytic reaction associated with that enzyme will also increase.

The pH or salt concentrations of a solution affect the shape of enzymes by altering the distribution of + and – changes in the enzyme molecules which, in turn, alters their substrate-binding efficiency. Temperature, within the physiological limits of 0° to 40° C, affects the frequency with which the enzyme and its substrates collide and, hence, also affects binding. All factors that influence binding obviously affect the rate of enzyme-catalyzed reactions. Some of these factors will be investigated during this laboratory.

Peroxidase

During this lab, you will study an enzyme called **peroxidase.** It is a large protein containing several hundred amino acids and has an iron ion located at its active site. Peroxidase makes an ideal experimental material because it is easily prepared and assayed. Turnips, horseradish roots, and potatoes are rich sources of this enzyme.

The normal function of peroxidase is to convert toxic hydrogen peroxide (H_2O_2), which can be produced in certain metabolic reactions, into harmless water (H_2O) and oxygen (O_2).

$$2 H_2O_2 \xrightarrow{\text{peroxidase}} 2 H_2O + 2 O:$$

The oxygen often reacts with other compounds in the cell to form secondary products.

The peroxidase reaction can be measured by following the formation of oxygen. The amount of oxygen present after the reaction can be measured in two ways: by the accumulation of gas in a closed system connected to a manometer or by the appearance of chemically active oxygen.

Many dyes will react with active oxygen by changing from a colorless to a colored state, and dye techniques are easier to perform than volumetric tests of gases in teaching laboratories. Such tests are called **dye-coupled reactions.** The enzyme-catalyzed reaction produces a product that enters into a secondary reaction with the dye. The enzyme, itself, does not bind with or affect the dye.

You will use the dye **guaiacol,** which turns brown when oxidized. The entire peroxidase reaction, including the measure of active oxygen through guaiacol, is as follows:

$$2 H_2O_2 \xrightarrow{\text{peroxidase}} 2 H_2O + 2 O:$$

$$O: + \text{guaiacol} \rightarrow \text{oxidized guaiacol}$$
$$\text{(colorless)} \qquad \text{(brown)}$$

To quantitatively measure the amount of brown color in the final product, the enzyme, substrate, and dye can be mixed in a tube and immediately placed in a spectrophotometer. As color accumulates, the absorbance at 500 nm will increase. The procedure for using a spectrophotometer was explained in lab topic 5. You should review those instructions before proceeding. (See fig. 5.5.)

LAB INSTRUCTIONS

In this exercise, you will determine the effects of several factors on the activity of peroxidase.

Preparing an Extract Containing Peroxidase

These steps will be done by the instructor before class to save time:

1. Weigh 1 to 10 g of peeled turnip, horseradish, or potato tissue.

2. Homogenize the tissue by adding it to 100 ml of cold (4° C) 0.2 M phosphate buffer at pH 7. Grind the mixture in a cold mortar and pestle with sand or blend it for 15 seconds at high speed in a cold blender. Filter extract to clarify. Alternatively, prepare a solution of peroxidase using an enzyme obtained from a biochemical supply house. The extract will keep overnight in a refrigerator.

Standardizing the Amount of Enzyme

The extract contains hundreds of different types of enzymes, including peroxidase. The activity of each enzyme will vary, depending on the size and age of the turnip, horseradish, or potato; the extent of the tissue homogenization; and the age of the extract. Only peroxidase, however, will react with H_2O_2.

To demonstrate that the amount of enzyme influences the rate of the reaction and to determine the correct amount of extract to use in future experiments, a trial run should be performed in which the amount of enzyme added is the only variable. Your instructor may do this section as a demonstration to show you how to best organize the procedures used in assaying an enzyme.

Before starting an experiment, a hypothesis should be stated. In this experiment, the hypothesis should relate the rate of reaction to the amount of enzyme added. An example will be given here, but you will make your own hypotheses in the other experiments done in this lab.

The amount of enzyme added to the reaction will have no effect on the rate of reaction.

To test your hypothesis, you should use the following directions to set up the chemical reactions and to conduct the experiment:

1. Label four 50 ml beakers as follows: *turnip (or horseradish or potato) extract containing peroxidase; buffer, pH 5; 10 mM H_2O_2 (substrate);* and *25 mM guaiacol.* Fill each about half full with the appropriate stock solution. Label four pipettes to correspond with the beakers. Alternatively, your lab instructor may have these reagents available in burettes or dispensers. If that is the case, you will be given verbal directions on how to add solutions to your test tubes.

2. Number seven test tubes from 1 to 7. The contents of the tubes will be:

   1: Control with all reactants except enzyme to be used in calibrating the spectrophotometer
   2: Substrate and indicator dye
   3: Dilute extract
   4: Substrate and indicator dye (same as 2)
   5: Medium concentration of extract
   6: Substrate and indicator dye (same as 2)
   7: Concentrated extract

   The exact quantities to be added to each tube are listed in table 7.1. Pairs 2 and 3, 4 and 5, or 6 and 7 will be mixed together when it is time to measure a reaction. Mix a pair only when you are ready to measure that reaction in the spectrophotometer.

3. Add stock solutions to each tube using the corresponding graduated 5 ml pipette or dispensing device. Use of the wrong pipette or dispenser will cross contaminate your reagents and introduce errors into your subsequent experiments.

4. Use the directions in figure 5.5 to adjust the spectrophotometer to zero absorbance at 500 nm. Pour

TABLE 7.1 — Mixing table for trial run to determine extract concentration (all values in milliliters)

Tube	Buffer (pH 5)	H_2O_2	Extract	Guaiacol (Dye)	Total Volume
1 Control	5.0	2.0	0	1.0	8
2	0	2.0	0	1.0	3
3	4.5	0	0.5	0	5
4	0	2.0	0	1.0	3
5	4.0	0	1.0	0	5
6	0	2.0	0	1.0	3
7	3.0	0	2.0	0	5

TABLE 7.2 — Results from trial run of enzyme activity (entries are absorbance units at 500 nm)

Time (Sec)	Tubes 2 and 3 0.5 ml Extract	Tubes 4 and 5 1.0 ml Extract	Tubes 6 and 7 2.0 ml Extract
20			
40			
60			
80			
100			
120			

the contents of test tube 1 into a cuvette (the special spectrophotometer test tube made of optical glass). This tube is used to "blank" the spectrophotometer, so that any color caused by contaminants in the reagents will not influence subsequent measurements.

5. If you are working as teams of students, one person can be a timer, another a spectrophotometer reader, and another a data recorder. Note the time to the nearest second and mix the contents of tubes 2 and 3 by pouring them back and forth twice. Mixing should be completed within ten seconds.

6. Add the reaction mixture to a cuvette by pouring or using an eye dropper, wipe the outside, and place the cuvette in the spectrophotometer. Read the absorbance at 20-second intervals from the start of mixing. If you are a little late in reading the meter, record the absorbance and change the table to show the actual time of the reading. Record your measurements in table 7.2. After two minutes (six readings) remove the tube from the spectrophotometer and visually note the color change. Discard the solution.

7. Mix the contents of tubes 4 and 5, transfer to a cuvette, and repeat your measurements for two minutes at 20-second intervals. Record the results in table 7.2.

8. Mix the contents of tubes 6 and 7, transfer to a cuvette, and record the absorbance measurements in table 7.2.

Analysis of Amount of Enzyme Data

4 Now plot the values in table 7.2 on one panel of graph paper provided at the end of this exercise. The abscissa should be the independent variable (time in seconds) and the ordinate the dependent variable (absorbance units). Explain why absorbance is considered the dependent variable.

Plot all three tests on the same coordinates using different plotting symbols. Note that since you adjusted the spectrophotometer with a blank containing the dye, the absorbance at zero seconds should be zero and a point may be plotted at the origin. Using a clear plastic ruler, draw the single straight line that best fits the points for each of the conditions. (Curves may plateau at the end.) Appendix B discusses how to make graphs.

⬦ In mathematical terminology, what characteristic of the graphed line is a measure of enzyme activity?

⬦ What are the units of activity in this experiment?

⬦ Which amount of enzyme gave a linear absorbance change from 0 to 1 in approximately 120 seconds? Use this amount in all subsequent experiments in this exercise.

⬦ Change the amount of enzyme called for in mixing tables 7.3, 7.5, 7.7, and 7.9 as necessary based on your first experiment. The standardizing procedure must be repeated for each batch of extract. Why?

Do you accept or reject your hypothesis regarding rate of reaction and amount of enzyme? Why?

Factors Affecting Enzyme Activity

If the scheduled laboratory period is short, your instructor may divide you into teams, each of which will test for the effects of one or more of the following experimental variables. The results will be shared at the end of the lab period and may be included in your report.

Temperature Effects

To determine the effects of temperature on peroxidase activity, you will repeat the enzyme assay in water baths at four temperatures:

1. In a refrigerator at approximately 4° C
2. At room temperature (about 23° C, but should be measured)
3. At 32° C
4. At 48° C

State a hypothesis that relates change in enzyme activity to the temperature of the solutions used.

5▷ To test your hypothesis, you should use the following directions to set up the reactions and conduct the experiment:

If constant temperature baths are not available, improvise with plastic containers, adding hot and cold water to adjust the temperature. Number nine test tubes in sequence 1 through 9. Refer to table 7.3 for the volumes of reagents to be added to each tube.

Preincubate all the solutions at the appropriate temperatures for at least 15 minutes before mixing. After reaching temperature equilibrium and adjusting the spectrophotometer with the contents of test tube 1, mix pairs of tubes (2 and 3, 4 and 5, 6 and 7, and 8 and 9) one pair at a time. After mixing one pair, measure the change in absorbance for two minutes at 20-second intervals for each temperature. The temperatures will not remain exact, but the effects can be overlooked. After measuring the absorbance changes, mix the second pair and measure the absorbance change, and so on.

Note: The room-temperature experiment can be performed immediately while the other tubes temperature-equilibrate. Be sure to measure the room temperature and record it in table 7.4.

Record changes in absorbance for the reaction mixture at each temperature in table 7.4.

Analysis of Temperature Data

6▷ These results should be graphed at the end of the laboratory period on one of the panels of graph paper at the end of the exercise. The slopes of the linear portions of these curves are a measure of enzyme activity. Does activity vary with temperature? _____ What is the optimum temperature? _____ °C

TABLE 7.3   Mixing table for temperature experiment (all values in milliliters)

Temperature	Tube	Buffer (pH 5)	H₂O₂	Extract	Guaiacol (Dye)	Total Volume
	1 Control	5.0	2.0	0	1.0	8
4°C	2	0	2.0	0	1.0	3
	3	4.0	0	1.0	0	5
Room temp = ____ °C	4	0	2.0	0	1.0	3
	5	4.0	0	1.0	0	5
32°C	6	0	2.0	0	1.0	3
	7	4.0	0	1.0	0	5
48°C	8	0	2.0	0	1.0	3
	9	4.0	0	1.0	0	5

TABLE 7.4   Temperature effects on peroxidase activity (entries are absorbance units at 500 nm)

Time (Sec)	Tubes 2 and 3 4°C	Tubes 4 and 5 Room temp = ____°C	Tubes 6 and 7 32°C	Tubes 8 and 9 48°C
20				
40				
60				
80				
100				
120				

Do you accept or reject the hypothesis stated earlier? What data do you have to support your decision?

(slope) at each temperature treatment from your linear graphs. Record below.

Temperature	Activity (ΔA/min)
4°	
23°	
32°	
48°	

On graph paper, plot the activity values as a function of temperature. Fit a curve to the plotted points. Your curve should be bell shaped. The top of the curve indicates the **temperature optimum,** the temperature at which the maximum rate is observed.

7 To show clearly this relationship, you should prepare a derivative graph after lab. This type of graph condenses the data into an easily understandable form. To prepare such a graph, first determine the absorbance change per minute

TABLE 7.5   Mixing table for pH experiment (all values in milliliters)

pH	Tube	Buffer	H_2O_2	Extract	Guaiacol (Dye)	Total Volume
5	1 Control	5.0 (pH 5)	2.0	0	1.0	8
3	2	0	2.0	0	1.0	3
	3	4.0 (pH 3)	0	1.0	0	5
5	4	0	2.0	0	1.0	3
	5	4.0 (pH 5)	0	1.0	0	5
7	6	0	2.0	0	1.0	3
	7	4.0 (pH 7)	0	1.0	0	5
9	8	0	2.0	0	1.0	3
	9	4.0 (pH 9)	0	1.0	0	5

TABLE 7.6   Effects of pH on peroxidase activity (entries are absorbance units at 500 nm)

Time (Sec)	Tubes 2 and 3 pH 3	Tubes 4 and 5 pH 5	Tubes 6 and 7 pH 7	Tubes 8 and 9 pH 9
20				
40				
60				
80				
100				
120				

pH Effects

Begin by stating a hypothesis that relates change in enzyme activity to the pH of the solutions used.

8 ▷   To determine the effect of pH on peroxidase, perform the following experiment.

Your instructor will supply buffers at pHs of 3, 5, 7, and 9. Number nine test tubes 1 through 9. Set up pH-effect tests by adding the reagents described in table 7.5.

After adjusting the spectrophotometer with the contents of test tube 1, mix pairs of tubes one at a time (2 and 3, 4 and 5, 6 and 7, 8 and 9). Measure absorbance changes at 20-second intervals for two minutes for each pair before mixing the next pair. Record the results in table 7.6.

Analysis of pH Data

9 ▷   The values in this table should be graphed at the end of the laboratory period on one panel of graph paper at the end of the exercise. The slopes of the linear portions of these curves are a measure of enzyme activity. Does activity vary with pH? What are the units of activity? What is the optimum pH?

Do you accept or reject the hypothesis regarding pH effects on enzymes? Why?

Temperature	Buffer (pH 5)	H_2O_2	Boiled Extract	Guaiacol	Total Volume
1 Control	5.0	2.0	0	1.0	8
2	0	2.0	0	1.0	3
3	4.0	0	1.0	0	5

TABLE 7.7   Mixing table for boiling extract (all values in milliliters)

10▶ To show clearly this relationship, you should prepare a derivative graph after lab. Determine the absorbance change per minute from your plots of absorbance versus time at each pH. Record below.

pH	Activity (ΔA/min)
3	
5	
7	
9	

Plot the activity values as functions of pH on graph paper. The resulting curve should be bell shaped with the top (peak) indicating the **pH optimum,** the pH at which the maximum rate is observed.

Demonstrating Effects of Boiling on Peroxidase Activity

Most proteins are denatured when they are heated to temperatures above 70° C. **Denaturation** is a nonreversible change in a protein's three-dimensional structure.

? If heating a protein to 100° C irreversibly alters its shape (denaturation), what do you predict will happen to measured enzyme activity?

State a hypothesis that relates heat treatment of an enzyme (boiling) to the expected effect on that enzyme's activity.

11▶ To perform the experiment that tests your hypothesis, follow these directions:

Add 3 ml of extract to a test tube and place it in a boiling water bath. After five minutes, remove the tube and let it cool to room temperature. Number three test tubes and add reagents as called for in mixing table 7.7.

Use the contents of tube 1 to blank the spectrophotometer. Mix the contents of tubes 2 and 3, pour the mix-

TABLE 7.8   Results from using boiled extract (entries are absorbance units at 500 nm)

Time (Sec)	Tubes 2 and 3 Boiler Extract	
20		
40		
60		
80		
100		
120		

ture into a cuvette, and read the absorbance at 20-second intervals for two minutes. Record the results in table 7.8.

Compare the activity of peroxidase after boiling to the activity of peroxidase kept at room temperature and pH 5 **?** (see table 7.4 or 7.6). How did boiling affect the activity?

? Do you accept or reject the hypothesis made at the beginning of this experiment? Why?

? Consult your text to see what kinds of chemical bonds are disrupted in proteins heated to 100° C. Describe these bonds and the effect on protein shape below.

TABLE 7.9 Mixing table for inhibitor experiments (all values in milliliters)

Tube	Buffer (pH 5)	H₂O₂	Extract	Hydroxylamine Treated Extract	Guaiacol	Total Volume
1 Control	5.0	2.0	0	0	1.0	8
2	0	2.0	0	0	1.0	3
3	4.0	0	1.0	0	0	5
4	0	2.0	0	0	1.0	3
5	4.0	0	0	1.0	0	5

In this experiment, there was no control. What should have been in another pair of tubes to act as an experimental control?

TABLE 7.10 Enzyme inhibitor results (entries are absorbance units at 500 nm)

Time (Sec)	Tubes 2 and 3 Normal Extract	Tubes 4 and 5 Hydroxylamine-treated Extract
20		
40		
60		
80		
100		
120		

After adjusting the spectrophotometer with the contents of test tube 1, mix pairs one at a time (2 and 3, 4 and 5) and measure the changes at 20-second intervals for two minutes. Record your measurements in table 7.10.

13▶ These data should be graphed at the end of the laboratory period. The slopes of the linear portions of the curves are a measure of enzyme activity. What are the units? Explain why the slopes differ.

Optional: The Effects of Inhibitors

Hydroxylamine ($HONH_2$) has a structure similar to hydrogen peroxide ($HOOH$). Calculate the molecular weight of hydrogen peroxide: _____. Calculate the molecular weight of hydroxylamine: _____. Hydroxylamine is small enough that it can enter the reactive site and bind with the iron atom in peroxidase. This prevents hydrogen peroxide from entering the site. What do you predict will be the effect on enzyme activity? State your predicted effects as a hypothesis.

12▶ To test this hypothesis, mix five drops of 2% hydroxylamine (neutralize to pH 7) and 2 ml of enzyme extract, letting the mixture stand for at least ten minutes. Then measure the peroxidase activity, comparing the activity in this treated enzyme preparation to that of the enzyme without the inhibitor. Table 7.9 lists the amounts of each solution to be used for the tests.

? Do you accept or reject the hypothesis made at the beginning of this experiment? Why?

Analysis

You have collected quantitative information about enzymes and how they work, using the enzyme peroxidase as an example. Your instructor will discuss the results in class and will indicate how to share the data from your experiments with the class. Directions for making graphs are given in appendix B.

? In all these experiments, tube 1 has been listed as a control. What is it controlling for? No reaction happens in this tube. Why was it included?

Learning Biology by Writing

This exercise provides material well suited for a written scientific report. You have tested five hypotheses and the collected results describe the properties of the enzyme peroxidase. Appendix D contains general directions on report writing. Your instructor may also have specific instructions for you to follow in writing your lab report. A suggested form follows.

Title
Purpose

A good lab report begins with a statement of purpose, which summarizes the hypothesis tested in the experiments. Write the purpose in the third person impersonal, concentrating on the scientific questions involved and not the teaching and learning objectives.

Techniques

Describe in about two paragraphs the techniques you used to extract the peroxidase and to measure enzyme activity.

Results

Use graphs to report your data. (See appendix B.) Time should be recorded on the abscissa and absorbance on the ordinate. A separate graph should be made for each factor—

pH, temperature, and inhibitors. All curves for any one factor should be on the same set of coordinates. Indicate the slope values for each curve, labeling curves and coordinates, and writing a descriptive legend (for example, "Peroxidase activity at different pH levels. Symbols used are . . .").

The results from the temperature or the pH experiments should be summarized in separate graphs in which you calculate the slope of your absorbance versus time plots at each temperature (or pH) and plot the slope as a function of temperature (or pH). These derivative plots make it easier to see what the temperature (or pH) optimum is for peroxidase.

Discussion

Read the section in your textbook about enzymes. Discuss how the experiments you performed can be interpreted in terms of the structural properties of a protein subjected to different chemical and physical treatments. What are the limitations of these experiments? Will all enzymes have the same pH and temperature optima? Will all denature at the same temperature or be inhibited by hydroxylamine? How would you do the experiments differently, if you were to repeat them? What are the sources of error?

As an alternative assignment, your instructor may ask you to complete the following lab summary and critical thinking questions.

Internet Sources

Use the WWW to locate information about the enzyme peroxidase.

Open your browser program and enter the following URL, http://www.expasy.ch/ to connect to the Expert Protein Analysis System (ExPASy) maintained by the Swiss Institute of Bioinformaties. On the screen that appears, click on Swiss-Prot in the Databases column. Type the words "turnip peroxidase" in the Quick Search box that appears. You will be presented with data retrieved from the database about the enzyme. Scan the entries. Under References find the citation to the original scientific paper published by Mazza and Welinder in 1980. Click NCBI for the PubMed numbered hypertext to read the abstract of the paper. Answer these questions:

What is the molecular weight of the enzyme? _____

How many amino acids does it contain? _____

Lab Summary Questions

1. Prepare graphs of the data in tables 7.4, 7.6, and 7.10. Follow the directions in appendix B and use the graph paper at the end of this exercise.
2. After reviewing your data plotted from table 7.4, give your reasons for rejecting or not rejecting the hypothesis: peroxidase activity does not vary with temperature. Describe what is meant by a temperature optimum for an enzyme.
3. Review the plots of the data in table 7.6 and state your reasons for rejecting or not rejecting the hypothesis:

peroxidase activity does not vary with pH. What is meant by a pH optimum for an enzyme?

4. What conclusion can you draw from the plots of the data in table 7.10? If hydroxylamine has a shape and size similar to hydrogen peroxide, propose a molecular-level explanation of why hydroxylamine changes the rate of (or stops) the conversion of hydrogen peroxide to oxygen and water.
5. What is protein denaturation? Propose a molecular-level explanation to explain why heating to 100° C and then cooling destroyed the ability of peroxidase to catalyse the conversion of hydrogen peroxide to oxygen and water.
6. What evidence do you have from these experiments that the ability of an enzyme to catalyze a reaction depends on its shape? Before answering be sure to consult your textbook and read the section about factors influencing rates of enzyme catalysed reactions and factors influencing shapes of proteins.

Critical Thinking Questions

1. Would performing the "temperature effects" portion of this experiment on a very hot day have any effect on the results?
2. If you used a cooked turnip as your extract source, what results would you expect? Why?
3. In the human digestive system, different enzymes have different pH optima. What happens when enzymes from the stomach which have an acid optimum enter the alkaline environment of the small intestine?
4. If you were hired to supervise a biochemical manufacturing plant that used enzymes to convert starches to sugars, how would you go about determining if the chemical reactions were operating at maximum rates?

Determining the Properties of an Enzyme **89**

Determining the Properties of an Enzyme

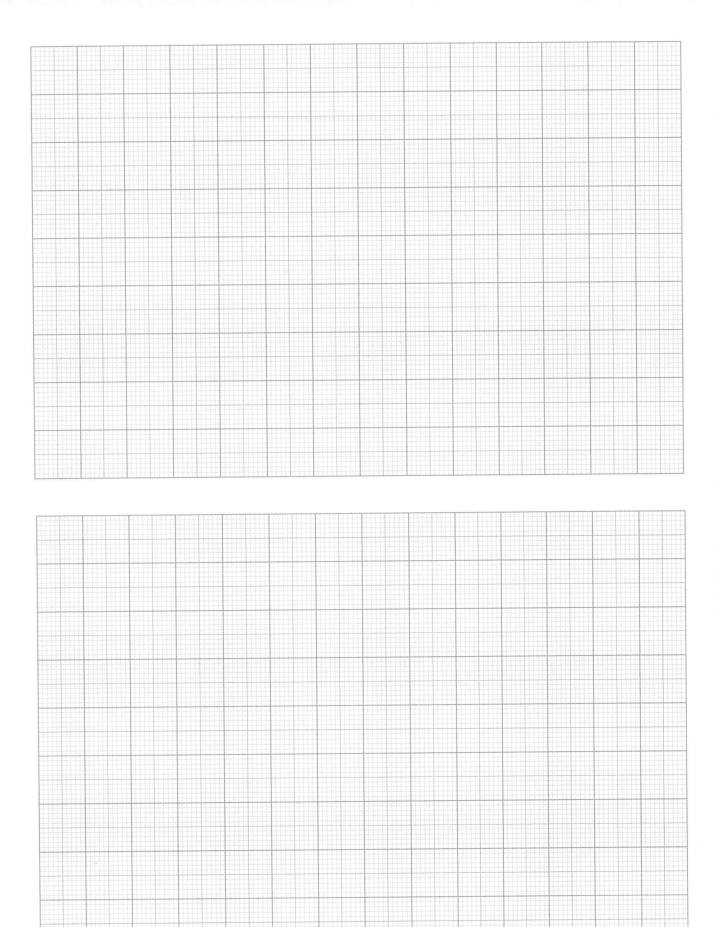

Determining the Properties of an Enzyme

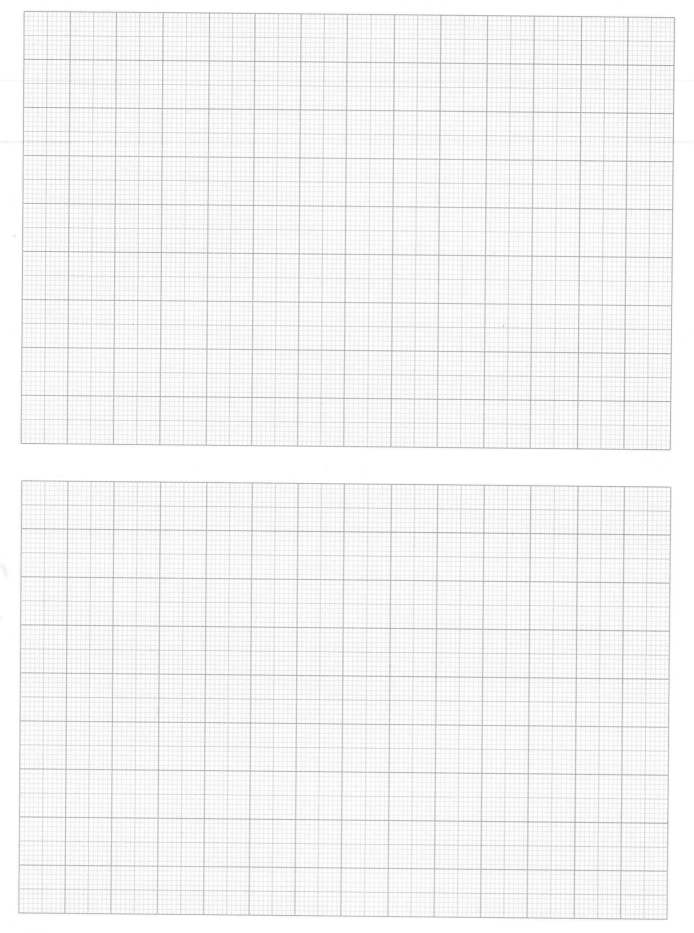

LAB TOPIC 8

Measuring Cellular Respiration

Supplies

Preparator's guide available at
http://www.mhhe.com/dolphin

Equipment

Barometer
Hot plates
pH meter

Materials

Distillation apparatus (see fig. 8.3)
Yeast packets
Four- to six-day germinating peas
Test tubes and racks
One-hole stoppers with 1 ml × 0.01 ml pipettes
    inserted (see fig. 8.4)
Test tube rack
Nonabsorbent cotton
Absorbent cotton
Miscellaneous beakers
Weighing pans
Pasteur pipettes

Solutions

Karo syrup or grape juice
0.1 M $Ba(OH)_2$
I_2KI (5 g I_2:10 g KI:100 ml H_2O)
1.5 M NaOH (*Caution:* caustic)
15% KOH in dropper bottles (*Caution:* caustic)
95% ethanol in dropper bottles

Student Prelab Preparation

Before doing this lab, you should read the introduction and sections of the lab topic that have been scheduled by the instructor.

You should use your textbook to review the definitions of the following terms:

Aerobic metabolism
Anaerobic metabolism
Cellular respiration
Electron Transport System
Ethanol
Fermentation
Glycolysis
Krebs cycle

You should be able to describe in your own words the following concepts:

What is ATP and why it is important
How ethanol is produced by yeast
How CO_2 is produced in cellular respiration
Where O_2 is used in respiration
Review appendix B

As a result of this review, you most likely have questions about terms, concepts, or how you will do the experiments included in this lab. Write these questions in the space below or in the margins of the pages of this lab topic. The lab experiments should help you answer these questions, or you can ask your instructor during the lab.

Objectives

1. To identify the by-products of aerobic and anaerobic respiration in yeast
2. To formulate and test a hypothesis regarding the effect of freezing on the aerobic respiration of peas
3. To measure the rate of oxygen consumption in germinating pea seedlings

Background

All organisms, whether plants, animals, bacteria, protists or fungi, perform **cellular respiration.** Respiration involves several complex enzyme-catalyzed reactions that break down organic molecules and yield energy as ATP that a cell can use to perform the work of growth, maintenance, and function. Biologists often use the metabolism of glucose as a model for respiration because the metabolic pathways are well understood and occur in most organisms.

After entering a cell, glucose can be broken down into two molecules of pyruvic acid by a series of approximately ten enzyme-catalyzed reactions known collectively as **gly-colysis.** The enzymes for these reactions float free in the aqueous portion of the cytoplasm. As the chemical reactions of glycolysis occur, energy contained in the covalent bonds of the glucose molecule is released. Some of this energy passes from the organism as heat, and some is used in the synthesis of the energy storage compound **ATP.** It is the form of energy that the cell uses to make other molecules and to perform work functions such as transport or movement.

To obtain energy from glucose, hydrogen atoms are removed from the glucose molecule as it is metabolized. These hydrogen atoms can be removed only by hydrogen (electron) carriers, such as the compound **NAD⁺** (nicotinamide-adenine dinucleotide), which acts as a coenzyme in the chemical reactions. Since a finite amount of NAD⁺ occurs in the cell and since each NAD⁺ molecule can combine with only two hydrogens, there must be a mechanism for removing hydrogen from NAD•2H complexes so that glycolysis can continue. Without such a mechanism, glycolysis would cease when all NAD⁺ molecules were saturated with hydrogen.

In many organisms, respiration can occur under anaerobic conditions where no oxygen is required to breakdown glucose. Many bacteria, yeast, and some animals, ferment glucose, producing lactic acid or ethanol. In these fermentations, hydrogens are removed from glucose, passed to the electron carrier NAD⁺, and then on to pyruvic acid, a product of glycolysis, converting it to lactic acid or ethanol that is excreted from the cell. Anaerobic respiration that produces an organic waste product is called **fermentation.** Fermentation allows cells to make ATP in the absence of oxygen.

Cells with the enzymes for forming lactic acid or ethanol in anaerobic metabolism can live in oxygen-deficient environments and still gain energy from glucose. Such cells do not, however, harvest all of the energy contained in the chemical bonds of glucose, since they excrete large organic molecules that contain substantial amounts of energy. Cells metabolizing glucose by fermentations harvest only about 5% of the available energy.

Many cells, when in an environment containing oxygen, are capable of metabolizing glucose by **aerobic respiration.** In this series of approximately 30 enzyme-catalyzed reactions, the glucose molecule is completely disassembled to yield CO_2 and H_2O. Of the usable energy contained in the chemical bonds of glucose, about half escapes as heat and about half is trapped in ATP during aerobic respiration.

In aerobic respiration, the initial series of reactions is the same as in anaerobic respiration; both begin with glycolysis. The products of glycolysis are also the same and include pyruvic acid, NAD•2H, and ATP. The difference between the two types of respiration lies in how the NAD•2H is regenerated and in the fate of pyruvic acid. In anaerobic respiration, the hydrogens were passed to pyruvic acid; in aerobic respiration, they are passed to oxygen through a series of compounds located in the cristae of the mitochondria or on the membranes of bacteria.

As the hydrogens pass through this series, known as the **electron transport system (ETS),** or **cytochrome system,** they lose energy. Some of this energy escapes as heat, but a large portion is trapped through the mechanism of **chemiosmosis** as usable energy in ATP. When the hydrogen finally combines with oxygen, water is formed. Thus in organisms that gain energy through aerobic respiration, the oxygen they consume leaves the body as the oxygen atoms in water molecules, not as the oxygen atoms in carbon dioxide.

Pyruvic acid, also produced in aerobic respiration, can be further metabolized. The covalent bonds of pyruvic acid are broken by the enzymes of the **Krebs cycle** found in the mitochondria, yielding CO_2 and additional hydrogen that combines with electron carriers like NAD⁺. These carriers are regenerated by means of the electron transport system, yielding more ATP by the mechanism of chemiosmosis.

Overall, 18 times more ATP is produced by aerobic respiration than is produced by anaerobic respiration. However, for aerobic respiration to occur, molecular oxygen is absolutely necessary. Organisms with aerobic metabolism are more efficient but are constrained to environments containing oxygen. Figure 8.1 summarizes the differences between these two types of respiration.

While the emphasis in this brief discussion has been placed on the electron carrier NAD⁺, remember that the important product of anaerobic or aerobic respiration is ATP. It is ATP that the cell needs to supply energy to its energy-consuming reactions of growth, maintenance, and movement. If ATP is not available to a cell, the cell dies. Because it is difficult to measure the amount of ATP in a cell, the experiments included here use indirect measurements to demonstrate anaerobic and aerobic respiration.

LAB INSTRUCTIONS

You will observe some of the properties of aerobic and anaerobic respiration. Because of the number of experiments included here, your instructor may omit some experiments, or do some as demonstrations.

Ethanolic Fermentation in Yeast

Yeast can break down glucose and obtain energy under both aerobic and anaerobic conditions. When oxygen is present, yeasts will break down glucose aerobically, using the metabolic sequence of glycolysis, Krebs cycle, and electron transport. Water and carbon dioxide are the waste products. When oxygen is not available, the Krebs cycle and electron transport system shut down. Glucose is metabolized to form pyruvate, which, in turn, is converted to two waste products, carbon dioxide and ethanol.

To test for ethanol production by yeast, your instructor will set up a culture about 24 hours before class as in figure 8.2. This was done by adding either diluted white corn syrup or grape juice to a 1-liter flask. A tablespoon of baker's yeast was added. This flask was labeled **anaerobic.**

Figure 8.1 Summary of reactions involved in anaerobic and aerobic metabolism of glucose.

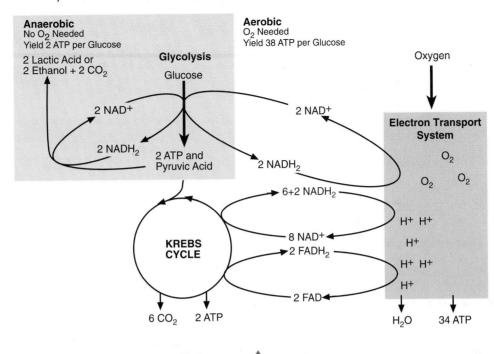

Anaerobic
No O_2 Needed
Yield 2 ATP per Glucose

2 Lactic Acid or
2 Ethanol + 2 CO_2

Glycolysis

Glucose

2 NAD^+

2 $NADH_2$

2 ATP and
Pyruvic Acid

Aerobic
O_2 Needed
Yield 38 ATP per Glucose

Oxygen

2 NAD^+

2 $NADH_2$

Electron Transport System

O_2

O_2 O_2

H^+ H^+

H^+

H^+ H^+

H^+

6+2 $NADH_2$

8 NAD^+

2 $FADH_2$

2 FAD

KREBS CYCLE

6 CO_2 2 ATP

H_2O 34 ATP

Figure 8.2 Culture setup for growing yeast anaerobically. Carbon dioxide produced in respiration forces all air (and oxygen) out of flask.

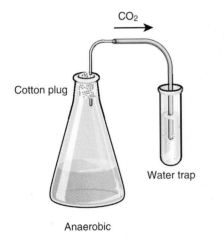

CO_2

Cotton plug

Water trap

Anaerobic

A second flask, without yeast, was stoppered completely and will serve as a control.

2 When you come to lab, you or your instructor will perform a simple classic chemical test to determine whether the culture is producing carbon dioxide. If the gas being emitted by the culture is bubbled through 0.1 M $Ba(OH)_2$, any carbon dioxide present reacts with the water in the test tube to form carbonic acid which in turn reacts with barium hydroxide to form barium carbonate, a white precipitate, according to the following equations:

$$CO_2 + H_2O \rightarrow H_2CO_3$$

$$H_2CO_3 + Ba(OH)_2 \rightarrow BaCO_3 \downarrow + 2H_2O$$

8-3 *Check water trap for pH*
(vs. control water)
by titrating w/ NaOH

? What are the results when the gas emitted by the culture is bubbled through $Ba(OH)_2$?

3 The other fermentation product should be ethanol. Ethanol can be isolated from the culture by distillation because it boils at 78° C, 22° C less than the boiling point of water.

Two distillation setups should be assembled (fig. 8.3). About 100 ml of anaerobic culture should be added to one, and 100 ml from the control to the other. Turn on the hot plates and collect about 30 ml of distillate from each still. Remember to turn off the hot plates! What you have just done is a scaled-down version of the process used in the liquor and gasohol industries.

? In which of the two flasks do you hypothesize you will find ethanol? _____

4 Ethanol in the distillates can be detected by a simple procedure called the iodoform test. Take four test tubes and number each one. First add 1 ml of strong I_2KI and then 1.5 ml of 1.5 M NaOH to each. These are the test reagents. Now add the following samples to each tube:

Tube

1   2.5 ml distilled water

2   1.25 ml distilled water and 1.25 ml 95% ethanol

3   2.5 ml of distillate from the **anaerobic** culture

4   2.5 ml of distillate from the **control**

Mix all solutions and let stand for five minutes. If ethanol is present, it will react with the iodine in the presence of NaOH to form **iodoform,** which will settle out as a

Figure 8.3 A reflux air-cooled distillation setup. The glass tube should measure about 30 cm from the flask to the curve to allow reflux action.

yellow precipitate. If no reaction occurs in tube 2 (the known sample of ethanol), add 1 ml more of NaOH to *all* tubes and mix. Record your results in table 8.1. Use *NR* to record where there is no reaction, and + to indicate where a precipitate is found.

Why were tubes 1, 2, and 5 included in the iodoform test procedure? What purposes do they serve?

Which tube(s) gave a positive test for ethanol? Is your hypothesis supported or falsified?

TABLE 8.1   Distillate tests from yeast cultures

Tube	Sample	Iodoform Test Results
1	Water	_____
2	Ethanol and water	_____
3	Distillate of anaerobic culture	_____
4	Distillate of control	_____

Aerobic Respiration in Peas

If living tissues or small organisms are placed in a closed chamber, they will consume oxygen and produce carbon dioxide as they metabolize sugars to produce ATP. If the CO_2 is chemically removed as it is produced, the pressure in the chamber will drop in proportion to the O_2 consumed. The simple device shown in figure 8.4 can be constructed to measure this change.

Potassium hydroxide can be used to remove CO_2. If CO_2 passes over a wet surface, it enters the water as carbonic acid. Carbonic acid will react with potassium hydroxide in the water to form potassium carbonate. Because potassium carbonate is a salt, it has an almost negligible volume compared to gaseous CO_2. This is summarized in the following equations:

$$H_2O + CO_2 \text{ (gas)} \rightarrow H_2CO_3 \text{ (solution)}$$

$$H_2CO_3 + 2\,KOH \rightarrow K_2CO_3 \text{ (crystalline)} + 2H_2O$$

If a small drop of dye is placed in the open end of the pipette in the apparatus shown in figure 8.4, it will move inward as the pressure in the chamber decreases due to oxygen consumption and to fluctuations in the classroom temperature and atmospheric pressure. The markings on the pipette allow direct readings as apparent volume units.

To convert apparent volume to actual volume of oxygen consumed, two corrections will be necessary. One correction will offset the effects of fluctuations in temperature and pressure during the experiment. The other will adjust the **apparent volume** to **standard volume** at a standard pressure of 760 Torr (mm of Hg) and standard temperature of 273 Kelvin. The first correction is made by using an experimental control called a **thermobar** and the second by a calculation using the **combined gas laws** that you learned in your chemistry classes.

Forming the Hypothesis

You will use the apparatus in figure 8.4 to determine the effect of freezing and thawing on pea seedlings. Peas are often planted very early in the spring and germinating peas may experience very cold frosts. You will investigate the question, Does freezing have any effect on germination?

Figure 8.4   Apparatus for measuring oxygen consumption in pea seedlings.

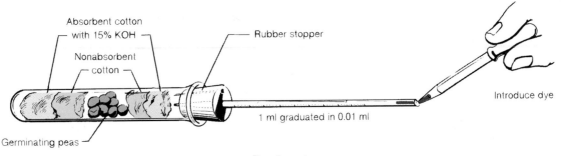

Respirometer

You will do this by measuring oxygen consumption as a measure of metabolic capability. If the peas are not aerobically metabolizing, they most likely will not germinate.

5▷   Convert this question into a testable hypothesis. State your hypotheses:

◆   Consider the experimental design for this experiment. How will you be able to control for temperature and pressure changes in the room during the experiment?

To be able to compare respiration between treated and untreated peas, you will add the same number of peas to two different tubes. However, all peas are not the same size and larger peas have more tissue that will consume more oxygen. How can you control for this?

Experimental Setup

Before lab, your instructor will have frozen and thawed some germinating peas, and will also have untreated germinating peas, all at room temperature. These are your experimental organisms. What is the variable being tested?

6▷   To begin, label and weigh two empty weighing pans. If you are using an electronic balance, adjust it to zero (tare) using the weighing pans. Now add about 8 to 12 four-to-six-day-old normal germinating peas to one pan and the same number of freeze/thaw-treated peas to the other. After weighing the pans with peas, subtract the empty pan weight to obtain the tissue weight.

	Normal	**Freeze/Thaw-Treated**
Weight of peas and pans	_____	_____
Weight of empty pans	_____	_____
Weight of peas	_____	_____

10 peas   5.84 g
sprouts   1.01 g

Record the pea weights in table 8.4.

7▷   Take three tubes and add a small ball of *absorbent* cotton 2 cm in diameter to the bottom of each tube. Hold the tube vertically and drop in four drops of 15% KOH, so that the drops fall directly on the cotton and do not run down the sidewalls. Cover the moistened cotton with a layer of *nonabsorbent* cotton. The nonabsorbent cotton protects the peas from the corrosive action of KOH, the active component of many drain cleaners. Add normal peas to one of the tubes, freeze/thaw-treated peas to another, and no peas to the third. This third tube serves as the *thermobar,* a control for random temperature and pressure changes in the room. Now add a second ball of *nonabsorbent* cotton to all three tubes and cover it with a ball of *absorbent* cotton. Moisten the absorbent cotton with two drops of 15% KOH.

In all of these procedures, care should be taken not to wet the sides of the tubes with KOH, since this will injure

TABLE 8.2   Raw oxygen consumption in milliliters

Tube	Contents	Reading Time (minutes)					
		0	**3**	**6**	**9**	**12**	**15**
1	Normal peas *sprouts*	-0.4 / -0.1	-0.4 / 2.0	0.0 / 0.9	0.2 / 1.6	0.3 / 2.2	0.4 / 2.6
2	Freeze/thaw-treated peas						
3	Thermobar						

TABLE 8.3   Thermabar-corrected data in milliliters

Tube	Contents	Reading Time (minutes)					
		0	**3**	**6**	**9**	**12**	**15**
1	Normal peas						
2	Freeze/thaw-treated peas						

plant tissue. Add dry stoppers with pipettes and syringes to all three tubes. The setups should look like that in figure 8.4. Be sure that the tip of the pipette does not poke into the cotton, which can plug it.

Measurement

8 ▶ Place the tubes on the table and let them equilibrate for five minutes. (Temperature equilibration after holding the tubes in your warm hands is absolutely essential for accurate results.)

◆ Is your body temperature above or below room temperature? What happens to the pressure exerted by a gas when it cools?

After 5 minutes, add a drop of dye into the end of each pipette by means of a Pasteur pipette. Try to get the front of the drop near the zero mark. When the front surface of the drop passes zero, record the time and the dye position in table 8.2.

Read the position of the front surface of the dye drop at two- to five-minute intervals (three minutes suggested but

should be modified according to rates observed), depending on the amount of activity. Record the readings in table 8.2. If fluid moves out from the 0 mark at the end of the pipette, estimate the volume change and use a minus sign to indicate readings less than zero.

Do not touch the tubes during the experiment because temperature increases will cause the gases in the tubes to expand and give false readings.

Analysis

9 ▶ Correct the raw data by subtracting the value of the empty thermobar. This tube corrects for any changes during the experiment due to changes in atmospheric pressure or temperature. Thus, changes in the thermobar tube reflect environmental variation, whereas those in the other tubes reflect changes due to metabolism plus environmental variation. Subtraction of the thermobar value, which may be either plus or minus, corrects for the effects of environmental variation. (Remember that if you subtract a negative number, you must actually add.) Enter the corrected values in table 8.3.

10 ▶ Now plot both sets of the apparent-volume-of-oxygen-consumed figures as a function of time on one piece of graph paper at the end of the exercise. Directions for making graphs are given in appendix B. Use different plotting symbols for each treatment and label all axes. Draw the straight line that best fits each data set.

11 ▶ Calculate the slope of each line and enter the slopes in table 8.4. Divide the slope values by the pea weights to arrive at a *specific rate of oxygen consumption per gram* of tissue for each treatment.

TABLE 8.4   Derived oxygen-consumption data

Tube	Contents	Slope (ml/min)	Pea Weight (g)	Specific Rate of O_2 Consumption (ml/min/g)
1	Normal peas			
2	Freeze/thaw-treated peas			

If you are going to compare rates of oxygen consumption between the samples, why is it necessary to calculate oxygen consumption on a per-gram-of-tissue basis?

12▶   These values are only apparent rates of consumption and must be corrected to standard conditions before they can be reported and compared to published data. By convention, all gas volumes in the literature are reported as volumes at 760 Torr (mm of Hg) pressure and 273 absolute temperature in Kelvin units.

The gas laws, mentioned earlier, are used to make this correction, as follows:

If V_1 equals apparent volume at P_1 (atmospheric pressure during the experiment) and at T_1 (temperature during experiment in Kelvins,° C + 273), and if V_2 equals corrected volume at standard pressure P_2 and at standard temperature T_2, then by the gas laws:

$$\frac{P_1V_1}{T_1} = \frac{P_2V_2}{T_2}$$

Rearrange to calculate the standardized rate of oxygen consumption,

$$V_2 = \frac{P_1T_2 \times V_1}{T_1P_2}$$

Add the standard values:

$$V_2 = \frac{P_1(273) \times V_1}{T_1(760)}$$

Add the experimental values.

Ask your lab instructor for the atmospheric pressure (in Torr) and temperature readings (in° C) in the room during the experiment. Record them below.   29.95 in

Atmospheric pressure (P_1) = _____ Torr
Temperature (T_1) = _22_ °C = _____ K

V_1 is the specific rate of O_2 consumption in ml/min/g, which you calculated in the last column of table 8.4. In doing the calculation, reduce the complex fraction to a constant multiplier.

What are the specific rates of oxygen consumption after adjustment to standard conditions for your experiments?

Normal peas _____ ml/min/g @ STP
Freeze/thaw-treated peas _____ ml/min/g @ STP

What was the percentage inhibition of respiration caused by freezing and thawing?

$$\% \text{ inhibition} = 1 - \left(\frac{\text{Normal} - \text{treated}}{\text{Normal}}\right)$$

Return to the hypothesis made at the beginning of the experiment. Must you accept or reject the hypothesis? _____

Learning Biology by Writing

The experiments in this laboratory exercise were chosen to demonstrate some of the properties of anaerobic and aerobic metabolism. The truly experimental portion of the exercise was the measurement of oxygen consumption in normal and treated pea seedlings.

Suppose you were working in the laboratory of a seed company that has developed a new strain of peas, the ones you just tested. They want to recommend planting in very early spring or maybe even fall in regions where frosts are common. They want to know if their seed viability will be affected by frosts. Assume you used oxygen consumption as a measure of viability and did the procedure in this manual.

Write a report that summarizes your observations and makes a recommendation to the company about early planting. Review the directions in appendix D for writing lab reports.

In the introduction to your lab report, briefly discuss aerobic respiration. Indicate how the experiment performed relate to these processes and outline the hypotheses tested by the experiments. Next, briefly describe the materials and methods used. Present your data as summary tables and a graph. Intermediate calculations need not be included unless your instructor requests them. In the conclusions section, advise your company about planting these peas.

As an alternative assignment, your instructor may ask you to complete the following summary and critical thinking questions.

Lab Summary Questions

1. What evidence do you have that both ethanol and carbon dioxide are produced in yeast fermentations?

2. Plot the data in table 8.3 on the graph paper at the end of the exercise and turn in the graph with this summary. What conclusion can you come to based on the data summarized in this graph and the hypotheses you made before doing the experiment?

3. Adjust the rate of O_2 consumption (ml/min/g) to standard temperature and pressure using the combined gas laws. Show the equations with substituted values. What is the percent inhibition of respiration due to the freezing treatment?

4. What are the advantages of aerobic respiration over anaerobic respiration? What are the advantages of anaerobic respiration over aerobic?

Critical Thinking Questions

1. One group of students in Denver, Colorado, and another group in New Orleans, Louisiana, performed the "aerobic respiration in peas" experiment. Are their results comparable? Explain.

2. Some students say that animals respire and plants photosynthesize. What evidence do you have to argue that this is a false generalization?

3. If you were looking at different types of cells through a transmission electron microscope and found certain cells that had no mitochondria, what prediction could you make about their metabolism?

4. In very deep lakes and several inches down in the muck on the bottom of swamps the dissolved oxygen concentration is zero. Many bacteria, fungi and some other organisms live in these environments and are able to metabolize glucose. How do they do it?

5. Does the pressure exerted by a gas increase or decrease when its temperature is increased but its volume is constant? When temperature decreases and pressure increases, what happens to the volume of a gas?

2. Plot data for all three plants.
(Use different colors for each plant).
~~Which plant used the most CO_2?~~
~~Did~~ respiration occur at the same rate for each plant?

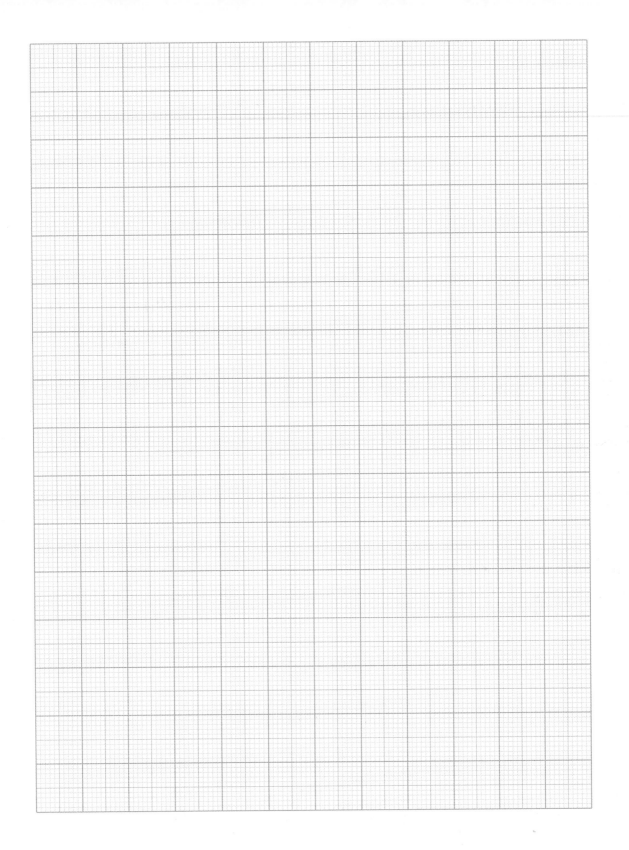

Measuring Cellular Respiration

LAB TOPIC 9

Determining Chromosome Number in Mitotic Cells

Supplies

Preparator's guide available at
     http://www.mhhe.com/dolphin

Equipment

Compound microscopes
Water bath or heating block at 60° C

Materials

Prepared slides of whitefish blastula
Prepared slide of sectioned onion root tip
Onion root tips four days old. Immerse green onions
     from grocery in water with aeration.
Glass pestles, 4 inches × ⅛ inch; round end in flame
     and then file flat area
Alcohol lamps
Small vials (15 to 20 ml) with caps
Small watch glass
Razor blades
Forceps
Slides and coverslips

Solutions

Fixative solution: one part glacial acetic acid to three
     parts 100% methanol made at start of lab
1 M HCl (Caution)
45% acetic acid (Caution)
Fresh Feulgen stain (Caution)

Student Prelab Preparation

Before doing this lab, you should read the introduction
and sections of the lab topic that have been scheduled
by the instructor.
     You should use your textbook to review the
definitions of the following terms:

Anaphase	Cytokinesis
Cell cycle	Daughter Cells
Centromere	Metaphase
Centrosome	Mitosis
Centriole	Prophase
Chromatin	Spindle
Chromosome	Telophase
Chromatid	

You should be able to describe in your own words
the following concepts:

     How and when chromosomes replicate
     Positions of chromosomes at stages of mitosis
     Significance of mitosis
     Review appendix C

     As a result of this review, you most likely have
questions about terms, concepts, or how you will do
the experiments included in this lab. Write these
questions in the space below or in the margins of the
pages of this lab topic. The lab activities should help
you answer these questions, or you can ask your
instructor during the lab.

Objectives

1. To identify stages of mitosis and cytokinesis on
   prepared slides of plant and animal cells
2. To stain chromosomes in dividing plant tissues
3. To determine the number of chromosomes in
   cultivars of onions
4. To test a hypothesis by using chromosome counting
   techniques and descriptive statistics

Background

An important characteristic of living cells is their ability
to divide, producing two daughter cells, which are genetically identical. Cell division in eukaryotes involves two
processes; **Karyokinesis** is the division of the nucleus by either mitosis or meiosis and **cytokinesis** is division of the cytoplasm. Prior to division, cells undergo a growth process in
which molecules, such as fats, proteins, and nucleic acids,
are synthesized from food molecules using energy derived
from respiration. Molecular synthesis alone, however, is not
sufficient to ensure proper growth. Molecules must assemble

Figure 9.1 How the cell cycle works.

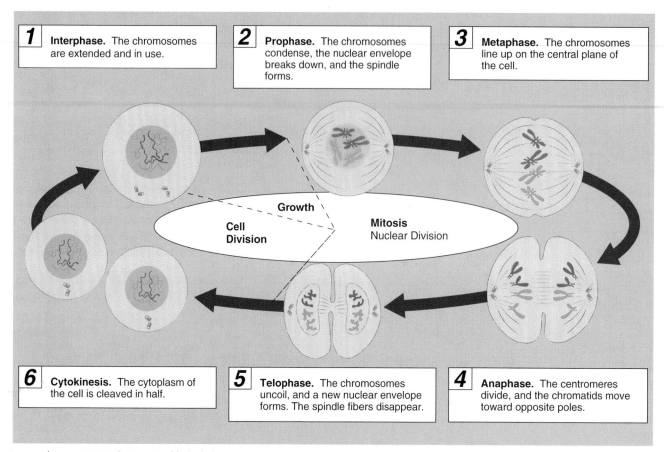

1 **Interphase.** The chromosomes are extended and in use.

2 **Prophase.** The chromosomes condense, the nuclear envelope breaks down, and the spindle forms.

3 **Metaphase.** The chromosomes line up on the central plane of the cell.

Growth

Cell Division

Mitosis Nuclear Division

6 **Cytokinesis.** The cytoplasm of the cell is cleaved in half.

5 **Telophase.** The chromosomes uncoil, and a new nuclear envelope forms. The spindle fibers disappear.

4 **Anaphase.** The centromeres divide, and the chromatids move toward opposite poles.

Source: Johnson, G. 2003. The Living World, Third edition.

or be assembled into eukaryotic cellular components, such as plasma membranes, ribosomes, mitochondria, and chromosomes. Before dividing, a cell contains hundreds of mitochondria, thousands of ribosomes, and literally trillions of small molecules, such as amino acids and sugars. Cells in balanced, continuous growth double their components and then divide these components in half, producing two equal daughter cells. Cells that continue to grow and divide are said to be going through the cell cycle.

Most eukaryotic cells have only one nucleus and its division involves a special mechanism. Important components of the nucleus are the **chromosomes,** the carriers of hereditary information. (Different organisms have different numbers of chromosomes in their nuclei: for example, the donkey has 66, humans have 46, and fruit flies, 8.)

During the growth period of the cell cycle, these chromosomes make copies of themselves. Each duplicated chromosome consists of two strands of genetic information called **sister chromatids** (fig. 9.1). Each sister chromatid consists of a single long DNA molecule that is coiled, folded, and wrapped around structures called nucleosomes. These are composed of proteins called histones. Each microscopic chromosome in an onion cell contains tightly coiled and folded DNA molecule that is over 10 cm long!

Mitosis is a karyokinesis process by which replicated chromosomes and other nuclear contents are sorted out to form two new nuclei. As a cell enters mitosis, its nuclear envelope breaks down, a spindle forms, and the chromosomes line up at the center, or **equator,** of the cell. Spindle fibers, composed of microtubules radiating from opposite poles of the cell, attach to each of the two chromatids. The microtubules attach to each chromatid at the **centromere,** a locally constricted region of a chromosome where the chromatids are held together.

As a cell progresses through mitosis, the centromeres split, and the microtubules of the spindle pull the separated sister chromatids to opposite poles of the cell. The cell then divides along the centerline, producing two daughter cells that each have a copy of all the chromosomes that were in the mother cell prior to growth and division (fig. 9.1).

Following karyokinesis, the cytoplasm divides. Fundamentally different cytokinesis mechanisms are found in animal and plant cells. In animal cells, the cytoplasm divides by constricting inward in a process called **furrowing.** In the furrow region, the protein actin, the same one involved in muscle contraction, encircles the cell. This contractile ring gradually pinches the cell in half, forming the daughter cells.

In plant cells, there is no constriction process. Instead, membrane vesicles containing cell wall components and derived from the Golgi apparatus migrate to the center of the cell and form a plate (phragmoplast) across the center of the mother cell. These vesicles fuse with each other and the plasma membrane to form the end membranes of two new daughter cells. Hence, the phrase cytokinesis by **cell plate** formation is used.

The alternating periods of growth (interphase) and division (phases of mitosis and cytokinesis) are called the **cell cycle.** Interphase can be divided into three subphases known as G_1, **S,** and G_2. During the S subphase, the DNA of the chromosome duplicates.

Mitosis in Animal Cells

Mitosis is most easily observed in growing, embryonic tissues, which have many cells dividing at the same time. The whitefish **blastula** is an early embryonic stage in the development of the whitefish from a fertilized egg. At this stage, the embryo is essentially a disc of dividing cells on top of a globe of yolk.

Biological supply houses "fix" the cells of the blastula; that is, they rapidly kill the cells and preserve them chemically. The disc is then embedded in wax or plastic to make it rigid and sectioned into thin slices. The sections are mounted on microscope slides. The tissue is stained with a dye to make the chromosomes visible. A resin is then placed on top of the stained specimen with a coverslip to make a permanent slide.

1▷ Obtain a prepared slide of a whitefish blastula and observe it under scanning power with your compound microscope. Note that several sections of the blastula are on the slide. Each contains many cells, some of which contain darkly stained chromosomes.

Center a cell containing chromosomes in the field of view and observe it first under medium power and then with the high-power objective. Sketch the cell, indicating such features as the **spindle, chromosomes, centromeres,** and **asters.** Though not visible through the light microscope, centrioles are found at the center of the asters in animal cells. The clear area in the center of the aster is called the **centrosome.** Centrioles have not been found in plant cells.

2▷ Now, locate another cell in which the chromosomes are visible. Chances are that the chromosomes are not aligned in the same patterns as in the previous cell. During mitosis, the chromosomes undergo choreographed movements that ensure each daughter cell obtains a full chromosome complement. In a blastula, the cell divisions are not synchronized, so different cells are in different stages of mitosis or may not have been dividing at the time the tissue was fixed and prepared. Furthermore, since you are looking at sectioned material, there may be cells in which neither chromosomes nor nuclei are visible because the section was taken from the edge of a cell and did not include any nuclear material. The absence of nuclear material in these cells is called an artifact of slide preparation. An artifact is a phenomenon that results from technique and not from some biological mechanism.

3▷ Over the next 15 minutes, look at various cells on the blastula slide and identify the mitotic stages. You should find examples of **interphase, prophase, metaphase, anaphase,** and **telophase.** Use figure 9.1 to help you identify the stages. Sketch each stage and label the structures.

Figure 9.2   Stages of mitosis in an onion root tip as seen in two types of preparations: (*a*) root tip sectioned and stained so cells remain intact, and (*b*) root tip stained with Feulgen reagent and then squashed.

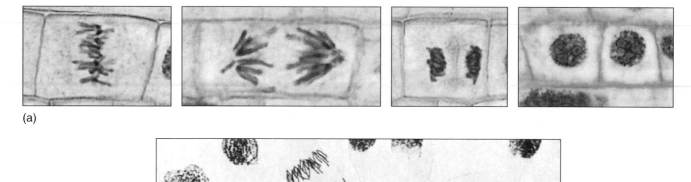

(a)

(b)

Mitosis in Plant Cells

4▷ If time permits, obtain a slide of a longitudinal section of an onion root tip. Cells just behind the tip divide by mitosis as the root elongates. Study this area on your slide and identify cells in interphase, prophase, metaphase, anaphase, and telophase (fig. 9.2). Sketch cells in these stages below.

◆ Did you see asters in any of the onion root tip cells?

Look at a cell in late telophase. Can you see a line of vesicles forming across the center of the long axis of the spindle? These will fuse together to form the **cell plate** that will separate the two daughter cells.

Mitosis lasts for about 90 minutes in onion root tip cells. Each of the four phases takes a different amount of time. The phase lasting the longest will be the most commonly observed. Why do you think this is so?

5▷ Look at 40 *dividing* onion root tip cells, and tally the frequency of occurrence of the phases in table 9.1.

Because frequency of occurrence is directly proportional to the length of a phase, multiply the percentage of the cells in a phase times the duration of mitosis to obtain an estimate of the time required for a phase in mitosis. Enter the time values in the last column of table 9.1.

◆ Which phase of mitosis is longest?

Staining Dividing Cells

Actively growing root tips, young leaves, flower buds, or other **meristematic** (dividing) plant tissues have a high percentage of cells in mitosis. These tissues can be prepared for study by the sectioning technique described earlier or by a squashing technique in which the tissue is softened, stained, flattened, and observed (fig. 9.3).

TABLE 9.1   Determining duration of mitotic phases

Phase	Number Seen	% of Total	Length in Minutes
Prophase	⎯⎯	⎯⎯	⎯⎯
Metaphase	⎯⎯	⎯⎯	⎯⎯
Anaphase	⎯⎯	⎯⎯	⎯⎯
Telophase	⎯⎯	⎯⎯	⎯⎯
Total	⎯⎯	⎯⎯	⎯⎯

Figure 9.3   Steps for preparing tissue by the Feulgen stain/squash method.

(1) Remove 2 tips.   (2) Add to fixative.   (3) Incubate at 60° C for 15 min.   (4) Pour fixative into watch glass.   (5) Add 5 ml of HCl.   (6) Incubate at 60° C (10 min).

(7) Pour HCl into watch glass.   (8) Add Feulgen stain (30–45 min).   (9) Transfer one tip to slide; add drop of acetic acid.   (10) Pulverize tip by tapping with glass rod.   (11) Add coverslip; press to squash cells.

The squashing method has two advantages over sectioning. First, intact cells are easy to identify in squashes, whereas in sectioned material, the plane of the sectioning may be tangential to the nucleus and may thus exclude some chromosomes. Second, with care, cells may be flattened without bursting. This allows easy observation of the chromosomes, although it does distort the original three-dimensional organization of the cell (see fig 9.2a and b).

The use of onion root tips is recommended, though it would be possible to use any other plant, bulb, or germinating seed that has vigorously growing roots.

Tissue Preparation Steps

6 ▸ 1. Pour 3 ml of *freshly* prepared methanol-acetic acid **fixative** into a small vial. Remove two roots and place them immediately in the fixative. The fixative rapidly kills the cells by denaturing proteins and extracting lipids. The roots should be fixed for 15 minutes at 60° C.

C A U T I O N

Methanol/acetic acid mixture is flammable, corrosive, and has noxious fumes.

---

2. To stain and squash the tissue, it is necessary to **hydrolyze** it, that is, to partially break down the cells and their components. After fixation, work in a ventilated area or fume hood and slowly pour off the fixative into a watch glass, being careful not to lose the roots. Discard the fixative in a waste container.

Add a few milliliters of 1 M HCl to the vial and incubate the vial in a water bath for ten minutes at 60° C to soften the plant cell walls and partially hydrolyze the DNA in the chromosomes. *The temperature and time in this step are critical.* Too short or too long a hydrolysis may result in poor staining in the next step.

CAUTION

HCl is hydrochloric acid which, if splashed, will damage your eyes (wear goggles) or clothing (wear apron). Feulgen stain appears colorless, but if it gets on your hands or clothing, it will stain them vivid pink. Be careful! Household bleach will remove stains.

---

3. After hydrolysis, pour off the acid into a watch glass, taking care not to lose the roots. Add a milliliter or so of **Feulgen stain** to the vial. The slide may be prepared after 30 minutes. Maximum stain intensity will be reached after 45 minutes to one hour at room temperature.

4. To make a slide, transfer a root to a very small drop of 45% acetic acid on a slide. Using a razor blade, cut off 1 to 2 mm of the root tip. Be sure to get the darkly stained area. Discard the remaining older portion. Look at the tip carefully. It should have a prominent purple to pink band of staining which is less than a millimeter wide. This is where the dividing cells are located.

CAUTION

Acetic acid is corrosive with noxious odor. Avoid contact. Flush skin with water if contacted.

---

5. Pulverize the root into a fine pulp on the slide by tapping on it with a polished glass rod about 100 times. The difference between making a superior and a mediocre slide lies in the pulping. It is important to use only a small amount of liquid so that the pulp is easily made. Now place a clean coverslip over the preparation. The amount of liquid on the slide should be just enough to flow out to the edges of the coverslip (fig. 9.3).

6. Lay the slide on the table and cover the slide with two thicknesses of paper towels. Put your thumb on the towel over the coverslip and press directly downward as hard as you can. Use care to prevent the coverslip from moving sideways—this will roll cells on top of one another and destroy the material. If this happens, make a new slide with the second root. If the tissue is especially tough, extreme pressure may be necessary.

7. Examine the slide using low power to locate pink-stained chromosomes on the slide. You should reduce the light intensity to see the chromosomes. Small quantities of 45% acetic acid may be added to the edge of the

coverslip to prevent the preparation from drying. If the cells are not sufficiently spread, they may be squashed further by repeating the pressing procedure. The Feulgen stain will gradually fade (in about two hours).

7 Study your slide and identify cells in the four mitotic phases and in interphase. Compare what you see to figure 9.3. How many chromosomes are in these cells? How many chromatids make up the chromosomes at different phases? Sketch some of the cells and label the stages.

If your microscope is equipped with an ocular micrometer and you calibrated it in lab topic 2, use it to measure the length of one of the chromosomes in a squashed onion cell. How long is an onion chromosome in millimeters?

Testing a Hypothesis

You have now learned to recognize the stages of mitosis and to carry out a classical chromosome-staining technique. You will now apply this knowledge in a simulation of real world situations.

A technique often used by taxonomists to identify closely related species is chromosome counting, also known as **karyotyping.** These scientists will often prepare cells using the same techniques presented in this exercise and then statistically analyze their data. In this portion of the exercise, you will estimate the diploid (2N) number of chromosomes for the species used to make your slide. The following describes what could be a real-life application of this procedure.

Imagine that you work in a biological laboratory that performs chromosome analyses for hospitals and anyone else who has an interest and is willing to pay the going rate.

A major horticultural firm has come to your lab with a problem it would like you to work on. The firm, Seeds Unlimited, has been selling a new strain of onion seed, which has been very successful. However, another horticultural firm, Gardeners Incorporated, claims that Seeds Unlimited stole the variety from them. Furthermore, Gardeners Incorporated wants damages because they spent years and thousands of dollars developing the new variety, which is protected under a horticultural patent. Your job is to prove in a court of law that the two varieties in question are genetically different and thus do not represent a patent infringement.

A literature search of horticultural patents yielded the basic karyotype information about the Gardeners Incorporated variety. It is a registered variety with a diploid number of chromosomes equal to 18. You must now determine the number of chromosomes in the Seeds Unlimited variety.

A hypothesis to be tested by collecting data on number of chromosomes would be: *The number of chromosomes in the Seeds Unlimited variety (the onions you have in lab) is equal to 18.*

As a head technician in the laboratory, you give the following directions to all other technicians:

Procedure

8▷ To count chromosomes, look only at mid- to late-anaphase cells. The number of chromosomes moving to one or the other pole of the cell represents the diploid number for the species. Select a cell in which the chromosomes are well spread by squashing, and count the number of chromosomes at one pole. Record the count in table 9.2. Now count those at the other pole and record. Find another cell and repeat until you have 10 counts.

When counting, be sure to use the high-power objective and to focus up and down as you count so as to include chromosomes that may lie above or below the plane of focus for the objective. It may be helpful to count the visible ends of chromosomes and then to divide by two to estimate the chromosome number. If this method is used, always round up to the next even number when an odd

TABLE 9.2 Number of chromosomes observed by you in onion root cells

Cell #		Cell #	
1. _____		6. _____	
2. _____		7. _____	
3. _____		8. _____	
4. _____		9. _____	
5. _____		10. _____	

number of ends is observed, because each chromosome has two ends and one may have been missed in counting.

A facsimile of table 9.3 will be on the blackboard or on a computer spreadsheet in the lab. Record your observations as directed. After everyone has recorded their observations, count the total number of cells seen by the group having four chromosomes, five chromosomes, and so on. Record these numbers in column B.

◆? If the number of chromosomes is constant for a species, why were different numbers of chromosomes seen in different cells?

Analysis

9▷ You will now statistically analyze the group data in table 9.3, using the following three steps.

1. Plot a frequency **histogram** of the data from the first two columns of table 9.3 on the graph paper at the end of the exercise. Appendix B contains graphing instructions. This histogram will give you a visual representation of the data variability.

2. Calculate the **average** number of chromosomes per cell from the class data by following the steps below:

$$\text{Avg. \# chromosomes per cell} = \frac{\Sigma(A \times B)}{\Sigma B}$$

$$\frac{\Sigma(\# \text{ chromosomes} \times \# \text{ cells})}{\Sigma(\# \text{ cells})}$$

(*Note:* Σ (Sigma) is a Greek letter meaning "sum of.")

a. In table 9.3, sum the values in column **B** and enter the value at the bottom.

b. For each row, multiply the value in column **A** by that in **B** and enter the product in column **C.**

Determining Chromosome Number in Mitotic Cells   **109**

TABLE 9.3   Summary and work table

A # chromosomes	B # cells seen	C A × B	D (A − Average)^2	E B(A − Average)^2
4	4	16		
5	5	25		
6	9	54		
7	1	7		
8	15	120		
9	2	18		
10	2	20		
11				
12				
13				
14				
15				
16				
17				
18				
19				
20				
	Σ 38 Total cells seen	Σ 260 Total chromosomes seen		Σ_____

c. Sum column **C** and enter value at the bottom.

d. Divide the sum of column **C** by the sum of column **B** to obtain the average number of chromosomes per cell.

Because the data in table 9.3 are already summarized into categories, the above equation can be translated into a series of arithmetic operations by rows and columns in the table.

To calculate the expression Σ (observation − average)^2:

a. For each row, subtract the average from the value in column **A.** Square this different and enter square in column **D.**

b. For each row, now multiply the value in column **D** by the value in column **B.**

c. Enter the product in column **E.**

d. Sum column **E** and enter the value at the bottom.

3. Calculate the **standard deviation** of the class data. The calculation of standard deviation is discussed at the end of lab topic 5 and in the beginning of appendix C. Remember that:

$$\text{Standard deviation} = \pm\sqrt{\frac{\Sigma(\text{observation} - \text{average})^2}{n - 1}}$$

To calculate the standard deviation, divide the sum of the fifth column by one less than the sum of column **B.** Take the square root of the quotient so obtained. What is the standard deviation for your sample?

What is your conclusion regarding the horticultural patent suit?

If your analysis showed that the two cultivars had the same number of chromosomes, does this prove they are the same cultivar? Do all species have different numbers of chromosomes?

Write the values for the average and standard deviation on your histogram.

Based on the class data summarized in the average, standard deviation, and histogram, what is your conclusion regarding the null hypothesis? Is the number of chromosomes in your class sample (obtained from Seeds Unlimited) the same or different from 18 (the number of chromosomes in the Gardeners Incorporated variety)? Must you accept or reject your hypothesis? Why?

Learning Biology by Writing

Continuing with the situation presented under Testing a Hypothesis, imagine that you must write a report summarizing your laboratory's studies of chromosome number in onion varieties. Your report will be entered as court evidence and must be concise, yet informative. Read appendix D for directions about writing lab reports.

The report should have the following components:

Descriptive title

Purpose of study including hypothesis

Methods
    Brief overview with lab manual cited for specifics
Results
    Histogram of lab data
    Average and standard deviation
        Give formula, substituted values, and answer
Discussion
    Accept or reject null hypothesis and cite evidence
    State potential sources of error
Literature cited

Determining Chromosome Number in Mitotic Cells

Internet Sources

Much genetic research is directed at building databases that contain what traits are located on what chromosomes. More detailed studies are actually trying to determine what is the nucleotide sequence in the single DNA molecule found in each chromosome. Such projects are now underway for a number of organisms, including the onion. To look for information on the chromosomes of onions use the search engine Google at http://www.google.com. When connected, enter *Plant Genome IV Abstracts*. These are abstracts of papers given at a conference in California in 1996. Submit the query.

When the listing of abstracts appears on the screen, several of them will be about the genetics of other plants besides onions. Use the Find feature of your browser (a button or under Edit at top of screen). When the dialog box appears, type *onion*. This will take you to the abstract that has onion in the title. Click on the abstract and read it. Close the window and repeat to find another abstract on onions. There should be at least four abstracts on onions.

In your own words, summarize the type of research that was reported at this conference. You can find more recent research by entering Plant Genome V (or VI or VII, *etc.*) in Google.

As an alternative assignment, your instructor may ask you to complete the following lab summary and critical thinking questions.

Lab Summary Questions

1. Describe the positions of the chromosomes in the following stages of mitosis:

Prophase
Metaphase
Anaphase
Telophase

2. For a species with a diploid number of chromosomes equal to six, draw a cell as it would look at metaphase.
3. Why are whitefish blastula and onion root tips used to study mitosis?
4. Plot the data in table 9.3 as a frequency histogram on the graph paper at the end of this lab topic and turn it in with this summary. Indicate with arrows the average and standard deviation for the distribution.
5. What is the average number of chromosomes observed by your class in onion root tip cells? What is the range of observations? What is the standard deviation?
6. Review the hypothesis that you made regarding the number of chromosomes in the tested variety of onions. Based on your data, must you accept or reject the hypothesis? Does the company, Gardeners Incorporated, have a valid claim?

Critical Thinking Questions

1. Can you think of any reasons why organisms rarely, if ever, have 2N chromosome numbers greater than 100?
2. If two organisms have the same 2N number of chromosomes, are they always members of the same species? Justify your answer.
3. Why do biologists say that mitosis produces genetically identical daughter cells? How can this be true?
4. Many protists, fungi, and plants have haploid cells that divide by mitosis. How many chromosomes would you expect to find in the cells resulting from such a division?
5. In table 9.1 you summarized your observations about the number of cells in each phase of mitosis and used this information to calculate the duration of each phase. What assumption concerning independence of cell divisions among the cells underlies the validity of this analysis?

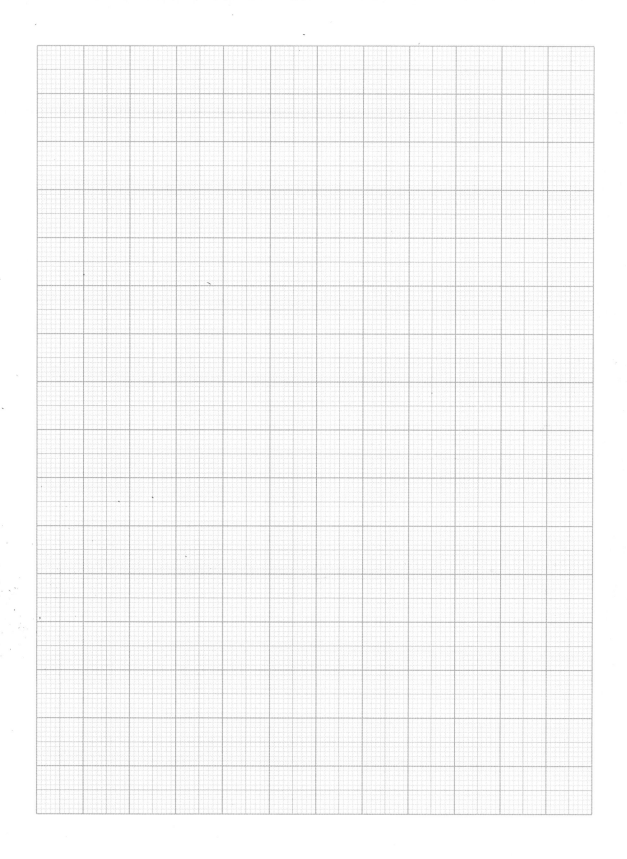

LAB TOPIC 10

Observing Meiosis and Determining Cross-Over Frequency

Supplies

Preparator's guide available at
   http://www.mhhe.com/dolphin

Equipment

Dissecting microscopes
Compound microscopes

Materials

Preserved female *Ascaris,* dissected (demonstration)
Prepared slides of *Ascaris* ovary-oviduct-uterus
Sordaria fimicola cultures on petri plates
   (Carolina/Biological Supply Biokit) and fresh
   plates for subculture
Slides and coverslips
Toothpicks or probes

Student Prelab Preparation

Before doing this lab, you should read the introduction
and sections of the lab topic that have been scheduled
by the instructor.

You should use your textbook to review the
definitions of the following terms:

   Ascus
   Crossing-over
   Diploid (2N)
   Haploid (N)
   Homologous chromosomes
   Spore
   Zygote

You should be able to describe in your own words
the following concepts:

   Chromosome positions during phases of meiosis I
   and II
   Genetic significance of crossing over
   How segregation occurs in meiosis I
   How independent assortment occurs in meiosis I
   General life cycle of an ascomycete fungus
   Review Chi square, appendix C

As a result of this review, you most likely have
questions about terms, concepts, or how you will do
the experiments included in this lab. Write these
questions in the space below or in the margins of the

pages of this lab topic. The lab experiments should
help you answer these questions, or you can ask your
instructor during the lab.

Objectives

1. To identify metaphase I and II of meiosis in *Ascaris*
   eggs
2. To determine experimentally the frequency of
   crossing-over in *Sordaria,* an ascomycete fungus
3. To perform a chi-square statistical test of a
   hypothesis

Background

Meiosis is a form of nuclear division which produces
daughter cells having half the number of chromosomes
found in the parent cell. In sexually reproducing species if
meiosis did not occur, the number of chromosomes would
double with each fusion of egg and sperm. For example,
humans have 46 chromosomes in their cells. If meiosis did
not occur in humans, a sperm and egg would each contribute 46 chromosomes to a fertilized egg, so that it would
have 92. Since mitosis always produces daughter cells with
the same number of chromosomes as the parent cell, all
cells in the new individual would also have 92 chromosomes. If this individual mated with another of the same
generation, the third generation of fertilized eggs would
have 184 chromosomes. Obviously, if this exponential progression continued, there soon would not be a cell large
enough to contain the increasing number of chromosomes
in each generation.

Meiosis prevents this problem in sexually reproducing
species by reducing the chromosome number by one-half at
some point in the life cycle. A **diploid** nucleus contains
pairs of chromosomes; a **haploid** nucleus has only one chromosome from each pair, or half as many. In mammals and
other higher animals, meiosis always results in either haploid egg or sperm production, but this is not the case in all

Figure 10.1 Schematic representation of meiosis. Note a double crossing-over during prophase I in the large pair of chromosomes. The positions of chromosomes at metaphase I and II are the key to understanding how reduction in chromosome number (segregation) occurs during meiosis because once they are lined up the pattern of separation is determined. Meiosis is not a random halving of chromosome number; it is highly ordered. Chromosomes are color coded to represent their origins: white are maternal and black, paternal. Random alignment of maternal and paternal chromosomes at metaphase I is the basis for independent assortment.

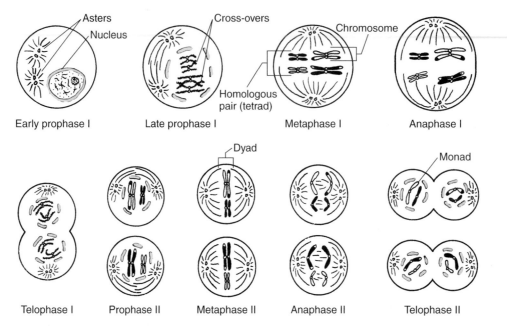

organisms. In some fungi, for example, meiosis occurs in cells soon after fertilization, forming haploid spores from diploid fusion cells. The haploid spores are then dispersed by wind. If they settle on a suitable substrate, they germinate and divide by mitosis, producing multicellular mycelia composed of haploid cells. During sexual reproduction in the next generation, cells of one mycelium fuse with those of another to produce the diploid **zygote,** and the cycle repeats.

As a result of fertilization, a zygote contains two similar sets of chromosomes: one inherited from the mother and one from the father. In humans, for example, if both parents are blue eyed, both the egg and the sperm will carry a gene for blue eyes on one of their 23 chromosomes, and the zygote will contain both of these genes on two different chromosomes. However, if one parent has brown eyes and the other blue, the zygote may carry one gene for brown eyes and one for blue. (The child that develops from this egg would have brown eyes, since brown is a physiologically dominant gene.)

These genetic examples illustrate that the chromosomes are in pairs in diploid cells, with one member traceable to each of the parents. Members of such pairs are said to be **homologous;** that is, they can be matched on the basis of the genes they carry. In the example just given, the eye-color trait in general, not the specific color, provides the basis for the homology. Recent work in the Human Genome Project has determined which genes are found on which chromosomes and is providing a complete map of all human chromosomes, as well as those of other species of economic or scientific interest.

Meiosis reduces the chromosome number by separating the members of each homologous pair in a diploid cell. In a human with 46 chromosomes, each egg or sperm produced by meiosis contains one member of each pair, or a total of 23 single chromosomes. However, although a single gamete contains *only one member of each pair,* it usually contains a mixture of the chromosomes derived from an individual's father and mother. There is no division mechanism that separates all maternally derived chromosomes from all paternally derived chromosomes. Therefore, meiosis produces gametes that each contain a mixture of maternal and paternal chromosomes. As you will see in lab topic 11, this leads to variation in future generations.

A stylized representation of meiosis for a hypothetical organism having four chromosomes is shown in figure 10.1. Note that there are two parts to meiosis. In **meiosis I,** the homologous chromosomes line up side by side during prophase I continuing into metaphase I, whereas in **meiosis II,** there are half the number of chromosomes and they line up singly at metaphase II. Understanding these differences in alignment is the key to understanding the differences between meiosis I and II. Why? Because once the chromosomes align as they do, the separation pattern is determined.

A cell that begins meiotic cell division has previously undergone a period of growth and **replicated** its chromosomes, so that each chromosome consists of two chromatids joined at the centromere. As this cell enters meiotic **prophase I,** the homologous chromosomes pair off and lie side by side as a **tetrad** of four chromatids, in much the

Figure 10.2 Three homologous chromosome pairs showing cross-over points in late prophase I chromosome pairs from a grasshopper (*Chorthippus parallelus*). Count the number of chromatids in each of the three pairs.

same way that one might hold four skeins of yarn clenched in one hand.

This pairing process, called **synapsis,** involves forming a connection between adjacent chromatids in tetrads. During this pairing process, the homologous chromosomes reciprocally exchange parts in a process called **crossing-over.** This means that parts of the maternally derived chromosome actually pass and bind to the paternally derived chromosome and vice versa. The net result is that each member of the homologous pair becomes a mosaic of parts derived from both the maternal and paternal chromosomes. Consequently, new gene combinations, or chromosomes, are created. Crossing-over has profound implications as an additional source of evolutionary variation in sexually reproducing species.

As prophase I ends, the recombined chromosomes unravel except at points of attachment called the **chiasmata** (fig. 10.2). The homologous chromosomes remain attached at the chiasmata as they line up at the center of the cell, eventually reaching the phase called **metaphase I.** While these events are occurring, the nuclear membrane breaks down and the spindle forms.

At the end of metaphase I, the homologous chromosomes in each pair separate from each other as the attachment points separate. Each mosaic homologous chromosome, consisting of two chromatids, moves to opposite poles of the cell during **anaphase I.** In most animals and many plants, the end of meiosis I is marked by the occurrence of cytokinesis, the division of cytoplasm. Each daughter cell thus produced has half as many chromosomes as the starting cell, but each of these chromosomes is a combination of maternally and paternally derived genes because of crossing-over and each chromosome consists of two chromatids.

In the second part of meiosis, these chromosomes again line up at the center of the cell in **metaphase II,** and

then proceed through **anaphase II** and **telophase II.** The result is the separation of the sister chromatids and the production of a total of four cells, each with the haploid number of chromosomes. Each chromosome at this point consists of one chromatid, in which the genes have been recombined as a result of crossing-over. Each cell, in turn, contains a mixture of chromosomes with some traceable to the individual's father and others to the mother.

LAB INSTRUCTIONS

You will identify cells in several stages of meiosis and experimentally determine the frequency of crossing-over during meiosis.

Meiosis in Ascaris

Experimental Organism

Ascaris is a genus of parasitic roundworms (nematodes) found in the intestines of swine. These worms reproduce sexually and have separate sexes, and internal fertilization. *Ascaris* is used to demonstrate meiosis because the diploid number of chromosomes is only four. This means that the chromosomes can be counted during the normal meiotic stages, an ideal characteristic for learning the stages of meiosis. The chromosomes are small, however, and are hard to see. You will have to work to see the meiotic phases.

Look at the demonstration dissection of a female *Ascaris* and refer to lab topic 20, page 262, for an explanation.

In *Ascaris,* as in most animals, meiosis occurs in the **gonads,** a term that refers to ovaries and testes. Diploid cells in these organs undergo meiosis, producing haploid **gametes,** or egg and sperm. (Because the chromosomes in *Ascaris* are small, crossing-over is not easily seen and will be studied separately.) In *Ascaris,* the production of **oocytes** starts in a tubular ovary (fig. 10.3).

As the developing oocytes descend from the ovary to the oviduct, meiosis occurs; different stages can be observed by looking at slides made from different regions of the reproductive system. As the developing oocytes enter the lower portions of the female reproductive tract or uterus, meiosis is completed and a sperm nucleus, which had penetrated during fertilization, fuses with the egg nucleus in an event called syngamy. A tough shell forms around the fertilized egg, or **zygote,** as these events occur. Inside the shell, the zygote eventually divides by mitosis, producing an embryo.

Procedure

1. Obtain a prepared slide containing sections of the ovary, oviduct, and uterus of *Ascaris.* You may have cross sections that will look round or longitudinal sections that will be longer than they are high. The ovary is the smallest section and should be examined first under medium power. You should see a mixture of round, maturing eggs, called

Figure 10.3 Developmental stages of eggs taken from different parts of the female reproductive tract in *Ascaris*. When the egg completes meiosis, the sperm and egg nuclei fuse to form a zygote. The zygote will develop into a worm embryo by mitosis.

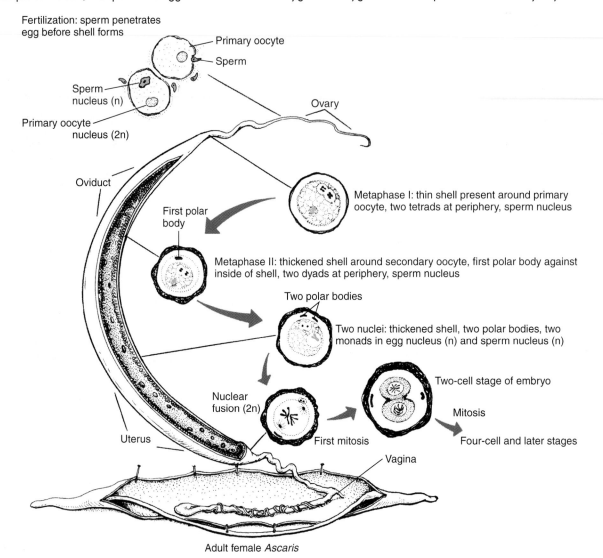

Fertilization: sperm penetrates egg before shell forms

Primary oocyte

Sperm

Sperm nucleus (n)

Primary oocyte nucleus (2n)

Ovary

Oviduct

First polar body

Metaphase I: thin shell present around primary oocyte, two tetrads at periphery, sperm nucleus

Metaphase II: thickened shell around secondary oocyte, first polar body against inside of shell, two dyads at periphery, sperm nucleus

Two polar bodies

Two nuclei: thickened shell, two polar bodies, two monads in egg nucleus (n) and sperm nucleus (n)

Two-cell stage of embryo

Mitosis

Nuclear fusion (2n)

Uterus

First mitosis

Four-cell and later stages

Vagina

Adult female *Ascaris*

primary oocytes, produced by the ovary and triangular sperm introduced when the female mated (fig. 10.4*a*).

After a sperm fertilizes a primary oocyte, the haploid sperm nucleus rests in the oocyte cytoplasm as the **sperm pronucleus.** A shell develops around the primary oocyte and its nucleus begins meiosis. The shelled oocyte now moves into the oviduct.

While the maturing egg is in the oviduct, the chromosomes condense, the nuclear envelope disappears, and the two pairs of homologous chromosomes each pair in synapsis, forming two tetrads. Although you cannot see it on your slide, what important genetic recombination process occurs in tetrads?

2 As meiosis continues into metaphase I, the paired, homologous chromosomes (tetrads) move to the center of a spindle that forms off center, or eccentrically. Look at the next larger section on your slide or get a new slide that is a section of an oviduct. Look at the oocytes in the oviduct to see if you can find any that look like figure 10.4b. These oocytes are in metaphase I and you can see that there are four chromosomes arranged as two pairs. The paired chromosomes will separate during anaphase I, moving to opposite poles. Since the centromeres do not separate, the sister chromatids remain together.

After the homologous chromosomes have separated, unequal division of the cytoplasm (cytokinesis) occurs. This results in one large, functional cell (the secondary oocyte) and one very small, nonfunctional cell called the

Figure 10.4   Photomicrographs of *Ascaris* eggs in stages of meiosis and the first mitotic division: (*a*) sperm penetration with primary oocyte nucleus caught in prophase I, (*b*) metaphase I, (*c*) cytokinesis I showing first polar body formation, (*d*) metaphase II, (*e*) fusion of haploid egg and sperm nuclei, (*f*) metaphase of mitosis as fertilized egg divides to produce two cells, (*g*) two-cell stage in *Ascaris* development.

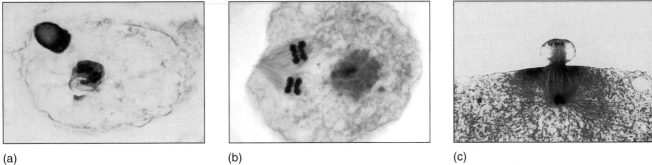

(a)   (b)   (c)

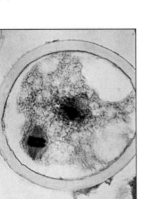  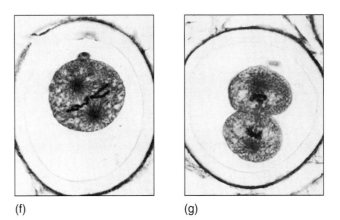

(d)   (e)   (f)   (g)

first polar body (fig. 10.4*c*). How many chromosomes are found in a secondary oocyte? In a first polar body?

You will be very lucky if you see a polar body forming as in figure 10.4*c*. Usually the polar body is seen flattened against the inside of the shell as a darkly staining body. If you see such, you know that the oocyte you are observing has completed meiosis I (telophase I).

After an interphase period, the second part of meiosis begins as the oocyte, now called a **secondary oocyte,** descends into the uterus. You should be able to observe metaphase II of meiosis II by looking at the largest section on your slide or by getting a new slide that contains a section of the uterus. Remember that you can tell if an oocyte has completed meiosis I by looking for the dark polar body against the inside of the shell. When you find such an oocyte look for the chromosomes, each consisting of two chromatids, lining up at the center of the spindle, which again forms eccentrically (fig. 10.4*d*). How many chromosomes are there at this point? _____ Metaphase II will end and anaphase II begins as the paired chromatids separate and move to opposite poles of the second eccentric spindle.

When the chromosomes reach the poles, the cytoplasm divides, resulting in a second polar body and a mature haploid egg. You can tell when an oocyte has completed meiosis II because there will be two darkly staining nuclei flattened against the inside of the shell.

In some of the oocytes from the uterus, you will see two large nuclei, called **pronuclei** (fig. 10.4*e*). One of these is the haploid egg nucleus produced by the meiotic steps you just studied and the other is the sperm nucleus, which has become active.

These nuclei will fuse into a single diploid nucleus. How many chromosomes will it have? _____

The cell containing the diploid nucleus is large and contains large amounts of stored energy. It is encapsulated in an eggshell and is thus a closed system. As this cell divides to form the embryo, it will draw on energy stored in the cytoplasm as yolk material. The meiotic production of eggs not only reduces the number of chromosomes in the cell but also ensures that a large amount of cytoplasm goes into the ovum for use by the embryo. The sperm essentially contributes only its nuclear material. The polar bodies, which you can see as dark bodies against the inside of the eggshell, are nonfunctional and eventually disintegrate, contributing to the metabolism of the embryo.

Figure 10.5 Stages in the life cycle of the fungus *Sordaria*: (a) diagram of sexual phases of life cycle; (b) scanning electron microscope photo of *Sordaria* mycelium.

(a)

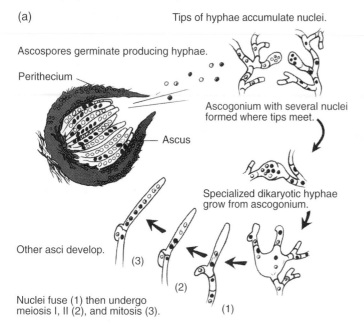

Tips of hyphae accumulate nuclei.

Ascospores germinate producing hyphae.

Perithecium

Ascus

Ascogonium with several nuclei formed where tips meet.

Specialized dikaryotic hyphae grow from ascogonium.

Other asci develop.

(3)

(2)

Nuclei fuse (1) then undergo meiosis I, II (2), and mitosis (3).

(1)

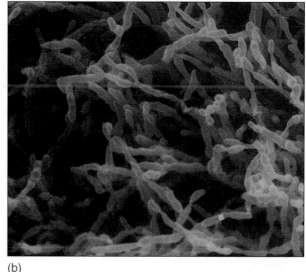

(b)

4 Make a series of drawings showing the stages of meiosis. Include eggs in the following stages: sperm penetration, metaphase I, and metaphase II. Label such structures as the eggshell, plasmalemma, spindle, tetrads, sperm nucleus, polar bodies I and II, and chromosomes. Refer to figure 10.4 to help identify the stages.

Measuring the Frequency of Crossing-Over

Background about Experimental Organism

The effects of crossing-over can be demonstrated in *Sordaria fimicola,* an ascomycete fungus. The life cycle of *Sordaria* is shown in figure 10.5a. In nature, the fungus grows in the dung piles of herbivorous animals. When mature, a fruiting body called a **perithecium** develops. It contains **ascospores,** small single haploid cells surrounded by a tough outer wall. The ascospores are forcibly ejected and stick to the leaves of grasses or other low-growing plants. Grazing herbivores ingest the spores. They pass through their digestive systems and are defecated. The ascospores germinate in the dung to produce thin haploid filaments called **hyphae.** The hyphae grow in length and branch by mitosis, eventually producing a network of filaments called a **mycelium** (fig. 10.5*b*). Cross walls in the hyphae separate individual cells.

Under appropriate conditions, not completely understood, an **ascogonium** having several nuclei is formed by the fusion of two hyphae. Special hyphae, called ascogenous hyphae with two nuclei per cell, are produced by mitosis from this structure. The ascogenous hyphae give rise to **asci,** saclike structures in which ascospores are produced. This process starts with the two nuclei in the cell fusing, the equivalent of fertilization. The diploid fusion nucleus then undergoes meiosis with crossing-over, forming four haploid nuclei. Each of these nuclei then undergoes mitosis, giving rise to eight ascospores in the ascus (fig 10.5*a*).

Several hundred asci will develop in the area of hyphal fusion. As they develop, the asci are surrounded by a growing mass of hyphae, forming the fruiting body or **perithecium** (fig. 10.6) with an opening at its apical end. As indi-

TABLE 10.1   Results from *Sordaria* ascospore counts

Type	Cross	Ascospore Pattern	Number Counted
(c) Nonrecombinant	Tan × black	4.4	_____
(d) Recombinant	Tan × black	2:2:2:2 or 2:4:2	_____

Figure 10.6   Perithecium of *Sordaria* has burst, releasing several asci, each containing eight ascospores. The perithecium forms from hyphal tissue that surrounds the dikaryotic cells.

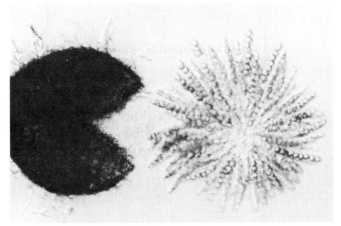

Open perithecium   Many asci each containing 8 ascospores

vidual asci mature, they elongate and forcibly eject their ascospores through the apical opening of the perithecium. Gelatinous materials on the surface of the ascospores attach them to the surfaces of leaves. When these ascospores are ingested by a herbivore, the cycle repeats.

The color of the ascospores is genetically controlled. Normal or wild strains of *Sordaria* produce black ascospores, while mutant forms of *Sordaria* produce tan ascospores. Consequently, in a mixed population of *Sordaria*, three types of fusions (matings) are possible: tan and tan, black and black, and tan and black. A conceptual model of the chromosome behavior in these matings would appear something like figure 10.7.

This figure is based on two experimentally verified assumptions: that the cells in a single ascus result from meiosis followed by a mitosis in each of the meiotic daughter nuclei; and that the eight ascospores in an ascus are arranged in linear order, reflecting the order of the steps in the meiotic and mitotic divisions that occur in the ascus. When the diploid stage is homozygous (both fusion members are genetically the same—for example, black and black), all haploid ascospores are the same. (See fig. 10.7*a* and *b*.) When the diploid stage is heterozygous (for example, black and tan), six possible results can occur. Two of these results are obtained only when crossing-over fails to occur and are called **nonrecombinants.** The two different chromosomes

duplicate, segregate, and undergo an additional mitosis forming the ascospore pattern shown in figure 10.7*c*. However, when crossing-over does occur, it produces the **recombinant** ascospore patterns shown in figure 10.7*d*. Because the ascospores cannot pass by each other in the ascus, they remain aligned in the order that they were produced. Thus, only the process of crossing-over can account for the patterns shown in figure 10.7*d*.

Though all of this may seem complex, it is really a simple biological system. The orderly patterns of colored ascospores in an ascus result from meiotic divisions in which crossing-over did or did not occur.

If you accept the chromosome model shown in figure 10.7 for fusion and ascus development in *Sordaria,* you can determine whether or not crossing-over has taken place by counting the number of asci having different arrangements of tan and black spores.

Procedure

5▷ Obtain a petri dish in which a cross was set up between two *Sordaria* strains about two weeks before your laboratory. One strain produced only black spores and the other only tan. Three types of perithecia will be present, corresponding to the three matings previously described.

Use a toothpick or dissecting needle to scrape a few perithecia from an area where the mycelia have merged, and make a wet mount on a microscope slide. Do not dig into the agar growth medium; just scrape off the perithecia! Add a coverslip and, using a pencil or probe, very gently press on the coverslip to squash the perithecia. When a perithecium bursts, it will release many asci, each containing eight ascospores (fig. 10.6). If you press too hard and rupture the asci, you will need to make another slide.

6▷   Observe the slide with a microscope on medium or high power. If the perithecia contain only asci with all black or all tan spores, make a new slide using material from a different region of the culture dish. Continue making slides until you find perithecia having asci with both types of spores. These are hybrid asci resulting from black by tan crosses. Use a marker to draw a circle on the outside of the plate where you found recombinant asci so it will be easier for others to get a sample.

The possible spore arrangements in the ascus are shown in figure 10.7.

7▷   Count all asci of types (*c*) or (*d*) that you can see on your slide. You should count at least 50 asci for statistical accuracy. Record the counts in table 10.1.

Figure 10.7 Alternative models of chromosome behavior during *Sordaria*'s sexual reproduction. Models (*a*) and (*b*) will be common. You must find areas where hyphae from tan and black strains have fused to form perithecia before you will observe asci of types (*c*) and (*d*).

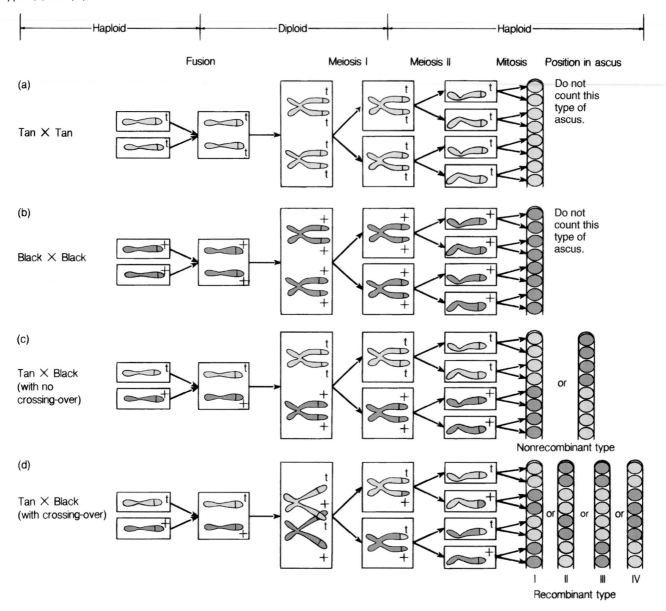

Analysis

The relative frequency of recombinant asci equals the number of recombinant hybrid asci observed divided by the total number of hybrid asci observed, or:

$$\% \text{ recombinants} = \frac{\# \text{ recomb.}}{\# \text{ recomb.} + \# \text{ nonrecomb.}} \times 100$$

8 Calculate the % recombinants found among the asci you observed.

9 Go to the chalkboard and write the % recombinants you observed. When all students have listed their values, calculate a class average. What is the average relative frequency of recombinants observed by the class? What is the standard deviation of the class data? (Refer to appendix C for method of calculation.)

Figure 10.8
Abbreviated "what if" models showing that cross-over must be in the chromosomal region between the spore color locus and the centromere to give a recombinant-type ascus.

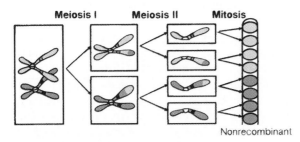

Situation A—Gene locus close to centromere; no cross-over in region between gene and centromere

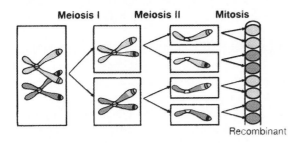

Situation B—Gene locus far from centromere; cross-over occurs between gene and centromere

Calculate the distance of the spore color gene locus from the centromere using the class average for percentage of recombinants.

What does the number you calculated describe?

What are the effects of crossing-over on total genetic variability in the next generation?

The relative distance in map units of the spore color gene from the centromere of its chromosome can be calculated from the percentage of recombinants (fig. 10.8). Geneticists apply a logical relationship to calculate map distances—the greater the distance between a gene locus and the centromere, the greater is the frequency of crossing-over. Thus, when the gene locus is at the centromere, no crossing-over will occur.

Why do you think there will be little or no crossing-over when a gene is located very close to the centromere or when two genes are very close together?

Map units are obtained by multiplying the percentage of recombinants by one-half because each recombinant ascus counted represents a tetrad of chromatids in which there has been an exchange involving only two of the four chromatids. So, if you find that 56% of the asci are the recombinant type, the gene for spore color would be said to be 28 units from the centromere.

Advanced Statistical Analysis

When students do this experiment, they often want to know what is the "right" answer. How many map units should they have? Researchers who have done this experiment agree that 28 map units separate the gene for spore color from the centromere, but this may vary with culture age and strains of fungi used.

If analysis of your class data does not yield this value, you are faced with a dilemma. Did your class perform the experiment in the wrong way or are you simply seeing random variation in results? Your class should discuss how the experiment was done and how they collected data to rule out any errors.

If all data are judged to have been collected by acceptable means, random variation may be the answer. To see if your class answer to map distance is within an acceptable limit of random variation, a statistical test called the **chi-square** (χ^2) test can be employed. This test compares actual results to expected results and will tell you with a certain probability whether differences may be due to chance deviations. You should read appendix C, which describes this test in more detail.

Before performing a chi-square test, you must formulate a **null hypothesis (H_o)** and an **alternative hypothesis(H_a)**. The H_o is that there is no difference between the observed results and the expected results other than due to chance. You found _____ map units between the centromere and the spore color gene using the class data and expected that number to be 28. The H_a is that there is a major difference, which may be due to error or to biological changes in the fungus. You would not be able to determine why the major difference was obtained without doing more experiments.

10 To perform the test, you must convert the map units back to the percentage of recombinant asci and the percentage of nonrecombinants. If the regions of the chromosomes are 28 map units apart, 56% of the asci counted should have been recombinants and 44% nonrecombinants. Enter the values you observed in the following table and calculate the difference by subtracting the expected.

	Recombinant	Nonrecombinant
Observed	_____	_____
Expected	56%	44%
Difference	_____	_____
(Difference)²/ Expected	_____	_____

Now calculate a χ^2 statistic that summarizes the variation in your data. χ^2 is defined as:

$$\chi^2 = \Sigma \frac{(\text{Difference})^2}{\text{expected}}$$

Square the difference for recombinants in the table and divide that number by the expected (56%). Repeat these calculations but use the values in the nonrecombinant column. Sum both these answers to calculate a χ^2 for your class data. What is the χ^2 value for the class data?

If you think about how χ^2 was calculated, its meaning should be obvious. When a calculated χ^2 is a large number, there is a lack of agreement between the observed and expected (the null hypothesis must be rejected); when χ^2 is small, the agreement is good (the null hypothesis would not be rejected). The problem you now face is to decide whether the χ^2 you calculated is large or small.

Statisticians use a standard table to aid in these decisions. This table provides χ^2 values for various experimental designs and levels of confidence where deviations are due to chance. Your experimental design has two categories: recombinant or nonrecombinant. You should use the value listed in the standard table under one degree of freedom (d.f.) because **degrees of freedom** are defined as being equal to one less than the number of categories. Next you must decide on a probability level. Biologists usually use 95% confidence level or a 5% chance that rejecting the null hypothesis will be a wrong decision. (An alternative way of expressing this is that you have 95% confidence in your decision to reject.)

11 If you look at table C.4 in appendix C, a critical value of 3.8 will be found at $\chi^2_{.95}$ and d.f. = 1. If your calculated χ^2 is less than or equal to this critical value, you should not reject the null hypothesis (i.e., your answer generally agrees with the value of 56% recombination). If your calculated χ^2 is greater than 3.8, you should reject the null hypothesis and infer the alternative hypothesis is true. What is your conclusion? Do you reject the null hypothesis?

$$\chi^2 = \_\_\_\_\_$$

Learning Biology by Writing

In a lab report entitled "Determining the Frequency of Crossing-Over in a Fungus," briefly describe the biology and life cycle of *Sordaria*, indicating why it is an ideal organism in which to study crossing-over. Describe how you measured the frequency of recombination and present the class results as a histogram. You can use the chi-square analysis to see if your class got the "right" answer. Discuss the significance of these results and the effect crossing-over has on genetic variability from generation to generation. Discuss how cross-over frequency can be converted to map units and indicate the value your class obtained. Describe why genetic variability is an important consideration in evolutionary theory. Finally, indicate possible sources of error in the experiment.

As an alternative assignment, your instructor may ask you to complete the following summary and critical thinking questions.

Internet Sources

Several labs work on the genetics of recombination using *Sordaria*. Use the Internet to locate three different labs that are doing such work.

Open your Internet browser program and use the search engine GOOGLE (www.google.com). When you get the query box in GOOGLE, type in the word *Sordaria*. You will get a long list back.

Look over the list and click on those entries that are research reports. Make a list of three labs that are doing research on *Sordaria* and briefly describe the research projects.

Lab Summary Questions

1. A hypothetical mammal has a diploid (2N) number of 6. Draw a picture of metaphase I and metaphase II as they would appear in this species. Add a third panel to your drawing in which you sketch metaphase of mitosis as it would appear in this species.
2. In *Ascaris,* meiosis and oocyte maturation occur at the same time. Describe the stages that oocytes go through as they pass from the ovaries to the uterus in these worms.
3. Describe how *Sordaria* grow and reproduce asexually. Without describing the details of meiosis and crossing-over, write a description of sexual reproduction in *Sordaria.*
4. You observed the results of crossing-over in crosses between black and tan *Sordaria.* Describe in your own words how crossing-over occurs and results in the ascospore patterns seen.
5. What frequency of crossing-over did you measure for *Sordaria?* Is this statistically the same as the value published in the literature? How did you reach this conclusion?

Critical Thinking Questions

1. Why do scientists use critical value tables in statistical analysis? Shouldn't each researcher decide what is an acceptable variation?
2. Why would you expect the diploid (2N) number of chromosomes always to be an even, not an odd, number in every species?
3. Crossing-over creates new combinations of alleles on chromosomes that are passed on to the next generation. Explain the evolutionary significance of such genetic variability. What are other sources of genetic variation in sexually reproducing species?
4. Can meiosis occur in a haploid cell? Why? Can mitosis occur in a haploid cell?

LAB TOPIC 11

Determining Genotypes of Fruit Flies

Supplies

Preparator's guide available on WWW at
   http://www.mhhe.com/dolphin

Equipment

Incubator
Compound microscopes
Dissecting microscopes

Materials

Prepared slide of giant chromosomes from fruit fly
   larva as demonstration
Drosophila cultures (Carolina Biological Supply)
   Wild type (sexes separated)
   Multichromosomal mutant for white eyes and
      apterous wings (sexes separated)
   Single chromosomal mutants (sexes separated)
      White eyes
      apterous wings
   Unknowns constructed by crossing wild type with
      any or all of above mutants (sexes separated)
Widemouthed vials for culture chambers
Cotton or sponge stoppers
White cards
Camel's hair brushes
Petri plate
Fly Nap® (Carolina Biological Supply) or carbon
   dioxide generator from bicarbonate or dry ice
Slides and coverslips
Dissecting needles and forceps
Dead flies for practice phenotyping
Alcohol lamp

Solutions

Instant *Drosophila* medium
   Pour dry mixture directly into culture chamber.
   Add water. Each chamber should be filled to a
      depth of 3/4 to 1 inch. Sprinkle several grains
      of dry yeast on the surface of the medium; the
      yeast will reproduce and act as a food source.
      Stop the chamber with cotton or foam plugs.
Jar of ethanol to kill and hold flies
Acetocarmine stain
0.8% saline

Student Prelab Preparation

Before doing this lab, you should read the introduction
and sections of the lab topic that have been scheduled
by the instructor.

   You should use your textbook to review the
definitions of the following terms:

allele	homozygous
autosome	locus
gene	mutant
genotype	phenotype
heterozygous	sex chromosome
homologous chromosome	wild-type

   You should be able to describe in your own words
the following concepts:

Mendelian dominants and recessives
Mendelian independent assortment
Mendelian segregation
Sex linkage
Use of a Punnett square

   As a result of this review, you most likely have
questions about terms, concepts, or how you will do
the experiments included in this lab. Write these
questions in the space below or in the margins of the
pages of this lab topic. The lab experiments should
help you answer these questions or you can ask your
instructor during the lab.

Objectives

1. To identify the life stages of the fruit fly, *Drosophila melanogaster*
2. To observe the giant chromosomes found in the salivary glands of larval fruit flies
3. To use Punnett squares to predict the results of genetic crosses
4. To breed fruit flies to determine their genotypes at one or more loci

Background

The discovery of the basic laws of heredity was made by Gregor Mendel, a Czech (then part of Austrian empire) monk and scientist, who in 1866 worked out a conceptual model of how seven characteristics are inherited in peas. However, the significance of Mendel's theories was not recognized by the scientific community until the early 1900s, after his death. Researchers reviewed his published papers and realized that this farsighted scientist had discovered the very principles of heredity that they were seeking in their own work.

Why was Mendel's work unappreciated when it was first published? At that time, the processes of cell division and fertilization were not well understood. The prevailing theories of inheritance suggested that liquids, not particles, were responsible for the inheritance of characteristics. In addition, Mendel's arguments were based on statistics from experimental crosses rather than from direct observation of biological structure. For all these reasons, no one paid much attention to the curious writings of the little-known monk-scientist.

The theories suggested by Mendel have been verified again and again. In fact, Mendel's ideas have held so true that they have been assigned the status of scientific laws by modern biologists. Furthermore, Mendel's methods are a perfect role model for beginning scientists: Ask simple questions, collect quantitative data, derive your explanations from data analysis, and publish your results.

What were Mendel's contributions? They included:

- The factors responsible for inheritance were particles not fluids;

- The particles occurred in pairs in the peas;

- One particle could dominate over the other when both were present;

- The particles segregated when pollen and eggs were produced (Law of Segregation);

- The particles governing an inheritable characteristic segregated independently of the particles governing other characteristics (Law of Independence).

In the years since the rediscovery of Mendel's work, additional principles of heredity have been discovered. For example, Walter Sutton demonstrated that the biological mechanism for Mendel's law of segregation was the separation of homologous chromosomes during anaphase I of meiosis. Sutton also showed that Mendel's law of independent assortment was controlled by the random alignment on the spindle of maternally and paternally derived chromosomes during metaphase I of meiosis. Sutton's work provided evidence that Mendel lacked: the biological mechanisms for the statistical results Mendel observed. As a result of the work of Sutton and others, the science of inheritance became a field of intense investigation.

Since chromosomes, meiosis, and the operation of Mendel's law of inheritance seemed universal in all eukaryotic organisms, it followed that mechanisms found in one eukaryotic organism were most likely applicable to all other eukaryotic organisms. Biologists then began looking for the best experimental organism to use in studying more complicated inheritance patterns.

For a number of reasons Thomas Hunt Morgan at Columbia University and others in the early 1900s decided that the common fruit fly was an ideal experimental animal. Fruit flies are small and feed on yeast, so that large populations can be kept in a small space and on a small budget. *Drosophila* lay hundreds of fertile eggs after mating and produce a new generation in 10 to 14 days, so results are quickly and easily obtained. And finally, the diploid number of chromosomes in the fruit fly is only eight, which makes it possible to determine on which chromosome a gene for a particular trait is located.

In his early studies of fruit flies as a model genetic system, Morgan was the first scientist to provide experimental evidence that a specific gene is part of a specific chromosome. He found that a gene determining eye color was located on a chromosome called the **X** chromosome, also called the sex chromosome. In fruit flies as in mammals, females have two **X** chromosomes (*i.e.,* a homologous pair), but males have only one. In place of the second **X** chromosome, males have another chromosome called the **Y** chromosome and it carries no gene determining eye color. When females produce haploid eggs by meiosis, every egg will contain an **X** chromosome. When males produce sperm by meiosis, half of the sperm contain an **X** chromosome and half contain a **Y.** Genes on the **X** chromosome, and the traits they cause, are said to be **sex-linked genes** or **traits.** The inheritance of a sex-linked trait can modify the Mendelian ratios expected from genetic crosses as you will see in your investigations.

LAB INSTRUCTIONS

In this lab you will first learn to identify normal and mutant traits that are inherited in fruit flies. You will also learn how to identify males and females. You will be given fruit flies of unknown genetic composition and will be asked to determine their genotypes through genetic crosses and use of Mendel's and Morgan's principles.

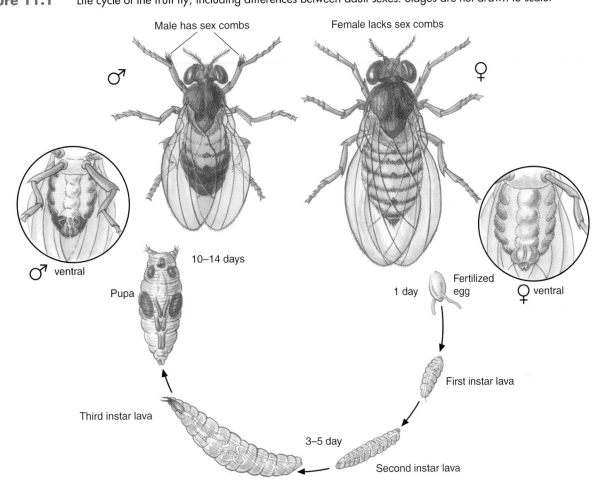

Life Cycle of the Fruit Fly

When male and female fruit flies mate, about 1,000 sperm are transferred in a packet, which is stored in the female's reproductive tract. As eggs descend the female's reproductive tract, sperm are released and fertilize them. A female may lay several hundred eggs during her reproductive period. Mating occurs about six to eight hours after the female emerges from the pupa, and eggs are first deposited on the second day after emergence. Each egg is about 0.5 mm long and has two thin filaments projecting from its anterior end (fig. 11.1).

After about a day at 25° C, a larva, or maggot, emerges from the egg. The maggot is eyeless and legless but has black mouthparts and conspicuous spiracles (or air pores) through which it breathes. In nature, the maggot pushes its way through rotting fruit, eating fungi and fruit tissue. As it grows, the maggot goes through three stages called **instars.** At the end of the first and second instars, the maggot molts and sheds its exoskeleton and mouthparts, allowing the organism to increase in size. When the third instar ends, a

pupa is formed. Before transforming into the pupal stage, the maggot stops feeding and crawls onto a dry surface.

Inside the pupa, major reorganizational events occur. The larval structures and tissues break down and are reabsorbed, and adult tissues begin to grow from imaginal discs (clumps of adult tissue producing cells) that have been dormant in the maggot. This tissue reorganization is completed in only four days. At the end of this period, the adult fly breaks out of the pupal casing.

When the fly emerges, it is elongated and light in color, and the wings are folded. During the next hour, the body becomes more rounded, the wings expand to their normal shape, and the body color darkens. Adult fruit flies live an average of 37 days but do not grow any larger.

About five to eight hours after emergence, courting behavior begins. The purpose of this behavior seems to be sex identification. A male initiates courtship with every fly, male or female, it contacts. Courtship gestures include tapping the other fly with his foreleg and circling closely around the other fly while rapidly vibrating the wing closest to it to stimulate mating behavior in a receptive female.

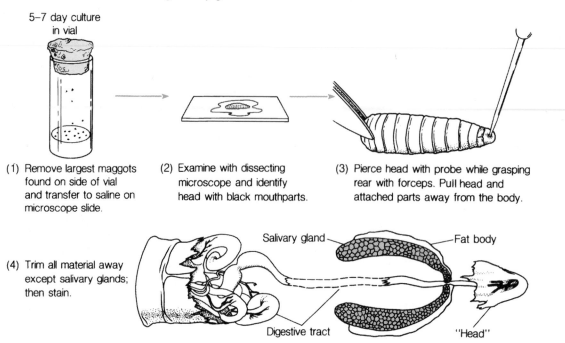

(1) Remove largest maggots found on side of vial and transfer to saline on microscope slide.

(2) Examine with dissecting microscope and identify head with black mouthparts.

(3) Pierce head with probe while grasping rear with forceps. Pull head and attached parts away from the body.

(4) Trim all material away except salivary glands; then stain.

Salivary gland

Fat body

Digestive tract

"Head"

If the female is not receptive or if the other fly is a male, a sharp kick ends the encounter. A receptive female, however, turns her abdomen and allows the male to mate.

Fruit Fly Anatomy

1 ▷ Obtain several fruit flies that have been killed by an overdose of anesthesia or by freezing. Place the flies on a white index card and observe them with a dissecting microscope using reflected light. Dead fruit flies always carry their wings at a 45° angle to the body.

2 ▷ Identify the following structures on several of the flies on your card. Compare different individuals to obtain some idea of the variation in these structures.

Head

    antennae (normal or aristopaedia)

    proboscis (mouthpart)

    ommatidia (compound eyes; may be red or white)

Thorax

    wings (normal and apterous)

    halteres (reduced second wing pair)

    legs (femur, tibia, foot)

Abdomen

    Note size and pigmentation.

3 ▷ Use a camel's hair brush to separate your flies into two piles by sex. Male and female flies are pictured in figure 11.1.

Males are generally a little smaller than females with five, rather than seven, abdominal segments. The male's abdomen also tends to be more rounded and blunt with darker markings at the tip, especially on the ventral surface. These dark markings do not fully develop until a few hours after emergence. Therefore, newly emerged males can easily be confused with females. The front legs of the male have dark hairlike projections called **sex combs,** which the female lacks. At the posterior end of the abdomen on the under surface, the male has two **claspers,** which the female also lacks.

After you have sorted your flies by sex, have your instructor check your identifications. Dispose of these flies as directed.

4 ▷ Now, take a jar in which there are larvae. Pick a few larvae out of the culture, and look at them under the dissecting microscope. Can you see the mouthparts and the outline of the digestive system?

5 ▷ Look for pupae on the sides of the container or on the dry paper wick above the medium. Larvae seek dry places in which to pupate so that the emerging adult will not become mired in sticky food materials. Remove a few pupae and observe them with your microscope. If you place a pupa on a slide in a drop of water and observe it with your compound microscope, you may be able to see the developing adult inside. Use low magnification and high light intensity.

Giant Chromosomes in Drosophila (Optional)

When the third instar larvae of *Drosophila* pupate, they secrete a proteinaceous material from their salivary glands that "glues" the pupa to a surface. In these larvae, the salivary gland cells contain giant chromosomes that are actively involved in RNA synthesis.

The large size of these chromosomes is due to three factors: (1) these chromosomes are partially condensed as in prophase; (2) the two homologous chromosomes lie side by side in permanent synapsis; (3) most important, each chromosome has been replicated a number of times (up to ten), and the copies are joined in register along their length. Though such giant chromosomes are not usual in most animals, their occurrence in fruit flies greatly aids genetic analysis, because it is possible to see chromosomal regions that nearly correspond to genes.

In order to view these chromosomes, you must dissect out the salivary glands, stain them, and observe them through your compound microscope (fig. 11.2). Alternatively, a prepared slide can be viewed on a demonstration microscope.

6▶ Locate a third instar larva crawling up the jar wall or the larger ones in the growth medium. Transfer one to a drop of saline on a microscope slide and observe under a dissecting microscope. Push a dissecting needle through the head just behind the black mouthparts and push a second needle through the midbody or hold with forceps. Pull the needles apart, and with a little luck, the mouthparts will separate from the body, trailing the salivary glands. The glands will be two elongated, grapelike, transparent clusters of tissue as shown in figure 11.2. Be careful not to confuse the salivary glands with the glistening fat bodies that stick to them.

7▶ Clean excess tissue off the glands with your needles and move the glands to a clean area of the slide that is devoid of saline. Cover the glands with a drop of acetocarmine or acetoorcein stain and let sit for five minutes. Gently heating the slide over a hot plate or lamp without boiling will speed staining.

8▶ When ready to observe, add a coverslip to the preparation, cover it with a piece of tissue, and press firmly and directly down, to squash the cells and nuclei. Do not let the coverslip move sideways, or a poor slide will result. Observe with your compound microscope using the medium power to find the tissue and the high power to observe chromosomes. If nuclei are not broken, remove the slide from the microscope and press again on the coverslip.

◆❓ How many chromosomes do you see? _____

Can you see banding in the chromosomes? _____

Sketch part of a chromosome below.

Theoretical Background for Crosses

We will now review some theoretical background before performing experimental genetic crosses with fruit flies.

Genes are located on chromosomes and chromosomes are found as homologous pairs in diploid organisms. The location on a chromosome where a gene is found is said to be the **locus** for that gene. Thousands of gene loci are found on each chromosome. If we could look at a single locus in a population of fruit flies, ignoring all other loci, we might find that some flies have exactly the same gene on both chromosomes in the homologous pair. These flies would be described as being **homozygous** for that gene. We could write this down in the following way.

Let **"A"** be the symbol for a specific gene. A homozygous individual would have two **"A"** genes present and this would be written as **AA.** When you write down the genes that are found in an individual, you are writing the **genotype.** However, as we studied a population of fruit flies, we might find some flies that had a variation of the gene, one that was not exactly the same. This would be called an **allele** of gene **"A"** and could be written as **"a."** Hypothetically, some flies could be homozygous for this allele and would have a genotype that was written **aa.** Some, however, could have the **"A"** form of the gene on one chromosome and the **"a"** form of the gene on the other chromosome in the homologous pair. The genotype of this individual would be written as **Aa,** and it would be described as being **heterozygous** at this chromosomal locus.

While the preceding discussion of genes was instructive, it is not possible to look at genes through a microscope and tell what kind are on a chromosome. The only way to determine what genes are present—short of doing DNA analysis, which is not easy—is to do genetic crosses and then to reason what genes must have been present to give the results. When you do genetic crosses, you deal with **phenotypes**—what an organism looks like, not with genotypes—what genes are present. Phenotypes are the result of gene expression (along with environmental influences) and the occurrence of a phenotype is usually a good indicator that a gene is present.

Fruit flies have eight chromosomes: six **autosomes** and two **sex chromosomes.** Females have two X sex chromosomes, so they have four homologous pairs. Males have X and Y sex chromosomes, so they have three homologous pairs plus an X and Y.

Because of these chromosomal differences between the sexes, one would expect to see differences in the inheritance of traits, depending on whether the traits were carried on autosomes or sex chromosomes. These differences are theoretically explained in the next two sections.

Autosome Model

If a gene locus for a trait with two different alleles is on one of the three pairs of autosomes, then a cross between two homozygotes for contrasting traits (alleles *A* and *a*) can be diagramed as follows:

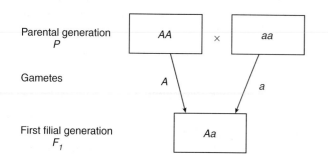

Obviously, one parent must be male and one female, but it makes no difference which parent is homozygous for one allele *(A)* and which is homozygous for the other *(a)*. The result will always be the same: progeny will always have two different alleles (heterozygous) regardless of whether they are male or female.

When Mendel did experiments like this, he had to use the concepts of **dominant** and **recessive** to explain his results. Because the parents are homozygous for different alleles, they look different; *i.e.,* they have different phenotypes. Hypothetically, we could say AA is blue because the "A" allele causes blueness and aa is white because the "a" alleles produce no color development. All progeny are Aa: they have both alleles. What will be their color? The only way to tell is to do an actual genetic cross. When Mendel did similar experiments, he found all progeny were the equivalent of blue. He explained it by saying the allele for blue dominates over white which is recessive.

If the heterozygous offspring from the first filial generation (F_1) are allowed to mate, the results will demonstrate that the recessive allele is present in the heterozygote because the recessive allele segregates from the dominant allele during meiosis and reappears in the next generation when the gametes combine. This can be demonstrated by using a **Punnett square,** a paper and pencil method of keeping track of the kinds of gametes that can be produced and the combination of gametes at fertilization as follows:

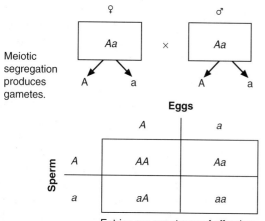

The genotypic results from a cross between two heterozygotes would be 1 *AA:* 2 *Aa:* 1 *aa,* or a phenotypic ratio of 3 dominant to 1 recessive if *A* were dominant to *a*. Because the parents have the same genotype, it makes no difference which one is male or which is female.

Sex Chromosome Model

Contrast the theoretical results of the autosomal model to those of a model in which the genes for the traits being studied are at loci on the sex chromosomes. Remember that a female has two X chromosomes but a male has only one plus a Y.

If you start with a gene that has two alleles (+ and *m*), then two types of matings can occur, as diagramed here in Cross A or B. (Note that the alleles are designated as superscripts to X, indicating that they are found at loci on the X chromosome.) In Cross A, the female is homozygous for X+, but in Cross B, she is homozygous for X^m.

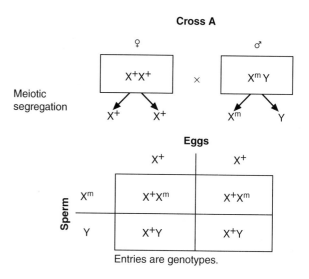

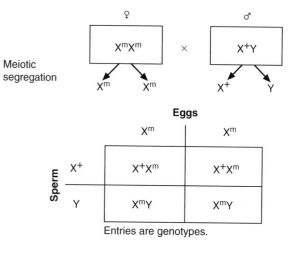

9▷ Summarize the phenotypic ratios, including sex, from the two crosses diagramed, assuming that X^+ is dominant and causes redness and X^m, recessive causing whiteness.

Hypothetical phenotypes from Cross A
Males

Females

Hypothetical phenotypes from Cross B
Males

Females

Obviously in the sex chromosome model, the genotypic and phenotypic ratios will vary according to which parent is homozygous for which trait. This contrasts with the autosomal model in which the sex of the individual is irrelevant to the result. Realize, however, that it is not sex per se that influences the result. It is the inheritance of the sex chromosomes that is important because males have only one X and females have two.

Applying Chromosomal Models to Crosses

In the lab there are several stocks of fruit flies. These flies have been specially bred with attention to two gene loci. These loci determine:

Eye color (red = W^+ or white = w)

Wing presence (normal wings = A^+ and apterous (no) wings = a)

Note: The notation for alleles indicates a dominant allele with a capital letter and a recessive with a lower case. The letter that has been chosen to represent the alleles reflects the trait. The + superscript is to indicate that it is a wild-type allele, *i.e.,* found in natural populations of fruit flies.

Problem to be solved: All of the fly stocks look alike. They have the dominant phenotypes, red eyes and normal wings. However, some of the flies are homozygous at one or both of the loci and others are heterozygous at one or both loci. Your job will be to perform a genetic cross to determine the genotypes at each of the loci for the unknown flies that are assigned to you. These will be called the **unknowns.** You will also have to determine whether any of the loci are carried on the X chromosome and are sex linked.

Method to be used: You will do this by performing a **test cross.** A test cross consists of taking an individual that has a dominant, but unknown genotype, and crossing it with one that is known to be homozygous for the recessive alleles at the loci of interest. If the unknown was homozygous for the dominant allele, all of the offspring from the test cross will show the dominant trait. If the unknown is heterozygous, half of the offspring will show the dominant trait and half will show the recessive. The appearance of the recessive phenotype among the offspring is the key observation for determining if the unknown was heterozygous. Figure 11.3 summarizes the reasoning you must use to determine the genotype.

In the lab, there are male and female flies that have the recessive phenotypes. Some of these flies have white eyes and normal wings, others red eyes and apterous wings, and the rest are double mutants with white eyes and apterous wings. These flies will be called the **testers.** You will use these to determine the genotype of the unknown that is assigned to you and your partner.

10▷ Before you actually perform the cross, you should engage in a little what-if thinking where you try to anticipate the results based on the autosomal and sex chromosome models described earlier. You will do a single locus by locus analysis. Do this by using the genetic notation given above, writing the genotypes for the parents, and then creating a Punnett square for each hypothetical cross. After you have filled in the Punnett squares, you will perform the actual cross and then compare the results to the what-if scenarios to determine the genotype of the unknown fly that you were given.

Figure 11.3 A test cross is a method for determining the genotype of an organism at a single locus when the organism has the dominant phenotype. It is based on comparison of results to hypothetical alternatives.

Test Cross

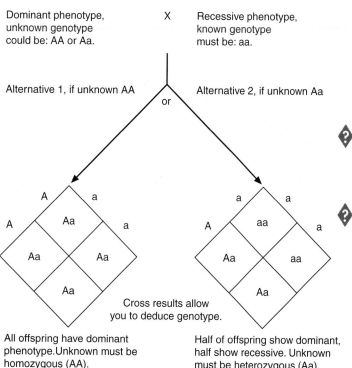

Dominant phenotype, unknown genotype could be: AA or Aa.   X   Recessive phenotype, known genotype must be: aa.

Alternative 1, if unknown AA   or   Alternative 2, if unknown Aa

Cross results allow you to deduce genotype.

All offspring have dominant phenotype. Unknown must be homozygous (AA).

Half of offspring show dominant, half show recessive. Unknown must be heterozygous (Aa).

Investigating Wings Locus

What if the locus for presence of wings is on an autosome?

What genotypic and phenotypic ratios would you expect among the offspring from your test cross if the unknown parent is homozygous at this locus?

Note that for this scenario, the required information has been printed to show you what to do. For the other scenarios, you will have to supply all of the information.

What if the unknown is homozygous at this locus?

Hypothetical parental genotypes: Unknown A^+A^+ × **Tester** *aa*

Note: *Tester flies will always be homozygous recessive.*
Possible gametes from unknown

Gametes from Tester	A^+	A^+
a	A^+a	A^+a
a	A^+a	A^+a

Predicted phenotypic ratios: *All will have normal wings.*
What if the unknown is heterozygous at this locus?

Hypothetical parental genotypes: Unknown A^+a × **Tester** *aa*

	A^+	a
a		
a		

Predicted phenotypic ratios:

What if the locus for presence of wings is on a sex chromosome?

What genotypic and phenotypic ratios would you expect among the offspring from your test cross if the unknown female parent is homozygous at this locus?

What if the unknown female is homozygous at this sex-linked locus?

Hypothetical parental genotypes: Unknown $X^{A^+}X^{A^+}$ × **Tester** X^aY

	X^{A^+}	X^{A^+}
X^a		
Y		

Predicted phenotypic ratios including sex:

What if the unknown female is heterozygous at this sex-linked locus?

Hypothetical parental genotypes: Unknown $X^{A^+}X^a$ × **Tester** X^aY

	X^{A^+}	X^a
X^a		
Y		

Predicted phenotypic ratios including sex:

But remember that you must also consider the reciprocal cross where the genotypes are reversed for the sexes.
What if the unknown is a male?

Hypothetical parental genotypes: Unknown $X^{A^+}Y \times$ Tester $X^a X^a$

	X^a	X^a
X^{A+}		
Y		

Predicted phenotypic ratios including sex:

Note: Males cannot be heterozygous for a gene on the sex chromosome. Why?

Investigating Eye Color Locus

 What if the locus for eye color is on an autosome?

What genotypic and phenotypic ratios would you expect among the offspring from your test cross if the unknown parent is homozygous at this locus?

What if the unknown is homozygous at this locus?

Hypothetical parental genotypes: _____ × _____

Note: *Refer to page 133 for allele notation to use.*

Predicted phenotypic ratios:

What if the unknown is heterozygous at this locus?

Hypothetical parental genotypes: _____ × _____

Predicted phenotypic ratios:

What if the locus for eye color is on a sex chromosome and the unknown female parent is homozygous at this locus?

Hypothetical parental genotypes: _____ × _____

Predicted phenotypic ratios including sex:

 What if the unknown female is heterozygous at this sex-linked locus?

Hypothetical parental genotypes: _____ × _____

Predicted phenotypic ratios including sex:

But remember that you must also consider the reciprocal cross where the genotypes are reversed for the sexes.

What if the unknown is a male and eye color is sex-linked?

Hypothetical parental genotypes: _____ × _____

Predicted phenotypic ratios including sex:

Remember males cannot be heterozygous for traits on X chromosome.

Determining Genotypes of Fruit Flies   **135**

Setting Crosses to Determine Genotypes

11 Get a culture chamber from the supply area. Following the directions given by your instructor, add flakes of growth medium, distilled water, and yeast to the vial. Let the medium hydrate for about 5 minutes. Label the vial with the code number you are given to identify the unknown and with your name.

When male and female flies are separated within four hours of emergence from their pupal casing, they are virgins and can be used in controlled mating experiments. Female fruit flies mate once and store sperm in a seminal receptacle. They will lay eggs for several days after mating. Each time eggs are produced some sperm are released from the seminal receptacle and fertilize the eggs.

Fruit flies can be anesthetized by treatment with carbon dioxide, with a proprietary chemical called Fly Nap, or with ether (***not recommended because of explosion hazard***). A small amount of dry ice put in an Erlenmeyer flask with a one-hole stopper is a safe carbon dioxide generator. A small rubber tube from the stopper can be inserted in a vial to anesthetize the flies. Your instructor will have two or more unknown types of anesthetized flies, separated by sex, available in the lab. Homozygous recessive tester strains will also be available for the two loci you are investigating.

12 Your instructor will dispense the unknown genotype flies first. Note the sex of the *unknown* you receive and add a *tester* fly of the opposite sex to your vial. Record the code number for the unknown strain below. Your instructor will compare your results later to a master list to determine if you correctly identified the genotype of your unknown.

Code number _____ Gender _____

What is the phenotype of your unknown?

What is the phenotype and genotype of your tester fly?

Once the flies are added, add a sponge stopper, but do not turn the vial upright. Doing so will cause the flies to fall on the surface of the wet growth medium and they will stick there. Keep the vials horizontal until the anesthesia wears off and the flies become active. After the flies have recovered, they may be incubated at room temperature or in a 25° C incubator. New flies will start appearing in the culture vials in about 10 days. You will harvest and count them in two weeks. Some care will be necessary in the interim.

Tending the Flies

13 During next week's lab, you should do the following.

1. Remove the parents, the flies you mated, and put them in a small vial containing a few drops of alcohol to kill them. Label the vial with your name and store it until the next lab.

2. Examine the growth medium. If it seems dry and shows flakes rather than being a homogeneous semi-solid, add a squirt of distilled water from a squeeze bottle. If, on the other hand, it seems wet, take a Chemwipe or piece of tissue and push this into the medium to wick away moisture. Leave the tissue in the vial.

3. Return the vials to the incubator. If pupae are not seen on the side of the vial by the ninth day, raise the temperature to 25° C, but not higher. High temperatures can cause abnormal development.

Counting the Offspring

14 Two weeks after setting the crosses, you will be able to collect your data to test your what-if models. Following instructions from your instructor, over-anesthetize your flies so that you kill them. Pour them out of the vial onto a white index card, which can be set on a block of dry ice for several minutes to be sure the files are dead. Examine them, using a dissecting microscope. Use a fine brush to separate them into two piles, males and females. Within each gender, separate the flies by phenotype for wing shape and eye color. Count the flies of each type and record in Tables 11.1 and 11.2.

Analysis

15 Calculate the ratios of the phenotypes among the offspring from your test cross.

16 Compare the ratios from your results to the hypothetical chromosome models for each of the what-if scenarios. Which one of the models best matches your results for each gene locus studied?

Phenotype	# of Males	# of Females	Total
Normal wings			
Apterous wings			
Red eyes			
White eyes			
Totals			

TABLE 11.2   Phenotypes of flies from reciprocal test cross with unknown: code # _____ gender _____

Phenotype	# of Males	# of Females	Total
Normal wings			
Apterous wings			
Red eyes			
White eyes			
Totals			

17▷ Using the notation given at the beginning of the experiment, give the genotype of your unknown at each of the four loci studied.

Code number _____

Phenotype for two traits _____

Genotype at two loci _____

Which of the traits is carried on the X chromosome? _____

❓ Can you think of any other chromosome models that would give you these same experimental results? Explain if necessary.

Your instructor may ask you to write your results on a piece of paper and hand them in for grading.

Learning Biology by Writing

Instead of simply turning in your results, you may be asked by your instructor to write a lab report describing how you determined the genotype of the unknown. Do this by first clearly stating the problem you are trying to solve. Discuss the alternative (what-if) scenarios very briefly, giving an example of how you did this for an autosomal and sex-linked trait alternative. Briefly summarize the methods used, citing the lab manual. Your results should be a facsimile of Tables 11.1 and 11.2. Your discussion should then explain how you determined the genotype of the unknown.

As an alternative assignment, your instructor may ask you to answer the following Critical Thinking or Lab Summary Questions.

Internet Sources

FlyBase is a large www database that compiles information on the genetics of *Drosophila*. It is located at http://flybase.harvard.edu:7081/

You can use this database to find additional information about the gene loci you studied in this lab. When connected to **FlyBase,** scroll to the query box at the bottom of the page. Enter the name of one of the phenotypes you studied. Click on *Search Everything.*

Look for information to answer the following questions:

   a) On what chromosome is the locus located?
   b) How many alleles have been described at this locus?
   c) What does the gene do?
   d) How can you relate this gene's function to the phenotypic effect of the mutation?

Lab Summary Questions

1. Describe the life cycle of *Drosophila* in detail and describe how to distinguish between male and female flies.
2. Describe a backcross and explain why it is a useful tool in genetics studies.

3. Discuss how Mendel's principle of segregation can be used to explain your results.
4. Give the code number, phenotype, and the genotype for the unknown fly assigned to you. Did everybody in the class who had a fly with the same genotype get the same progeny ratios from the test cross? How do you explain any differences?
5. The genes for which trait studied in this lab were on the X chromosome? Explain how you arrived at this answer.
6. If you were studying the inheritance of both eye color and wing shape in the same flies, how would you use Mendel's principle of independence to predict the results?

Critical Thinking Questions

1. In humans, brown eyes are dominant to blue and are autosomal. A child who is brown eyed wonders whether she is homozygous or heterozygous. She has a sister who is blue eyed. What is the "unknown's" genotype?
2. In humans, the ability to metabolize the sugar galactose is genetically determined. The gene locus is on an autosome. The allele for normal metabolism is dominant, and the recessive allele in the homozygous condition cause galactosemia—an uncomfortable but not deadly disease. A woman who is galactosemic marries a normal man and they have one child who is normal. Is this convincing proof that the man is homozygous for the normal gene?
3. Hemophilia, a hereditary disease where the blood does not clot well, is determined by a recessive allele that occurs at an X chromosome locus. A man with this condition marries a normal woman who had no history of the disease in her family over 5 generations. He is concerned about having children because he thinks all of their sons will have the disease—though none of their daughters will. Is he correct in his thinking? Explain.
4. Read the classic scientific papers written by Mendel and Sutton. Type are available on the WWW at:

   http://www.mendelweb.org/Mendel.html
   http://www.esp.org/foundations/genetics/classical/wss-02.pdf

INTERCHAPTER

An Outline of Sterile Technique

Lab Topics 12, 13, and 14 will require the use of sterile technique as you work with bacteria. This section outlines what is involved.

Sterile or aseptic technique is a lab procedure to minimize the chances that foreign bacteria and fungi will be introduced into a pure culture. Bacterial and fungal spores are virtually everywhere and so small that they drift around in air currents. When you work with bacterial cultures in a lab, you want to avoid unintentionally letting these spores enter the growth media that you are using. Spores are single cells surrounded by heavy cell walls that make them resistant to drying, heat, and often chemicals, such as acids, but spores will germinate and produce living cells when exposed to growth conditions. In medical labs, food and water quality labs, and research labs, sterile technique is used to prevent getting false readings. The unintended introduction of bacteria into a culture is called **contamination.** Although you will not work with pathogenic (disease-causing) bacteria, those who do such work use sterile techniques like these to protect themselves.

Sterile techniques starts with sterile solutions and tools. The most common way to sterilize these is to heat them in a pressure cooker-like device called an **autoclave.** It has a large chamber that can be loaded with materials to be sterilized. The chamber is sealed and steam is introduced. The pressure builds up to 15 lb/in^2, which yields a temperature of 120° C. This is held for 15 minutes and then cooled back to room temperature. Any living organisms or their spores are killed by this treatment. This sterility is forever, unless the materials get contaminated.

To maintain sterility as you work with sterile materials requires constant diligence and common sense. There are seven things to consider to prevent your cultures from getting contaminated and also to prevent you from getting contaminated.

1. Choose to work in an area that is free from blowing air. Sometimes it is convenient to work in a clear plastic hood to minimize drafts. Disinfect your work area with a wipe down of 10% bleach. Lay out what you will need: source culture, packaged sterile transfer tools, new growth medium, *etc.* Inspect any new growth medium to see if it is cloudy and possibly contaminated. Be prepared to work quickly and efficiently in what you have to do; *i.e.*, mentally go through the steps of the procedure before doing it, so that you will not have long pauses as you try to recall what to do.

2. Use only sterile tools to transfer materials. Never introduce into a sterile medium a tool that you suspect is not sterile and never touch the tip of the tool to anything unless you know it is sterile. Micropipetters,

inoculating loops, and sterile toothpicks are the primary tools for transferring bacteria in labs. Loops may be reusable metal loops or disposable plastic ones. To reduce likelihood of contamination, open sterile packs just prior to use. If a metal loop is being used, it should be flamed until it just starts to glow red and then allowed to cool before use.

3. In the labs included in this manual you will be asked to make transfers from one liquid to another, from liquid to agar, from agar to liquid, or from agar to agar. The specific techniques are described below and will be demonstrated by your instructor just before you do a procedure.

4. *Liquid-to-liquid transfer:* Loosen all caps, mentally review the steps you will be performing, and arrange materials to ensure they are easily accessible. Anytime you have a container open that is to remain sterile, avoid drafts and control your breathing to minimize air currents. Transfers can be made with a loop as directed here or with a pipette as discussed in #5.

   a. Remove a sterile transfer tool from its package, touching just the handle, or flame your wire loop.
   b. Hold it like you would a pencil and do not allow it to touch any nonsterile surface.
   c. If your source material is a culture in a capped container, grasp the cap between your little finger and the palm of the same hand holding the transfer tool. Remove the cap/plug, pulling the tube down and away. Do not put the cap down. Normally what would be done at this step is to pass the mouth of the tube through a flame to heat it just a small amount. This causes convection currents that keep air flowing away from the opening. However, if plastic tubes are used and you are not experienced, this step can be skipped.
   d. Holding the open tube at an angle (to minimize cross-sectional area for bacteria or fungal spores to fall into), insert the transfer tool and dip it into the culture. The tiny bubble that forms in the loop will hold thousands of bacteria.
   e. Replace the cap and open a fresh tube using the same technique.
   f. Dip the transfer loop holding the suspended drop into the new medium to transfer the bacteria.
   g. Replace the cap and mix.
   h. Clean area as directed and dispose of contaminated materials in biohazard bags for autoclaving. If a transfer loop was used, reheat it to kill any residual bacteria.

5. *Liquid-to-agar transfer:* Loosen caps and mentally review your procedure before starting. Avoid drafts and control your breathing to minimize air currents.

Transfers can be made with a loop as described above or with a pipette as outlined below.

a. Select an appropriate sterile pipetter for the volume to be delivered, and set the volume to be transferred. Open the sterile pipetter tip box by tilting the lid to one side. Insert the pipetter in a sterile tip and exert pressure to secure the tip to the pipetter. Remove the tip from the box and reclose the box immediately.

b. Hold the pipetter so that the thumb operates the plunger and you keep your little finger of the pipetting hand free for grasping the cap of the tube.

c. Pick up specimen tube with other hand, grip cap with little finger of pipette hand and twist off, pulling the tube down and away from the cap. Do not lay the cap down. You can skip the flaming step usually done here.

d. Insert just the tip of the pipette into the fluid in the tube and withdraw the required volume.

e. While avoiding air currents, open the petri dish lid just far enough to insert the pipette tip and expel fluid in a drop. Close the dish and dispose of the pipette tip as directed.

f. Take a sterile loop or a sterile bent glass rod and again lift the plate lid just enough to insert tool. Spread the drop of fluid evenly on the agar. Turn plate 90° and repeat spreading. See figure 12.6.

g. Clean area as directed and dispose of contaminated materials in biohazard bags for autoclaving.

h. Once agar has been inoculated, invert the petri plate for incubation.

6. ***Agar-to-liquid or agar-to-agar transfer:*** Mentally review your procedure before doing it. Transfers are to be made with a sterile loop.

a. Remove a sterile transfer tool from its package just before use.

b. Hold it like you would a pencil and never touch anything nonsterile with tip.

c. Touch the surface of the loop to a well-isolated colony and lift off a sample without getting agar. Remember that bacteria are very small and you need not have a large amount on the loop. See figure 12.4.

d. If transferring a sample to liquid, open a tube containing the growth medium as you would in a liquid to liquid transfer as in #4. Insert the transfer tool and swish it in the medium. Replace the cap.

e. If transferring a sample to another agar plate, open the petri plate lid just far enough to insert the sterile loop and steak the plate in a zig-zag pattern. Turn the plate 90° and repeat this zig-zag pattern. Remove the loop, and invert sample for incubation.

f. Clean area as directed and dispose of contaminated materials in biohazard bags for autoclaving.

7. Always wash your hands after you have finished working with bacteria. Many, but certainly not all, bacteria are pathogenic and you do not want to inadvertently contaminate yourself or others. Also, wipe down the area where you have worked with 10% bleach. All tools that have been used in transfers and all old cultures should be autoclaved or treated with bleach before disposal.

LAB TOPIC 12

Isolating DNA and Working with Plasmids

Supplies

Preparator's guide available on WWW at
    http://www.mhhe.com/dolphin

Equipment

Water baths at 42° C and 60° C
Incubator at 37° C
Refrigerator
Spectrophotometer

Materials

Carolina Biological Supply Colony Transformation Kit
(materials marked with asterisk [*] below come in this kit)

*Plasmid-free culture of a competent strain of *E. coli*
    (DH5 α) Gibco 18265017
*Petri plates
*Pipettes and suction device
*Bacteriological loops
*Glass hockey sticks
*Test tubes
Hot plate and 800-ml beaker with boiling water
Beakers, 30 ml
Marker pens
Bunsen burner
Glass rods
Capped plastic tubes, 9.5 cm × 1.5 cm for transformation
Test tubes
Cuvettes
Aluminum foil
Autoclave bags for waste
Medium-sized onion

Solutions

10% household bleach
Woolite/enzyme/salt solution
95% ethanol (stored in freezer)
Ice
DNA dissolved in 4% NaCl
4% sodium chloride
Diphenylamine solution
*Luria broth
*Luria agar, plain and containing 0.1 mg of ampicillin
    per ml
*50 mM CaCl$_2$
*Plasmid DNA (0.01 μg/μl)

Student Prelab Preparation

Before doing this lab, you should read the introduction and sections of the lab topic that have been scheduled by the instructor.

You should use your textbook to review the definitions of the following terms:

    competent cells
    genomic DNA
    plasmid
    spectrophotometer
    transformation

You should be able to describe in your own words the following concepts:

    Solubility of DNA in polar and nonpolar solvents
    How you could "genetically engineer" a bacterium
        to produce a protein normally found only in
        humans
    The difference between genomic and plasmid
        genes
    How a spectrophotometer is used to construct a
        standard curve (Lab Topic 5)

As a result of this review, you most likely have questions about terms, concepts, or how you will do the experiments included in this lab. Write these questions in the space below or in the margins of the pages of this lab topic. The lab experiments should help you answer these questions or you can ask your instructor for help during the lab.

Objectives

1. To isolate genomic DNA from an onion
2. To measure the amount of DNA in a sample
3. To conduct a plasmid transformation for ampicillin resistance in *E. coli* and interpret the results to test the following hypothesis: There will be a significant difference between plasmid-treated and untreated bacteria in their ability to grow on a growth medium containing ampicillin.

Background

Genetic engineering is one of the newer applications of basic understandings in biology. It involves the isolation of specific genes from one organism and the insertion of the genes into a second organism of the same, or a different species. The development of these techniques has raised much excitement and many questions. Scientists see genetic engineering as a way to explore how genes are regulated, to cure diseases caused by genetic deficiencies, and to add desirable traits to agricultural crops, such as nitrogen fixation to corn, thus eliminating the need for fertilizers.

Private companies have spent millions of dollars to build factories for growing bacteria that have been modified by these techniques to produce human hormones, various antigens/antibodies to diseases, and other useful proteins. Because the courts have ruled that new strains produced by these techniques are patentable, stockbrokers, lawyers, and business executives have had to learn biology as a new industry develops. What discoveries led to this tremendous interest?

For almost 100 years it has been possible to isolate DNA from cells, but it has been only for the last 40 years that its function was understood. The history of molecular biology is discussed well in most textbooks and will not be repeated here. However, the work of one set of investigators is worthy of mention. In classic experiments performed in the 1940s, Oswald Avery, improving on earlier experiments of Fred Griffith, demonstrated that if DNA was isolated from a pneumonia-causing strain of bacteria and added to cultures of nonvirulent bacteria, the recipient cells were transformed into virulent types that caused pneumonia. This **transformation experiment** was the first experiment to show that genes could be artificially transferred in the laboratory.

Since the time of those experiments, our knowledge of gene structure, function, and control has increased astronomically. Modern genetic engineering techniques depend upon four critical factors: (1) techniques that allow scientists to isolate single, whole genes, (2) a suitable host organism in which to insert the foreign gene, (3) a vector (transmission agent) to carry foreign genes into the host, and (4) a means of isolating the host cells that have taken up the foreign gene.

The bacterium *Escherichia coli* is widely used in genetic engineering as a host organism. It has a single large circular DNA molecule, called the **genomic DNA,** as a chromosome containing about two and a half billion base pairs (fig. 12.1). We now know how these base pairs are arranged into the approximately 4,300 genes that are found in *E. coli* as well.

A **vector** in molecular biology is a substance used as a vehicle to carry genes from one organism to another. It may be a virus, a naked DNA molecule, or a process involving heat or electrical shock. DNA molecules may be modified by gene-splicing techniques so that they carry genes of interest.

A **plasmid** is a small circular DNA molecule containing 1000 to 200,000 base pairs that naturally exists in the

Figure 12.1   Electron micrograph of a ruptured *E.coli* cell showing the genomic DNA spewing from cell.

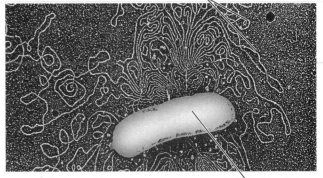

Figure 12.2   Plasmid DNA is small in relation to the host cell: see size of genomic DNA in figure 12.1.

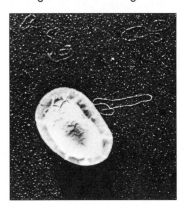

cytoplasm of many strains of *E. coli* (fig. 12.2). Independent of the genomic DNA, plasmids replicate as the cell grows and pass to each of the daughter cells during cell division. Plasmids that carry genetic information that is beneficial to the host cell are maintained in a given population. Plasmids frequently carry genes that confer antibiotic resistance, an obvious benefit.

When *E. coli* are exposed to a plasmid, many cells, but not all, will take up the DNA; *i.e.,* the plasmid somehow crosses the cell membrane. Cells that take up DNA this way are said to be **competent cells.** Competency can be induced by treating cells with divalent cations, such as calcium ions, followed by a heat shock. Although the exact mechanism is not understood, it is often described as making membranes leaky so that large molecules that would not normally pass through the plasmalemma can enter the cell. This treatment does not make all cells competent, but it greatly increases the chances that plasmids will enter some of the treated cells. Once in a cell, the plasmid is integrated into the host cell's normal metabolism. It replicates and is passed to the daughter cells as they divide. Even though every cell does not take up DNA when exposed to it,

E. coli has another useful characteristic. It grows rapidly, dividing every 20 minutes, so that a single cell can give rise to a billion descendants in a matter of ten hours. Thus, a few transformed cells can provide a huge number within a day.

E. coli carrying a plasmid can be selected from the population by using a **selection medium.** If the plasmid carries a gene for resistance to the antibiotic ampicillin, only those cells that have taken in the plasmid will grow (be selected) when placed in a medium containing ampicillin, and in a matter of a day, billions of ampicillin-resistant cells can be grown.

Genetic engineers use a modification of this technique to insert a foreign gene into *E. coli*. Although the following description is a gross simplification, it outlines the basic strategy used by many groups. The plasmid is modified by inserting a specific foreign gene into the plasmid DNA, perhaps the gene for human insulin. The modified plasmid is then mixed with competent *E. coli* and the *E. coli* are placed in a medium containing ampicillin. Only those cells having the plasmid grow in the medium. They also are the cells that carry the gene (in their new plasmid) for insulin. These cells can then synthesize insulin, which can be isolated from the bacterial cultures and used for insulin therapy of diabetics. Essentially any protein for which we can isolate a gene can be manufactured using this technique.

LAB INSTRUCTIONS

You will explore some basic techniques in molecular biology. You will isolate genomic DNA from an onion and estimate the amount of DNA obtained. You will also do a plasmid transformation of *E. coli* so that the cells are changed from ampicillin sensitive to ampicillin resistant. Your instructor may choose to do this lab in two parts.

Isolation of Genomic DNA

For most students, DNA is an abstract substance, one that they have never seen or handled. In this portion of the exercise, you will isolate DNA from an onion and be able to observe it.

The isolation procedure is fairly direct. You will rupture onion cells, releasing their cell contents: proteins, DNA, RNA, lipids, ribosomes, and various small molecules. You will then precipitate the DNA from the suspension by treatment with ethyl alcohol.

Rupturing Onion Cells

The following recipe should make enough DNA for 6 to 8 student groups.

1. Take a medium-sized onion and chop it into ten or so pieces. Add the pieces to a blender along with 100 ml of a detergent/enzyme/salt solution made up by diluting the commercial product Woolite in 1.5% sodium chloride solution. This solution contains enzymes that degrade proteins, a detergent that emulsifies fats, and salt that creates a polar environment to dissolve the DNA.

2. Briefly blend (20 to 30 seconds) until the onion is completely disintegrated. If the homogenate is too foamy, add another 50 ml of the detergent/enzyme/salt solution.

3. Pour the homogenate into a beaker and incubate in a 60° C water bath for exactly 15 minutes. The time and temperature are critical in this step. Higher temperatures or longer incubations will denature the DNA. A DNA double helix is composed of two strands held together by hydrogen bonds. Heat can break hydrogen bonds. What do you think happens to the double helix when DNA is heated?

4. Cool the mixture in an ice bath for five minutes and then filter the cool homogenate through four layers of cheesecloth to remove pieces of the onion.

5. Each student group should take 6 ml of the filtrate in a test tube.

DNA Precipitation

DNA is not soluble in ethyl alcohol and will precipitate when mixed with this substance. Because DNA is a long linear molecule, the precipitate is stringy.

6. The next step is a bit tricky and requires a steady hand. The technique is outlined in figure 12.3. Six milliliters of ice-cold ethanol (stock stored in freezer) must be slowly added from a beaker so that it flows down the side of the tilted test tube. If done correctly, the less-dense ethanol will layer on top of the aqueous solution of DNA.

7. Let the tube sit for two to three minutes without disturbing it and you will see a whitish, stringy substance starting to form at the interface of the two liquids. What do you think this is?

Ethanol is a less polar solvent than a 1.5% saline solution. DNA is a polar molecule. As the ethanol diffuses into the saline at the interface, DNA comes out of solution as a gelatinous precipitate.

8. Take a glass stirring rod and extend it into the tube past the interface of the two layers. Do not stir, but roll the

Figure 12.3 Technique for spooling DNA.

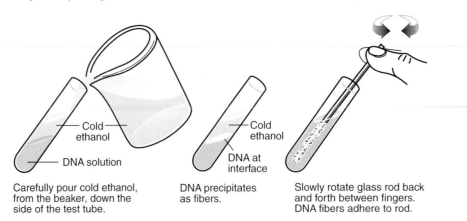

Carefully pour cold ethanol, from the beaker, down the side of the test tube.

Cold ethanol
DNA solution

DNA precipitates as fibers.

Cold ethanol
DNA at interface

Slowly rotate glass rod back and forth between fingers. DNA fibers adhere to rod.

TABLE 12.1 Mixing table for DNA standard curve (all values in milliliters)

Tube	Diphenylamine	4% NaCl	Known DNA 500 µg/ml	Unknown DNA from onion
1	3.0	3.0	0.0	0.0
2	3.0	2.8	0.2	0.0
3	3.0	2.5	0.5	0.0
4	3.0	1.5	1.5	0.0
5	3.0	0.0	3.0	0.0
6	3.0	0.0	0.0	3.0

stirring rod between your fingers. The DNA fibers will wrap around the rod. The technique is called spooling.

9. Remove the rod and press it against the side of the tube above the liquids. This will expel excess ethanol from the spooled DNA. The DNA can be redissolved by stirring the rod with the spooled DNA in another tube containing 4% NaCl. This sample can now be tested by the following colorimetric technique that will allow you to quantify the amount of DNA present.

Determining the Amount of DNA in Solution

DNA chemically reacts with diphenylamine to give a blue reaction product in proportion to the amount of DNA present. The amount of blue color can be quantified by reading the absorbance at 600 nm in a spectrophotometer.

To determine the amount of DNA that you have isolated, you will first prepare a standard curve measuring the absorbance of known amounts of DNA. Then you will measure the absorbance of your sample and compare it to the values obtained for the known amounts.

1. Number six test tubes. Add the solutions called for in mixing table 12.1.

2. Cap all six tubes with aluminum foil and place them in a boiling water bath for five to ten minutes. Remove and place the tubes in a beaker of tap water to cool.

3. Calculate the number of micrograms of DNA added to tubes 2 through 5 (*hint:* quantity = concentration of known stock multiplied by the volume used) and record in table 12.2.

4. Review the directions for using a spectrophotometer (fig. 5.5). Set it to 600 nm and use tube 1 to adjust the absorbance to zero, so that any color due to contaminants is blanked. Now read the absorbance of the tubes 2 through 6 and record in table 12.2.

5. Plot absorbance as function of concentration for the known solutions of DNA. Draw a straight line that best fits the plotted points. Find the absorbance for your unknown on the Y axis and read the amount of DNA present by dropping down to the X axis and reading the amount of DNA that gives that absorbance. This represents the yield from the DNA isolation procedure for the volume of DNA solution you started with.

TABLE 12.2 Absorbance of DNA solutions at 600 nm

Tube	Amount of DNA (μg/Tube)	Absorbance
1	0	_____
2	_____	_____
3	_____	_____
4	_____	_____
5	_____	_____
6	Unknown	_____

What is the concentration of DNA in your unknown? _____

Transformation by Plasmids

In this experiment you will make bacterial cells competent and then expose them to a plasmid that contains a gene conferring resistance to the antibiotic ampicillin. Your assay to tell whether the cells contain the plasmid will be to determine if they can grow on a selection medium containing ampicillin.

4 ▷ Before starting the experimental procedures that follow, you should formulate a testable hypothesis that relates plasmid treatment to the ability of treated and untreated cells to be grown on nutrient agar containing ampicillin. State your hypothesis:

Note: All procedures are to be performed using sterile technique and sterile solutions and glassware. Read the Interchapter on pages 139 and 140 before doing the experiment. Disposable pipettes, loops, and tubes should be placed in specially marked bags after they are used so they can be autoclaved before disposal. Why?

Before starting this procedure swab the table where you will work with 10% household bleach. Why?

Making Competent Cells

5 ▷ 1. Mark two capped plastic tubes: one + *plasmid* and the other – *plasmid.*

2. Remove the caps one at a time. Use sterile technique and a sterile pipette to add 250 μl of sterile 50 mM $CaCl_2$ to each. After recapping, set the tubes in a beaker containing crushed ice. Discard the pipette.

3. Obtain a petri plate containing Luria agar on which a plasmid-free strain of *E. coli* has been growing for 24 hours at 37° C.

4. Flame a bacteriological loop and let it cool, so that it does not melt the agar when the tip touches it. If you use sterile disposable plastic loops, do not flame! With the loop, scoop up a visible mass of *E. coli* from the petri plate (fig. 12.4). Be very careful and collect only bacteria and *not any of the agar* on which they are growing. If you collect agar, you will prevent the cells from becoming competent. A small dab will contain millions of cells.

5. Transfer the mass to the – *plasmid* tube so the cells are suspended directly in the drop of $CaCl_2$ at the bottom of the tube. Be sure to transfer only cells and not agar.

6. Use a sterile Pasteur pipette to aspirate the suspension until a single-cell suspension is achieved with no cell clumps. If the suspension is not somewhat cloudy, add another cell mass and repeat the suspension procedure. Put on the cap and return the – *plasmid* tube to the ice bath.

Repeat procedures 4 through 6 to add cells to the + *plasmid* tube. Discard the pipette and flame the loop.

Transformation Procedure

If the *E. coli* are to take up the plasmid DNA, the temperature and timing in these steps must be followed exactly.

7. Now add plasmid DNA (0.01 μg DNA/μl containing the ampicillin-resistance gene) to the + *plasmid* cell suspension by flaming an inoculating loop, letting it cool, and dipping it in the stock solution of plasmid DNA so that a bubble forms across the loop. This will transfer about 10 μl.

8. *Carefully* insert the loop into the + *plasmid* tube without touching the sides or you will lose the bubble and the plasmid DNA will not mix with the cells. Swish the loop in the cell suspension. After recapping, tap the side of the tube to mix in the plasmid DNA with the cells (fig. 12.5). Return the tube to the ice bath. Nothing is added to the – *plasmid* tube.

(a) Flame loop until red hot; let cool.

(b) Tilt lid, insert loop, touch to agar surface to cool, then scrape colony off agar.

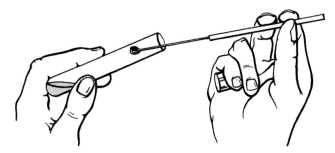

(c) Open tube as shown and swirl loop in medium.

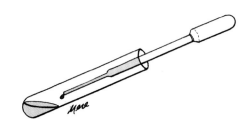

(d) With sterile pipette, aspirate fluid to disperse cells; recap tube.

9. After 15 minutes, remove both tubes from the ice bath, and heat shock the cells by immediately suspending both tubes in a 42° C water bath for 90 seconds.

10. After they have been in the water bath for 90 seconds, return the tubes immediately to the ice bath and let them sit for 20 to 30 minutes.

Selective Growth

11. Use a sterile pipette to add 250 µl of Luria broth to both tubes and cap the tubes. Mix well by tapping the bottom of the tubes (fig. 12.5). Discard the pipette and put the tubes in a beaker.

12. Obtain two petri plates containing *only* Luria agar and label them **norm** for normal growth medium. Obtain two plates containing Luria agar *plus* ampicillin and label them **amp** indicating they contain the antibiotic.

13. Follow the directions below to transfer cells from the – *plasmid* tube to one of each type of plate. Repeat the procedure for cells from the + plasmid tube.

   a. Take one of each type of plate (norm and amp) and label them *No Plasmid* to designate the type of cell suspension to be added. Label the remaining two plates *Plasmid-Treated.*

   b. Use a sterile pipette to add 100 µl of the – *Plasmid* cell suspension to a plate labeled *norm.* Repeat the procedure to add cells to a plate labeled *amp.* Use

Figure 12.5   Technique for mixing solutions added to closed tubes.

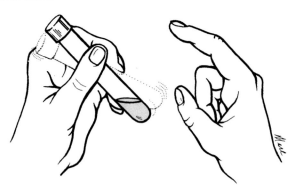

a sterile glass "hockey stick" to spread the cell suspension uniformly over the surface of the agar (fig. 12.6). When not using the hockey stick, place it in a beaker of 70% ethanol. When using it again, remove it and pass it through a flame to burn the ethanol off. After it has cooled, use it to spread the bacteria. A cool "hockey stick" will not melt agar when touched to its surface. A hot one will kill the bacteria.

   c. Use a fresh sterile pipette to repeat step b but now use the +*Plasmid* cell suspension. Discard the pipette. Spread with a glass "hockey stick."

Figure 12.6 Technique for spreading bacteria from plasmid incubation on surface of agar. After placing drop near center of plate (a), spin plate while holding hockey stick on surface to spread bacteria evenly on the surface (b).

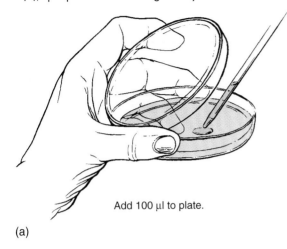

Add 100 µl to plate.

(a)

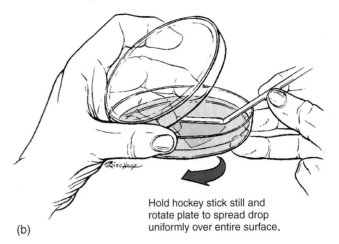

Hold hockey stick still and rotate plate to spread drop uniformly over entire surface.

(b)

d. Initial the four petri plates.

e. Incubate the plates upside down in a 37° C incubator for 12 to 24 hours. Do not overincubate. When the *Amp/Plasmid-Treated* plate has visible individual colonies, analyze all plates or put them in a refrigerator for up to one week before the analysis is performed.

6▶ Discard all disposable materials according to the directions from your lab instructor. Wash the bench area where you worked with 10% household bleach.

Analysis

7▶ After 1 to 2 days when visible colony growth occurs, open the plates and look at the bacterial growth on the surface of the agar. Each colony is the product of one cell that went through the heat shock treatment. Three results are possible: *no* growth, *selected* colony growth, and bacterial *"lawn"* growth. Colonies are produced when a single cell grows and divides to produce a ring of whitish cells. Lawn growth occurs when there are so many bacteria growing and dividing that they grow into their neighbors, making a continuous "lawn" on the surface of the agar.

8▶ Look at your plates and score them according to the above classification. On the plate showing selected colony growth, observe the colonies through the bottom of the culture plate and count the number of colonies present, using a marker to put a dot over each colony as it is counted. This technique keeps you from counting the same colony twice. Record the result in table 12.3.

Sit down with your lab partners and systematically discuss the purpose of each plate and cell suspension combination in this experiment.

Which plate demonstrates that the heat shocked *E. coli* are able to grow on the Luria medium?_____

TABLE 12.3 Number of colonies counted in selection experiment

	Norm Medium	Amp Medium
No plasmid	_____	_____
Plasmid-treated	_____	_____

Which plate shows that the plasmid treatment did not kill the cells?_____

Which plate shows that the plasmid DNA was taken up by the cells?_____

Which plate shows that the antibiotic was active?
_____

Which pair of plates shows that transformation is a rare event?_____
_____

Review the hypothesis that you made at the beginning of this experiment. Come to a conclusion on whether to accept or reject the hypothesis. What is your decision? Why?

Isolating DNA and Working with Plasmids   **147**

Genetic engineers often speak of transformation efficiency and express it as the number of transformed colonies per μg of plasmid DNA. The concentration of plasmid DNA solution used in this transformation should be 0.01 μg/μl. To calculate the amount of plasmid DNA used, multiply the number of μl (about 10 μl) used by the concentration in μg per μl. This is the amount of plasmid DNA in the transformation tube.

9 To calculate the number of transformed cells in the transformation tube, you must take into account the fact that you used only part of the solution. You pipetted 100 μl onto each plate from a total volume of 500 μl in the tube. Therefore, the tube contained five times more transformed cells than were observed on the plate. What was the transformation efficiency for this experiment?

After you have finished analyzing your plates, flood the surface of the agar with 10% household bleach solution and place the plates in a bucket for disposal. Why?

Learning Biology by Writing

If you did the transformation experiment with the plasmid, you have a well-controlled experiment for a lab report. State the hypothesis that the experiment tests. Indicate the technique used. Give the data obtained and draw a conclusion from the experiment.

Instead of a report, your instructor may ask you to answer the Lab Summary and Critical Thinking Questions.

Internet Sources

The transformation of competent *E. coli* cells is something that is done routinely in research laboratories. For this reason, many researchers are concerned with optimizing this procedure. Use a browser to search the Internet for information on the optimization of transformation.

In your Internet browser, choose a search engine such as **GOOGLE.** Type in *efficiency of E. coli transformation.* You will get a long list back.

Read a few of the articles. What are the authors suggesting be done to improve the efficiency of transformation?

Lab Summary Questions

1. Explain the difference between genomic and plasmid DNA in bacteria.
2. Why does DNA precipitate when ethanol is added? Why is the precipitate "stringy"?
3. Explain how you used a standard curve to determine the amount of DNA in a sample. Do not restate the technique. Present the reasoning behind the procedure.
4. Why were the cells placed in $CaCl_2$ and heat shocked during the transformation experiment?
5. Discuss the experimental design of the transformation experiment, describing the purpose of each petri plate that was inoculated.
6. What evidence do you have that the bacteria were transformed in this experiment?

Critical Thinking Questions

1. Three critical factors required for genetic engineering techniques are described in this exercise. How was each of these factors met in your transformation experiment?
2. Design a procedure that theoretically would allow you to take the gene for insulin and insert it into the cells of a person with diabetes. What would be the major problems encountered in doing such a procedure?
3. What are some possible sources of error in your transformation experiment?
4. What do the "control" plates in the transformation experiment control for?

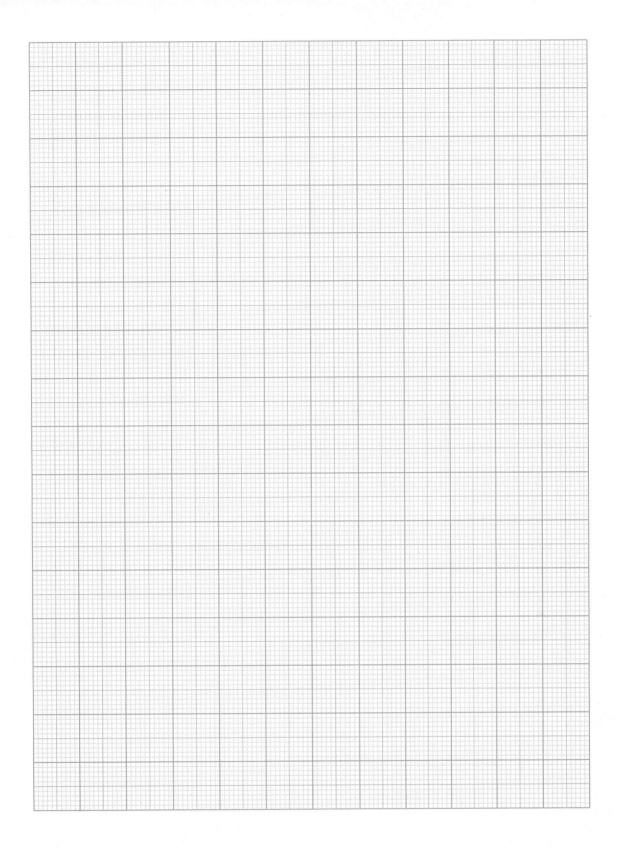

Isolating DNA and Working with Plasmids

LAB TOPIC 13

Testing Assumptions in Microevolution and Inducing Mutations

Supplies

Preparator's guide available on WWW at
http://www.mhhe.com/dolphin

Equipment

Incubator at 25° C
Ultraviolet light box (see fig. 13.1)
Desktop computers with Hardy-Weinberg simulation
program (EVOLVE in BioQuest Collection
recommended)

Materials

White 3″ × 5″ index cards cut in half with letter **A** on
half and **a** on half, one of each kind for each
member of class. Six additional cards will be
needed, 3 with letter **a** and three with α.
Sterile
    Petri dishes with King's agar or Tryptic Soy Agar
    Automatic pipetters and tips
    0.85% saline, 9.9 ml in capped tubes
    Glass rod bent into "hockey stick"
Cultures of *Serratia marcescens* D1
Wax pencils
Alcohol in covered beakers
Alcohol burners
Household bleach
Ultraviolet-shielding safety glasses
Rubber gloves
Brown paper bag

Student Prelab Preparation

Before doing this lab, you should read the introduction
and sections of the lab topic that have been scheduled
by the instructor.

    You should use your textbook to review the
definitions of the following terms:

allele	homozygote
dominant	phenotype
genotype	population
heterozygote	recessive

    You should be able to describe in your own words
the following concepts:

Allele frequency
Gene pool
Genetic drift
Hardy-Weinberg equilibrium
Mutation
Random mating

    As a result of this review, you most likely have
questions about terms, concepts, or how you will do
the experiments included in this lab. Write these
questions in the space below or in the margins of the
pages of this lab topic. The lab experiments should
help you answer these questions, or you can ask your
instructor during the lab.

Objectives

1. To simulate the conditions of the Hardy-Weinberg
   equilibrium.
2. To test hypotheses regarding the effects of selection
   and genetic drift on gene frequencies in
   populations
3. To test a hypothesis that mutations can be induced
   by ultraviolet light
4. To interpret results from bacterial mutagenesis
   experiments

Background

Populations, not individuals, evolve by gradual changes in
the frequency of alleles over time. These changes result
from mutation, selection, migration, or genetic drift. Collectively, these processes comprise **microevolution.** The
mechanisms of microevolution are well understood. In fact,
one of these mechanisms, selection, has been used for cen-
turies to increase the productivity of crops and livestock.

    Models of microevolution have been developed since
1858 when Charles Darwin published his version of the theory of evolution. Darwin's work was monumental in that it
established the role of natural selection in evolution. However, Darwin worked 50 years before Mendel's ideas about
genetic mechanisms became widely known; thus, Darwin
was ignorant of the basic concepts that you know about

genes, inheritance, and DNA. Researchers in the last 75 years have established the hereditary mechanisms that support Darwin's theory.

According to the microevolution model, a population of organisms can be considered to be a **gene pool,** which theoretically is composed of all the copies of every allele in the population at a given moment. In diploid organisms, the genes in the gene pool occur as pairs in each individual, and individuals may be homozygous or heterozygous for a particular trait. The genes found in an individual comprise its genome, and it is all of the genomes in a population that comprise the gene pool. Population genetics, the basis of microevolution, deals with the frequency of alleles and genotypes in a population's gene pool and attempts to quantify the influences of mutation, selection, and other factors causing evolution.

In the early 1900s, when biologists first started to think about the genetics of populations, a common misconception was that a dominant allele would drive a recessive allele out of a population after several generations. People holding this view observed that the recently rediscovered Mendelian genetics indicated that every time two heterozygotes mated, 75% of the offspring expressed the dominant trait. Furthermore, every time a homozygous dominant individual mated with a homozygous recessive individual, all offspring had the dominant trait. They reasoned that the dominant allele over several generations would become the only allele in the population.

In 1908, two mathematicians, G.H. Hardy and G. Weinberg, independently considered this concept and proved it was wrong. They showed that no matter how many generations elapsed, sexual reproduction in itself does not change the frequency of alleles in a gene pool. Changes, if they occur, must be due to the action of other factors—the real agents of evolution.

To understand the insights of Hardy and Weinberg, imagine a hypothetical population of plants that have red flowers and white flowers. Red flowers result from a dominant allele, and white flowers are found only in individuals homozygous for the recessive allele. A field study of 10,000 plants of this species indicated that 84% of the plants had red flowers and 16% had white. Laboratory analysis of the red-flowered plants indicated 36% of the total plants were homozygous for the red allele and 48% of the total plants were heterozygous. Thus, the genotypic frequencies in the population were:

36% *AA*	where *A* = red allele, dominant
48% *Aa*	*a* = white allele, recessive
<u>16% *aa*</u>	
100%	

The frequencies of alleles in the population's gene pool can be obtained by a series of simple multiplications from this information:

Every individual with the genotype *AA* contributes two *A* alleles to the gene pool and every individual with the genotype *Aa* contributes one *A* allele. Given 10,000 individuals, the frequency of the *A* allele in the gene pool equals:

$$2 \times 36\% \times 10,000 + 48\% \times 10,000$$

$$= 12,000 \ A \text{ alleles}$$

In the same population, the frequency of *a* in the gene pool can be calculated by similar reasoning (*aa* individuals contribute two *a*; and *Aa*, one *a*). Therefore,

$$2 \times 16\% \times 10,000 + 48\% \times 10,000$$

$$= 8000 \ a \text{ alleles}$$

On a percentage basis, $60\% \left(\dfrac{12,000}{12,000 + 8000} \right)$ of the alleles

in the gene pool are *A* and $40\% \left(\dfrac{8000}{12,000 + 8000} \right)$ are *a*.

This information describes the gene pool at the time of the study. What happens to the allele frequencies when this population reproduces sexually?

The answer to this question can be determined by creating a theoretical model as did both Hardy and Weinberg. They proposed that random mating occurred in a population; *i.e.,* every individual of one sex could mate with any individual of the opposite sex. This means that the alleles of the gene pool can randomly combine with each other. If no selection or mutation occurs in this large, isolated population and every pollen grain has an equal opportunity to fertilize every egg, the frequencies of all genotypes in the next generation can be predicted by common sense reasoning.

To do this, we create a modified Punnett square for the population to show all the possible combinations of gametes in producing the next generation:

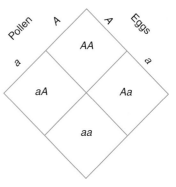

To this calculation, we can add the concept of probability. If 60% of the gene pool is the *A* allele, then 60% of the eggs and sperm will carry the *A* allele and 40% will carry the *a* allele. The *multiplication law of probability* can be used to predict the frequency of each type of fertilization and thus the genotypes of the next generation. This law states that *the probability (as a %) of two independent events occurring together is equal to the arithmetic product of their individual relative frequencies.* To apply this law to populations, you need only to recognize that egg and sperm production are independent events and that when egg and sperm occur together, fertilization results.

If we now add probability information to the modified Punnett square we get:

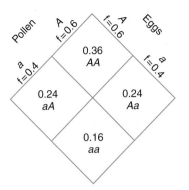

Summarizing the table, we are predicting that the new generation will have 36% *AA*, 48% *Aa,* and 16% *aa,* or 84% red flowers and 16% white. Compared to the parent generation, the genotype and allele frequencies are the same. In fact, the frequencies of alleles and genotypes would remain the same if we repeated these calculations over a hundred generations. The dominant allele does not drive the recessive allele out of the gene pool. The two reside in the gene pool in equilibrium, called **Hardy-Weinberg equilibrium**—as long as no selection, mutation, or other agents of evolution are acting on the population.

The hypothetical situation just described can be generalized. If we say that the frequency of the dominant allele is **p** and that of the recessive allele is **q,** in any population where there are only two alleles at one gene locus:

$$p + q = 1 \text{ and } p = 1 - q \text{ or } q = 1 - p$$

Because of these relationships, we need only measure the frequency of one allele and we can calculate the frequency of the other.

Once the frequency of alleles in a gene pool is known, the frequency of genotypes can be predicted by:

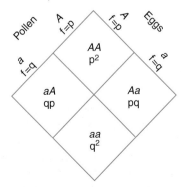

This model indicates that for a population in Hardy-Weinberg equilibrium, the homozygous dominant will be found **p²**% of the time, the heterozygote **2pq**% of the time,

and the homozygous recessive **q²**% of the time. This mathematical representation is called the Hardy-Weinberg law. In the earlier hypothetical flower population, the red flowers would have had a frequency of

$$p^2 + 2pq = 84\%$$

and the white flowers of

$$q^2 = 16\%$$

If we had known this relationship then we could have quickly determined allele frequencies because:

$$q = \sqrt{q^2} = \sqrt{0.16} = 0.40$$

Since

$$p + q = 1$$

then

$$p = 1 - 0.40 = 0.60$$

This mathematical model is very useful in studies of microevolution. It is used as a null hypothesis—a baseline against which to measure populations. If a population is mating randomly and no other factors change the frequencies of alleles, the frequencies of genotypes in the population should be those predicted by the Hardy-Weinberg equation. If the frequencies are quite different, then this is taken as strong evidence that an agent of change is influencing the population. It is then up to the biologist to seek what is causing the change. This is done by asking such questions as: Is natural selection occurring against a genotype? Is mutation occurring? Is the population small? Is mating not random? Are individuals migrating into or out of the population? By understanding how these factors change gene frequencies in populations, we can understand how evolution occurs on a small scale. Brief discussions of the effects of natural selection, population size, and mutation follow:

Natural selection, as a cause of evolution, acts on the total phenotype of an organism, not directly on its genes or genotype. In essence, it is a test by the environment of the organism's hereditary phenotype. In sexual reproduction, no two offspring are genetically alike, because of biparental inheritance, crossing over, and mutations. This can lead to subtle differences in how well some offspring function. Some function better than others, and produce more offspring. Because offspring tend to resemble their parents more than other members of a population, the next generation in this sequence will be better adapted to its environment. This selection process repeated over generations increases the population's fitness as certain alleles, and the genotypes in which the alleles occur, accumulate.

Not all changes in allele frequencies are directed by natural selection. Some are random changes, especially in small populations. Biologists call this concept **genetic drift.** This can be explained using statistical sampling theory. Let's consider a classical example, coin tossing. The probability of getting a heads in a coin toss is one half: likewise for tails. Does this mean that when tossing a coin twice, you will get only one heads followed by one tails? It does not. There are equal chances that you will get a tails first followed by a heads, or two heads, or two tails. However, if you were to toss the coin 100 times, there is a better chance that you will get very close to having heads one half of the time and tails one half of the time. The probability of getting the theoretical result increases even more if you were to toss the coin 1,000 times. Statistical sampling theory tells us that the bigger the sample size, the more likely we are to get the predicted result. When Hardy and Weinberg said that populations must be large in order for their theorem to apply, they were addressing the sample size concern. However, all populations are not large. Many endangered species consist of only a few breeding individuals. Consequently, we would expect to see random changes in gene frequencies in these populations due to statistical sampling error.

Mutation is a never ending process that occurs when random mistakes happen in gene replication before cell division. When mutations occur in cells that will become eggs or sperm, the gametes carry the mutation. If an egg or sperm carrying a mutation participates in a fertilization, then the individual thus created has the mutation as part of its genotype even though the parents did not. Modern molecular biology has shown that many mutations are neutral and have no discernable effect on an organism's phenotype. Other mutations are harmful but, because they are also recessive, most offspring do not express the mutation and only act as carriers. In later generations if two carriers mate, then the mutation is expressed as a phenotype and natural selection will act on it. Some mutations are beneficial and improve the organism's phenotype when tested against the forces of natural selection.

Students often ask, Where do alleles come from? The answer is that mutation creates them. As mutation is always happening at a low rate, alleles are always being created during DNA replication.

Other factors causing evolutionary change are migrations into or out of the population, and nonrandom mating. Migration can add alleles to a gene pool or it can remove them, depending on the genotypes of those that enter or leave. Nonrandom mating occurs when a species practices mate selection or consists of subpopulations that breed only among themselves.

LAB INSTRUCTIONS

In this lab, you will investigate several aspects of the microevolution model. You will start by playing a game that simulates the conditions of the Hardy-Weinberg equilibrium, demonstrating the effects of random mating, selection, migration, and population size. The long-term effects of natural selection and population size will be then investigated using a computer simulation. The lab ends with an experiment investigating how mutations can be induced in bacteria.

Mating Game

I would like to acknowledge the ideas of Dr. David Robinson, who suggested this activity.

In this group activity involving all of the students in the lab, you will model the conditions of the Hardy-Weinberg equilibrium. Your instructor has prepared two small cards for each member of the class. Half of the cards have the capital letter **A** on them and half have the lower case **a.** These are to represent the gene pool of a population where the allele frequencies are known. The cards should be distributed two to a student with one-fourth of the students receiving **AA;** one-half **Aa;** and one-fourth **aa.** The two cards represent the genotype of individuals of a hypothetical species, and the population is in Hardy-Weinberg equilibrium. Everyone should note their starting genotype and record it here: _____

We will now set up some conditions for the hypothetical organisms that the students represent. The species, like many insect species, is an annual one that mates once a year and then dies with the new generation emerging in the following growing season. Mating is random, with every gamete having a chance to combine with any other gamete. The population is at the carrying capacity for its environment, so each season the number of offspring is the same as the number in the parental generation. The one factor that we cannot adequately control in this simulation is population size. Most lab sections will have fewer than 30 students. In small populations like this statistical sampling errors may enter in. Nonetheless, we can still develop some understanding of microevolution through the planned activities.

2 ▶ Random Mating

Now for the fun part. It is time to mate! To simulate random mating everyone should throw their cards into a box, and after all cards have been added and stirred, everyone should draw two out to reconstitute a new generation. Shake the box occasionally to ensure randomness. This is similar to corals on a reef spawning where clouds of eggs and sperm are released, randomly combining to give the next generation. Count the number of individuals in the new generation of each genotype and record below.

	Number	**Ratio**
AA		
Aa		
aa		
Totals		

Has the ratio of genotypes changed from the starting parental generation? Is this population in Hardy-Weinberg equilibrium? Explain.

Count the number of **A** and **a** alleles in the gene pool of the new generation and determine their relative frequencies; *i.e.,* what percent of the gene pool is **A** and what percent is **a**? Have they changed from the first generation? Explain.

Count the number of **A** and **a** alleles in the gene pool of the new generation and determine their relative frequencies. Have they changed from the first generation? Why?

What do you think would happen to allele and genotype frequencies if you repeated the mating game? Why? If you cannot imagine what would happen, repeat the game and experimentally determine what would happen.

What do you think would happen if this scenario repeated over and over again through several generations? Can you see a trend? What is it?

3 ▶ Natural Selection

Everyone should exchange cards in order to return to their first genotype. Before mating again, we are going to have a selection event happen. Those with the genotype **aa** cannot compete in life as well as those with a dominant allele in their genotype. Imagine that **aa** can barely get enough food to survive and that they are susceptible to disease. Consequently, when mating time comes around, half of the homozygous recessives cannot mate; they are sick. Decide who among the **aa** genotypes will be sick and have them go to the side. They have died without leaving any offspring. All others get to play the mating game and can throw their cards in the box. When all have been added, those who added cards can draw them back out to form the next generation.

Count the number of individuals in the new generation of each genotype and record below.

	Number	**Ratio**
AA		
Aa		
aa		
Totals		

Has the ratio of genotypes changed from the starting parental generation? Why?

4 ▶ Migration and Bottlenecking

Everyone should swap cards to get back to their original genotype. About one-third of the class, based on name or major but not genotype, should separate themselves from the others. They have just migrated to a new locality. Perhaps a hurricane blew them from their homeland to an island. Each population should randomly mate within itself, using the box technique as before.

For the *population remaining* on its homeland, count the number of individuals in the new generation of each genotype and record below.

	Number	**Ratio**
AA		
Aa		
aa		
Totals		

Has the ratio of genotypes changed from the starting parental generation? Why?

⬥ Count the number of **A** and **a** alleles in the gene pool of the new generation and determine their relative frequencies. Have they changed from the first generation? Why?

Count the number of individuals in the new generation of each genotype and record below.

	Number	**Ratio**
AA		
Aa		
aa		
Aα		
aα		
αα		
Totals		

5> For the *separated population,* count the number of individuals in the new generation of each genotype and record below.

	Number	**Ratio**
AA		
Aa		
aa		
Totals		

⬥ Has the ratio of genotypes changed from the starting parental generation? Why?

⬥ Has the ratio of genotypes changed from the starting parental generation? Why?

⬥ Count the number of **A, a,** and **α** alleles in the gene pool of the new generation and determine their relative frequencies. Have they changed from the first generation? Why?

⬥ Count the number of **A** and **a** alleles in the gene pool of the new generation and determine their relative frequencies. Have they changed from the first generation? Why?

6> Mutation

Everyone should return to their original genotype. Simulate a mutation occurring in which the **A** allele mutates to **a** in three cases and to **α** (alpha) in three cases. When everyone puts their cards in the mating box the instructor will reach in and replace six of the **A** cards with three **a** cards and three **α** cards, simulating mutations occurring in **A.** Now everyone should draw two cards to simulate the next generation.

Computer Simulation of Microevolution

While the mating game was instructive for short-term changes from one generation to the next, it did not address the effects of changes over many generations, say 50 or more. Several excellent computer programs have been written to do just this. In this section, you will use a computer program to test hypotheses related to how allele frequencies will change over time as a result of selection against different phenotypes and as a result of varying population size; *i.e.,* genetic drift. Your instructor will describe the program and the computers that you will use. In this section, the experiments that you should conduct will be described.

7 Before starting, check the computer program that you use to be sure it has the following starting values/conditions set. The simulated population should be:

Large, about 8,000;

No migration should be occurring into or out of the population;

No mutation should be occurring;

Mating should be random;

And initially no selection should occur.

In addition, the program should be set up to:

Have graphic output;

Have the simulation run 50 or more generations;

Plot allele frequencies.

8 Effect of No Natural Selection in a Large Population

The program is now set to simulate a Hardy-Weinberg equilibrium over 50 generations. Run the simulation and note what happens to gene frequencies during the elapsed time.

Did the allele frequencies change? Describe the change and suggest what were the causes.

9 Effect of Natural Selection

You will now change the starting parameters for the program by introducing selection into the equation. You should first look at selection against the recessive allele (as expressed through the homozygous recessive phenotype) and then against the dominant allele (as expressed through the homozygous dominant and heterozygous phenotypes). Your instructor will tell you how to introduce selection into the program that you are using. This can be done in two ways. Adults of a certain phenotype may die and thus not be able to breed, or breeding pairs with certain genotypes will simply produce fewer offspring.

10 Before running the selection program, develop a hypothesis to be tested. If selection is against the recessive allele, what do you think will happen to the frequency of the recessive allele over 50 generations? Will it increase, decrease, or remain the same?

State this prediction as hypothesis.

Run the simulation and study the graphic output. Must you accept or reject your hypothesis? _____

11 Your instructor will tell you how to switch the graphic output of your simulation program from plotting the frequency of alleles to plotting the frequency of genotypes under the same selection conditions as before. Run the program. Describe what happens to genotypic frequencies as the allele frequencies change.

12▶ Now remove the selection against the recessive allele. Instead, add selection against the dominant allele. Predict what you think will happen.

State these predictions as a testable hypothesis.

14▶ Run the simulation and describe the results. Must you accept or reject your hypothesis? _____
Run this same simulation again. Did you get the same result? _____
Run it again. Did you get the same result? _____

 Run the program to test your prediction. Describe what happened. Was your prediction supported?

 How do you explain the variation in results that you are seeing?

13▶ Genetic Drift

Change the starting conditions of the simulation program back to those you used at the beginning for graphic output and no selection. Now change population size from 8,000 to 40.

Before running the program, predict what you think will happen to allele frequencies over 50 generations in a small population compared to the large one you investigated first.

15▶ Genetic Drift and Natural Selection Together

As you did before, introduce a natural selection factor against the homozygous recessive phenotype, using exactly the same value. Do not change the starting population size. Leave it at 40. Before running the simulation, predict what you think will happen to the frequencies of the dominant and recessive alleles.

 Run the program to test your predictions.
Describe what the results look like. Do they support your prediction? Explain what happened.

Experimental Induction of Mutations

In nature, mutation is the process that creates new alleles. **Mutations** are mistakes that happen during DNA replication before cell division, or abnormalities that develop in chromosomes during cell division. Whatever the source, mutations contribute to genetic variability in populations. However, their effect is usually much less than that of genetic recombination that occurs as a result of crossing over and biparental inheritance. Mutations can be neutral, having no effect, or be detrimental or beneficial. Usually, but not always, mutations create recessive alleles. This means that the trait would not be expressed until it appeared in a homozygous individual, a process that could take several generations.

Natural selection is the agent in nature that determines whether a mutation is "good" or "bad." A detrimental mutation (allele) would be selected against; *i.e.,* organisms with the mutation would not function as well in the environment and would leave fewer offspring. Over time, the frequency of genotypes carrying a detrimental mutation should decrease in the population. The opposite would be true for a beneficial one.

In this part of the lab topic, you will experimentally induce mutations in a bacterium, *Serratia marcescens*. Bacteria are good organisms to use in mutation studies because they are haploid. If a mutation occurs, it is expressed because there is not a second allele present to mask it. You will study mutations affecting viability in *Serratia*. Mutations can be induced by several means. Chemicals, called mutagens can change an organism's DNA, causing changes in hereditary information. Ultraviolet radiation has similar effects and will be used in the experiment. Because mutations caused by UV exposures are random, many different mutations will be induced in *Serratia*. Some may affect pigment synthesis. This bacterium is normally red, but when a mutation occurs in the genes producing the enzymes involved in pigment synthesis, no pigment is made and the bacteria are white. Other mutations will affect viability because the UV exposure damages genes that are essential to life.

State a hypothesis to test in your experiment. It should relate the length of exposure to UV light to the amount of mutation expected.

Figure 13.1 Ultraviolet light irradiation box. Insert petri dish into box and remove lid for appropriate time (table 13.1). Replace lid and remove from box. When working at the box, protect your eyes by wearing glasses made of ultraviolet-absorbing glass. Protect your skin by wearing a long-sleeve shirt and rubber gloves.

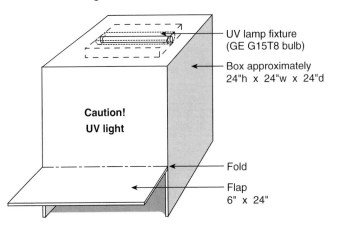

UV lamp fixture (GE G15T8 bulb)

Box approximately 24"h x 24"w x 24"d

Caution! UV light

Fold

Flap 6" x 24"

Procedure

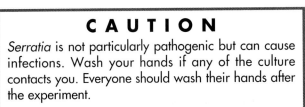

CAUTION

Serratia is not particularly pathogenic but can cause infections. Wash your hands if any of the culture contacts you. Everyone should wash their hands after the experiment.

16▶ About 18 to 24 hours before the lab, *Serratia* was transferred from a stock culture into 100 ml of nutrient broth and cultured at room temperature. You or your instructor should take this culture, and make a sterile, serial dilution. Add 0.1 ml of the culture to 9.9 ml of 0.85% sterile saline. Cap and shake the container 10 times. Take 0.1 ml of this dilution and add to a second 9.9 ml of sterile saline; cap and shake. Again take 0.1 ml of this second dilution and add it to a third 9.9 ml of sterile saline. Shake it well. The cell suspension in the third tube represents a millionfold dilution of the original culture. The first and second dilutions will not be used and should be autoclaved before disposal.

17▶ About 30 minutes before you are going to use it, a germicidal UV lamp enclosed in a box should be turned on and allowed to stabilize (fig. 13.1).

CAUTION

Never look at a UV lamp, because it can destroy cells in your cornea (surface of the eye).

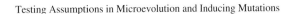

Testing Assumptions in Microevolution and Inducing Mutations

TABLE 13.1 Results from uv irradiation of *Serratia*

Plate	Sample Vol.(µl)	Cumulative UV Exposure (sec)	Total No. Colonies	Corrected No. Colonies	% Surviving
1	25	0 (control)	___ × 4*	___	___
2	50	10	___ × 2*	___	___
3	50	20	___ × 2*	___	___
4	100	30	___	___	___
5	100	40	___	___	___
6	100	60	___	___	___
7	200	80	___ × 0.5*	___	___

*Note: Number of colonies should be adjusted to compensate for differences in sample-size plated.

Figure 13.2 Inoculation of petri plates. Label the bottom of the petri dish with a marker. Open the dish from one side and add culture liquid containing bacteria. Use a sterile glass rod bent like a hockey stick to spread bacteria over agar surface. Turn plate 90° and repeat to ensure even spreading.

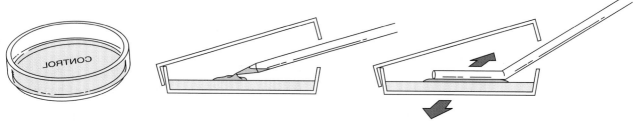

Mark dish with sample identification.

Add appropriate volume of bacterial suspension to dish.

Spread bacteria uniformly over agar; incubate at 23° for 24 hrs.

The lamp should be in a box, and anyone near the box should wear safety glasses that will filter out ultraviolet light. Anyone reaching into the box should wear a rubber glove and a long-sleeve shirt to protect their skin from the UV light. Why?

18▶ Take seven petri plates containing a growth agar and number the bottoms 1 through 7.

19▶ Shake the third dilution tube well to distribute the cells. Refer to table 13.1 to determine the volume of the *Serratia* dilution to be added to each plate. Note that the plates to be irradiated longer receive a larger volume of cells to compensate for the lethal effects of UV exposure. Use a sterile pipette to add the required amount to the agar surface. Then use a sterile inoculation loop or sterile bent glass rod ("hockey stick") to spread the culture evenly across the surface of the agar (fig. 13.2). The "hockey stick" can be sterilized between the times you spread bacte-

ria on each plate by putting it in a beaker of 70% ethanol. Before using it again pass it quickly through a flame to burn off the ethanol. Let the rod cool for 15 to 20 seconds before spreading the bacteria on a new plate.

Plate 1 is a control and nothing more should be done to it. Plates 2 through 7 will be irradiated with increasing amounts of UV light.

20▶ Refer to table 13.1 and determine the exposure time for each plate. Take one plate at a time and slide it into the UV light box. Reach in with a gloved hand and *remove the top* for the time indicated. This is done because glass and many plastics absorb UV light and little UV would reach the bacteria if the lids were left on during the irradiation period. Replace the top and remove the plate. Immediately place the plate in a paper bag or drawer to prevent photoactivation of DNA repair mechanisms. Repeat the procedure for the other plates for the appropriate times.

21▶ When all plates have been irradiated they should be placed in a refrigerator until 48 hours before the next lab. When taken out of the refrigerator, they should be incubated for 48 hours at room temperature. They may be incubated longer if growth is slow. (If incubated at temperatures higher than 25° C, no color develops.) When colonies are clearly visible, the plates should be counted.

Analysis

22 When you examine the plates, count the total number of colonies on each plate. Record the results in table 13.1. Note that the plates receiving short UV exposures received lesser volumes of the cell suspension, and the numbers should be corrected by multiplying the colony counts times a volume correction factor. This calculation makes all counts for all plates directly comparable. Record the corrected number of colonies for each treatment in table 13.1.

Calculate the percent surviving by dividing total colony count from plate #1 into colony counts for all other plates and multiplying by 100. On the panel of graph paper at the end of the exercise, plot the percent surviving (from total counts) as a function of irradiation time.

? Why do you think there are fewer total colonies formed in those samples that were irradiated longer?

? Return to the hypothesis you made and come to a conclusion about accepting or rejecting your hypothesis.

Learning Biology by Writing

In this lab you did two experiments. One was a computer simulation to test hypotheses related to selection and population size, and the other was a classical "wet lab" experiment demonstrating how mutations can be induced with UV light. Although related, these two experiments are not directly comparable. Consequently, your instructor may ask you to write a lab report on only one of the activities.

For either lab report, follow the directions in appendix D. When biologists sit down to write a scientific paper, they often start by writing an abstract of the work they are reporting. This helps them organize their thoughts. You might try to do the same. Start your report with a clear description of the hypothesis that were tested and why these are important ideas. The methods section should be brief and can cite the lab manual as a source for more detail. The results section should contain any tables or graphs of data. Usually one or the other is included for any experiments performed, but not both. The conclusions section should state what was learned from the work and how that relates to what is already known. Future experiments can be suggested to test the ideas that were presented.

As an alternative, your instructor may ask you to turn in answers to the Critical Thinking or Lab Summary Questions that follow.

Internet Sources

Use one of the search engines available through your Internet browser to find information on the topic of *Serratia* mutations (or mutants). Google at http://www.google.com is a good search tool. Compare your results to those of others who have done similar experiments. Are your techniques and results similar to those of others? Do any of these sites contain information that would be useful in writing your report? If you use this information, be sure to cite the source. See appendix D.

Lab Summary Questions

1. Define natural selection. Describe what happens to the frequency of a dominant allele in a population when selection is against the homozygous dominant phenotype? Against the heterozygote? Against the homozygous recessive phenotype?

2. How can you determine the frequency of a recessive allele in a population? Of a dominant allele at the same locus, assuming there are only two alleles at the locus in a population?

3. What is genetic drift, and why is the concept important in understanding evolution? Based on your computer simulation work, give comparative examples of what happens to allele frequencies over many generations in small populations compared to large ones.

Testing Assumptions in Microevolution and Inducing Mutations **161**

4. Plot the data from table 13.1 as directed in the **Analysis** section. What is it about these plots that suggests ultraviolet light is mutagenic?
5. Hand in your graphs of the effects of UV irradiation on survival in *Serratia marcescens*.

Critical Thinking Questions

1. Huntington's disease is a fatal nervous system disorder causing degeneration of the brain. It is caused by a defective gene on chromosome #4. This is an autosomal dominant trait affecting 1 in 20,000 people. However, it is a delayed action gene that is not expressed until the affected individual is in his or her late 30s or 40s. The homozygous condition is lethal to the fetus. Using the Hardy-Weinberg equation, determine (1) the frequency of carriers in the population, (2) the frequency of fetuses affected by the lethal homozygosity, and (3) the frequency of unaffected individuals.

2. In the Lake Maracaibo region of Venezuela, there is a family of about 3,000 people who are descendants of a German sailor who had Huntington's disease. Would sampling this population reflect the whole population of Venezuela? What mechanism of microevolution is occurring? Explain.

3. Present prenatal screening tests (amniocentesis and chorionic villi sampling) allow parents to screen babies for possible genetic defects. As well, reproductive technologies using surrogacy and sperm donation are becoming more common with couples seeking sperm donors or seeking surrogate mothers with "desirable" traits such as high IQ, tallness, and so on. Speculate on the implications these procedures have on the population as a whole with regard to microevolution.

4. Natural selection can be viewed as the sum total effects of the environment on the organism and includes both abiotic (physical) and biotic (other organisms) components. Does natural selection have its effects by acting directly on the genes, genotypes, or phenotypes? Explain your answer.

5. Explain how bottlenecking, genetic drift, and natural selection might lead to allopatric speciation if a single population was divided into two isolated populations by a natural event.

6. Ultraviolet rays in sunlight are responsible for tanning of human skin. Many see this as a healthy activity. Are you aware of any evidence that is contrary to this opinion?

7. Ultraviolet lamps are sometimes called germicidal lamps. Why?

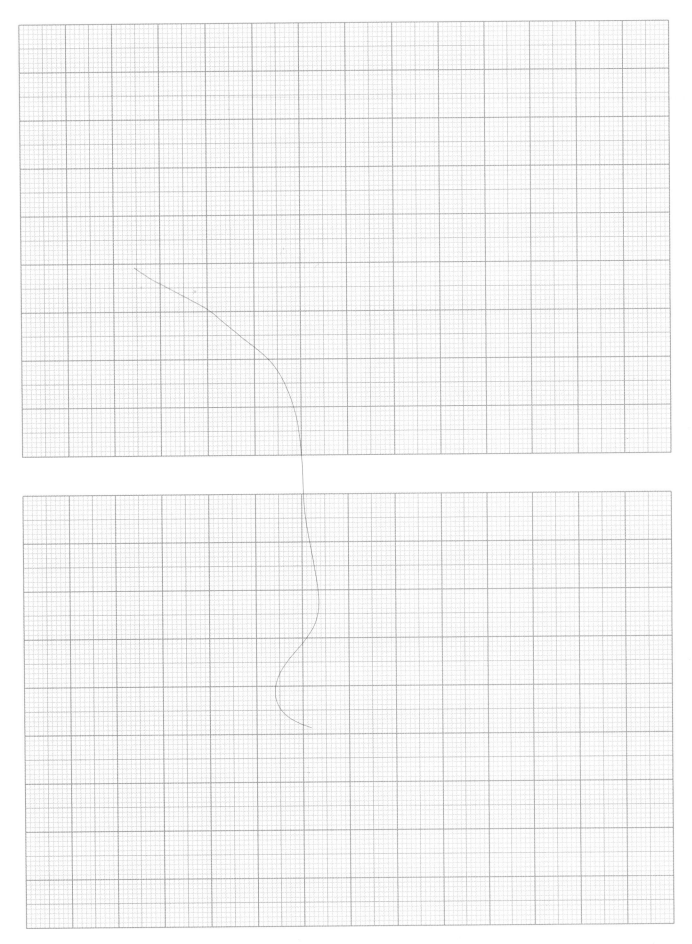

LAB TOPIC 14

Working with Diverse Bacteria

Supplies

Preparator's guide available on WWW at
    http://www.mhhe.com/dolphin

Equipment

Compound microscopes
Dissecting microscopes
Oil immersion objectives
Incubators or water baths at 37° C and 42° C
pH meter or pH tape
Balance, 0.1g sensitivity

Materials

Alcohol lamps
Grease pencil
Bacterial loops
Sterile pipettes and suction bulbs
Microscope slides and coverslips
Natural yogurt (about a 1-to-10 dilution)
Whole milk
Living cultures
    Bacillus megaterium or *Bacillus subtilis*
    Pseudomonas fluorescens
Mixed cyanobacteria cultures containing any of:
    Anabaena
    Gloecapsa
    Merismopodia
    Oscillatoria
Soil samples
Immersion oil
Prepared slides for demonstration
    Composite of bacterial types (cocci, bacilli,
        and spirilla)
Autoclave bag for disposal of cultures

Solutions

Gram's stain kit with Gram's iodine, crystal violet,
    safranin, and counterstain
Nutrient agar with 0% and 6% NaCl in petri dishes
Sterile water
Sterile 0.85% saline, packaged 9.9 ml per capped
    tube
India ink
70% ethanol
Test substances, such as antibiotics, metal salts, or
    pesticides; filter disks, 1 cm diameter

Student Prelab Preparation

Before doing this lab, you should read the introduction
and sections of the lab topic that have been scheduled
by the instructor.

You should use your textbook to review the
definitions of the following terms:

bacillus   Domain
bacteria   eukaryote
coccus   Gram's stain
colony growth   prokaryote
Cyanobacteria   spirilla

You should be able to describe in your own words
the following concepts:

Agar and growth media
Bacterial diversity
Differential growth
Sterile technique (See figures 12.4 and 12.6)

As a result of this review, you most likely have
questions about terms, concepts, or how you will do
the experiments included in this lab. Write these
questions in the space below or in the margins of the
pages of this lab topic. The lab experiments should
help you answer these questions, or you can ask your
instructor during the lab.

Objectives

1. To recognize the diversity of prokaryotic cell types
   among bacteria
2. To investigate where bacteria live
3. To determine the number of bacteria in soil and
   meat samples

Background

Bacterial diversity is staggering. Although there are only about 5,000 species that have been described, it is estimated that the total number of different types might be 100-fold more. We simply have not had enough time or effort put into finding and describing these new species. Clearly, there is much work to be done to understand what types of bacteria there are and what they are capable of doing.

A few years ago the term bacteria was synonymous with prokaryote. Recent studies, however, indicate that a group called the Archeae should not be considered bacteria. They are so different that they should be assigned to their own domain. Taxonomists now recognize three domains that include all living organisms: Bacteria, Archaea, and Eukarya (which include protists, plants, fungi, and animals). This lab topic will focus on the bacteria. Several of the other chapters in this manual look at the different types of Eukarya. The Archeae are not covered in this lab manual because they are a difficult organism to work with. They live in extreme environments such as thermal vents or concentrated salt lakes such as the Dead Sea, or they are strictly anaerobic and are killed by oxygen.

Our lives are intertwined with those of bacteria. By far, most bacteria are beneficial as important members of ecosystems where they function as decomposers. Without bacteria (and fungi) to recycle carbon from dead organic matter as carbon dioxide, plant photosynthesis would be greatly reduced. Of course, as they decompose materials, bacteria grow and reproduce becoming the basis for many food chains. Their decomposition capabilities also recycle a number of other materials such as nitrogen, phosphorous, and sulfur. Their ecological roles cannot be overstated. Bacteria living in association with plant roots convert atmospheric nitrogen into ammonia that the plants can use in their metabolism to make amino acids and nucleotides. In the intestines of vertebrates, symbiotic bacteria make vitamins that the host cannot. About half of our antibiotics (*e.g.*, streptomycin, neomycin, and tetracycline) come from bacterial sources. In nature, the soil bacteria producing these use them to fend off competitors. In today's world of bioengineering, researchers are genetically engineering bacteria to produce drugs and proteins. Environmental scientists are using bacteria for bioremediation projects. For example, modern gold-mining methods use cyanide to extract gold from crushed rock. A bacterium was found that could metabolize the highly toxic cyanide, producing harmless carbon dioxide and nitrogen compounds. By spreading this bacterium on cyanide contaminated sites, they could be returned to normal conditions. What other kinds of applications are there for what I will call bacterio-technology? The limits are our knowledge and maybe our imaginations.

The bacteria that attract most of the attention are disease-causing organisms. Although the list of such pathogenic bacteria is small relative to the total number, the impact can be significant. The Bubonic plague of the 1300s was caused by bacteria. Cholera, Lyme disease, syphilis, and tuberculosis are other examples of diseases caused by parasitic bacteria. Some bacteria produce toxins. Botulism, a type of food poisoning, occurs when the bacterium *Clostridium botulinum* grows anaerobically on improperly canned foods. As little as one microgram (one millionth of a gram) of the exotoxin produced by this organism can kill a human.

True bacteria and the cayanobacteria are prokaryotes: they lack a nucleus and membranous cytoplasmic organelles, such as mitochondria, endoplasmic reticulum, and vacuoles. The relatively simple cells are surrounded by a cell wall composed of peptidoglycans, a type of polymer not found in the cells of organisms in the other domains. In addition, bacterial DNA is arranged in circular molecules and is not combined with proteins to form chromosomes as in the other domains.

Bacteria reproduce asexually by binary fission. This involves a doubling of cellular components including DNA and then a dividing in half. Under good conditions, this might happen every half hour, creating the potential for explosive population growth. Under adverse conditions many bacteria produces spores, called endospores. These contain a quiescent cell that can live for years in a sort of suspended animation, but when they fall into good conditions, they grow and divide to start a new population. Bacteria do not sexually reproduce, but there are natural mechanisms for exchanging genetic materials between individuals. Some bacteria conjugate by forming cytoplasmic tubes that link two cells together so that DNA can be exchanged. Transformation occurs when bacteria take up genes released into the surrounding environment by the death and fragmentation of others. Bacteria are also invaded by viruses. When a virus leaves one bacterium and enters another, it often carries a piece of the first cell's DNA with it to the new cell in a process called transduction. It is an understanding of these genetic exchange processes as well as of gene function that has made bacteria favorite experimental organisms for genetic engineers.

As a group, bacteria are more metabolically diverse than all other types of organisms. In the group, there are two kinds of autotrophs and two kinds of heterotrophs. Some bacteria are obligate anaerobes that must have oxygen to live. Others are obligate anaerobes and will die if exposed to oxygen. Yet others are facultative aerobes and perform one type of metabolism when oxygen is present and another when it is absent. However, what is truly amazing is their metabolic flexibility as a group. There is probably no naturally occurring organic compound that bacteria cannot metabolize, and they also have enzyme systems that can break down many synthetic molecules. The U.S. Environmental Protection Agency funds research on the use of bacteria in bioremediation, the restoration of contaminated sites to near natural conditions by the action of living organisms. Their hope is that someday we will have identified a mixture of bacteria that could be applied to an oil spill site or an old industrial dump containing toxic organic compounds. The bacteria would then degrade the toxin to

Figure 14.1   Cell wall structures in Gram + and Gram – bacteria. The additional outer second membrane and small amount of peptidoglycan in Gram – bacteria greatly reduce stain binding.

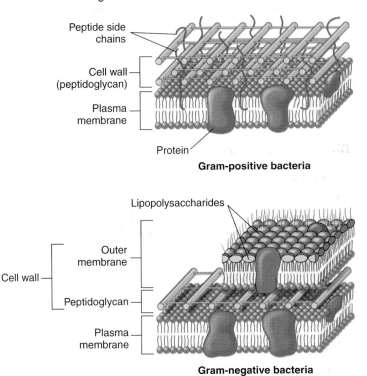

Peptide side chains

Cell wall (peptidoglycan)

Plasma membrane

Protein

Gram-positive bacteria

Lipopolysaccharides

Cell wall

Outer membrane

Peptidoglycan

Plasma membrane

Gram-negative bacteria

apart or may be more or less permanent and characteristic for a species, in which case they are called colonies.

1 Obtain a prepared microscope slide containing bacteria with all three shapes of cells. Look at it first with your 10× objective, switching to 43× for a better view. If your microscope is equipped with an oil immersion objective, review the directions on page 17 for using this. Because such objectives provide magnifications up to 1000×, you should be able to see the cell shapes more clearly. However, because such objectives are also expensive, your instructor may have this as a demonstration slide. Compare what you see to the scanning electron micrograph of bacteria in figure 3.1. Sketch the basic shapes of bacteria below.

harmless carbon dioxide, water, and maybe ammonia, rendering the site suitable again. As we discover new species of bacteria and understand their metabolic capabilities, we take a step toward achieving such a goal.

LAB INSTRUCTIONS

You will look at representative bacteria and explore some of the techniques used in microbiology.

Bacterial Cell Shapes

Bacteria are small, although not the smallest living organisms. They are very difficult to observe even through the finest light microscopes, let alone those that are normally found in teaching laboratories. Therefore, the identification of bacteria is based on a combination of characteristics. One characteristic used is cell shape. Some bacterial cells are **cocci,** small spheres, while others are rod-shaped **bacilli** or cork-screw-shaped **spirilla** or **spirochaetes.** Most bacteria are unicellular, although some species typically consist of cell aggregations. These may be transient and break

Gram's Staining

Bacteria are also identified by their staining response to certain dyes. One such staining protocol is known as the **Gram's stain.** One of its dyes is crystal violet. It binds irreversibly to cell wall components of only some bacteria and is used as a lab test to identify bacteria as belonging to one of two groups. Bacteria that retain Gram's stain when washed with alcohol are said to be **Gram positive;** those that are decolorized by the alcohol wash are said to be **Gram negative.** Cell walls from Gram positive bacteria are rich in peptidoglycan, a polymer consisting of polysaccharides and short polypeptides. Gram negative bacteria have less peptidoglycan, and a second plasma membrane covers the cell wall, blocking the stain's access (fig. 14.1). The Gram's staining reaction is one important tool in bacterial identification. As a child you may have had a throat culture when you were sick. Among other things, the doctor wanted to determine if you were infected with Gram + or – bacteria, because certain antibiotics are effective only against Gram + bacteria.

2 You will perform the Gram's stain on two 18- to 24-hour cultures of bacteria available in the laboratory. (Alternatively, scrapings from between your teeth could be used as a source. You should find both Gram + and – bacteria, but

Figure 14.2   Directions for Gram's staining.

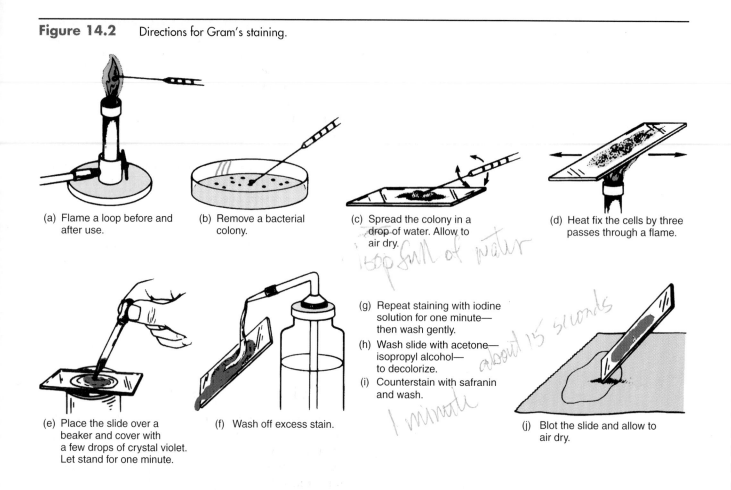

(a) Flame a loop before and after use.

(b) Remove a bacterial colony.

(c) Spread the colony in a drop of water. Allow to air dry.

loop full of water

(d) Heat fix the cells by three passes through a flame.

(e) Place the slide over a beaker and cover with a few drops of crystal violet. Let stand for one minute.

(f) Wash off excess stain.

(g) Repeat staining with iodine solution for one minute— then wash gently.

(h) Wash slide with acetone— isopropyl alcohol— to decolorize.

(i) Counterstain with safranin and wash.

about 15 seconds

1 minute

(j) Blot the slide and allow to air dry.

will not know the species.) One culture contains *Bacillus megaterium,* and the other *Pseudomonas fluorescens.* Both species are large, but one is Gram positive and the other Gram negative. You will determine which bacterium is Gram + by using the following procedures (fig. 14.2):

1. Wash a microscope slide with soap and water to remove oils. Dip the slide in a beaker of alcohol and let it air dry. Use a diamond pencil or wax pencil to put a "B" on the side of the slide where you will put *Bacillus* and a "P" on the *Pseudomonas* side.

2. If you start with colonies from petri plates, put two tiny drops of distilled water on the slide. If you start from a liquid culture, water is not needed. Flame a bacterial loop and let it cool. Dip the loop into the culture or scoop a small amount, not a whole colony, off the agar plate surface. Alternatively, a sterile toothpick can be used in place of a wire loop. Use the loop to spread one species of the bacteria evenly on one-third of the slide and the other species on another third with a blank area between. Let the slide air dry at room temperature. Gently heat the slide by passing it through a low flame three times. This makes the cells adhere to the surface of the slide, a process known as **heat fixing.**

3. Put the slide on top of a beaker and flood the surface with crystal violet. After one minute, carefully pick up the slide and wash off the stain with a gentle flow of water. Wearing rubber gloves will protect your skin from the stain.

> ### BE CAREFUL!
> This stain is very difficult to remove from your skin and clothing.

4. Now the slide should be flooded with Gram's iodine reagent that enhances color development. After one minute, gently wash the slide again with water.

5. Take a squeeze bottle containing 25% acetone in isopropyl alcohol and *gently* squirt it on the surface of the slide until the solvent running off the slide is colorless. This will take less than a minute. If you squirt too hard, the bacteria will be washed off. Gently wash the slide with water.

6. Flood the surface of the slide with safranin, a **counterstain** that helps you see Gram-negative cells. After 30 seconds, wash off the counterstain with water.

7. Blot the excess at the edge of the slide onto paper toweling and allow the slide to air dry. Examine with a high dry objective, or an oil immersion objective if available.

Which species of bacterium is Gram positive? Which is negative? Sketch a few of the cells below and describe in words how they look.

4 There is a corollary chemical test that can be used to confirm the Gram's stain results. Because of the differences in the cell wall structure between Gram + and Gram − bacteria, they react differently when treated with KOH. Take two small beakers and add about 3 ml of 3% KOH to each. Add the *Bacillus* to one beaker and *Pseudomonas* to the other and stir. Dip a toothpick into the cultures. For the Gram − bacteria, but not the Gram +, a mucus-like material should adhere to the toothpick, confirming a difference in cell wall chemistry.

Applied Microbiology

Fermentation is a term used to describe a collection of metabolic reactions that occur in microorganisms and release energy from organic molecules. No oxygen is required and an organic molecule is used as the terminal electron acceptor. The organic molecule that is produced is excreted from the cell and accumulates in the environment.

Fermentation historically has been used as a means of preserving dairy products, resulting in a "sour milk" product. Yogurt is one such product produced when *Streptococcus thermophilus* ferments the sugar lactose in milk to lactic acid (souring) and *Lactobacillus bulgaricus* produces the flavors and aroma of yogurt. As lactic acid accumulates, the pH of the milk drops and the proteins in the milk curdle. Eventually, an acid concentration is reached that prevents the growth of other microorganisms and the yogurt-producing bacteria as well. When this happens, the food is preserved and can be kept at room temperature for extended periods. However, it will not keep indefinitely and the growth of other acid-tolerant microorganisms may eventually cause spoilage.

5 In the lab is yogurt that has been diluted in water. Place a drop on a microscope slide and add a coverslip. Observe to see the bacteria used to make this product.

6 To test for lactic acid, measure the pH of the yogurt and of whole milk. Record the values below.

pH of whole milk _____

pH of yogurt _____

Determinative Microbiology

Identification of bacteria involves looking at a combination of structural, physiological, and ecological characteristics of an unknown microbe. Cell shape, selective staining by one or more dyes, and motility can be used to narrow the possibilities, but then other tests must be employed. Often this involves growth on specialized media or growth under special conditions.

Differential Growth

7 To illustrate how different species grow under different conditions, a simple experiment can be performed. You or your instructor should take a gram of rich organic soil and add it to 10 ml of sterile 0.85% saline. The suspension should be agitated to extract and suspend the bacteria. About 0.1 ml of the resulting soil extract should then spread onto the surfaces of each of four petri plates containing nutrient agar. Two plates have nutrient agar made up in 6% sodium chloride, and the other two in 0% sodium chloride. One of each pair of plates should be incubated at room temperature, and the other at 42° C for 12 to 24 hours. This is a 2 × 2 experimental design which combines conditions of high and low salt with high and low temperatures to determine if different kinds of bacteria grow under different conditions.

8 Additional variables may also be tested within this design. You could add to the plates small discs of filter paper that were soaked in antibiotics, such as penicillin, or in salts of heavy metals, such as mercuric chloride, or in organic compounds, such as insecticides. Materials in the discs will diffuse into the agar and affect the growth of bacteria in a circular area around the discs. Bacteria some distance away

TABLE 14.1 Differential growth results for soil bacteria

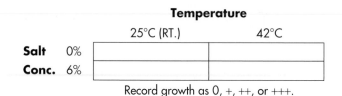

		Temperature	
		25°C (RT.)	42°C
Salt	0%		
Conc.	6%		

Record growth as 0, +, ++, or +++.

from the discs should not be affected and can serve as a control for growth under the basic four conditions.

Look at these plates and note the differences between them. Do all growing colonies look the same despite the growth conditions? Is the same number of colonies found on all plates? What other differences can you see? Record your observations below and in table 14.1.

Explain why these differences were found if all plates were inoculated with an extract from the same soil sample. What does this experiment tell you about the general ecological requirements (or tolerances) of different species?

Bacterial Population Counts

Biologists frequently need to estimate the number of individual bacteria in a complex mixture of materials, such as in soil, in food, or in body fluids.

This is done by taking a measured sample of the material and diluting it with sterile saline. A small amount of that dilution is spread on the surface of nutrient agar in a petri dish. For example, if 1 g of soil is mixed with 99 ml of sterile saline, and 0.1 ml of the mixture is placed on the agar, then 1/1000 of the soil sample has been "plated." Any bacteria in the sample will be spread on the agar surface. They will draw nutrients from the agar medium and each single bacterium will repeatedly divide to form visible colonies. The colonies can be counted easily, and by multiplying by 1000, the number of bacteria in the original gram of soil is obtained.

If the bacteria are plentiful in the soil, a 1/1000 dilution may be insufficient; when a petri plate is covered by thousands of colonies, counting is tedious and subject to error. To avoid this, **serial dilutions** are often made. In this technique, a sample is suspended first in 99 ml of saline, and then 1 ml of that suspension is added to a second 99 ml of saline to further dilute it: in this case to 1/10,000. One milliliter of this suspension may then be diluted in a third 99 ml to give a 1/1,000,000 dilution and so on. Describe how you would make 1/10,000,000 serial dilution below.

In groups of 3;

9ᐅ   In the lab are samples of various foods: milk, ground beef, or fresh vegetables. The problem you must solve is to conduct an experiment that will allow you to estimate the number of bacteria in the food sample.

In order to reduce the number of necessary plates and bottles, each pair of students may be assigned a dilution to use according to table 14.2. In that case, the directions below should be followed with each pair performing one of the serial dilutions.

To reduce the volumes required, all sample dilutions will be based on 0.1 g of sample in 9.9 ml of saline, a 1 to 100 dilution.

TABLE 14.2   Bacterial counts from food sample

Sample was _____

Dilution	Am't plated (ml)	Colonies/plate	Multiply by Dilutions	Bacteria/g sample
1:100	0.5		$(10 \times 20 \times 100 =)$	
1:10,000	0.5		$(10 \times 20 \times 10,000 =)$	
1:1,000,000	0.5		$(10 \times 20 \times 1,000,000 =)$	

10▶ Weigh 0.1 g of ground beef or other food sample and add it to 9.9 ml of sterile saline in a test tube. Mix well. Cap the tube and label it as a 1 to 100 dilution. Use a sterile pipette to take 0.1 ml of this dilution and add it to a second tube containing 9.9 ml of sterile saline to make a 1 to 10,000 dilution. Cap and label the tube. In turn, use a fresh sterile pipette to take 0.1 ml of this new dilution and add it to 9.9 ml of fresh sterile saline to make a 1 to 1,000,000 dilution. Cap and label the tube.

11▶ Take the 1 to 100 dilution and shake it well or vortex mix it to disperse the bacteria evenly. Take a fresh sterile pipette and transfer 0.5 ml of the dilution to a sterile petri plate containing nutrient agar. Label the plate with the date, dilution, amount added, and your initials. Repeat this procedure for each of the other two dilutions.

Once the plates are inoculated, allow them to sit covered for several minutes. Then invert the plates (so condensation will not drop onto growing culture) and incubate for one to two days at 37° C. (If growth is too rapid, the plates can be placed in a refrigerator and held until the next lab.)

Analyzing Results

12▶ At the next laboratory period, you will count the number of colonies you see on each plate. Some will have too many colonies to count and you should just say "too numerous to count." Record the results in table 14.2.

To arrive at the number of bacteria per gram of sample, remember that each colony you see represents a single bacterium in the food sample. You spread dilutions of these invisible cells evenly on the agar surface and each single cell developed into a colony that you could see. Usually, results are expressed as number of bacteria per gram of sample. You started with 0.1 g of the sample to keep volumes of reagents low, so you will have to multiply all of your results by 10 to get to a per gram basis. In addition, you did not plate the entire amount of each dilution. You plated only 0.5 ml. 1/20 of the diluted sample. Therefore, the number of colonies you count will have to be multiplied by 10×20 and then by the dilution factor to arrive at the number of bacteria per gram of food.

Where Do Bacteria Live?

13▶ Your instructor will provide you with a sterile petri plate of nutrient agar, a sterile tube of saline, and sterile cotton swabs. Dip the swab and wipe a surface that you think is "clean." For example, a desk top, a soda machine, or a pencil you use. After collecting the sample, open the dish and rub the swab on the surface of the nutrient agar. If you use only half the plate, a second sample can be collected with a second swab. Incubate overnight and look at the plates during the next lab.

Cyanobacteria

Cyanobacteria (once called blue-green algae because their prokaryotic cell structure was not known) exist as single cells or as colonies or as filaments of cells, depending on the species (fig. 14.3). Because they contain chlorophyll, they are greenish in color and perform photosynthesis. Like many bacteria, they produce spores that are resistant to drying. Spores allow cyanobacteria to survive unfavorable seasonal conditions. They may also be windblown and distribute the species. Cyanobacteria live in freshwater, marine, and terrestrial environments, and they flourish when given water, sunlight, carbon dioxide, nitrogen, and certain inorganic salts. Their gelatinous capsules and the toxins they produce make them poor food for many heterotrophs. They can be a nuisance in water supplies where they cause taste and odor problems. However, on a global scale they are probably more important than rain forests in terms of photosynthetic productivity, the oxygen geochemical cycle, and in nitrogen fixation. They are also a relatively unexplored source of bioactive secondary products.

14▶ Go to the supply area and make a wet mount slide from the mixed cyanobacteria culture. Before adding a coverslip, dip a dissecting needle in India ink and then touch the tip to the drop of culture on the slide. This will transfer a small amount of carbon particles to the drop and will allow you to see the gelatinous sheaths that surround the cells.

After adding a coverslip, look at the preparation with the 10× objectives, switching to the 43× objective when you find something of interest. Because this is a mixture,

Figure 14.3   Common cyanobacteria (blue-green algae).

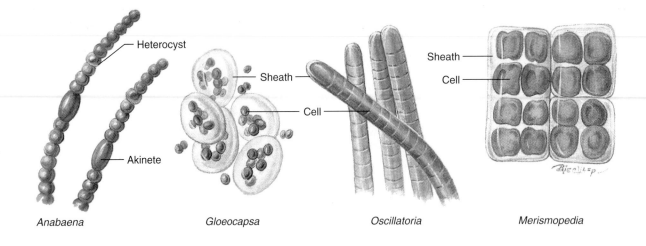

Anabaena   *Gloeocapsa*   *Oscillatoria*   *Merismopedia*

you should see most of the species seen in figure 14.3, plus there may be others. The idea here is to see the diversity of colonial organization found in the cyanobacteria. Read about each genus as you see it and draw an example in the circle below based on your observations.

Anabaena sp. has colonies of cells arranged as filaments. Filamentous organization may be a defensive mechanism making it difficult for microscopic animal grazers to ingest the colony. You may be able to see some differentiation along the filament. Interspersed with numerous vegetative cells, you may see some elongated cells called akinetes. They are spores and will survive desiccation. Other cells are more spherical and are called heterocysts. They have enzymes that allow them to convert atmospheric nitrogen gas into compounds that can be used in amino acid and nucleic acid synthesis. Some species of

Anabaena produce neurotoxins that inhibit neuromuscular junctions, and others produce toxins that interfere with liver function.

Gloeocapsa sp. The about 40 species in this genus form colonies. Cells form new mucilaginous sheaths as they divide, so that sheaths develop within sheaths. Unlike *Merismopodia* that has cells arranged in regular arrays, the cells are arranged in a more random fashion. This species is commonly found growing on moist rocks and flower pots in greenhouses. Often the dark stains seen on rock or buildings that are intermittently damp are due to dried cells of this species. Some species of *Gloeocapsa* are found associated with fungi in lichens as the photosynthetic partner.

Merismopodia sp. has colonies of single cells in a regular array within the gelatinous sheath. Usually you can see a unit that is 4 cells wide by 4 cells tall within a colony and this may be repeated several times.

Oscillatoria sp. also has a filamentous organization but there are no real cell specializations as in *Anabaena*. Filamentous cyanobacteria often trap oxygen released in photosynthesis in mats of criss-crossed filaments so that the mats float at the surface of the water. This genus is well known for its ability to move. If you do not touch your slide, you may be able to see the filaments move back and forth. If you scan along the length, you will see where there are a few dead cells in the filament. Typically, the filament will fragment at this point, producing two filaments that will continue to grow. People swimming in tropical oceans where there have been blooms of these organisms have suffered skin rashes. The Red Sea takes its name from the fact that a species of *Oscillatoria* blooms there and produces a red pigment that periodically colors the water red.

As you look at these different species of cyanobacteria, answer these questions:

- Do you see nuclei in any of the cells?

- Is their green color due to a concentration of chlorophyll in any region of the cell, or is the chlorophyll generally dispersed throughout the cell?

- Would you expect to find any of these cells dividing by mitosis or meiosis? Why?

Your instructor may bring some scrapings from flowerpots in the greenhouse to the lab or may have collected some blue-green algae from a lake or stream. Take a few minutes to make a slide and look at these "wild" cyanobacteria.

Learning Biology by Writing

Devise a procedure that would allow you to determine the number of bacteria found in potato salad collected from a salad bar in a restaurant. Describe how you would use sterile techniques to avoid false readings.

As an alternative assignment, your instructor may ask you to answer the questions in the Lab Summary and Critical Thinking Questions that follow.

Internet Sources

Use the **Google** search engine (http://www.google.com) in an Internet browser to locate additional information on cyanobacteria. When connected type in the phrase, *cyanobacteria in drinking water*. Scan the list that is returned and answer Critical Thinking Question #4.

Lab Summary Questions

1. List the characteristics that separate organisms in the Domain Bacteria from those in Domain Eukarya. Which ones were you able to see in this laboratory?

2. What evidence do you have from this laboratory that bacteria are diverse? Cite specific examples as you explain this diversity.
3. Describe how to do a Gram's stain. Why is this important?
4. Explain how the number of colonies seen on an agar surface in a petri plate can be related to the number of bacteria in a sample.
5. What is a serial dilution and why is the technique used?
6. What evidence do you have from your experiments that bacteria are metabolically different?
7. What quantitative evidence do you have that bacteria are found in foods? *Which foods had the most bacteria?*
8. Describe the diversity of shapes and sizes seen in cyanobacteria.

Critical Thinking Questions

1. "To make strawberry jam, mash 4 cups of berries with 7 cups of sugar (sucrose). Bring to a boil and continue boiling for twenty to thirty minutes. Ladle into sterilized jars and cover with melted paraffin wax or sterilized lids."

Relate jam making to the growth requirements of different bacteria. Would these steps be adequate for preserving (canning) vegetables? Meats? What other methods could be employed to preserve foods?

2. Should a plant nursery that plans to grow pea and bean seedlings use a sterilized potting soil mix?

3. It is well known that many pathogenic bacteria are developing resistance to antibiotics. Explain why this is happening and formulate a recommendation that you would give your physician regarding when antibiotics should be used in the treatment of disease.

4. Why should you be concerned if cyanobacteria are found in your drinking water? Use Internet Sources to locate the information that you need.

5. Why are bacteria, rather than say human cells, used as experimental organisms in molecular biology studies?

6. Bioremediation is a process that uses bacteria to remove pollutants from soil or water. What characteristic(s) of bacteria make them excellent organisms for doing this?

7. Do you think of bacteria as being "bad" and just as human pathogens or do you think of them as being "good," essential components of all ecosystems. Justify your position.

8. You may have heard of green algae. Are they related to blue-green algae? Make a list of the similarities and differences.

LAB TOPIC 15

Diversity Among Protists

Supplies

Preparator's guide available on WWW at
http://www.mhhe.com/dolphin

Equipment

Compound microscopes
Dissecting microscopes

Materials

Cultures
Mixed culture of marine dinoflagellates
Volvocales mixed algae culture
Mixed ciliate culture
Mixed amoeba culture
Euglena
Saprolegnia culture kit
Physarum (slime mold) culture kit
Cultures or preserved specimens of *Polysiphonia,
Bangia,* or *Callithamnion, Porphyra, Rhodymenia,*
or *Gigartina Corallina*
Plastomount of *Ulva* and other marine algae
Herbarium sheets with examples of brown, red, and
green algae
Prepared slides
Giardia lamblia
Trichomonas vaginalis
Oedogonium
Spirogyra conjugating
Trypanosoma sp.
Radiolarian strew
Diatomaceous earth
Slides and coverslips
Sample food labels listing carrageenan or agar additives

Solutions

0.85% saline
Protoslo or methyl cellulose
1% water agar for slime molds in petri dish
Unprocessed (old-fashion) oatmeal

Student Prelab Preparation

Before doing this lab, you should read the introduction
and sections of the lab topic that have been scheduled
by the instructor.

You should use your textbook to review the
definitions of the following terms:

algae	heterotrophic
autotrophic	isogamy
cilia	multicellularity
clade	phylum
colonial	protist
flagella	protozoa
genus	species
heterogamy	taxonomic kingdom

You should be able to describe in your own words
the following concepts:

Differences between bacteria and protists
Diversity among the protists
How ancestral protists might have evolved into
multicellular organisms

As a result of this review, you most likely have
questions about terms, concepts, or how you will do
the experiments included in this lab. Write these
questions in the space below or in the margins of the
pages of this lab topic. The lab experiments should
help you answer these questions, or you can ask your
instructor during the lab.

Objectives

1. To observe the diversity among several clades of
   protists
2. To learn the life cycles of biologically important
   protists
3. To illustrate evolutionary trends among the protists

Background

With this group, the protists, you will begin your study of
diversity among eukaryotic organisms in the Domain Eukarya. Protists inhabited the earth billions of years before

plants, fungi, and animals: but considerably after bacteria originated. Some fossil evidence indicates that protists might have been present over 2 billion years ago. The ancestors of this group were the first to have a true nucleus, chromosomes, organelles such as chloroplasts, mitochondria, endoplasmic reticulum, cilia, and cell division by mitosis or meiosis. The endosymbiotic chimera theory suggests that two or more types of prokaryotic cells fused together to give rise to this eukaryotic cell plan.

Most protists are unicellular, although several algae are colonial or truly multicellular. By studying the colonial and multicellular forms, we can gain information that allows us to speculate how multicellularity might have developed in fungi, plants, and animals. Although protists are often described as being simple organisms, their cellular organization and metabolism are every bit as complex as those found in the so-called higher organisms. In fact, higher organisms are often much simpler at the cellular level because their many cells are specialized to perform particular functions while single protistan cells perform all functions necessary for life as independent organisms.

Protists live everywhere there is water: as plankton in the oceans and freshwater, in puddles, in damp soils, and, as symbionts, in the body fluids and cells of multicellular hosts. Some are autotrophic, making their own food materials through photosynthesis, while others are heterotrophic, absorbing organic molecules or ingesting larger food particles. They are ecologically very important in food chains, especially the algae (phytoplankton) which are the energy base for most aquatic ecosystems. Most protists can reproduce asexually by mitosis while some reproduce sexually as well, involving meiosis and nuclear exchange, or in some cases, cell fusion called syngamy. A cyst stage is found in the life cycle of many protists; it allows the species to lie dormant during temporarily harsh conditions.

Protists can be placed in three broad categories. Those that are photosynthetic are commonly called algae. Protists that ingest food are commonly called protozoa. Another group of protists are fungus-like and absorb food materials. They lack a common name. None of the common names have taxonomic status.

The classification of the protists is in a state of flux. The Kingdom Protista, originally proposed by Robert Whitaker in 1969, included eukaryotic organisms that did not fit into Kingdoms Animalia, Planta, or Mycota. It was a category of convenience and not one that represented evolutionary relationships. This means that we are somewhat ignorant about the ancestry of an estimated 60,000 or so species of protists. Modern scientific investigation of cell structure and nucleic acid analysis is starting to show us common characteristics among the 30 or more phyla included in Whitaker's Kingdom Protista. Biologists now reject Protista as a kingdom. What is not clear is what should replace it.

Modern taxonomists are using a clade approach to classify organisms where relationships are not definite. A clade is a tentative grouping of organisms based on available evidence. As more evidence accumulates the tentative grouping may change or remain constant. When all biologists agree about a grouping then the clade may be given a taxonomic name. In this lab topic I will use the clade approach. The clades are listed below with brief descriptions of the protists included. Students should recognize, however, that by the time this revision of the lab manual appears in print, new information may be available, resulting in a reorganization of the protists. While this is frustrating, it also demonstrates that science is an ongoing, dynamic process that is not locked into beliefs that betray the facts. Those phyla marked with an asterisk will be studied in lab.

(Protozoa-like Organisms)

Clade Diplomonadida*—Unicellular, anaerobic protists lacking mitochondria with multiple flagella and two nuclei; *Giardia lamblia* is a human parasite transmitted as cysts in drinking water. Infections lead to intestinal cramping and diarrhea.

Clade Parabasala*—Unicellular, anaerobic, flagellated protists lacking mitochondria in which the nuclear envelope persists during mitosis. They are characterized by a modified Golgi apparatus located at the base of the flagella called a parabasal apparatus. Includes *Trichonympha sp.*, a mutualistic symbiont found in termites' guts that aids in cellulose digestion, and *Trichomonas vaginalis*, a human parasite, that can be sexually transmitted.

Clade Euglenozoa*—All members of the group are unicellular and have an anterior surface chamber containing two flagella: one short and one long. Movement of the longer one draws the organisms through the water.

Phylum Euglenophyta* (Euglena): Includes about 800 species with a proteinaceous pellicle covering the cell membrane. Although euglenoids are typically thought of as photosynthetic they are descended from heterotrophic ancestors which had a feeding pocket adjacent to the flagellar opening. The chloroplasts are surrounded by three membranes suggesting that they originated from green algae that were ingested and became symbionts. Reproduction is by mitosis. Polysaccharides are stored as paramylon, a glucose polymer.

Phylum Kinetoplastida*: Includes thousands of species commonly called kinetoplastids, a name derived from a single mitochondrion in the cell, the kinetoplast. It contains DNA and is located near the base of the flagella. *Trypanosoma* is the causative agent of African sleeping sickness in humans and nagana in cattle. Another species causes leishmaniasis.

Clade Alveolata*—Most members of this group are unicellular and have small indentations in their surface membranes, called alveoli, beneath the material covering the outer surface.

---

*Those marked with an asterisk will be studied in lab.

Phlyum Dinoflagellata* (also called Pyrrhophyta): Includes over 2,000 photosynthetic and heterotrophic species with two flagella and an outer covering of cellulose plates. They are abundant in oceans and freshwaters where they form the basis for many food chains. Zooxanthellae are members of this group that are symbiotic in the tissue of corals, contributing photosynthetic products that supplement the corals' feeding.

Phylum Ciliophora* (Ciliata): Includes about 8,000 species commonly called ciliates. All have many cilia that are used as locomotor organelles. Two types of nuclei are found in cells: a macronucleus that directs metabolism and a micronucleus that is exchanged in conjugation during sexual reproduction.

Phylum Apicomplexa: Includes about 3,900 species with a special apical complex used to invade animal host cells. Commonly called apicomplexams or sporozoans, these parasitic protists have complex life cycles, including the causative agent of malaria (*Plasmodium* sp.), which is transmitted to humans by *Anopheles* mosquito bites. Some form spores, single haploid cells capable of developing into adult without fusion with another cell.

(Algae-like Organisms)

Clade Stramenopila*—Includes both photosynthetic and heterotrophic species which may be unicellular, colonial, or multicellular. All have unique flagella with hair-like projections along the length of the shaft. For many species, reproductive cells are the only flagellated stages in the life cycle.

Phylum Bacillariophyta: Includes about 11,500 photosynthetic and unicellular species commonly called diatoms that are common in freshwater and marine environments. Siliceous cell walls are often highly patterned. Some authors place these organisms with the golden algae in the Phylum Chrysophyta.

Phylum Oomycota: Includes about 500 species commonly known as water molds, white rusts, and downy mildews, which were once thought to be fungi. They are multicellular with cellulosic cell walls, not chitinous walls as in fungi. They have heterotrophic nutrition as parasites and decomposers. Body form is filamentous with differentiation into reproductive structures. Members of the genus *Phytophora* are important plant parasites.

Phylum Phaeophyta: Includes about 1,500 photosynthetic species commonly called brown algae because of a brown pigment fucoxanthin. They are the most complex and largest of all algae. Most are marine and multicellular in cooler coastal waters, such as kelps which can reach 100 m in length. Body shows division of labor into specialized parts such as holdfast, stalk or stips, blades, and bladders or floats. Life cycle includes an alternation of generations between a sporophyte and gametophyte stage.

Clade Rhodophyta* (red algae)—Over 5,000 described species which may be uni- or multicellular, all lacking flagella. Most species are marine, growing as encrustations or as miniature "tree-like" organisms. The red color comes from phycoerythrin (a phycobilin-type pigment). However, not all species appear red. Life cycles include an alternation of generations between a sporophyte and gametophyte stage. Cell walls consists of cellulose embedded in a matrix of other polysaccharides which include agar and carrageenan, that are commercially used as sources of thickening agents in lab cultures or food. Some, called coralline algae, secrete $CaCO_3$ around their cell walls, which can contribute to coral reef formation.

Clade Chlorophyta*—About 7,000 unicellular, colonial, or multicellular species which are photosynthetic with bright green chloroplasts (Chlorophylls a and b) Molecular studies show a close affinity to plants and some authors include them with plants in a new clade called Kingdom Viridiplantae. Most are aquatic but others can live on damp surfaces.

Phylum Chlorophyta*: Includes green algae that may be unicellular, colonial, or multicellular. About 90% of the species are found in freshwater.

(Amoeboid-like Organisms)

Amoeba Clade*—Hundreds of species found in aquatic and moist soil habitats. They lack cell walls and any form of sexual reproduction; multiply by division into two cells. They move by extension of the cytoplasm and pulling the body forward; called amoeboid movement. The phylogenetic placement of this group is quite a puzzle. Includes organisms with such names as naked amoebae, shelled amoebae, foraminiferans, actionopods, heliozpoans, and sun animacules.

Clade Mycetozoa*—About 1,200 species with restricted mobility

Phylum Myxogastrida*: Includes about 500 species commonly known as plasmodial slime molds which were once thought to be fungi. Body consists of a non-walled mass of multinucleate cytoplasm which streams or flows like slime, giving group its common name.

Phylum Dictyostelidae: Includes about 70 species commonly known as cellular slime molds. Despite their name they are not closely related to the plasmodial slime molds. They live in damp soils ingesting bacteria. Body form is single amoeboid cells that can aggregate during times when food is short to form a slug phase; amoebae in the slug can develop spore walls and be air-carried to new locations.

LAB INSTRUCTIONS

To develop an appreciation of protist diversity, you will study representatives of the groups listed above that are marked with *.

Figure 15.1 *Giardia lamblia:* a diplomonad human intestinal pathogen.

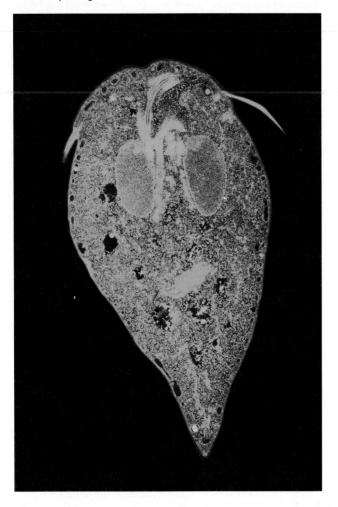

we had cysts

Giardia divides by binary fission and huge numbers can build up. As trophozoites are carried into the large intestine, they begin to make cyst walls which protect them from dehydration. It was estimated that a single diarrhetic stool from an infected person contained 14 billion cysts. You can get the disease by eating a single cyst in contaminated food or water. Infection is not fatal, but is extremely discomforting. It is accompanied by severe diarrhea, flatulence, intestinal pain, dehydration, and weight loss. Infection usually runs its course in two weeks. There are drugs to speed recovery. It is the most commonly identified protozoan parasite found in fecal samples examined by the U.S. Public Health Laboratories. In the U.S.A. it has estimated an estimated prevalence of up to 7% in adults and more in children. There is evidence that *Giardia* can be transmitted to humans through the feces of dogs, beavers, and bears. Campers often pick up the infection when drinking untreated, unfiltered water, hence its common name "beaver fever."

In preparing some of the materials for this lab, your instructor may mix cultures of two or more organisms so that you will have to prepare only one slide instead of two or three. This will save time and materials.

Clade Diplomonadida

1▷ Obtain a microscope slide of *Giardia lamblia* from the supply area. Look at it first with the 10× objective to locate a specimen and then study it through your 40× objective. Compare what you see to figure 15.1.

Giardia is an intestinal parasite of vertebrates. It has a feeding stage called a trophozoite in its life cycle and a cyst stage. You are looking at a trophozoite stage.

The trophozoite has four flagella and two nuclei. The nuclei look like eyes in the cells. Careful examination of several specimens may reveal a depression on the cell surface that acts like a suction disk allowing the organism to attach to the intestinal wall lining. No mitochondria are found in these organisms. Sketch a *Giardia* in the circle.

Clade Parabasala

2▷ Obtain a prepared slide of *Trichomonas vaginalis* from the supply area. Observe the slide first with the 10× objective, switching to high power when examples are found. Compare what you see to figure 15.2.

Transmitted during sexual intercourse, this parasite resides in the lower genital tract of infected women and in a man's urethra and prostate. Commonly known as "trich," it causes irritation, and in severe cases purulent discharges from the vagina or inflammation of the male's urethra and prostate. It is estimated that infection rates may be as high as 20%. The parasite divides by binary fission and infections develop within a month of exposure. It is treatable with antiprotozoal drugs although resistant strains are appearing. It does not form any cyst stages and can live only briefly outside the body. *Trichomonas* feeds by pinocytosis and phagocytosis as well as by absorbing nutrients from the host's body fluids across its membrane. They lack mitochondria, and energy is obtained by anaerobic metabolism.

label flagella and nuclei

Figure 15.2 *Trichomonas vaginalis:* a parabasalid human pathogen.

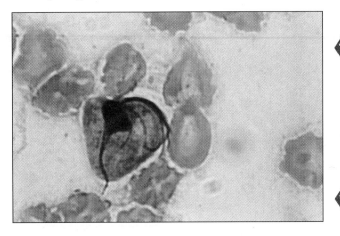

Other species are parasites of the urinary, reproductive, or digestive tracts of other vertebrates.

Note the prominent nucleus. Identify the flagella. How many can you count? ___/___ Near the nucleus you should be able to see the parabasalar body, a modified Golgi apparatus, which characterizes this group. Running the length of the cell is the axostyle composed of many microtubules. It should be visible. Note the undulating membrane which appears a fold of membrane-like material along the cell body. All three of these structures allow the organism to move through mucus linings of the vagina or urethra. Sketch a few cells below.

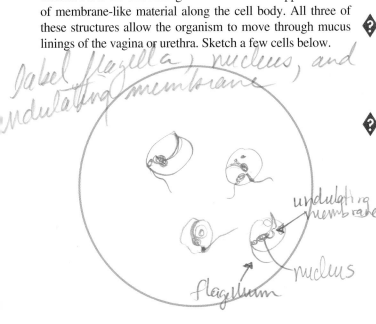

label flagella, nucleus, and undulating membrane

undulating membrane

flagellum

nucleus

Clade Euglenozoa

Phylum Euglenophyta

This small group of about 800 species contains flagellated, autotrophic protists that lack cell walls, having instead a flexible outer covering called a **pellicle.** Species are common in waters polluted with organic matter and on the surfaces of wet soils.

3▷ Make a wet-mount slide of *Euglena* from the stock culture and observe through your compound microscope. If the organisms are swimming too fast to be studied, make a new slide but add methyl cellulose, a thickening agent, or shred a small piece of lens paper or cotton into the drop so as to trap the organisms. Reducing the light intensity will help you see structures in the organism.

? *Euglena* is usually pear-shaped with the blunt end anterior. Does the flagellum push or pull the organism through the water?

pull

? Watch the organism closely. What evidence is there that the surrounding pellicle is flexible?

some bending/flexing

As you study *Euglena*, find the structures indicated in figure 15.3.

? What color is the **pigment spot,** also called the stigma, near the base of the flagellum? How many flagella does *Euglena* have?

pink (coral)

? What evidence do you have before you that *Euglena* is photosynthetic?

green

? Excess sugars produced during photosynthesis are converted into **paramylon,** a unique form of storage starch. Is the chlorophyll of *Euglena* localized in structures, or spread throughout the cell as in cyanobacteria?

Some in structures, but whole cell is pale green

Phylum Kinetoplastida

Members of this phylum are unicellular and characterized by the presence of a unique organelle, the **kinetoplast,** located at the base of the flagellum. Electron microscope studies show that it is a modified mitochondrion containing a large amount of DNA. Organisms in this group are free living or parasitic on animals and plants. Three cause serious human disease (African sleeping sickness, Chagas disease, and leishmaniasis).

4▷ Obtain a prepared microscope slide of a blood smear from an individual with trypanosomiasis (African sleeping sickness or Chagas disease). Look at it first with your 10× objective, and when a specimen is located, with the 43× objective. You should see round red blood cells, irregular shaped white blood cells with large nuclei, and long, thin cells which are the trypanosomes (fig 15.4). These organisms live and reproduce in the host's blood stream.

Figure 15.3   Anatomy of *Euglena* sp.

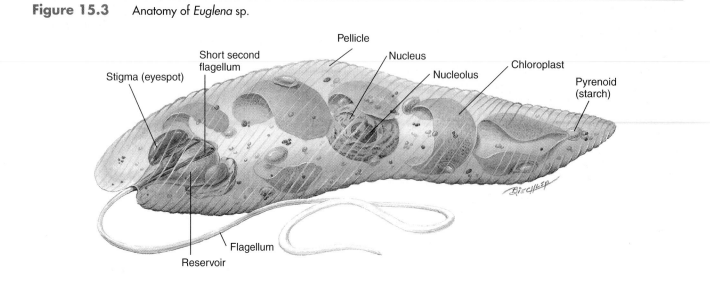

Figure 15.4   A blood smear showing trypanosome parasite against a background of red blood cells.

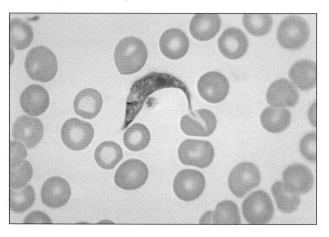

Draw a few of the cells below. Can you see the flagellum? The kinetoplast? Note the undulating membrane that aids in locomotion.

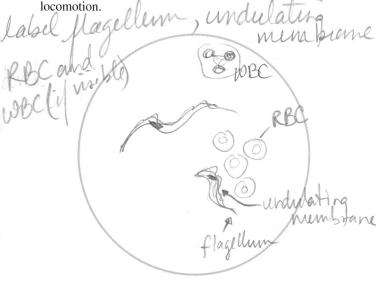

African sleeping sickness and Chagas disease are called trypanosomiasis, a reflection of the genus name for the causative agent. Chagas disease affects up to 18 million people in the rural tropical Americas. It is transmitted from one infected individual to another by "kissing bugs, biting insects that take blood meals, by blood transfusions, and transplacentally from mother to fetus. The protozoan is contained in the feces of the bug and is transmitted when a person rubs the fecal material into the wound bite or into the eyes. It can also be transmitted by ingesting food contaminated with the bug's feces. Most people who are infected do not seek medical attention and the infection will run its first course in 4 to 8 weeks. Symptoms include fatigue, fever, rash, enlarged liver/spleen, sometimes vomiting. Often 10 to 20 years later, people then develop the most serious symptoms which include cardiac problems and difficulty in swallowing and defecation, leading to reduced lifespan. African sleeping sickness is a serious disease and strikes almost a quarter of a million people annually in tropical Africa and can affect cattle as well, in which case it is called nagana. It is transmitted from an infected individual to others by the blood feeding tsetse fly. Initial symptoms are fever, headaches, rash, and edema progressing to meningoencephalitis and death within several weeks. It is treatable by appropriate anti-trypanosomal drugs.

Trypanosomes have attracted the interest of molecular biologists. Studies show they survive in the blood stream by tricking a mammal's immune system. Normally when a foreign cell enters the blood, an antibody to it is made and the cell is killed. For the immune system to do this, it must "recognize" the chemicals found on the foreign cell's surface. Biologists have found that every few days trypanosomes are able to change the types of molecules found on their surfaces and thus always stay one step ahead of the body's natural defenses.

Clade Alveolata

Phylum Dinoflagellata (Pyrrhophyta)

Dinoflagellates, along with diatoms, make up most of the phytoplankton found in marine and freshwater ecosystems. They form the basis of many aquatic food chains as they are gathered by filter-feeding animals which are, in turn, consumed by other animals. Over 61 species of dinoflagellates produce nasty toxins. During "algal blooms," which turn coastal waters red or brown, incredibly dense populations of dinoflagellates can produce neurotoxins that kill fish. Filter-feeding shell fish (clams, mussels, and oysters) can accumulate large amounts of the algae before or after a bloom and, if eaten by humans, can cause poisoning. Another group of dinoflagellates, collectively called zooxanthellae, live in the tissues of coral polyps and some mollusks where their photosynthetic capability contributes to the energy budgets of the animals. About 1,000 species of dinoflagellates have been described (fig. 15.5).

5▷ Make a wet mount of the mixed marine dinoflagellate culture in the supply area. Observe first with the 10× objective and then with the 43× objective of your compound microscope. Focus up and down to see the outlines of the surrounding plates, called a **theca.** The sculpted patterns of the theca are used by taxonomists to identify different species.

Try to locate the flagella as the organism swims. Dinoflagellates have two: one is in a groove that encircles the cell body, and the other runs in a longitudinal groove. The beating of the flagella give dinoflagellates a distinctive spiral swimming motion. Draw a few dinoflagellates below, indicating the differences between the species that you can see on your slide.

label flagellar grooves
sketch as many varieties as you see

groove

groove

What color are the dinoflagellates on your slide? _____While some dinoflagellates are heterotrophic, most are photosynthetic but the green color of their chlorophyll is often masked by accessory pigments.

Figure 15.5 Representative dinoflagellates showing the characteristic outer cellulose plates and flagella originating from grooves between plates.

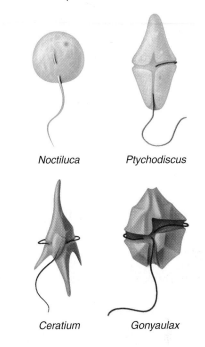

Noctiluca   *Ptychodiscus*

Ceratium   *Gonyaulax*

Sexual reproduction is known in only a few species. Most reproduce asexually when a cell divides into two halves. In the autumn as temperatures decrease, dinoflagellates shed their theca and form cysts that sink to the bottom where they remain dormant. As temperatures warm in the spring, they exit the cyst, form a new theca, and become active. The cyst wall contains sporopollenin, a chemical compound also found in the walls of pollen grains. Cysts are often preserved as fossils and can be used to reconstruct ancient climates and ecological situations.

Phylum Ciliophora

About 7,000 species of ciliates have been described. They are all unicellular. Most are free-living heterotrophs although a few are parasitic. The phylum is characterized as having the cell surface covered by **cilia** used in locomotion and feeding. They also have a complex surface covering, called a **pellicle,** and compound ciliary structures called **cirri.** There is a complex feeding apparatus with an oral groove that leads to a fixed area where food is ingested into **food vacuoles** via a mouth or **cytostome.** Freshwater species have a **contractile vacuole.** Most ciliates have two types of nuclei. **A macronucleus,** which regulates the normal physiological functioning of the cell, has diverse shapes, resembling a sphere, a beaded ribbon, or a horseshoe. One or more **micronuclei** are found; these are involved with sexual reproduction, a process called **conjugation.** During conjugation, two paramecia, tem-

Figure 15.6   Types of ciliates likely to be found in mixed culture: (a) *Blepharisma* sp.; (b) *Colpidium* sp.; (c) *Euplotes* sp.; (d) *Paramecium* sp.; (e) *Spirostomum* sp.; (f) *Stentor* sp.

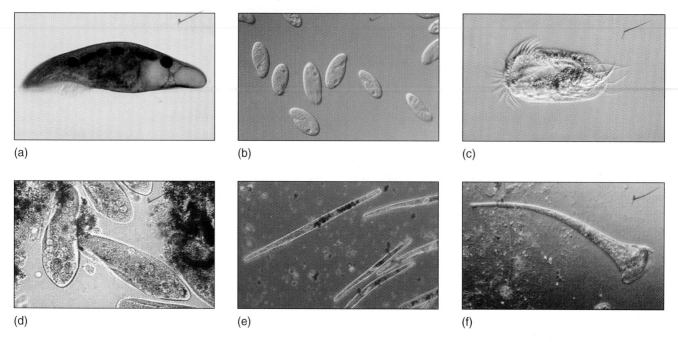

(a)

(b)

(c)

(d)

(e)

(f)

porarily fuse along their oral areas, and exchange micronuclei. Then the cells separate, reorganize their nuclear material, replace their macronuclei with the newly organized genetic material, and subsequently divide. Asexual reproduction in ciliates occurs by **binary fission** in a transverse plane.

6 ▶   Make a wet-mount slide of the mixed ciliate culture found in the supply area. Look at it first with 10× objective and switch to the 43× when you find something of interest. Compare to figure 15.6. Continue to switch back and forth as you scan the slide. How many different types of organisms can you see? _≥8_ You may find it useful to adjust the condenser height and diaphragm opening on your microscope to get different views of these organisms. Can you see evidence for cilia in all of these? _yes_

Observe the different locomotory movements: forward, rotating, and swerving. Now, make a second slide, but this time add a drop of methyl cellulose to the culture drop before you add a cover slip. This chemical is very viscous and slows the ciliates so that you can see details of cellular structure. There will be several different species of ciliates on the slide (fig. 15.6). Check them off as you are able to identify them. These may include the following:

Blepharisma sp. These medium-size ciliates have a rose color (but not in strong light) that makes them easy to spot. *Blepharisma* feeds by sweeping small bacteria into the cytosotome using a thin membrane (not visible) and fused cilia called cirri that you should be able to see beating in waves. At the base of the cytostome, the

bacteria are ingested as food vacuoles. A contractile vacuole may also be seen. It controls the volume of water in the cell.

Colpidium sp. These are probably the smallest ciliates on your slide. Along with other ciliates, they are important in food chains in aquatic ecosystems. They eat bacteria which are probably living on dead and decaying materials. In turn, the ciliates are consumed by small animals and their energy passes along to top predators such as fish or birds.

Euplotes sp. This genus is characterized by rows of fused cilia called cirri which resemble spines. A freshwater inhabitant, *Euplotes* uses its cirri for swimming and also to "walk" along a substrate.

Paramecium sp. Several different species in this genus will be on the slide. They are medium-size cells. Some will be green because of symbiotic algae. Others will be translucent. Known as slipper animals because of the body shape, they feed on bacteria. The stiff outer surface of the cell is composed of several layers of membrane, which along with other structures, make up the pellicle. Beneath the pellicle are numerous rod-like trichocysts which are thought to be used in defense. They can discharge to produce a sticky mass of threads. You should be able to see the contractile vacuole(s). How often do they contract? _5-6 times/min_ On the next page, sketch a *Paramecium,* as well as other representatives of the group. On the sketch indicate the macronucleus, micronuclei, oral groove, gullet, contractile vacuoles, and cilia. Describe the major similarities of *Paramecium* and *Blepharisma*. What are the differences?

Spirostomum sp. These will be very large, long, tubular cells with a macronucleus that extends the length of the cell. It looks almost like a string of beads. One of the remarkable things of *Spirostomum* is the way it can contract. The organism can contract to 1/4 of its length in 6–8 millisec. Contraction can sometimes be induced by touching the coverslip. *Spirostomum* feeds on bacteria which are swept into the cytostome with a row of specialized fused cilia.

Stentor sp. These are large, trumpet-shaped ciliates that can be up to 2mm long! They feed on bacteria that are swept into a gullet by a crown of cilia. *Stentor* has remarkable regenerative powers; a small fragment one-hundredth the volume of an adult will grow back to a complete organism. When free floating, the cilia on the surface give *Stentor* mobility, but it will attach to substrates such as dead leaves of aquatic plants. Tap on the cover slip and watch it change from a trumpet shape to a blob of cytoplasm. Some species have symbiotic green algae that use the *Stentor's* waste products in photosynthesis and supply the host with carbohydrates. You may be able to see the macronucleus in the cytoplasm. A contractile vacuole should also be visible, and if you spend some time observing a single specimen, you should see it cycle.

Clade Stramenopila

Phylum Bacillariophyta

The 11,000 species of diatoms share a common characteristic: a cell wall consisting of two valves made of silica. They are often golden-yellow in color because of an excess of carotenoid and xanthophyll pigments, which mask the green of the chlorophylls that are also present. When the diatom cell dies, the siliceous valves do not disintegrate but accumulate as sediments (fig. 15.7). In California, some deposits of diatoms are 300 feet deep. These are mined to produce diatomaceous earth, a fine powdery material used as a filtering material in swimming pools and as a fine abrasive in silver polishes and toothpastes.

Figure 15.7 Diatoms exist in an exquisite variety of geometrical shapes. Here an artist has arranged several species into a beautiful composition on a microscope slide using a single hair to drag each one into position.

7▷ Make a wet-mount slide from the diatomaceous earth available in the laboratory. Place a drop of water on the slide and then add a very small amount of diatomaceous earth to the drop. Stir well before adding a coverslip and viewing.

Note the exceedingly delicate patterns of the diatom valves. These are the skeletal remains of cells that lived thousands of years ago. In a top-down view, some valves will be round, others triangular, ovoid, and irregular. When viewed from the side, these same valves appear rectangular or ovoid.

The ornamentation of the valves is often used as a test of the resolution of microscopes. In a poor microscope, only the outline of the valve will be visible, while in very good microscopes the fine indentations and perforations will be apparent. Sketch a few different types of diatom valves below. Indicate the location of the **girdle,** the region of overlap between the two valves.

Diversity Among Protists   **183**

Phylum Oomycota: Water Molds

At one time these organisms were considered to be fungi but are now considered protists in the clade Stramenopila. The characteristics that set water molds apart from fungi include: (1) asexual reproduction by biflagellated zoospores; (2) cell walls that contain cellulose; (3) diploid "body" cells with meiosis occuring in gametangia; and (4) the production of heterogametes.

Water molds live in freshwater, marine, and moist terrestrial environments. They are important decomposer organisms. Several species are economically important. One species causes potato blight which led to famines in Ireland during the nineteenth century and another causes grape downy mildew which nearly wiped out the French wine industry during the same time period.

The "body" form of water molds superficially resembles that of fungi. Long filaments without cross walls are produced and are called **hyphae,** a term from when they were included with the fungi. The tips of the hyphae may differentiate into zoosporangia which asexually produce motile diploid zoospores. The zoospores swim short distances or are carried by currents and colonize new locations. Under harsh conditions some hyphae differentiate into **gametangia** in which meiosis occurs to produce haploid eggs and sperm. The diploid condition is restored at fertilization. The zygote may become encased in a hard covering and dispersed to produce a new individual. With the water molds we encounter the first protist in this lab topic that is multicellular. The hyphae have normal body cells and others that produce gametes: a primitive specialization that you will see again later in the lab.

Saprolegnia is an easily cultured member of this group in which asexual reproduction can be studied (fig. 15.8). Cultures of this mold may be ordered from biological supply houses or grown in the laboratory as follows. Two weeks before class, a few dead insects were placed in petri dishes containing pond water. By lab time, *Saprolegnia* hyphae should be apparent. (Keep culture aerobic, or it will putrefy.) Carefully cut the ends of some of the hyphae from the culture and use a pipette to transfer them into a drop of water on a slide. Add a coverslip and observe with a compound microscope. Sketch what you see.

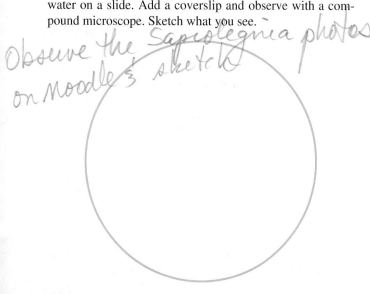

Figure 15.8 Life cycle of the water mold *Saprolegnia* sp.

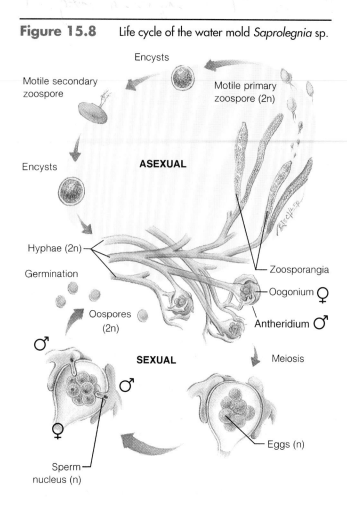

The nuclei in the hyphae of water molds are diploid. Note the absence of cross walls in the hyphae except at the edges of the **zoosporangia.** They should contain a number of motile, asexual spores. If you are lucky, you may find a zoosporangium in which the flagellated spores are starting to escape. The spores will disperse and, if a suitable substrate is found, will germinate to produce new hyphae. Under adverse conditions, spores will encyst and remain quiescent until favorable conditions develop. The asexual spores are called **zoospores** because of their motility.

Sexual reproduction involves production of nonmotile gametes by meiosis inside reproductive structures. Eggs are produced in spherical **oogonia.** Tips of hyphae of the same individual function as male gametangia, or **antheridia.** These grow into the oogonium, the sperm nucleus fuses with the egg nucleus, which then becomes a **zygote.** The zygote develops a thick wall and becomes an **oospore.** The oospore produces a new mycelium on germination.

Clade Rhodophyta

Most red algae grow at great depths in the oceans where the red accessory pigment phycoerythrin aids in photosynthesis. It is associated with the external membranes of their chloroplasts and absorbs blue light while reflecting red. Be-

Figure 15.9 Red algae from the clade Rhodophyta are often delicate in appearance. They live at great depth but are often washed up on beaches.

cause blue light penetrates deeper in water than red, red algae can live at greater depths than most other algae. Some species have been found at depths greater than 600 feet. Because they are not exposed to wave actions, their body forms can be delicate (fig 15.9).

9 ▶ In the lab, look at the herbarium sheets with dried red algae and at the preserved or living specimens available. Depending on the collection at hand, you may be able to observe species from the following genera:

Polysiphonia sp., *Callithamnion* sp., or *Bangia* sp. are highly branched and feathery.

Porphyra sp. or *Rhodymenia sp.* are flattened sheets, only a cell or two thick.

Based on general body size and shape, would you say that these are unicellular or multicellular? Do you see any examples of cell specialization?

cells at tips look different

some cells have green structures inside (larger browner algae)

Describe any morphological evidence for a division of labor among the cells in this organism.

Look at the example of *Corallina* sp. Species in this genus are coralline red algae whose cell walls are impregnated with secreted calcium carbonate. These algae are important reef-building organisms and, together with corals, are responsible for forming what we generally refer to as coral reefs, such as the Great Barrier Reef, in Australia.

In the lab will be labels from various recognizable food products such as salad dressings, ice cream, *etc.* Look at the labels to see if they contain agar or carrageenan, thickening agents isolated from the cell walls of red algae.

Clade Chlorophyta

Phylum Chlorophyta

The 7,000 or so species of green algae can be grouped to create a natural progression from single cells to multicellularity. Three lines of evolution are apparent: (1) the formation of colonies; (2) the formation of multicellular filaments; and (3) the formation of definite multicellular organisms.

Students often ask, what is the difference between colonial and multicellular organisms? After all, both consist of many cells. Why the use of different words to describe this organizational plan? The difference is subtle but significant. Colonial organisms consist of the same types of cells that aggregate to produce the organism. All of the cells perform the same functions. There is not any specialization of cells to perform unique functions as there are in our bodies. Multicellular organisms also consist of many cells. However, cells are specialized to perform specific functions. Consequently, there is a division of labor among the cells, and the shape and functioning of the total organism is due to the sum of the individual contributions of each cell type. Any account of the evolutionary development of plants, fungi, and animals must offer hypotheses as to how multicellularity might have developed.

Colonial Series

The volvocine series consists of flagellated green algae which may be single cells or colonies. Colonies may be plate-like, ellipsoid, or spherical. Colonies may reproduce asexually, producing smaller versions of the parent. In the colonies, there may be differentiation into vegetative and reproductive cells suggesting multicellularity, but it is not well developed. The molecular similarities between the green algae and plants suggest to some that plants may have developed along these lines. Sexual reproduction varies from **isogamy** where both genders product flagellated gametes of equal size, which cannot be distinguished from each other, to **anisogamy** where males and females produce different-sized gametes. Generally this takes the form of **oogamy** where the female produces a much larger, non-motile egg and the male produces a flagellated sperm.

10 ▶ In the lab there is a mixed culture of green algae called a volvocales culture. It contains several species ranging from unicells to colonies of varying shapes and sizes (fig. 15.10). First make a wet-mount slide of the culture without a coverslip to see the colonies. Later, add a coverslip to observe more detail. This single slide will illustrate an evolutionary trend from simple single cells to more complex colonies. The culture may have any of the following species:

Chlamydomonas sp.—This genus contains about 500 species of unicellular photosynthetic algae found worldwide in soils, fresh water, oceans, and even in

Figure 15.10 Several types of colonial green algae likely to be found in mixed culture: (*a*) *Eudorina* sp.; (*b*) *Gonium* sp.; (*c*) *Pandorina* sp.; (*d*) *Volvox* sp.

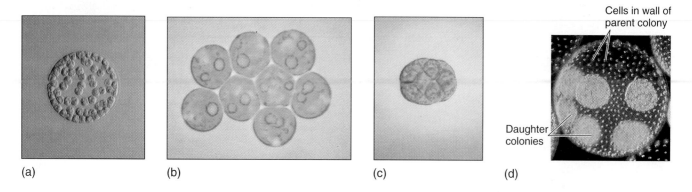

(a)　　　　(b)　　　　(c)　　　　(d)

Figure 15.11 Life cycle of the green alga *Chlamydomonas*.

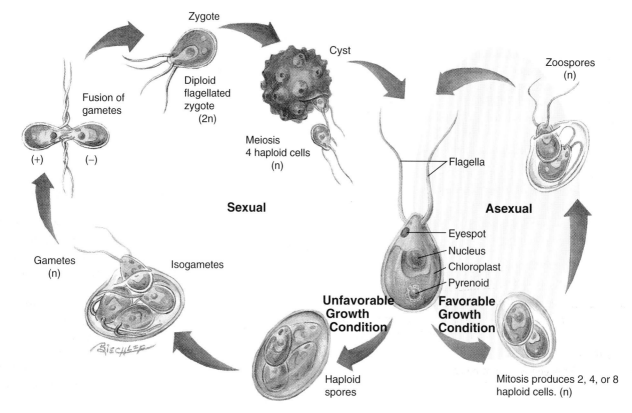

snow on mountaintops. The cells have a single cup-shaped chloroplast and two flagella that pull it through the water in a breast-stroke fashion. The cells can divide asexually by mitosis or can reproduce sexually (fig. 15.11). Gametes look alike and are haploid with mating types. Gametes fuse to form a diploid zygospore with an outer wall that protects it from adverse environmental conditions. When conditions improve, the diploid zygote undergoes meiosis and releases four haploid cells that resume the vegetative life cycle. *Chlamydomonas* can be said to have an alternation of generations life cycle. The zygote represents the sporophyte (2N) stage and the haploid stage is represent by the gametophyte (N) stage. *Carteria sp.* are similar in appearance to *Chlamydomonas* but are smaller.

Figure 15.12   Sexual life cycle of *Volvox*.

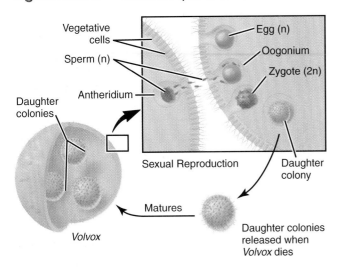

Vegetative cells
Sperm (n)
Antheridium
Daughter colonies
Egg (n)
Oogonium
Zygote (2n)
Sexual Reproduction
Daughter colony
Matures
Volvox
Daughter colonies released when *Volvox* dies

In the space below, draw examples of *Chlamydomonas, Eudorina,* and *Volvox.* Discuss how the progression from simple to more complex, with the change from isogamy to anisogamy, could be considered a model for the development of multicellularity.

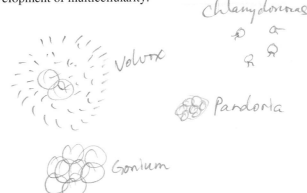

Eudorina sp.—Organisms in this genus have a colonial organization of 16 to 64 cells (mostly 32) covered by a gelatinous envelope in which the cells are quite distinct and separated from each other. The cells individually resemble *Chlamydomonas*. The two flagella found in each cell project to the outside of the colony's envelope and provide mobility. Sexual reproduction is anisogamous.

Gonium sp.—Organisms in this genus consist of nearly spherical cells, usually 8 to 16, arranged in a flat plate. Each cell is similar in appearance to *Chlamydomonas*. Because the cells are flagellated, the colony is able to swim. It is usually found in water which has a high organic content.

Pandorina sp.—Usually 4 to 32 closely pressed cells are found in the gelatinous envelope of the spherical colonies of these algae. Again the cells resemble the base structure of *Chlamydomonas*. New colonies can be formed by asexual reproduction. Sexual reproduction is isogamous.

Volvox sp.—This genus cannot be confused with any of the others. The spherical colonies, consisting of 500 or more cells, are the largest of this grouping and can be seen by the naked eye. Again the cells are similar to those of *Chlamydomonas* and the flagella allow the colony to be mobile. Most of the cells are vegetative cells, but some are specialized for reproduction. As you observe these colonies, you may see daughter colonies inside a parent colony. During asexual reproduction, some cells of a *Volvox* divide, bulge inward, and produce new colonies. When the parent colony dies, the daughter colonies are released. Sexual reproduction in *Volvox* is oogamous; nonmotile eggs and motile sperm are formed (fig. 15.12).

Filamentous Algae

The filamentous algae have a very different body form from the green algae studied so far. Transverse cell divisions add new cells to the growing tip of a single filament. In some of the species, not only are different gametes produced, but the gametes are produced in specially differentiated regions of the organisms that are called gametangia.

11▶   Obtain a prepared slide of *Oedogonium* and look at it through your compound microscope. Scan along the filament and find a vegetative cell, which should be examined under high power (fig. 15.13). It contains a net-shaped chloroplast that surrounds the cytoplasm and nucleus.

Some of the cells in the filament will be dark colored and swollen. These are **oogonia** and contain a single egg. In fact, the genus name *Oedogonium* means enlarged egg cell. Find other cells in the filament that are short and disk-shaped. These are **antheridia** and each produces motile sperm. Mature sperm have a crown of flagella on the anterior end. When released, sperm swim to and enter the oogonium where fertilization occurs. Is *Oedogonium* isogamous or anisogamous?

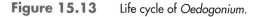

Figure 15.13   Life cycle of *Oedogonium.*

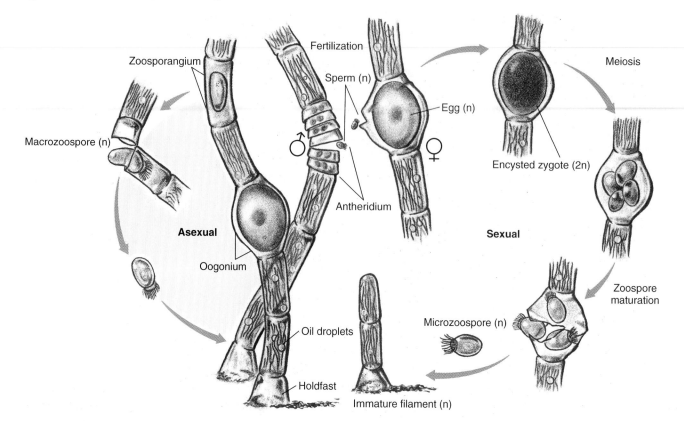

Fertilization

Zoosporangium

Sperm (n)

Meiosis

Egg (n)

Macrozoospore (n)

Encysted zygote (2n)

♂

♀

Asexual

Sexual

Oogonium

Zoospore
maturation

Oil droplets

Microzoospore (n)

Holdfast

Immature filament (n)

Zygotes are identifiable by a thick cell wall that surrounds the oogonia. If you think you see one on your slide, ask your instructor to confirm and then share the view with others. Zygotes eventually divide by meiosis to produce four haploid **microzoospores.** When they escape from the oogonium, they form new filaments which consist of haploid cells. When *Oedogonium* produces eggs and sperm, which type of cell division is involved? Mitosis? Meiosis?

Meiosis

Asexual reproduction occurs when vegetative cells differentiate into large **macrozoospores,** which leave the filament and give rise to new filaments.

12 Before leaving the filamentous green algae, you should examine one other species. Obtain a prepared slide of *Spirogyra.*

Also known as green silk, the 400 or so species are usually found in cool, clear, running water. Vegetative cells are cylindrical and contain a single, spiral-shaped chloroplast (fig. 15.14). *Spirogyra* is isogamous and does not produce motile gametes. Gametes are transferred by the process of **conjugation.**

Scan your slide until you find two adjacent filaments where tubes are growing outward toward one another (fig. 15.14). One filament is of mating type + and the other −. When the tubes meet they will fuse, forming a connection between the

Figure 15.14   Conjugation between mating types in *Spirogyra:* (*a*) vegetative cells grow as filaments and have a spiral-shaped chloroplast; (*b*) sexual reproduction (conjugation) starts when tubes grow outward from adjacent filaments; (*c*) condensed protoplast leaves male filament and enters female filament; (*d*) zygotes are formed when protoplasts fuse.

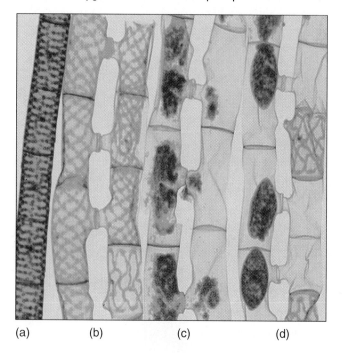

(a)   (b)   (c)   (d)

Figure 15.15 *Ulva,* the sea lettuce: (*a*) mature individuals in the ocean; (*b*) life cycle of *Ulva.*

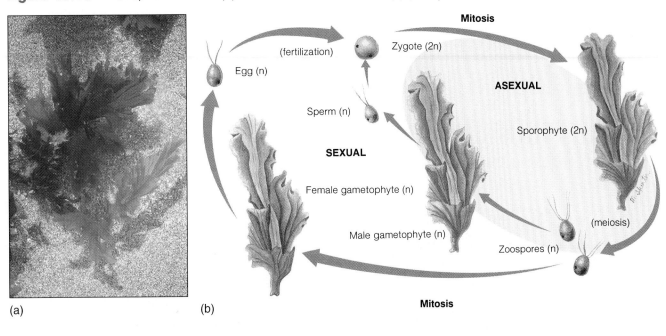

(a)   (b)

two cells. The contents of one cell will move through this conjugation tube and fuse with the other cell to form a diploid zygote. The zygote overwinters and then will divide by meiosis to produce four haploid nuclei. Three will degenerate and the fourth will grow to produce a new filament composed of haploid cells. Is *Spirogyra* isogamous or anisogamous?

Leafy Series

13▶ Examine either a living or plastic mounted specimen of *Ulva,* the sea lettuce (fig. 15.15). It typifies another line of evolution among the green algae. The leaflike body of this alga, the **thallus,** is multicellular and is only two cells thick. Some of the cells at the base are modified into a holdfast, which anchors the thallus in place against tidal currents.

If the number of chromosomes in mature individual *Ulva* are counted, some individuals are diploid and others haploid. The diploid individuals result from the formation of a zygote by the fusion of isogametes. The isogametes are formed by the haploid individuals. Haploid individuals grow from zoospores, motile cells produced by meiosis in the diploid individuals. Thus, *Ulva* shows a clear-cut pattern of **alternation of generations.** The diploid individuals represent the **sporophyte** generation and the haploid individuals are called **gametophytes.** This life cycle pattern is found in all plants and will be studied in detail in lab topics 16 and 17.

The series of green algae you have examined show some interesting trends toward the characteristics seen in plants. Your lab instructor will lead a discussion about these. Be prepared to discuss:

Isogamy/heterogamy

Colonial/Multicellular

Alternation of Generations

Amoeba Group

The Rhizopods

Commonly known as amoebas, all members of this phylum are unicellular and move by means of pseudopodia. Meiosis and sexual reproduction do not occur in this clade. All reproduction is asexual. Amoebas are common in soils, freshwater, and marine habitats where they feed on detritus or nonmotile bacteria and algae.

14▶ Go to the supply area and locate the culture of mixed amoebae. Unlike the other cultures you have used today, the amoebae do not swim; they crawl on the bottom. Use a pipette to pick up some of the "scum" from the bottom and put a small drop on your slide. Add a coverslip and observe first with the 10X objective to find the plane of focus and then with the 43X objective as you spot something of interest. In viewing this slide, it is best to be patient. When amoebae are disturbed, they contract into a blob or into their shells. When they are not disturbed, they begin to spread out and to move about. When this happens, the viewing can be quite interesting.

Several different kinds of amoebae will be found in this mixture (fig. 15.16). These include:

Actinosphaerium sp.: Known as sun animalcules, these amoebae have long **axopodia** that project from a central body, looking something like a pin cushion. The axopodia attach to prey which can be larger than the amoeba. The spherical cytoplasm is divided into an outer ectoplasm containing numerous large vacuoles and an inner endoplasm where the nucleus is located. They live attached to vegetation in freshwater. Cysts are formed during adverse conditions.

Figure 15.16 Types of amoeba likely to be seen in mixed culture: (a) *Actinosphaerium* sp.; (b) *Amoeba* sp.; (c) *Arcella* sp.; (d) *Centropyxis* sp.; (e) *Difflugia* sp.

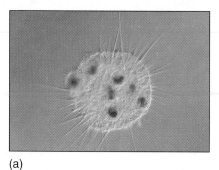

(a)

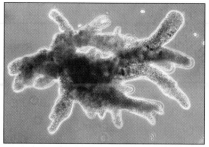

(b)

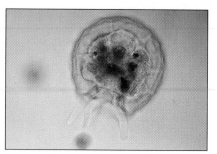

(c)

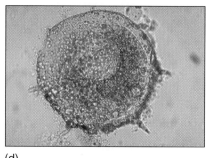

(d)

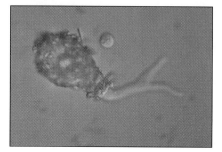

(e)

Amoeba sp.: This is a large "typical" amoeba that you probably saw in your high school biology class. There are numerous species in the genus and they live in fresh- or saltwater. Species in this group produce lobose cytoplasmic extensions called **pseudopodia** which are used in locomotion and to engulf prey. There usually is a single nucleus.

Arcella sp.: These freshwater amoebae use fine sand grains which they bind together to build a **test.** The amoeba hangs in the chamber inside the test. An aperature at the bottom allows it to extend pseudopodia for feeding and locomotion. Tests can be yellowish to brown. Some host symbiotic algae and are green. These amoebae form cysts in adverse conditions. *Arcella* is a member of the group called testate amoebae.

Centropyxis sp.: Members of this genus are also testate amoebae but often the test has spines. The test surface is rough, often encrusted with diatom valves. The pseudopods will often be branched. These amoebae live in water drops that accumulate in damp areas or in freshwater streams and lakes.

Difflugia sp.: Another testate amoeba, members of this group build a test containing both large and small particles and parts of diatoms, giving it a rough appearance. The organic material that binds the rings together is laid down in rings. They live in wet bottom sands and marshes.

Clade Mycetozoa

At one time this group was included in kingdom Fungi but they are now considered to be protists.

Phylum Myxogastridia

Plasmodial slime molds are common in moist leaf litter and decaying wood. There are about 500 known species. During the growth phase of its life cycle, it has a body form called a **plasmodium,** a nonwalled mass of cytoplasm containing hundreds of nuclei. Plasmodia may be bright orange or yellow in color (fig 15.17). The cytoplasm in a plasmodium streams back and forth and allows the organism to locomote, moving several meters until it "finds" food. They are able to pass through a mesh cloth. If you can imagine a large cytoplasm several inches long "crawling" about, you recognize where the common name slime mold originates. They are not closely related to the cellular slime molds found in another phylum in this grouping. If food or moisture are reduced, plasmodial slime molds differentiate to form **sporangia** which produce spores that are haploid. The spores are resistant to harsh conditions but will germinate in a moist environment with food. Cells emerging from spores may fuse to start a new plasmodium.

Your instructor will have *Physarum* growing in the lab on 1% water agar with a few flakes of oatmeal for food. Several hours ago a piece of the slime mold was transferred to a fresh petri dish and was set up under a demonstration

Figure 15.17 Plasmodium of slime molds are colorful and capable of moving in one direction by cytoplasmic streaming.

microscope so that you can see the cytoplasmic streaming. Watch for a few minutes and note how it changes direction. What purpose might this serve in a large organism that feeds by secreting digestive enzymes and absorbing the products of digestion?

Moves nutrients to various regions, especially growing/reproducing regions

16 If it is available, your instructor may provide you with a petri dish containing water agar and a few flakes of oatmeal. Use a clean spatula to transfer a small amount of large slime mold in the lab to your petri dish. This is an inoculum. Take it back to your room and keep it in a warm (not hot) place. For the next week, look at it each night and record your observations. Bring this "journal" to lab next time.

Learning Biology by Writing

Your instructor may ask you to answer the questions in lab summary 15 and turn in the answers at the next lab period.

Internet Sources

On the World Wide Web there are several databases that list information about diseases caused by protozoa. The U.S. Center for Disease Control (http://www.cdc.gov) is one and the World Health Organization (http://www.who.int/tdr) is another. The WHO lists ten major tropical diseases. Of these, four are caused by protists that are transmitted by biting insects. The diseases are Chagas disease, leishmania, malaria, and African sleeping sickness. Connect to one of these sites and read about how the diseases are transmitted, what the symptoms are, how they are treated, and where the diseases are common.

Lab Summary Questions

1. How do protists differ from bacteria and cyano-bacteria?
2. Use the summary table on the next page to make a list of the genera (plural of genus) observed in this lab. Indicate the clade into which each is classified.
3. Describe what is meant by isogamy and anisogamy. Which is considered more advanced? Why?
4. What is alternation of generations? Sporophyte? Gametophyte?
5. Describe how the volvocine series of algae (*Chlamydomonas, Pandorina,* and *Volvox*) can be interpreted as representing the evolutionary developmental stages leading from unicellular species through colonial ones to multicellularity.
6. Describe how the filamentous alga *Oedogonium* and the leafy alga *Ulva* may be interpreted as representing evolutionary developmental stages leading to green plants.
7. Why are protists considered to be important members of ecosystems?
8. Name several human diseases caused by protists.

Critical Thinking Questions

1. Describe the different adaptations that protists exhibit for withstanding a harsh environment. Compare these adaptations to those of bacteria and cyanobacteria.
2. Diatomaceous earth is classified as an "insecticide" by gardeners and is used to control such pests as ants, snails, slugs, and caterpillars. Explain its insecticidal properties.
3. Why aren't protist cells larger? Relate the shape of protist cells to their function and their environment.
4. Speculate on how colonial aggregates of cells may have led to multicellularity.
5. Because the classification of protists is a work in progress, compare the classification used here with that in your textbook. How are they similar? How do they differ?
6. Describe what captured your attention in this lab. Explain why.

Diversity Among Protists   **191**

Clade	Phylum if known	Representative Genus	Common Name	Distinguishing Characteristics	Ecological Significance
Diplomonadida					
Parabasala					
Euglenozoa					
Alveolata					
Stramenopila					
Rhodophyta					
Chlorophyta					
Amoeba-like					
Mycetozoa					

LAB TOPIC 16

Investigating Plant Phylogeny: Seedless Plants

Supplies

Preparator's guide available on WWW at
http://www.mhhe.com/dolphin

Equipment

Compound microscopes
Dissecting microscopes

Materials

Living plants
- *Chara* or *Nitella*
- *Marchantia* with gemmae
- Moss gametophytes with sporophytes
  (*Polytrichum*)
- Moss spores germinated on agar
- Mature ferns
- Where possible, examples of lower seedless
  plants
- Fern spores
- C-fern sexual differentiation kit from Carolina
  Biological or fern prothalli growing
  on agar

Prepared slides
- *Chara* or *Nitella* w.m. with reproductive
  organs
- *Marchantia* thallus, cross section
- *Marchantia* antheridia, longitudinal section
- *Marchantia* archegonia, longitudinal section
- *Marchantia* sporophyte, longitudinal section
- Moss sperm, demonstration slide
- Mature moss capsule, longitudinal section
- Mature moss antheridia and archegonia,
  longitudinal sections
- Fern antheridial and archegonial prothalli
- Fern sori, w.m.
- Fern sporophyte growing from archegonium

Slides, coverslips, razor blades
Pasteur pipettes

Student Prelab Preparation

Before doing this lab, you should read the introduction
and sections of the lab topic that have been scheduled
by the instructor.

You should use your textbook to review the
definitions of the following terms:

antheridium	protonema
archegonium	Pterophyta
Bryophyta	rhizoid
gametangium	Sphenophyta
gamete	sporangium
gametophyte	spore
heterogamy	sporophyte
lignin	thallus
Lycophyta	xylem
phloem	

You should be able to describe in your own words the
following concepts:

Alternation of generations
Vascular tissue (xylem and phloem)
Reproduction in seedless plants
Adaptations to land

As a result of this review, you most likely have
questions about terms, concepts, or how you will do
the experiments included in this lab. Write these
questions in the space below or in the margins of
pages of this lab topic. The lab experiments should
help you answer these questions, or you can ask your
instructor during the lab.

Objectives

1. To recognize that some species of green algae are
   closely related to plants
2. To observe the body plans of liverworts, mosses,
   and ferns
3. To learn the life cycle stages in the alternation of
   generations found in mosses, liverworts and ferns

TABLE 16.1   Taxonomic divisions of plants

Division	Common Name	Estimated Living Species
Nonvascular Plants		
Seedless plants		
Bryophyta	Mosses	10,000
Hepatophyta	Liverworts	6,500
Anthocerophyta	Hornworts	100
Vascular plants (Tracheophyta)		
Lycophyta	Club mosses	1,000
Pterophyta	Ferns and Horsetails	12,000
Seed plants		
Gymnosperms		
Coniferophyta	Conifers	550
Cycadophyta	Cycads	100
Ginkgophyta	Ginkgo	1
Gnetophyta	Gnetae	70
Angiosperms		
Anthophyta	Flowering plants	250,000

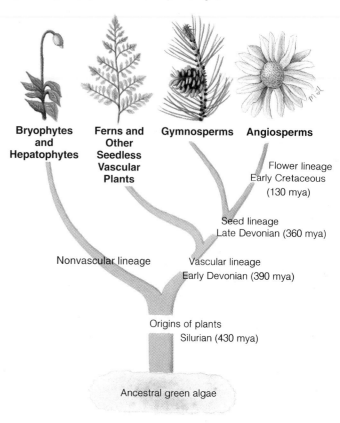

Figure 16.1   A summary of plant evolution. Note the early origin of both Bryophytes and Tracheophytes from the green algae (mya = millions of years ago).

4. To collect evidence that tests the following hypothesis: Seedless plants can be arranged in a phylogenetic sequence demonstrating increasing adaptation to a terrestrial environment.

Background

Plants are a diverse group of multicellular, photosynthetic organisms ranging from simple mosses to complex flowering plants. All plants, and some green algae, share certain characteristics, which include: (1) chloroplasts with thylakoid membranes stacked as grana and containing chlorophylls a and b, (2) starch as a storage polysaccharide in the chloroplasts, (3) cellulose in cell walls, (4) cytoplasmic division by cell plate (phragmoplast) formation, and (5) complex life cycles involving an alternation of generations in which a diploid sporophyte stage alternates with a haploid gametophyte stage; (6) heterogamy with gametes produced in organs called gametangia.

Botanists use the term "division" rather than phylum for the major taxonomic categories within the plant kingdom and, in turn, divide the divisions into classes, orders, families, genera, and species. Some botanical taxonomists divide the kingdom Plantae into two divisions, the Bryophyta (mosses and relatives) and the Tracheophyta (vascular plants), while others recognize ten divisions (table 16.1).

Phylogeny is the study of evolutionary relationships among groups of organisms. The evolutionary story of plants is linked to the colonization of the terrestrial environment, an event that started about 400 million years ago (fig. 16.1). Most botanists agree that plants arose from an ancestral green algae. For the first 3.5 billion years, no life existed on the land, but at the end of the Silurian period, the first land plants appeared and rapidly diversified. The appearance of producer organisms other than cyanobacteria in the terrestrial environment established a strong basis for terrestrial food chains. Plants in the terrestrial environment were then rapidly exploited by the evolution of terrestrial fungi and animals from aquatic ancestors.

As do many algae, all plants show an **alternation of generations** in their life cycles. In plants, there are two alternating multicellular stages. The diploid **sporophyte** generation produces haploid spores by meiosis. These, in turn, produce plants of the **gametophyte** generation, which produce haploid gametes by mitosis. When gametes fuse at fertilization (syngamy), the fertilized egg (zygote) is the first cell of a new sporophyte generation (fig. 16.2). Remember gametophytes are multicellular, haploid and produce gametes by mitosis. Sporophytes are multicellular, diploid and produce spores by meiosis.

Figure 16.2   Generalized idea of alternation of generations.

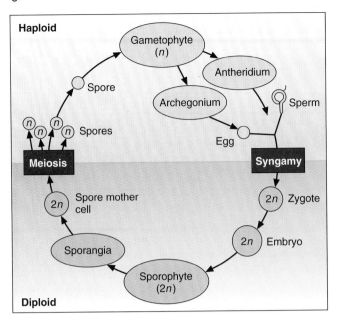

Haploid
Gametophyte (n)
Antheridium
Spore
Archegonium
Sperm
(n)(n)(n)(n) Spores
Egg
Meiosis
Syngamy
(2n) Spore mother cell
(2n) Zygote
Sporangia
(2n) Embryo
Sporophyte (2n)
Diploid

Plant Adaptations to Land

The critical steps for plants in making the transition from aquatic to terrestrial environments involved the development of adaptations to live in a dry environment. In land plants the reproductive organs are covered by a protective covering, the **gametangium,** a jacket of sterile (nonreproductive) cells that prevents the drying of gametes and embryos. Furthermore, land plants have modified basal organs (roots) that absorb water and minerals from the substrate to replace what is lost by evaporation and waxy surface **cuticles** that prevent water loss. Hereditary variations of this type arising in ancient shoreline algae would have allowed them to enter an environment that, though harsh in its physical characteristics, was free from competitors for resources. Such an environment would have allowed the colonizers to establish themselves and to further diversify.

Once the transition to land occurred, two major lines of evolution developed among the land plants. The direction the nonvascular plants (table 16.1) took saw the gametophyte generation become the predominant photosynthetic stage in the life cycle. Following fertilization, the gametophyte retained the embryo and it developed into a multicellular sporophyte that was dependent on the gametophyte. The sporophyte produced **spores** that were resistant to desiccation and that became the primary means of dispersal.

The other evolutionary alternative, the tracheophyte line, gave rise to vascular plants (table 16.1). In these plants, cells located in the central portion of the shoots became elongated and specialized for transport and support, the progenitors of vascular tissues. A vascular cambium developed, which allowed secondary growth, increasing the girth of the stem and root. These tissues allowed tracheo-

phytes to grow to large sizes, which, in turn, favored development of more efficient anchorage and absorptive systems. At the same time, reproductive adaptations occurred, giving rise to the seed plants, which you will study in Lab Topic 17.

In seed plants, such as conifers and flowering plants, the gametophyte stage of the life cycle became dependent on the sporophyte stage whose spores developed into a microscopic gametophyte encased in reproductive organs borne on the sporophyte. Male gametophytes became desiccation-resistant airborne **pollen** which carried sperm that no longer needed water to swim to the eggs. Eggs were retained in a protective organ where, following fertilization, the zygote and surrounding tissues developed into seeds containing an embryo and food reserves. In this group, **seeds,** rather than spores, became the effective dispersal mechanisms. Seeds allowed the plant embryos to be carried overland by wind, water, or animals, allowed a period of dormancy, and provided initial nourishment for germination and growth in the new location.

The efficiency of the adaptations of plants to land over eons of time allowed them to invade many different environmental niches. Relatively simple changes allowed the colonization of swamps and bogs, while more complex adaptations were necessary for life in desert environments. As you study the trends outlined in this lab topic, remember that contemporary plants in different divisions have not descended from one another. They have similarities because they share common ancestors. Present species are related as cousins or more distant relatives, not as you are to your parents or siblings.

Algal Preadaptations to Land

We will start our investigation of plant diversity by looking at a group of aquatic green algae (Clade Chlorophyta) commonly called stoneworts. The "stone" part of the name comes from the fact that they secrete $CaCO_3$ on their surfaces and "wort" means herb or small plant. This small group of about 250 species is distributed worldwide in freshwater streams and lakes. What makes them interesting is the fact that they have certain structures considered necessary for life on land as well as many other shared characteristics with plants. Were organisms like these the progenitors of land plants? When you are finished looking at them, I think you will be asking why these are not considered plants. Why do I say this?

Stoneworts (in the taxonomic order Charales) have several plantlike characteristics. These include:

- Multicellular, consisting of many cells which are specialized into types that perform only certain functions

- Gametangia, organs formed from sterile nonreproductive cells that surround and protect the gametes as they develop

- Heterogamy, the production of two different types of gametes, one motile and other immotile, so that the egg is retained in the female gametangium where fertilization occurs

Investigating Plant Phylogeny: Seedless Plants

Figure 16.3 *Chara sp.* is a green alga (Chlorophyta) with remarkable plantlike characteristics. (*a*) whole plant; (*b*) enlarged view showing gametangia.

(a)

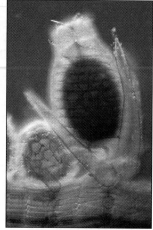

(b)

- Embryophyte condition, the retention of the zygote for a period of time after fertilization
- Sporopollenin, a waterproofing compound found in the walls of plant spores and pollen
- Parenchymal-like cells which retain the ability to produce other types of cells during development
- Plasmodesmata, continuity of cytoplasm by small strands of cytoplasm connecting adjacent cells
- Cellulose cell walls
- Chlorophylls a and b in chloroplasts similar in structure to those of plants
- Similar mechanisms in mitosis and cytokinesis involving microtubules
- DNA sequencing has shown many homologies between these organism and plants
- Terminal growth from an apical meristem-like cell

While this impressive list of shared characteristics argues persuasively for stoneworts being closely related to plants, one essential characteristic is missing. Stoneworts do not show an alternation of generations with distinct multicellular gametophyte and sporophyte stages.

LAB INSTRUCTIONS

You will study first a group of green algae that are very plantlike. Then you will look at the terrestrial adaptations of liverworts, mosses, and ferns. You will learn the distinctions between the sporophyte and gametophyte stages in the alternation of generations that characterizes plants.

Anatomy and Life Cycle of a Stonewort

Our representative of this group will be the stonewort *Chara* sp. The growth form of this is shown in Figure 16.3(*a*) and possibly in whole specimens available as demonstrations in the lab. Note the "stem" with large internodes and nodes that produce whorls of what look like "leaves." These are neither stems nor leaves as your microscopic observations will prove. Some species can grow to be 30 cm tall. The base of the central stalk has branched rhizoids, long thick cells that anchor it in soft mud or sand. If the specimen is in reproductive condition, some of the nodes will have **gametangia,** organs where the gametes develop.

1 ▷ Obtain a slide from the supply area of a whole mount of *Chara* sp. with reproductive organs and look at it with your scanning objective. Switch to 10× to observe structures in more detail. Note that the long internodal areas consists of a single, large **internodal cell.** Cytoplasmic streaming in these cells and plasmodesmata connections between cells allow for long-distance transport. In some cases, the internodal will also be surrounded by smaller pericentral cells that do not extend from node to node. As internodal cells mature, they elongate as the large central vacuole enlarges and repetitive nuclear divisions occur making the cell multinucleate. In some species, the single cell may divide lengthwise to produce several parallel internodal cells.

At the nodes are several small **nodal cells.** These resemble parenchyma tissue in higher plants because they (1) are connected to each other by small cytoplasmic bridges called plasmodesmata; and (2) they retain the ability to divide and form other cell types. Division of nodal cells gives rise to the whorls of **laterals** at each node. Note that they have the same cellular construction as the main axis of the plant and are not leaves. This stage could be considered a gametophyte stage in a life cycle because it is haploid and also bears specialized reproductive structures called **gametangia.**

Look at several nodes to locate gametangia. Male gametangia, called **antheridia,** are round and orange in color at maturity. Inside a jacket of nonreproductive (sterile) cells, other cells will divide by mitosis to produce biflagellated sperm cells. If the sperm cells are produced by mitosis, what must be the ploidy level of the sperm? _haploid_

Gametangia that are somewhat oval, not round, are female gametangia called **oogonia** (sing. oogonium). Each contains a single egg produced by mitosis. What is its ploidy level? _haploid_ The sterile cells in the jacket around the gametangium have a helical shape and give a spiral surface pattern to the organ.

Sperm released from the antheridia swim to the oogonium. The crown of cells at the tip of the oogonium have openings between them allowing the sperm to enter and fertilize the egg. The resulting zygote is diploid. It is retained in the oogonium and during maturation synthesizes a thick wall composed of sporopollenin around itself. Internal fertilization of an immobile egg followed by zygote retention is a characteristic of plants called embryophytes. All higher plants are embryophytes. The oogonium then releases the mature zygote. Under the right conditions, it will germinate. First it divides by meiosis to produce four haploid cells. Three degenerate, leaving a fourth that divides by mitosis to produce a filament of cells called a protonema. It develops into a new individual. What is its ploidy level? *haploid*

Earlier it was stated that stoneworts do not have an alternation of generations lifecycle. What is missing is a multicellular diploid sporophyte stage. The only diploid stage of the lifecycle is the single-celled zygote. It is, in part, this omission in the life cycle that keeps us from calling stoneworts plants. The diploid phase ends with the meiotic division of the zygote. The multicellular haploid stage that follows, however, is a fine candidate for a gametophyte. In the next two sections, you will study liverworts, mosses, and ferns that clearly show not only haploid multicellular gametophyte stages but also diploid multicellular sporophyte stages.

Bryophytes

The term "bryophyte" has been used traditionally to describe plants belonging to three divisions: Hepatophyta—the liverworts, Anthocerophyta—the hornworts, and Bryophyta—the mosses. The bryophytes illustrate a clear alternation of generations in which a multicellular gametophyte is the predominant stage of the life cycle. The multicellular sporophyte stage of the life cycle is reduced in size and function, becoming nutritionally dependent on the gametophyte for its survival. Reproduction occurs sexually by gametes and asexually by fragmentation or dispersal by spore formation. Although the bryophytes are land plants, they are not completely adapted to a terrestrial way of life. Motile sperm require water films to travel from the male gametangium to the eggs held in the female gametangium in order for fertilization to occur. Most bryophytes are small, less than 10 cm tall, due in part to their lack of vascular tissues.

one slide of each

Division Hepatophyta (Liverworts)

The name of this division is derived from the fanciful resemblance of the shape of some of these plants to the human liver and from the root word *wort* meaning herb. Two body types are found among the species of liverworts: the thallose type and leafy type.

Gametophyte Stage

2 ▶ Examine living specimens of *Marchantia*, a thallose-type liverwort. It is commonly found on the surfaces of damp rocks and soil. Individual plants consist of a "leafy" body called a **thallus** (fig. 16.4). You are looking at the haploid gametophyte stage of the two stages found in the life cycle. The surface of the thallus has a diamond-shaped pattern, which corresponds to air chambers located beneath the upper epidermis. Note the numerous hair-like **rhizoids** on

Figure 16.4   Thallus of the liverwort *Marchantia* in background with stalked sexual reproductive structures.

the lower surface. These are an adaptation for anchorage and their surface area allows water and mineral absorption although they lack the specialized conducting cells found in true roots. Projecting above the upper surface may be tiny **gemmae cups.** These produce asexual **gemmae,** small, multicellular discs of green tissue, which develop into new gametophytes when dispersed to suitable locations. If the cells of gemmae develop from thallose tissue by mitosis, what is their ploidy level?_____

3 ▶ Obtain a slide of a cross section of a thallus and look at it with your compound microscope (fig. 16.5). Sketch the cross section in the circle that follows. Indicate the upper and lower **epidermis.**

Is there evidence of a **cuticle?** _____ Note the **air pores** on the upper surface, which open into **air chambers** and allow carbon dioxide and oxygen gas exchange. Beneath the chambers is a layer of **chlorenchyma** tissue where the cells contain chloroplasts. Beneath this layer is the lower **storage tissue.** Are chloroplasts present in these cells? _____ Based on your observations, which of these layers is photosynthetic? _____ Projecting from the lower epidermis note the **rhizoids,** which extend into the

Figure 16.5 Stages in the life cycle of the liverwort *Marchantia*.

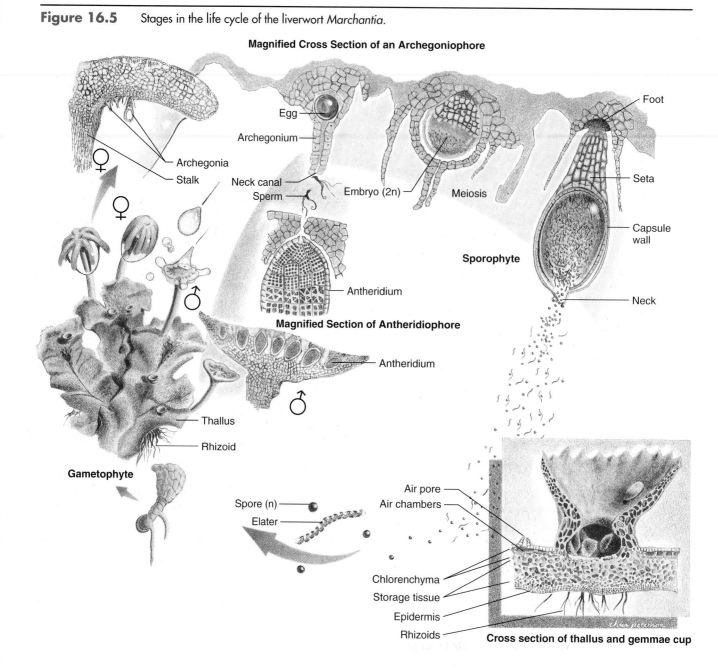

Magnified Cross Section of an Archegoniophore

Egg
Archegonium
Neck canal
Sperm
Embryo (2n)
Meiosis
Foot
Seta
Capsule wall
Neck
Archegonia
Stalk
Antheridium
Sporophyte

Magnified Section of Antheridiophore

Antheridium

Thallus
Rhizoid

Gametophyte

Spore (n)
Elater

Air pore
Air chambers
Chlorenchyma
Storage tissue
Epidermis
Rhizoids

Cross section of thallus and gemmae cup

soil, absorbing moisture and minerals. A rhizoid is composed of how many cells? _____ Your slide may show ventral scales, which enclose longitudinal bundles of rhizoids that aid in the capillary transport of water. How do you think this happens?

Note the significant similarity here with the green alga *Chara*. Gamete formation is in a sex organ (gametangium) that is surrounded by a protective layer of nonreproductive tissue. What could be the advantage of this structural development to a land plant?

Sexual Reproduction in Gametophyte
Sexual reproduction in *Marchantia* involves separate male plants that bear gametangia (sex organs) called **antheridia** (singular-antheridium) and female plants that bear gametangia called **archegonia** (singular-archegonium).

In the proper season, a new body part appears, Small, stalked **antheridiophores** or **archegoniophores** grow from the notches in the tips of thalli (fig. 16.5). The gametangia develop on the caps of these stalks. Antheridiophores are male and will bear antheridia. They look somewhat like an inverted large, flat cone. Archegoniophores are female and will bear archegonia. They look like fleshy minature palm trees.

4▷ Look at a slide of a longitudinal section of an antheridiophore, the male reproductive structure. Note the general shape and the multicellular nature of the organ. Beneath the upper surface are several ovoid **antheridia.** Sperm develop by mitosis inside the sterile jacket and are released through the canal that leads to the upper surface (fig. 16.5). Several hundred sperm are produced by mitosis in each antheridium.

5▷ Now look at a slide of a longitudinal section of an archegoniophore, the female reproductive structure. In rows on the underside are flask-shaped **archegonia** (fig. 16.5). At the base of each archegonium is a swollen area containing the **egg.** Eggs are produced by mitosis, as are sperm. Are eggs and sperm haploid or diploid? _____ Why?

Extending downward from the egg in the archegonium is a **neck canal.** Sperm must swim up this canal to fertilize the egg. Fertilization requires water. Raindrops falling on the antheridia splash sperm onto adjacent archegonia. Using their flagella, the sperm swim to the egg and fertilize it. The zygote (2N) develops into the sporophyte stage, embedded in the parent's tissue on the underside of the archegoniophore's cap (see yellow area of fig. 16.5).

Sporophyte Stage

Obtain a slide with maturing sporophytes on the under sur-
6▷ face of an archegoniophore. Following fertilization, the stalks of the archegonia elongate. The zygote divides repeatedly, forming a multicellular sporophyte stage within the tissue of the gametophyte. Find the sporophyte stage on the slide and study it using low power. Most of the cells in the sporophyte form a **capsule,** which surrounds other cells that become **spore mother cells** and **elaters,** the latter being slender, elongate cells (fig. 16.5). Together these cells make up the **sporangium,** the structure in which spores develop. Thus, the sporophyte generation of the thalloid liverwort is a small sporangium that developed from the zygote attached to the lower surface of the archegoniophore growing from the gametophyte thallus.

Inside the sporangium, each spore mother cell undergoes meiosis, forming four haploid cells that mature into spores with thick cell walls. They have sporopollenin in their walls, which prevents drying. When the sporangium matures, it ruptures, exposing a cottony mass of spores and elaters. The elaters, intertwined in this mass, are hygro-

scopic. Spiral thickenings in their cell walls cause the elaters to twist and coil as they dry or take up moisture, dislodging spores from the mass. The spores are disseminated by the wind and, if they land in a moist area, will germinate and divide by mitosis to form a new gametophyte generation. What will be the ploidy level of these gametophyte's cells?

When you finish this section, quiz your lab partner about the life cycle of *Marchantia* with emphasis on the alternation of generations.

Division Bryophyta (Mosses)

Many organisms commonly called "mosses" are not even plants. Reindeer moss is a lichen, Spanish moss is a vascular plant, and Irish moss is an alga. True mosses are members of the division Bryophyta and have the anatomical and reproductive features which you will study in this section.

Observe Protonema on Moodle

Gametophyte Stage

About a week before the lab period, haploid moss spores were placed on agar medium in petri plates. By the time of the lab, the spores should have germinated and divided by mitosis, producing **protonema,** algalike filaments of cells.

7▷ Remove a protonema from the culture, make a wet-mount slide of it, and observe through your compound microscope. Alternatively, you may use a prepared slide of a protonema and a young gametophyte. Each protonema is a branching thread of single cells joined end to end. Does this remind you of the general appearance of a filamentous alga? _yes_ The colorless branches will develop into **rhizoids** (fig. 16.6), which are analogous to the roots of higher plants. Leafy, green **gametophytes** will develop from **buds** on the protonema. Protonema can spread to cover an area 40 cm in diameter in a matter of months. Sketch what you see on your slide. What is the ploidy level of the cells in a protonema? _haploid_

extensive tangled tube-like filaments

Figure 16.6   Stages in life cycle of a moss.

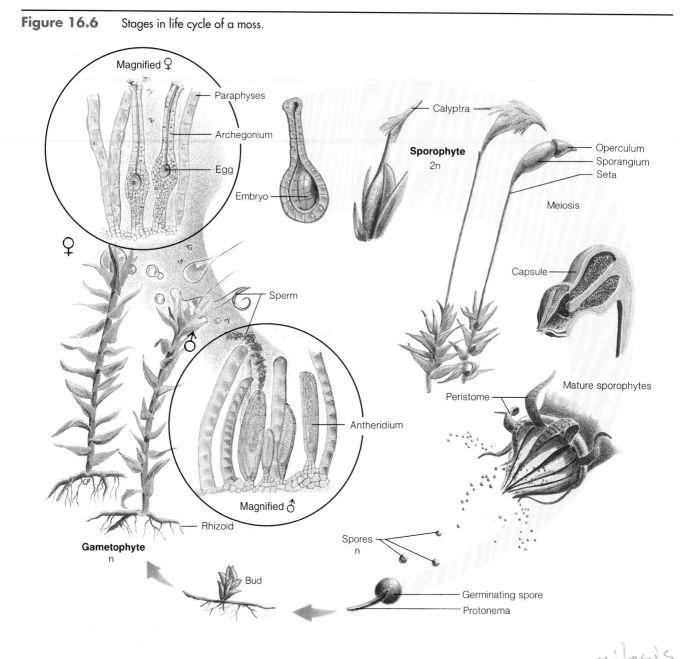

As the gametophyte matures, gametangia differentiate at the tip. **Archegonia,** the egg-producing organs, will develop on one gametophyte and **antheridia,** the sperm-producing organs, on another. The gametophytes of many moss species can bear both.   *2 slides*

Sexual Reproduction in Gametophyte

8 ▶ Examine a prepared slide of a longitudinal section of a female gametophyte. Look at it first with your lowest power objective, increasing magnification where necessary to see detail. At the tip look for the bowling-pin-shaped structures (fig.16.6), called **archegonia,** between thin filamentous paraphyses. Note that the archegonium is surrounded by a sterile jacket of cells. They protect the egg during development. The single egg is located in the swollen base. The cells of the gametophyte are haploid. Does the egg develop by mito-

sis or meiosis from the gametophyte cells? *mitosis* In the narrow upper region of the archegonium you may be able to see the **neck canal.** Sperm must swim down this canal to fertilize the egg.   *2 slides*

9 ▶ Now obtain a prepared slide with a longitudinally sectioned tip of a male gametophyte. At the tip, find the oval-shaped **antheridia** between the thin filamentous paraphyses (fig.16.6). The antheridia are surrounded by a thin layer of sterile jacket cells that protect the sperm during development. The tightly packed angular cells inside the jacket are sperm in various stags of development. Are sperm formed by mitosis or meiosis? *mitosis* If a water film develops on the tip of sexually mature male gametophyte, the sterile jacket of cells splits open releasing sperm into the water film. After a few minutes they begin to actively swim, using two flagella.

10 Moss sperms must pass from the antheridium through a film of water to the eggs in the archegonium. Often raindrops will splash sperm from the antheridium on one gametophyte to the archegonium on another. A prepared slide of moss sperm should be in the lab as a demonstration. Be sure to look at it.

Sporophyte Stage

Zygotes develop into embryos surrounded by a protective jacket of sterile archegonial cells. The developing diploid embryo marks the start of the next generation, the sporophyte. The growth of the zygote into a multicellular diploid sporophyte stage by mitosis differs from the situation seen in the green alga *Chara.* It represents an evolutionary progression from the ancestral condition where the diploid zygote alone represents the sporophyte.

from moss specimen

As the embryo grows by mitotic division, differentiation occurs, and a long stalk of sporophyte tissue, the **seta,** grows upward (fig.16.6). The end of the seta differentiates into a **capsule** covered by the **calyptra,** derived from clinging archegonium tissue. The top of the capsule is closed by a lid, or **operculum.** Inside the capsule, meiosis occurs, producing haploid spores. These are the same types of cells that grew and produced the protonema described at the beginning of this section. When the capsule matures, the operculum drops off. Some species have a **peristome,** a ring of toothlike units that flex with changes in humidity releasing the capsule's spores during periods of dryness.

11 Spores remain viable for several years and are the primary agents of dispersal for the species. If mature capsules are available, observe them under a dissecting microscope, removing the calyptra and operculum to release the spores. Also observe a longitudinal section of a mature capsule on a prepared microscope slide. *4 slides*

In what ways are mosses adapted to land? In what ways are they still dependent on water?

Adaptations:
jacketed gametangia
waxy cuticle
stomata
rhizoids to draw water from soil

Dependent on water:
— no vascular systems, so depend on diffusion
— sperm must swim

What similarities were there in the life cycles of mosses to the life cycle of *Chara?* What differences were there?

— Chara lack sporophyte, so zygote undergoes meiosis
— Both have multicellular gametophyte that makes gametes for sexual reproduction
— Both have sperm that swim

Read Tracheophytes

As a group, the tracheophytes (table 16.1) exhibit a number of additional adaptations to living in a terrestrial environment not seen in the bryophytes. There is a general trend across the group whereby the plant body is differentiated into a subsoil root system for the absorption of water and minerals, and an aboveground stem that can be quite tall in comparison to bryophytes. The leaves are often greatly enlarged and adapted to high rates of photosynthesis. This regional specialization of the plant body required the development of new types of plant tissues, the **vascular tissues.** They consist of **xylem** and **phloem** that form a transport system. The cylindrical cell walls of the xylem function in carrying absorbed water and minerals to the aboveground parts of the plant, while the elongated phloem cells conduct the products of photosynthesis from the leaves to all other parts of the plant. The cell walls of the xylem are reinforced with lignin, so that the xylem assumes the additional function of being a support tissue. Consequently, most tracheophytes are larger than the nonvascular plants.

The tracheophytes include both seedless and seed-bearing plants. The seedless tracheophytes include the club mosses, horsetails, and ferns (see fig. 16.7). Your instructor may have brought living specimens from the greenhouse into the lab. If so, compare them to the photos and note the following features.

Psilotum, a whisk fern, is unique among the tracheophytes because it lacks roots and leaves. Root function is performed by a stem that grows underground and endomycorrhizal fungi that absorb water and minerals. Leaf function is carried out by the photosynthetic stem.

Club mosses, such as *Lycopodium,* are low-growing herbs. They have a belowground stem that gives off roots and aerial branches. Horsetails, such as *Equisetum,* have underground stems called rhizomes that give rise to aerial branches that have jointed stems bearing whorls of leaves. Various silicon compounds are found in the stem and a common name in colonial times was "scouring rushes" because

Figure 16.7 Sporophyte stages of the fern allies: (*a*) *Psilotum*, a whisk fern, has an aboveground and belowground stem that functions in water and mineral absorption; (*b*) *Lycopodium*, a club moss sometimes called a ground pine or cedar, has a belowground stem that produces roots and an aboveground stem that bears leaves and sporangia; (*c*) *Equisetum*, a horsetail, has jointed aboveground stems that bear whorls of leaves and a belowground true vascular root. Some stems bear a terminal reproductive structure called a strobilus (foreground) which produces spores that give rise to the gametophyte generation.

(a) (b) (c)

the abrasive nature of the stems made them useful in cleaning pots.

Division Pterophyta (Ferns and Horsetails)

In this section, you will study the life cycle stages of ferns as the example of seedless, vascular plants. The sporophyte generation is the most conspicuous stage in the life cycle of ferns, but they also have an independent gametophyte stage. In Lab Topic 17 you will study the vascular plants that produce seeds.

Gametophyte Stage

12▶ Make a wet-mount slide of some fern spores and look at them with a compound microscope. The haploid spores are produced by meiosis and by differentiation of cells in **sori** (singular *sorus*), organs on the underside of the sporophyte leaf.

13▶ If they are available, examine spores that germinated on an agar surface. Do these protonema remind you of filamentous algae? The protonema will differentiate into a very small gametophyte stage called a **prothallus** (fig. 16.8c) that has a heart-shaped leaf and rootlike **rhizoid** areas. This stage is haploid and will grow by mitotic cell division with energy input from photosynthesis.

Sexual Reproduction in Gametophyte

Under appropriate conditions, certain cells in the prothallus stage will differentiate into gametangia, producing eggs and multiflagellated sperm. While many textbooks and figure 16.9 show that a single prothallus will produce both

egg and sperm, this may not be the case in nature. In many species, the earliest developing prothalli (plural) produce eggs and the slower developing ones produce sperm. A hormone produced by the faster growing prothalli may influence sex organ development in the slower growing ones. A mating system such as this promotes cross fertilization. What is the evolutionary advantage of cross fertilization?

more genetic variability

14▶ Obtain a slide showing the gametangia of the fern gametophyte (prothallus). Sperm-producing gametangia, called the **antheridia,** are located on the lower surface in the center of the heart-shaped, leaflike thallus near the rhizoids. **Archegonia** that produce eggs are located near the cleft or notch of the thallus. A water film must cover the lower surface of the prothallus for sexual reproduction to occur. The sperm swim through this film to the archegonium, where one sperm swims down a canal to fertilize the egg (fig. 16.9).

Figure 16.8   Ferns: (*a*) fern fronds represent the sporophyte stage of life cycle; (*b*) sori on undersurface of the frond contain sporangia, which produce spores; (*c*) the gametophyte (prothallus) stage of fern life cycle develops from a spore.

(a)

(b)

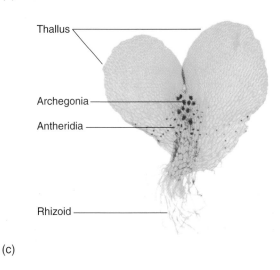

Thallus

Archegonia

Antheridia

Rhizoid

(c)

This zygote is the start of the sporophyte generation in the alternation of generations. As the diploid sporophyte grows by mitotic division and photosynthesis, the gametophytic prothallus withers and dies. What is the ploidy level of the sporophyte's cells? _diploid_

Gametophyte Stages (Optional)

Stages in the development of the gametophyte may be available in the lab. About two weeks ago and then again last week, your instructor inoculated some spores of the fern *Ceratopteris* sp. (C-fern) on a growth medium. As a result you should have gametophytes in various stages of development, including some that are sexually mature.

15 Gametophyte stages should be removed from the culture and placed lower surface upward in a small amount of water in a depression slide and observed through your microscope. Sketch a series of developmental stages below.

If the oldest gametophytes are sufficiently mature, it may be possible to observe motile sperm and fertilization. As you study the oldest gametophytes, identify the **gametangia** (sex organs), which project from the surface of the leaflike prothallus. When a gametophyte with gametangia is found, study it.

Antheridia may be anywhere on the surface of the gametophyte, and often are most frequent on small, irregular-shaped plants. If it appears to be constructed of three cells—a basal cell, a ring cell, and a cap cell—surrounding a mass of small cells, it is an antheridium (fig. 16.9). You may see motile sperm escaping.

Archegonia are composed of many cells, some embedded in the thallus and others projecting from the thallus as a neck. The neck is composed of four rows of cells with each row consisting of four cells (fig.16.9).

Fertilization

16 To observe fertilization, you must look at the gametangia immediately after placing the gametophytes in water. Select several mature gametophytes of different sizes and place them on a slide. Add a drop of water and immediately observe through your compound microscope.

Figure 16.9   Stages in the life cycle of a fern.

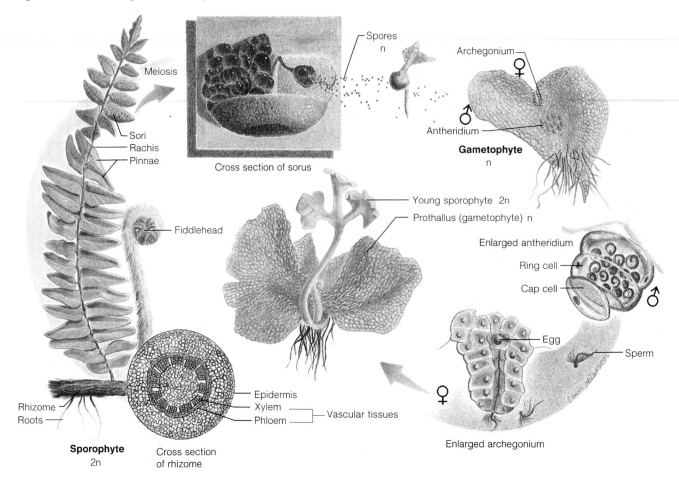

The water causes the cap cells to pop off the antheridia, releasing sperm. Water also causes the apical neck cells to break open, creating a channel to the egg at the base of the canal. Sperm swim to the neck and down the channel and fertilization occurs. You will have to scan several gametophytes to see all of these events. Everyone may not be fortunate enough to see sperm or fertilization.

Sporophyte Stage   3 slides

17 ▶ Obtain a whole-mount slide of a young sporophyte growing from the archegonium of a prothallus and observe it under low or medium power with your compound microscope and compare to figure 16.9.

The sporophyte stage will eventually develop into well-defined leaves, stems, and roots with an interconnecting vascular system. The stems of most ferns are horizontal just beneath the soil surface and are called **rhizomes.** However, these structures are quite different from the rhizoids found in the gametophyte fern stage or in the mosses. Rhizoids are single, elongated cells joined end to end, which do not connect with vascular tissue. Rhizomes, on the other hand, are multicellular and contain vascular tissue.

The rhizome extends under the surface of the soil and sends up leaves from its buds. These leaves break from the

soil in a coiled position called a **fiddlehead.** It gradually unrolls to produce the fern **frond,** a single compound leaf with **pinnae** projecting from a central **rachis.** Roots with root hairs also develop from the rhizome and provide the sporophyte stage with water, minerals, and anchorage. Since all of the cells making up these structures arise from the zygote by mitosis, they are diploid.

On the undersides of the pinnae, small **sori** develop (fig. 16.9). They are composed of clusters of **sporangia.** Cells in the sporangia divide by meiosis to produce haploid cells that differentiate into spores. These spores are resistant to desiccation and aid in the dispersal of ferns. They are often sold as fern "seed." How do spores differ from seeds?

Spores are single cells. Seeds are multicellular and contain an embryo, nutrients, and seed coat.

18 ▶ Obtain a prepared slide of sporangia from the supply area. Look at it and compare it to figure 16.10. Can you see the annulus? yes Are some of them open, releasing spores? yes — on one slide no on two other slides

As the sporangia mature, they dry and split on the side opposite the **annulus** (fig. 16.10). The annulus with its thickened cell walls snaps back explosively releasing the spores. If ferns with mature sporangia are available, try this experiment. Remove a pinna (leaflet) from a leaf and place it on a piece of white paper. Shine a bright, hot light on it while observing through a dissecting microscope. You may see explosive movements as the sporangia open.

The life cycle of a fern is summarized in figure 16.9. Recall the stages you have observed as you study this illustration. The adaptations of the fern to the terrestrial environment include the desiccation-resistant, dispersible spores; the well-developed vascular system; the root system; and the erect, desiccation-resistant photosynthetic fronds. Most of these adaptations are in the sporophyte stage, which, in contrast to the mosses, is the dominant stage in the life cycle.

When you finish this section, pair off with someone in the lab and quiz one another about how *Chara,* liverworts, mosses, and ferns are similar and how they differ.

Figure 16.10   Spores are forcibly released by snapping back of the reinforced annulus on one side of the sporangium as humidity changes.

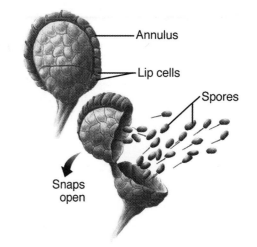

Learning Biology by Writing

Write a one or two page essay that describes alternation of generations life cycle found in plants. Describe what a sporophyte is and how it produces spores. Describe why spores are genetically dissimilar and their role in dispersal. Mention how spores increase genetic outcrossing. Describe how a gametophyte develops and what are gametangia. Discuss the genetic consequences of producing gametes by mitosis. Finally, discuss how a sporophyte develops and how genetic diversity develops among sibling sporophytes. Include examples from the plants studied in this lab.

Internet Sources

Read more about an exciting new experimental system for investigating the biology of the life cycle of ferns at *http://www.bio.utk.edu/cfern/index.html* You might want to use a search engine to look for other research work that is being done using ferns as experimental organisms. What are the advantages of ferns as experimental organisms in the study of developmental genetics?

Lab Summary Questions

1. How is reproduction in *Chara* similar to that in a moss? How is it different?
2. How is reproduction (life cycle) in a moss similar to that in a fern? How are they different?

3. How do spores differ from seeds?
4. Write a short letter to a child describing how to tell the difference between a moss, liverwort, and fern. You could include some sketches.
5. Devise a diagram that explains alternation of generations in plants.
6. What adaptations do mosses and ferns have to a terrestrial way of life?
7. List five reasons why the plants studied in this lab would not be found growing in hot, dry environments.
8. As you studied the plants in today's lab, what captured your imagination? Why?
9. What event starts the sporophyte stage of the life cycle? When and where does it occur in liverworts? Mosses? Ferns?
10. Fill in table 16.2 on the next page.

Critical Thinking Questions

1. Few, if any, mosses and ferns are found growing in dry biomes. Based on your studies, why do you think this happens?
2. Were plants or animals first to invade terrestrial environments? Why do you say so?
3. Two roommates with little botanical knowledge thought they could make a quick profit by growing ferns for the commercial florist market. They thought it would be very easy. All they had to do was collect the thousands of spores produced by a fern on their desktop and plant these like tiny seeds in individual pots. If they placed the pots in the sun, fertilized and watered the spores, and waited about a

year, they would have beautiful plants that they could sell for several dollars apiece. Why are they doomed to failure?

4. Would you expect to find separate male and female sporophyte-stage mosses growing in the wild? Why, or why not?

5. A person interested in studying crossing-over during meiosis in ferns prepared microscope slides of an archegonium and antheridium at different stages of development. Why will this study fail?

6. A species of fern has a diploid number of chromosomes equal to 18. How many chromosomes would you expect to find in a spore? In a zygote? In a cell from a leaf? What would you expect if the species was a moss?

7. Explain how vascular tissues are an advantage in terrestrial plants.

TABLE 16.2   Summary: Seedless Plant Observations

	Chara sp.	Heptophyta	Bryophyta	Pterophyta
Common name				
Habitat				
Produces spores (Y/N)				
Distinct gametophyte (Y/N)				
Gametophyte develops by				
Gametophyte ploidy				
Gametes produced by Mit/Mei				
Gametes same or different				
Name where egg produced				
Egg ploidy				
Name where sperm produced				
Sperm ploidy				
Where zygote formed				
Zygote ploidy				
Sporophyte develops by				
Sporophyte ploidy				
Name where spores produced				
Spores produced by Mit/Mei				
Spore ploidy				
Spores same or different				

LAB TOPIC 17

Investigating Plant Phylogeny: Seed Plants

Supplies

Preparator's guide available on WWW at
    http://www.mhhe.com/dolphin

Equipment

Compound microscopes
Dissecting microscopes

Materials

Living plants
    Gladiolus flowers
    Dicots in flower
    Where possible, examples of gymnosperms
        other than pines
    Various grocery store fruits and vegetables—to
        include apples, peppers, beans, and peas;
        specimens cut in longitudinal and cross sections
Plant specimens
    Needles of various conifers
    Staminate and ovulate cones, various species
    Hardwood trees herbarium set
    Gymnosperm herbarium set (or locally collected
        tree specimens)
    Pine pollen
Prepared slides
    Pine needle, cross section
    Pine stem, cross section
    Pine pollen cone, longitudinal section
    Pine pollen, w. m.
    Pine ovulate cone, longitudinal section
    Pine embryo
    Lily ovary, cross section
    Germinating pollen, whole mount
    Lilium anther, cross section
Slides, coverslips, razor blades

Solutions

10% raw sugar solution (do not use white or brown
    sugar)

Student Prelab Preparation

Before doing this lab, you should read the introduction
and sections of the lab topic that have been scheduled
by the instructor.

You should use your textbook to review the
definitions of the following terms:

angiosperms	locule
anther	monocot
Anthophyta	ovary
carpel	ovule
dicot	pistil
endosperm	pollen
gametophyte	seed
gymnosperms	sporophyte
integument	zygote

You should be able to describe in your own words
the following concepts:

Alternation of generations
Double fertilization
How seeds are formed
How seeds differ from spores
Plant adaptations to land

As a result of this review, you most likely have
questions about terms, concepts, or how you will do
the experiments included in this lab. Write these
questions in the space below or in the margins of the
pages of this lab topic. The lab observations should
help you answer these questions, or you can ask your
instructor during the lab.

Objectives

1. To observe the life cycle stages in the alternation
   of generations found in gymnosperms and
   angiosperms
2. To appreciate how pollen, ovarian development,
   seeds, and fruits are reproductive adaptations to
   the terrestrial environment
3. To continue to collect evidence (started in previous
   lab) that tests the following hypothesis: Plants can
   be arranged in a phylogenetic sequence that
   demonstrates increasing adaptation to a terrestrial
   environment.

Background

In the previous lab topic, you saw how bryophytes could evolve from green algal ancestors that were preadapted for life on land. Reproductive adaptations were critical for living on land as it is a dry environment. The significant development was the gametangium that protected gametes from drying during development, although water was required for the sperm to reach the egg. Furthermore, because the egg was retained in the gametangium and fertilization occurred internally, there was the potential for the zygote to develop into an embryo in a protective environment.

Bryophytes took advantage of this potential opportunity, evolving a multicellular sporophyte stage that developed by mitosis from the zygote retained in the gametangium. This sprophyte stage was nutritionally dependent on the gametophyte. Furthermore, in the sporophyte stage a new organ appeared, the sporangium, where haploid spores were produced by meiosis and released. The microscopic spores were surrounded by a desiccation resistant wall made of sporopollenin. They became dispersal agents that could be air-carried great distances without drying (and dying) before settling to produce a gametophyte stage, thereby completing the life cycle.

In ferns these two life cycle stages, the gametophyte and sporophyte, became separate stages. Each was capable of an independent existence, although their life cycle linked the two independent stages in an obligatory sequence where one stage produced the other. Starting with haploid spores that divided by mitosis to produce gametophytes bearing gametangia, the fern life cycle resembles that of a bryophyte. In both groups, sperm must move through a water film to fertilize the egg held in the female gametangium and the zygotes develop by mitosis into mature sporophytes. However, in ferns the gametophyte stage dies off leaving the sporophyte stage to live and mature on its own. The independent sporophyte develops sporangia that produce spores by meiosis, completing the life cycle.

In this lab you will continue your study of plant life-cycle evolution and plant adaptation to land by looking at a group of plants called the seed plants. They have become quite successful in terrestrial environments. If we look at the life cycle strategies of seed plants, what is striking is the change in the role of the gametophyte compared to its role in the life cycles of seedless plants. In bryophytes the gametophyte predominates, and although reduced in ferns, it is still a recognizable stage. In seed plants you will discover that the gametophyte stage has become reduced to a few cells that are nutritionally dependent on the sporophyte: a reversal of roles for life cycle stages compared to the bryophytes. In seed plants there is also a change in the role of spores. No longer are they agents of dispersal. Instead, they are retained in the sporangium where they develop into the gametophyte.

Let's look at seed plants in more detail. Seed plants are heterosporous, producing two kinds of spores from different sporangia. The spores are retained in the sporangium so that the gametophyte initially develops inside the sporangium (male gametophytes finish their development outside the sporangium). One type, called **microspores,** develops into male gametophytes which are the individual pollen grains of seed plants. One or two cells in the pollen act as sperm. Pollen is resistant to desiccation due to sporopollenin in its wall. Consequently, pollen can travel long distances on air currents and avoid drying, which would kill the sperm. With the development of pollen, plants were no longer dependent on water films and splashes to transfer sperm from male gametangia to female gametangia. Looking at it another way, in seed plants pollen became mobile gametophytes capable of delivering sperm over long distances to remote females. This is a significant adaptation to life in terrestrial environments.

A second type of spore in seed plants, a **megaspore,** is also produced. It is microscopic, developing into a multicellular female gametophyte inside a megasporangium. One of the gametophyte's cells becomes a haploid egg by mitosis while others take on supporting roles. If pollen successfully transfer sperm to the egg, the resulting zygote marks the start of a new sporophyte generation. What is interesting is that the zygote is inside the gametangium of a gametophyte growing in the tissues of a sporangium of a previous sporophyte generation. Confusing? Read the paragraph again and recognize that three generations are being spanned, initial sporophyte, gametophyte, and new sporophyte. The other gametophyte tissues surrounding the zygote became a source of nutrition allowing the zygote to undergo limited development to an embryo stage of the new sporophyte. After the embryo grows to a certain size, growth stops. The outer walls of the sporangium then harden, encasing the embryo and nutritive tissues. This composite structure is the **seed.**

Seeds, themselves, represent adaptations to life in the terrestrial environment. They are dispersal agents that are resistant to mechanical damage and drying. Seeds can travel substantial distances without injury to the embryo. They resist desiccation, remaining viable from growing season to growing season. When seeds are released into suitable environments, the embryos resume growth, producing mature sporophytes to complete the life cycle.

To summarize, the additional life cycle adaptations to a terrestrial environment that we see in seed plants are:

- Gametophyte stage is reduced to a microscopic stage dependent on the sporophyte.

- The male gametophyte is mobile, encased in a pollen grain and capable of delivering sperm to distant females independent of water availability.

Figure 17.1 Cladogram of plant evolution showing the sequential development of reproductive adaptations.

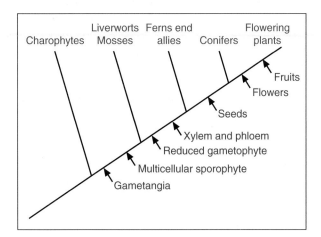

- Development of heterospory allowed development of separate male and female gametophytes, a necessary condition for seed development.
- Retention of the female gametophyte embedded in the sporophyte and retention of the embryo in the female gametophyte allowed seed development.
- Seed development provided an effective means for sporophyte dispersal.

This sequential progression of evolutionary events can be represented in a cladogram (fig. 17.1).

In addition to the life-cycle strategy changes found in seed plants, there are other adaptations to a terrestrial environment. Having well-developed vascular tissues allowed seed plants to develop large "bodies," with some species reaching 300 feet in height. This eliminated the competition with bryophytes and ferns for sunlight. Secondary growth allowed the stems to increase in girth so that they could carry large numbers of leaves and not break under the strain. Root systems also became highly developed to anchor the stems and to provide the needed water for metabolic activities.

The sporophyte stage is the predominant stage in the alternation of generations life cycle found in seed plants. What selective pressures other than desiccation resistance might have lead to the decrease in the prominence of the gametophyte and to the dominance of the sporophyte? Sporophytes are diploid. Diploid organisms have two copies of every gene. If one fails, the other is a backup. In terrestrial environments, the ultraviolet component of sunlight can cause mutations. Because most mutations are harmful, those plants having an exposed stage that was resistant to the effects of mutations and a "hidden" stage that was susceptible might have had an advantage over others.

If this is true, then we could also ask why has the gametophyte stage been retained in the life cycle of these plants? Some authors think that this might be related to two things. First, haploid cells, as found in gametophytes, are a good allele-screening mechanism. If an essential gene becomes nonfunctional as a result of mutation, the haploid stage will not be able to perform the essential function and die, removing the genotype from the population. The second line of reasoning addresses the mobility of male gametophytes encased in pollen. For immobile organisms such as plants, the ability to now combine gametes with distant sexual partners would greatly increase genetic variability in populations, providing new gene combinations for natural selection to act upon. It could be that these two advantages were sufficient to allow the gametophyte stage to continue in the life cycle, although much reduced in prominence.

LAB INSTRUCTIONS

You will study the life cycles of pines and flowers in this lab, learning to apply to these common plants the concept of alternation of generations with distinct sporophyte stages but less conspicuous gametophyte stages. You will also learn how seed plant structures represent adaptations to a terrestrial environment.

Gymnosperms

The term *gymnosperm* means "naked seed" and is used today to describe a clade (grouping) of four divisions (phyla) that all have this characteristic. It is not a taxonomic category. Present-day gymnosperms include: cycads, ginkgo, gnetophytes and conifers (fig. 17.2). All bear naked seeds at the base of the scales found in the characteristic cones or similar structures of these plants. **Cycads** are tropical plants. They look somewhat like palms from a distance and have a stem that may be 18 m high bearing leaves in a cluster at the top, hence the common name "sago palms." Only one species of **ginkgo** survives today, and all plantings are derived from a few trees found in the Imperial Gardens of China. It is widely planted as an ornamental in the United States. It has fan-shaped leaves with parallel venation and drops its leaves in the fall as do many dicots. The seeds are born on stalks and have a fleshy coating. The **gnetophytes** are a diverse group found in moist tropical areas and arid areas, including North America. Some species are treelike: others, shrubs; and yet others, vines. The **conifers** are by far the largest and most diverse group and include the familiar pines, as well as many others.

In this section you will look at pines as examples of the gymnosperms. The sporophyte stage is the conspicuous stage in the life cycle and the gametophyte stage is much reduced to a few cells that give rise to functional eggs and sperm.

Figure 17.2   Examples of gymnosperms: (*a*) Cycad; (*b*) Ginko; (*c*) Gnetae.

(a)

(b)

(c)

Division Coniferophyta (Conifers)

The conifers are an economically important group of over 500 species of plants. They include pine, fir, spruce, hemlock, redwood, sequoia, cypress, and cedar; the species are used for building lumber and as a source of fiber for paper. Most, but not all, have needlelike leaves, tree or shrublike body plans, well-developed vascular tissues, and secondary growth. Cones are common reproductive structures, but may be modified berry-like structures as in yews and junipers.

Sporophyte Stage of Pine

You will start your study of conifers by looking at the anatomy of structures found in the sporophyte stage.

1 ▶ *Leaves*   Look at the demonstration specimens of conifer needles (pines, firs, spruces, yews, *etc.*). Needles are produced in bunches of 1 to 8, collectively called a **fascicle,** with the number being characteristic for the species. Note the shiny appearance to the needles. This is caused by the presence of a thick, waxy **cuticle,** which prevents evaporation of water, adapting the plants to dry environments. The reduced surface area of the needles and the pyramidal shape of most conifers are adaptations that enhance shedding of snow in cold climates and allow these plants to hold their leaves year round. Needles are usually retained for 2 to 4 years, although the bristlecone pine may retain its for 45.

2   Obtain a slide of a cross section of pine needle (fig. 17.3) and look at it with your compound microscope. The cuticle may not be visible because it was extracted by solvents used in the slide preparation process. Note the **epidermis,** which secretes the cuticle, and the **stomata,** pores on the needle surface that connect to spaces inside the leaf. **?** What is their function?

Figure 17.3   Cross section of a single pine needle.

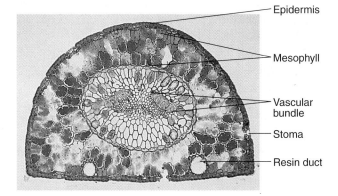

- Epidermis
- Mesophyll
- Vascular bundle
- Stoma
- Resin duct

Beneath the hypodermis is the **mesophyll,** consisting of cells that seem to fit together like pieces of a jigsaw puzzle. These are the photosynthetic cells and contain large, darkly staining chloroplasts. In the center, find one or two **vascular bundles.** Each bundle contains xylem with thick-walled tracheids and thinner-walled sieve cells in the phloem. Xylem brings water to the needles and the phloem conducts photosynthetic products to the other parts of the plant. Near the flat surface of the needle, two or more **resin ducts** should be visible. Resins seal breaks in the needles and may protect the plant from herbivores.

Stem   Vascular tissues are highly developed in the gymnosperms, allowing these plants to grow to great heights. The stems of pines and other conifers are woody, and annually increase in girth due to secondary growth. This leads to the formation of substantial amounts of secondary xylem toward the center of the stem and secondary phloem around the circumference (fig. 17.4). The pine you buy at a lumberyard is secondary xylem.

3   Obtain a cross section of a pine stem and look at it with your compound microscope. The very center of the stem contains **pith.** It is surrounded by elongated **tracheids**

Figure 17.4   Cross section of a pine stem.

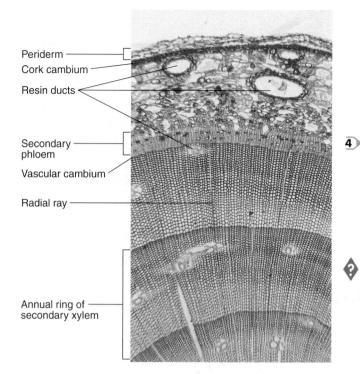

Periderm
Cork cambium
Resin ducts
Secondary phloem
Vascular cambium
Radial ray
Annual ring of secondary xylem

Male Gametophyte   The small male cones appear singly or in clusters near the bases of terminal branch buds in the spring of each year (fig. 17.5*a*). They are called **staminate** or **pollen cones**. Pollen cones usually occur on lower branches and female cones on the upper. Since pollen falls when it is released, it is carried away from female cones on the same tree by the wind, promoting genetic outcrossing. Examine the specimens of pollen cones available in the lab. Each cone is composed of whorls of **microsporophylls,** which each contain two **microsporangia** where pollen are produced.

4   Examine a slide of a longitudinal section of an immature staminate cone and locate the swollen microsporangia on the undersurface of the microsporophylls (fig. 17.5*b*). **Microspore mother cells** in each microsporangium divide by meiosis, to produce four haploid microspores that each develop into a pollen grain. Each of these is surrounded by a thick wall, the outer layer of which expands to produce two air-filled bladders in mature pollen. (fig. 17.5*c*).

?   Can you speculate on what might be the function of these two bladders so typical of pine pollen?

helps float on wind

5   If mature staminate cones are available in the lab, remove a microsporophyll and place it in a drop of water on a slide. Tease it apart, add a coverslip, and use your compound microscope to look for pollen. Alternatively look at a prepared slide of pine pollen. Draw a few pollen grains below.

pine pollen

cedar pollen

While in the microsporangium, the pollen grain matures to its final multicellular state of only four cells: two **prothallial cells,** a **tube cell,** and a **generative cell.** The prothallial cells will degenerate and the generative cell will divide by mitosis to produce two sperm. These four microscopic cells in the pollen grain represent the complete **male gametophyte** stage in the pine life cycle. The sequence of events in the origin of the male gametophyte is summarized in figure 17.6.

of the **secondary xylem** positioned so that their long axes are parallel to the long axis of the stem. You are looking at them in cross section. New tracheids are laid down in **annual growth rings.** At the outer edge of the secondary xylem is the **vascular cambium** layer. Its cells divide to form new **secondary xylem** inwardly and **secondary phloem** outwardly. Xylem and phloem are vascular tissues, Xylem conducts water and minerals to the needles, Phloem conducts carbohydrates from the needles to the roots as well as in the reverse direction. Both tissues are traversed by **rays** that conduct materials across the radius of the stem. Resin ducts should be visible in the layers of xylem. Many conifers produce resins that seal injuries, preventing fungal and bacterial infections. The stem is surrounded by a thick bark or **periderm,** which originates from a **cork cambium** layer outside of the phloem (fig. 17.4). It protects the trunk from invasion by microorganisms, from desiccation, and from damage by fire. The roots of gymnosperm sporophytes are also highly developed but will not be studied.

Gametophyte Stages

The familiar pine, fir, and spruce trees are all examples of the sporophyte generation of the gymnosperms. To locate the gametophytes, the reproductive structures must be examined. For pines, these are the young cones that develop in the spring of the year. Pine trees develop two types of cones: small **staminate** or pollen cones, which are on the tree for only a few weeks, and the larger, more familiar **ovulate** seed cones ("pine cones"), which can remain on the tree for two or more years.

Investigating Plant Phylogeny: Seed Plants   **211**

Figure 17.5   Pine reproductive structures: (*a*) staminate (male) cones; (*b*) Longitudinal section of pollen cone; (*c*) characteristic pollen of pine.

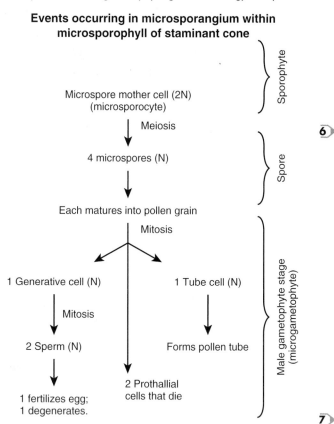

(a)

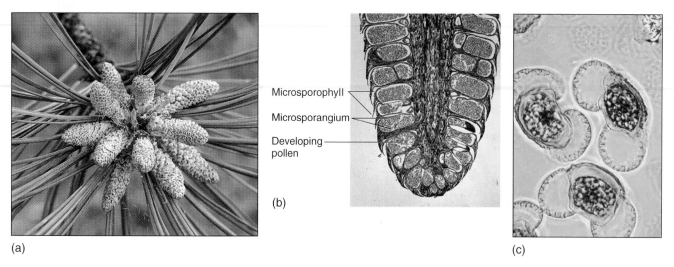

Microsporophyll

Microsporangium

Developing pollen

(b)

(c)

Figure 17.6   Sequence of spore formation and development of male gametophyte generation in gymnosperms.

Events occurring in microsporangium within microsporophyll of staminant cone

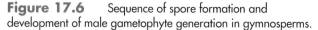

Microspore mother cell (2N) (microsporocyte) — Sporophyte

↓ Meiosis

4 microspores (N) — Spore

↓

Each matures into pollen grain

Mitosis

1 Generative cell (N)   1 Tube cell (N)

↓ Mitosis   ↓

2 Sperm (N)   Forms pollen tube

↓

1 fertilizes egg; 1 degenerates.   2 Prothallial cells that die

Male gametophyte stage (microgametophyte)

When the pollen are mature, the microsporangia, now called pollen sacs, rupture, releasing enormous amounts of pollen into the air. After releasing pollen, the staminate cones dry and fall off the tree.

The development of pollen was a major adaptation in land plants. The male gametes became highly resistant to desiccation. No longer was a water film required for the sperm to reach the egg. Pollen could carry the sperm through the air. Furthermore, the possibilities of genetic outcrossing beyond the individuals in the immediate vicinity was greatly increased. The wind could carry pollen (the male gametophyte) long distances where its sperm could fertilize eggs borne by individuals outside the neighborhood of the pollen-producing tree.

Female Gametophyte   The familiar pinecone is the female ovulate cone. In it you will find several female gametophytes with their eggs.

6   Obtain a young ovulate cone from the supply area. Ovulate cones are borne on the tips of young branches. The cone at first grows with the pointed end upward, but as it matures it inverts. Each cone is made up of whorls of heavy **scales.** Two ovules are on the upper surface of each scale. An **ovule** is a small, complex reproductive organ produced by the sporophyte. It surrounds and protects at first the sporangium in which spores are produced. Because only a single megaspore is produced and is not released, it also protects the much-reduced female gametophyte stage as it develops. After the egg is fertilized, the coat of the ovule will surround and protect the embryo.

Ovules can be observed in young cones by cutting away the tip and about one-third of the scales. While observing through a dissecting microscope, remove scales from the large remaining piece. Look for two oval structures on the surface of the scale that faces the smaller end (upper surface when the young cone is on tree).

7   To study the microscopic structure of an ovule, obtain a prepared slide of a longitudinal section of an ovulate cone and look at it with your scanning objective. After determining which end of the section is the tip of the cone and what structures are scales, find the swollen ovules at the base of the scales (fig. 17.7). Identify the **integument,** the outer layer of the ovule surrounding the female gametophyte. It may be in any stage of development, depending on when the section was made. Because this is sectioned material

Figure 17.7 (*a*) Longitudinal section through an ovulate (female) pinecone. (*b*) Magnified section through ovule showing micropylar opening and megaspore mother cell.

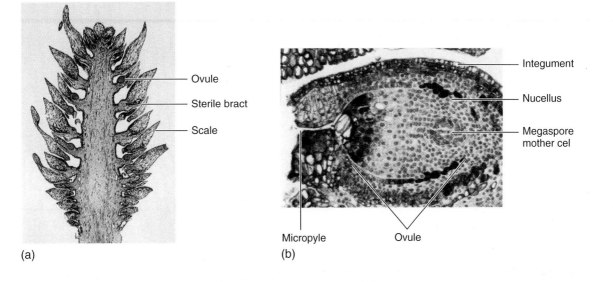

(a)

(b)

and all slices do not contain the same structures, you may have to look at several ovules to see the following. If you find one of these structures, share the view with others.

Inside a young ovule is a mass of cells that makes up the megasporangium, called the **nucellus** in gymnosperms. Embedded in the nucellus tissue is a single diploid **megaspore mother cell.** It divides by meiosis to produce four haploid **megaspores,** three of which degenerate. The remaining haploid megaspore undergoes repeated mitotic divisions, yielding the haploid **multicellular female gametophyte.** As the gametophyte continues to develop, it draws nutrients from the surrounding nucellus tissue. At a later stage of development, two haploid **archegonia** form at the end of the ovule that faces the central axis of the cone. Each will contain a functional egg. The sequence of events in the development of the female gametophyte is summarized in figure 17.8.

Look again at the slide of the ovulate cone and examine each ovule carefully. When the plane of the section is correct, you should be able to see an opening in the integument on the end of the ovule facing the central axis of the cone (fig. 17.6). Called the **micropyle,** this opening provides sperm with access to the egg.

To summarize, two eggs are held in a female gametophyte which is embedded in the megasporangium produced by the sporophyte. The composite structure is called an ovule.

Fertilization and New Sporophyte Generation
Fertilization in pines involves a series of events that may take over a year to complete, whereas in spruces, firs, and hemlocks, it may be completed in a few weeks. At the time of pollination, the ovulate cones are small, project upward from the branch tips, and have open spaces between the scales. Windborne pollen enters the cone openings and is trapped in a droplet of fluid secreted at the micropyle of the naked ovules on the upper surface of the scales. As this droplet dries, pollen is drawn toward the micropyle. The

Figure 17.8 Sequence of spore formation and development of female gametophyte generation in gymnosperms.

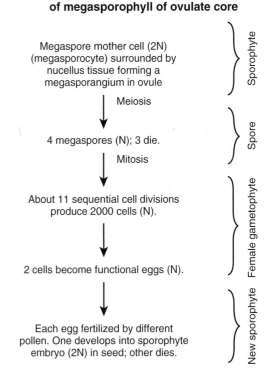

scales of the cone now press tightly together (close) and over the next year fertilization and seed development occur.

As the pollen reaches the micropyle, a tube grows down the canal toward the egg (fig. 17.9). Two sperm pass down the tube and enter the egg cytoplasm. The nucleus of one fuses with the nucleus of an egg, signaling the start of a new diploid sporophyte generation. The second sperm nucleus

Investigating Plant Phylogeny: Seed Plants **213**

Figure 17.9 Life cycle of a pine.

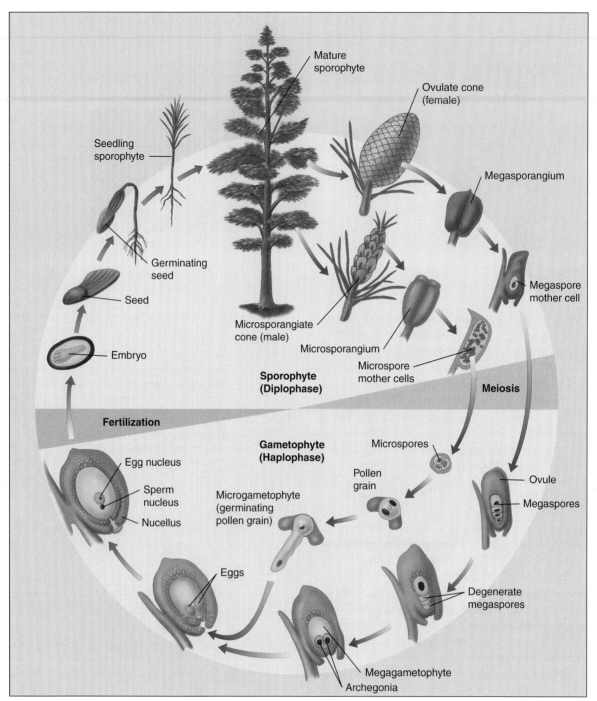

degenerates. The second egg in the ovule also may be fertilized by another pollen, but only one of the zygotes will develop. The zygote that divides to form the embryo pushes into the surrounding nucellus tissue, drawing nutrients to sustain growth.

As the embryo forms, the integument covering the ovule develops into the seed coat. It may take up to an additional year for the seed to develop. In essence, the seed is a mature ovule containing the embryo of the next sporo-

phyte generation. These seeds are truly remarkable structures containing cells from three generations in the life cycle of the pine: (1) the parent sporophyte contributes the seed coat; (2) the embryo is the sporophyte of the new generation; and (3) the stored food materials in the seed are haploid female gametophyte tissues.

8 ▷ Examine dry, mature pinecones for evidence of seed. When air- or water-borne, or transported by an animal to a new location, the seed can germinate, allowing the spread of the species. Thus in gymnosperms, when compared to mosses and ferns, seeds have replaced spores as the means of dispersal.

As you finish this section, pair off with someone in the laboratory and quiz one another about gamete production, fertilization, and where you have seen the sporophyte and gametophyte generations. Use figure 17.9 as a guide.

Angiosperms

Angiosperms are seed plants that produce flowers. They are the most diverse of the plant groups, living in mountain, desert, freshwater, and seawater habitats. Nearly 250,000 species have been described. They range in size from the tiny duck weeds of about 1 mm tall to the 100 m tall eucalyptus trees in Australia. All angiosperms are classified in Phylum **Anthophyta.** There are two taxonomic classes of anthophytes: **monocots** include the grasses, lilies, bamboos, palms, bromeliads, and orchids; and **dicots** include most herbaceous plants, broad leaf shrubs, and trees, vines, and cacti.

The sporophyte stage dominates the angiosperm life cycle as in gymnosperms, but unlike gymnosperms the ovules are contained in a tissue-encased ovary that matures into a fruit with seeds after fertilization. The name angiosperm comes from the Greek words angeion (=container) and sperma (=seed) and together emphasize the role of the fruit as a seed container. Like gymnosperm, the term *angiosperm* does not have taxonomic status and is used like a common name for the clade.

A colorful floral structure in many species attracts animal pollinators that promotes genetic diversity through cross fertilization. Fertilization in angiosperms involves a **double fertilization** where one sperm fuses with the egg to produce the next generation and another sperm fuses with a second type of cell to produce a nutrient-laden tissue called the **endosperm.** The endosperm fuels the early development of the embryo and seedling during germination of the seed.

Division Anthophyta (Flowering Plants)

The life cycles and vegetative structures of flowering plants have many of the same adaptations to terrestrial environment as do gymnosperms. Additional changes include the development of flowers and fruits, and it is these structures that we will focus on in our study. Greater detail on the biology of the anthophytes is covered in Lab Topics 24 through 27.

Gladiolus is recommended as a readily available flower to dissect, but other species will work.

Sporophyte Floral Structure

The flower is a sexual reproductive structure in the sprorophyte stage of anthophytes. Within the male and female floral organs, the gametophyte stages develop in sporangia. As in the gymnosperms, spores produced in the sporangia are not released and the gametophyte develops internally and is very small. Some species have flowers that contain both male and female structures while others have flowers with only male or female parts. Those species producing separate male and female flowers may have individual plants with flowers of one

sex or may have flowers of different sexes on the same plant. *Gladiolus* has both male and female parts in the same flower. Flowers may be single or in clusters as in *Gladiolus*. The flower aggregation is called the **inflorescence.**

9▶ Break a single *Gladiolus* flower from an inflorescence in the supply area and compare it to figure 17.10. The stalk of the inflorescence is called the **peduncle** and that of an individual flower is the **pedicel** which you broke to get your specimen. Note that the floral parts arise from a green **receptacle** at the end of the pedicel. The brightly colored **petals** are obvious. These modified leaves function to attract pollinators. Many flowers are pollinated by insects, birds, bats, or other small mammals. Animals are attracted to specific flower species as a result of co-evolution of visual and olfactory attractants in the plants and corresponding receptors and behaviors in the animals. Consequently, the evolution of many flowers and animals are intertwined. In flowering plants that are wind pollinated, such as the grasses, the flowers lack these attractants and are inconspicuous.

10▶ If you pick off the petals, you will see the sex organs inside the flower. The central stalk-like structure, known as the **pistil,** is the female portion of the flower. It may consist of one or more **carpels.** Carpels are thought to have evolved by the folding and fusion of specialized ancestral leaves called **megasporophylls** that bore female sporangia along their margins. The base of the pistil contains the ovary where the ovule(s) is (are) located. Egg development, fertilization, and seed development will occur here. Above the ovary is a stalk, the **style,** which ends in an expanded tip of glandular tissue, the **stigma.**

Alongside the pistil are long filaments with expanded tips. The structures at the tips are **anthers,** the male portion of the flower that will produce microspores. They will develop into the male gametophyte stage (=pollen). The anthers are thought to have evolved by the fusion of specialized leaves called **microsporophylls** which bore pollen-producing structures.

11▶ Coordinate this part of the dissection with another person at your lab table so that one of you cuts a cross section of the ovary and of the style on one flower and the other cuts a longitudinal section through all parts of the pistil from another flower. Look at the sections. You may want to use a dissecting microscope to see more detail.

In the cross section you should be able to see three open chambers in the ovary that are called **locules** (fig. 17.10). Projecting into the locules from the central shaft are pairs of little bead-like structures, the **ovules.** As in gymnosperms, each ovule contains a megasporangium that produces a single megaspore. It will develop into the female gametophyte that will produce a single egg. Fertilization occurs in the ovule, and each ovule develops into a seed.

Now look at the longitudinal section of the pistil. Note that there are several ovules in vertical arrays in the locules. Try to count the number of ovules in one locule. How many did you find? 15-18 Estimate how many ovules are in the ovary of your plant. 45-54 How many seeds could this flower have produced? 45-54

Investigating Plant Phylogeny: Seed Plants **215**

Figure 17.10   The sporophyte generation of an angiosperm bears flowers that contain a few cells that act as the male and female gametophyte generation. Male gametophytes are the pollen produced by the anthers of the stamens. Female gametophytes are the embryo sacs contained in the ovules located in the ovaries. Nuclei of the cells in the gametophyte generation are haploid. The diploid number of chromosomes is restored with formation of the new sporophyte generation at fertilization.

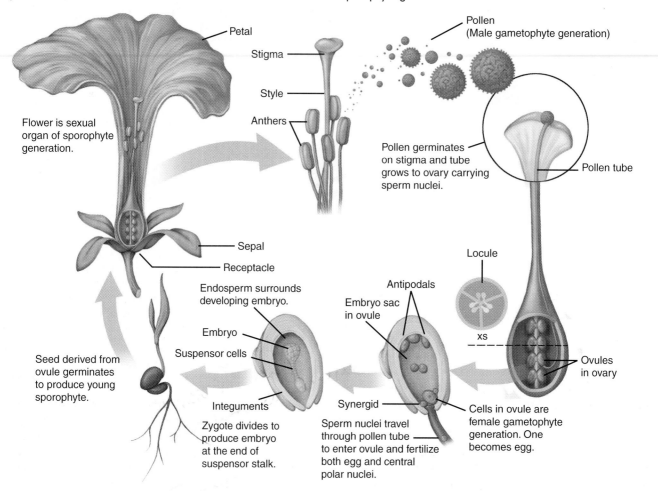

To understand how the carpels might have evolved from megasporophylls, visualize three long narrow leaves that bore ovules on their margins (fig. 17.11). Fold the margins inward in a curling manner so that the ovules are now inside a tube formed by the megasporophyll. Push three of these structures together with the folded edges innermost. You now have a structure that resembles the cross section of the ovary. Over time, the upper parts of the leaf differentiated into the style which had no ovules or locules and into the stigma with its characteristic moist, glandular surface. Examine the surface of the cross section of the style and the surface of the stigma for evidence of an ancestral fusion of three megasporophylls.

The surface of the stigma is moist with sugary secretions from underlying glandular tissues. When mature pollen land on this surface, they germinate, producing a pollen tube. Pollen tubes grow down through the surface of the stigma and follow a pathway toward the ovules, guided by tissue in the style. What is the distance from the stigma to the lowermost ovule in the *Gladiolus* that you dissected? **~80 mm**   For the egg in that ovule to be fertilized, a pollen tube must grow this distance, from a microscopic pollen grain caught on the surface of the stigma.

Male Gametophyte

12▶ Take a *Gladiolus* anther and place it in a drop of water on a slide. Use a dissecting needle to tease it apart, add a coverslip, and observe with your compound microscope. Sketch what you see.

Figure 17.11   Hypothetical model for evolutionary development of ovary from carpels, ancestral leaves that bore ovules along edge.

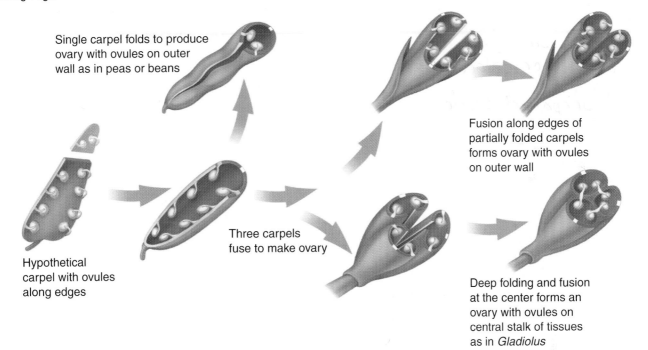

Single carpel folds to produce ovary with ovules on outer wall as in peas or beans

Hypothetical carpel with ovules along edges

Three carpels fuse to make ovary

Fusion along edges of partially folded carpels forms ovary with ovules on outer wall

Deep folding and fusion at the center forms an ovary with ovules on central stalk of tissues as in *Gladiolus*

13▶   Examine a prepared slide of a cross section of an anther. Four **pollen sacs** should be readily visible. They serve as **microsporangia** in which thousands, if not more, **microspore mother cells** divide by meiois to produce haploid **microspores.** Depending on the age of the anther used to make your slide, you may find cells in meiosis I or II or differentiating into mature pollen. In mature pollen, three cells are found: a **tube cell** and two **sperm.** These three cells together represent the entire **male gametophyte** generation of an anthophyte. Because they are contained in the pollen grain, they are mobile and can be carried to distant locations. Often the male gametophyte does not fully develop until pollen have been transferred to the stigma. The steps in the formation of the male gametophyte are summarized in figure 17.12. Sketch a cross section of an anther below, drawing detail for only one of the pollen sacs. See figure 27.4.

looks like after 1st meiotic division

Figure 17.12   Sequence of spore formation and development of male gametophyte generation in anthophytes.

Events occurring in microsporangia within anther

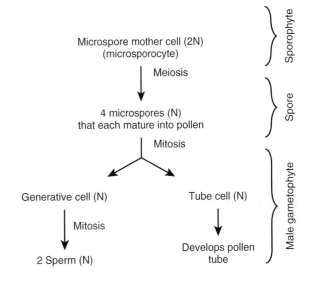

Microspore mother cell (2N) (microsporocyte)

Meiosis

4 microspores (N) that each mature into pollen

Mitosis

Generative cell (N)

Mitosis

2 Sperm (N)

Tube cell (N)

Develops pollen tube

Sporophyte

Spore

Male gametophyte

14▶   It is possible to artificially induce pollen germination, *i.e.,* to activate the male gametophyte. Your instructor may set this up as a demonstration or you may do the procedure. Harvest some mature pollen or split open an anther from a flower that is shedding pollen. Take a depression slide and add a drop of germinating solution which contains 10% raw sugar. Transfer some pollen to the drop and add a coverslip. Observe this preparation for the next hour or so at approximately fifteen-minute intervals for evidence of pollen tube

Investigating Plant Phylogeny: Seed Plants   **217**

growth. Record your observations below as drawings with marginal time notations. Alternatively, you may look at a prepared slide of a germinating pollen tube. See figure 27.5.

After 3 hours, no pollen tubes for lily or gladiolus.
Used prepared slide

Female Gametophyte

15▶Obtain a prepared microscope slide of a cross section of a lily ovary and ovules. Look at it with low power through your compound microscope. Orient yourself by finding the three locules that you saw in your dissection. Each should contain two ovules, although some ovules may be missing because the plane of the section did not cut through where they were located. Compare what you see to figure 27.6.

The anthophyte ovule is similar to those you observed in gymnosperms. Identify the central **megasporangium** surrounded by **integuments.** At one end of the ovule, the integuments have an opening, the **micropyle,** that allows a pollen tube containing the sperm to reach the egg.

The female gametophyte develops in the following way. Within the megasporangium, a **megaspore mother cell** divides by meiosis to produce four haploid **megaspore** cells. Three disintegrate, leaving one functional megaspore. It develops into a multicellular female gametophyte stage within the ovule. Its nucleus divides three times by mitosis to produce eight nuclei contained in the original megaspore's cytoplasm. What is the ploidy level of each of these nuclei? *haploid* Three of the nuclei migrate to the end of the megaspore away from the micropyle, two migrate to the center, and three migrate to the end nearest the micropyle where one enlarges to become the egg nucleus. Cell membranes form around each nucleus separately except for the pair in the center which are surrounded by a single membrane and are contained in a single cell. These seven cells with eight nuclei comprise the entire **female gametophyte** stage of the flowering plant. Collectively they are referred to as the **embryo sac.** The events in the development of the female gametophyte are summarized in figure 17.13.

Fertilization and the New Sporophyte Generation

Germinating pollen grains are the active male gametophytes of flowering plants. They consist of only three cells.

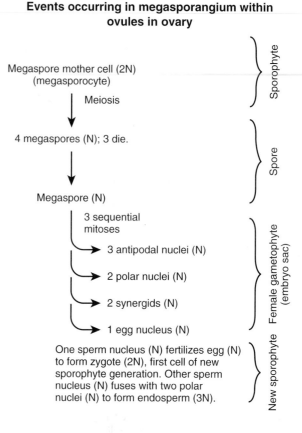

Figure 17.13 Sequence of spore formation and development of the female gametophyte in anthophytes.

Events occurring in megasporangium within ovules in ovary

One forms the pollen tube that penetrates the tissues of the pistil navigating to the micropyle of the ovule. The other two are sperm. **Fertilization** occurs when the growing pollen tube enters the micropyle of an ovule and penetrates the embryo sac. Two sperm cells leave the pollen tube. One fuses with the egg to form a diploid zygote, the next sporophyte generation of the plant. The other combines with the two polar nuclei in the single cell at the center of the embryo sac to form a special triploid tissue, the **endosperm.** It nourishes the developing plant embryo. Anthophytes are said to have **double fertilization** because two sperm are involved, but only one diploid embryo is formed.

Fruits

Following fertilization, the embryo develops inside the ovule as the ovule matures into a seed. The integuments surrounding the seed become the hard seed coat. As the seed develops, the ovarian wall enlarges to become the fleshy part of a fruit. Fruits are another example of how plants and animals have co-evolved. Many fruits are colorful, sweet, and fragrant. They attract animals that eat them. The seeds encased in their protective coats pass through an animal's digestive system unharmed. They are often deposited along with a little fertilizer at some distance from the parent plant where they germinate and continue the life cycle.

16 In the lab are several fruits that have been sectioned either across or longitudinally. Study the fruits and make some quick sketches below in which you show the arrangements of the locules and the seeds in the sections. Estimate the number of seeds in some representative fruits, such as a pepper, bean, or pea (estimate the number in one locule and multiply by the number of locules). Record these numbers next to your drawings. Using your seed data, estimate the number of ovules that would be found in a flower of the same species. Are any of these numbers impressive?

cucumber- 4 locules x ~80 seeds → 240 seeds
Strawberry - 8 locules x 80 seeds → 640 seeds
green beans ; 4-6 seeds per pod (locule)

Learning Biology by Writing

A major theme in this lab topic and in the one before was the phylogenetic trend in the plant kingdom that allowed plants to inhabit terrestrial environments. In a short, two-page essay give all of the observations that you made in both labs that support or contradict the following hypothesis: there is strong evidence that plants can be arranged in a phylogenetic sequence that corresponds to increasing adaptations to a terrestrial environment. Indicate whether you accept or reject this hypothesis based on your observations.

Internet Sources

The Wollemi pine (*Wollemia nobilis*) is a newly discovered (1994) pine. Forty individuals were found growing in canyon lands to the west of Sydney, Australia. Use a search engine such as Google to locate information on this newest discovery of an ancient group. Why is it considered to be a significant discovery?

Your instructor may ask you to turn in answers to the lab summary questions below.

Lab Summary Questions

1. What similarities do gymnosperms and anthophytes have in their mechanisms of gamete formation, fertilization, and dispersal? What differences?

In later labs you will look at the tissues, vascular anatomy, and leaf structure of the sporophyte stage of flowering plants. You and your lab partner should now quiz one another about the anatomy of a flower and the life cycle of a flowering plant. Make a list of how gymnosperms and anthophytes are similar as well as different in the details of their life cycles.

apple - 6 locules x 2 seeds = 12 seeds
pepper - 4 locules x 60-70 seeds = 240-280 seeds
mango - 1 locule x 1 seed = 1 seed
peanuts - 2-3 seeds per locule
tomato; 4 locules x ~25 seeds = 100 seeds

2. Explain where you would look for the gametophyte generation of a conifer and what it would look like.
3. Explain where you would look for the gametophyte generation of an anthophyte and what it would look like.
4. Explain how a seed is an important adaptation to the terrestrial environment.
5. Explain how pollen is an important adaptation to the terrestrial environment.
6. What did you find most interesting about this lab and why?
7. Summarize your observation from the previous lab (#16) and this one by filling in table 17.1 on the next page.

Critical Thinking Questions

1. How does a seed differ from a spore?
2. If a species of flowering plant has a diploid number of chromosomes equal to 30, how many chromosomes will be found in the embryo in a seed? In a pollen nucleus? In a megasporocyte? In endosperm tissue?
3. If a species of pine has a diploid number of chromosomes equal to 40, how many chromosomes will be found in an embryo in the seed? In a pollen nucleus? In a megasporocyte? In nucellus tissue?
4. If all flowering plants depended on the wind to carry pollen from one individual to another, do you think there would be any colorful flowers and fruits? Why?
5. Describe co-evolution and how the concept applies to flowering plants and animals.
6. Why do grasses and conifers lack any showy, sweet, fragrant reproductive structures?
7. How does pollination differ from fertilization?

Investigating Plant Phylogeny: Seed Plants **219**

TABLE 17.1 Life cycle comparisons. Summarize your observations by adding yes or no to indicate if a feature is found in a group.

Characteristics	Bryophyta (Mosses)	Pterophyta (Ferns)	Coniferophyta (Conifers)	Anthophyta (Flowers)
Airborne spores				
Haploid gametophyte				
Egg and sperm produced by mitosis				
Flagellated sperm				
Water-dependent fertilization				
Diploid sporophyte				
Dependent sporophyte				
Independent photosynthetic gametophye and sporophyte				
Dependent gametophyte				
Spores produced by meiosis				
Ovules				
Pollen				
Embryo protected by seed coat				
Seed within fruit				
Vascular tissues				

LAB TOPIC 18

Observing Fungal Diversity and Symbiotic Relationships

Supplies

Preparator's guide available on WWW at
http://www.mhhe.com/dolphin

Equipment

Compound microscopes
Dissecting microscopes

Materials

Living Specimens
Mushrooms (from grocery store)
Rhizopus zygospore culture kit
Lilac leaves infected with powdery mildew (If out of
season, try freezing leaves in sealed containers
for future labs.)
Miscellaneous fungi samples: puffballs, bracket
fungi, molds, others as locally available
Lichen set (crustose, foliose, and fruticose)
Arthrobotrys (nematode-catching fungus) culture kit
with worm, fungus, and growth medium
Microscope slides
Allomyces gametophyte
Allomyces sporophyte
Peziza section
Rhizopus combination slide with sporangia and
zygospores
Coprinus mushroom, section of gills
Mycorrhiza and root sections
Lichen thallus section
Cross section of leaf infected with powdery
mildew
Field guide to mushrooms
Slides and coverslips
Loaf of bread with wrapper showing sodium
propionate added
Sealable plastic bags
Fresh bread with no preservatives

Student Prelab Preparation

Before doing this lab, you should read the introduction
and sections of the lab topic that have been scheduled
by the instructor.

You should use your textbook to review the
definitions of the following terms:

Ascomycota
ascus

Basidiomycota
basidium

Chytridiomycota
coenocytic
dikaryon
fruiting body
gametangia
haustaria
hyphae

lichen
mycelium
mycorrhizae
septa
sporangia
spores
Zygomycota

You should be able to describe in your own words
the following concepts:

Fungal life cycle from spore stage to spore stage
General body form of a fungus when growing and
at time of sexual reproduction
In addition, do some field work and bring fungi to
the lab that you find on campus.

As a result of this review, you most likely have
questions about terms, concepts, or how you will do
the experiments included in this lab. Write these
questions in the space below or in the margins of the
pages of this lab topic. The lab observations should
help you answer these questions, or you can ask your
instructor during the lab.

Objectives

1. To learn the anatomy and life cycles of
representative fungi from each phylum
2. To observe the symbiotic relationships that fungi
have developed with various plants and algae

Background

Known as yeasts, mildews, rusts, blights, mushrooms, and
puffballs, the fungi are common organisms in all ecosystems.
At one time, fungi were considered degenerate plants that had
lost their photosynthetic capability and adapted a saprophytic
mode of existence. In recent years, this hypothesis has been
rejected in favor of the theory that the fungi are a separate
kingdom, the **Fungi (Mycota).** Recent molecular evidence
shows that fungi may be more related to animals than they are
to plants. The oldest fossils are from 400 million years ago.

Figure 18.1 Scanning electron micrograph of interconnected fungal filaments called hyphae that form a mycelium.

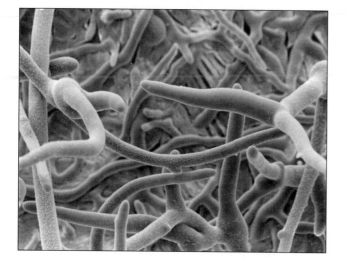

Fungi are found everywhere. They are extremely important in the recycling of minerals in ecosystems as they break down dead plant and animal matter. It is estimated that an acre of forest soil ten inches deep can contain over two tons of fungi. About 5,000 species are pathogens of crops and 150 species cause animal and human disease. A 1984 study showed that 40% of the deaths from infections acquired while in hospitals were from fungal infections, not bacteria.

The typical body of a fungus, except in yeasts, is long filaments called **hyphae.** The hyphae branch to form interconnecting filaments collectively called a **mycelium** (fig. 18.1). No cell in a fungus is far from its surrounding environment because the mycelium is not solid but more like a tangled mass of filaments. No tissues are found in a mycelium: all cells are generalized to perform the functions of the entire organism. The hyphae of most fungi are surrounded by cell walls composed of the polymer **chitin,** the same material found in the exoskeletons of arthropods. This polymer is not found in the bacterial, protistan, or plant kingdoms.

In many fungi, the hyphae are divided by cross walls called **septa,** but the septa can be perforated, allowing cytoplasm to flow from one compartment to the next. This allows the bulk transport of material from one part of the organism to another without crossing cell membranes. New hyphal growth is achieved as new compartments form at the tips. Many fungi, except for the yeasts, are **coenocytic;** meaning many nuclei are found in a single large cytoplasm.

Fungi reproduce both asexually and sexually by releasing **spores,** single cells encased by a tough cell wall. They are small enough to be airborne to virtually any location. If they land on suitable substrates, the spores germinate and produce hyphae by mitotic nuclear divisions to continue the life cycle. Spores are haploid, as are hyphae. Spores are produced in specialized structures called **sporangia.** Asex-

ual sporangia produce spores by mitosis. These asexual spores are genetically alike to the parent and to each other. They are produced when conditions are good for growth. The life-cycle strategy can be stated something like this: If the parent genotype can grow under the present conditions, then asexual spores allows that genotype to produce genetic clones rapidly to take full advantage of the situation. The process of producing sexual spores is a bit more complicated and usually occurs in response to changing environmental conditions when growth slows.

Sexual reproduction in most fungi (excluding the chytrids) involves two distinct processes: plasmogamy and karyogamy. **Plasmogamy** occurs when the cytoplasms from the hyphae of two parents fuse. Fungi do not have sexes. Instead there are **mating types** designated + and –. The differences are not obvious and involve the surface chemistries of hyphae. When the cytoplasms of two mating types unite, a **heterokaryon** is formed, meaning that there is a single cytoplasm containing nuclei from at least two different parents. If only two parents are involved, then the heterokaryon is called a **dikaryon.** The dikaryotic hyphae can continue to grow for years following plasmogamy with each type of nucleus dividing by mitosis during the growth period.

Usually under conditions of poor nutrition, the haploid nuclei in the dikaryon will fuse to produce a single diploid nucleus in a process called **karyogamy.** This diploid nucleus is only a transient stage. It will quickly divide by meiosis to produce four haploid nuclei. These nuclei are then packaged into spores that are released. Given suitable conditions, they produce new hyphal growth. During this brief diploid phase and meiosis, genetic recombination occurs through crossing over and independent assortment so that sexually produced spores are genetically different from each other. This mixing of genes allows the organisms to test new combinations against the natural selection factors in the environment. Often the events of karyogamy occur in what is called a **fruiting body,** the familiar mushrooms, bracts, and cups that are apparent to the casual observer of nature.

Fungi are heterotrophic; they are completely dependent on preformed carbon compounds from other sources. Consequently, fungi live as saprobes, parasites, and symbionts of dead and living plants, animals, and protists.

Because of their chitinous cell walls, the fungi cannot engulf food materials. Instead, the hyphae secrete digestive enzymes that break down polymeric organic matter into small organic molecules that the hyphae absorb. Hyphae are never more than several micrometers thick; thus, they have a great surface-area-to-volume ratio for efficient absorption. Cytoplasmic streaming, in addition to diffusion, provides efficient transport to nonabsorbing areas. Some parasitic fungi produce specialized hyphae called **haustoria,** which penetrate a host's cells and absorb food materials produced by the host.

Fungi are important components of the geochemical cycles in the biosphere. Their decomposing action releases inorganic compounds and carbon dioxide that would otherwise be tied up in dead organic matter and unavailable to other organisms. Fungi also enter into symbiotic mycorrhizal relationships with higher plant roots, increasing the absorptive capacity of the host plants. Lichens (fungi in association with algae) are important colonizing organisms on rock faces, creating an environment for other organisms and building soil. Parasitic fungi destroy many crops in the field or in storage. A few species are important in manufacturing; yeasts are used in baking and brewing and others in manufacturing antibiotics, such as penicillin. Remarkably, the fungi also are able to colonize synthetic environments. For example, some fungi live in jet fuel, on photographic plates, and on other manufactured hydrocarbon materials.

Some fungi that grow on foodstuffs produce extremely powerful toxins. Aflatoxins produced by fungi that grow on stored grain are carcinogenic at concentrations of a few parts per billion. Ergot is an LSD-like hallucinogen produced by some fungi that grow on stored grains. Some historians think that the Salem witch hunts of the 1600s were induced by ergot contamination of grain supplies, causing mass hallucinations.

Nearly 100,000 species of fungi have been described. It is estimated that over ten times that number await discovery. Mycologists group the fungi into four phyla and one fungus-like clade. These are:

Phylum Chytridiomycota—chytrids: aquatic organisms that link the protists and fungi; saprobes and parasites; about 1,000 species.

Phylum Zygomycota—Zygomycetes: zygote fungi living in soils, or on decaying plant and animal matter, or as parasites of insects, or in mutualistic association as mycorrhizae of plant roots; about 1,100 species.

Phylum Ascomycota—Sac fungi living in a variety of habitats and in symbiotic relationships; about 60,000 species.

Phylum Basidiomycota—Club fungi are important in decomposing wood and in forming mutualistic and parasitic relationships with plants; about 25,000 species.

Deuteromycetes—a nontaxonomic clade; imperfect fungi with asexual reproduction only. This group will not be studied in this exercise; about 17,000 species.

Traditionally, the lichens are studied with the fungi. Lichens, however, are not single organisms. Instead, they are a unique association of two species: one an alga (or a cyanobacterium) and the other a fungus.

LAB INSTRUCTIONS

In this lab topic, you will examine life cycle stages of several representative fungi and observe the anatomical basis for several symbiotic relationships.

---

air, leaf, skin

Observation of Field Samples

1 You should start this lab by looking at locally collected samples of fungi. On your way to class, look under bushes on campus or in the back of your refrigerator for fungi. Bring them to the lab.

Look at your samples under a dissecting microscope and tease them apart. Make slides of small pieces and look at them with your compound microscope.

Describe where you got your sample, what it is growing on, and what it looks like. Make some sketches of it. Write any questions you have about how it grows, what it feeds on, or how it reproduces. Use the space below.

Phylum Chytridiomycota

The members of this division are thought to be a link between the protists and the fungi. Based on recent molecular data from protein and nucleic acid analysis, the chytrids have been moved from being protists and are now considered

Observing Fungal Diversity and Symbiotic Relationships **223**

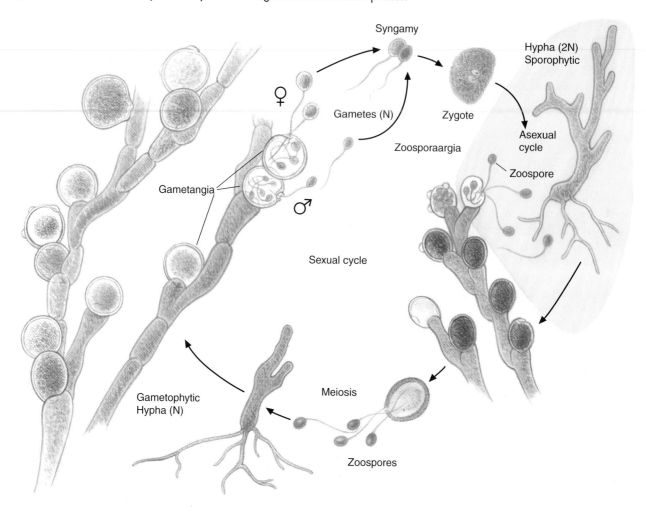

Syngamy

Gametes (N)

Hypha (2N)
Sporophytic

Zygote

Asexual
cycle

Zoospore

Zoosporaargia

Gametangia

♀

♂

Sexual cycle

Gametophytic
Hypha (N)

Meiosis

Zoospores

fungi. They are aquatic fungi important in decomposition and as parasites of plants and invertebrates.

Chyrids tell us that the origin of the fungi was probably in water with the transition to the terrestrial environment coming later. They are the only fungi that have flagellated cells (**zoospores**) in their life cycles which suggest an algal ancestor. Their cell walls contain chitin, a complex carbohydrate polymer also produced by animals, suggesting that animals and fungi are closely related.

2 ▶ Go to the supply area and get a prepared slide of a chytrid from the genus *Allomyces*. These organisms live in freshwater, salt water and soil environments where they are important decomposers. Look at the chytrid slide first with the 10× objective and then switch to higher power to see detail.

The cells of *Allomyces* are organized into filaments called **hyphae.** A mass of hyphae is called a **mycelium.** The filaments essentially have unlimited growth potential and divide in a branching pattern. Cells in the hyphae are separated from each other by partial cell walls which allow free circulation of intracellular materials along the filaments. Essentially, the cytoplasm in the hyphae is one large

interconnected cell with many nuclei, called **coenocytic** organization. Cytoplasmic extensions from the hyphae secrete extracellular enzymes that digest organic materials.

Look for the reproductive structures located at the tips of the hyphae (fig. 18.2). The diploid mycelium can reproduce asexually or sexually. Asexual reproduction occurs when diploid flagellated zoospores are produced by mitosis and swim to new locations where they settle to produce new hyphal growth. Sometimes the flagellated cells will encyst, forming a stage that is resistant to adverse environmental conditions. When conditions turn favorable, the cyst germinates to produce a new individual. This is called the sporophytic life cycle.

There is also a **gametophytic** phase in the life cycle. If available, get a slide showing this phase. In this phase, the diploid mycelium can produce haploid zoospores by meiosis. These scatter and form haploid mycelia. At the ends of these haploid mycelia, haploid male and female **gametangia** can form. The male gametangium releases flagellated sperm and the female releases flagellated female gametes (eggs). **Syngamy,** the fusion of gametes, occurs to produce a zygote. The zygote develops by mitosis into a new diploid mycelium.

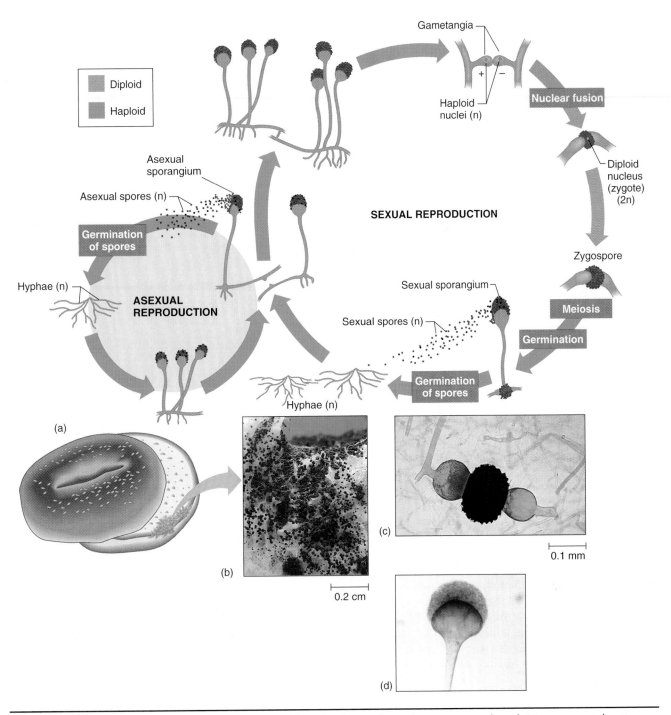

Figure 18.3 *Rhizopus,* the black bread mold: (*a*) life cycle; (*b*) mycelia and zygospores on bread; (*c*) zygospore photo; (*d*) asexual sporangium photo.

Phylum Zygomycota

The 1,100 members of this division are saprophytic, terrestrial fungi that live in the soil on dead plants or animals. Many species form mycorrhizal associations with roots of higher plants. The zygomycetes do not produce flagellated spores nor do they produce a distinct egg or sperm. In sexual reproduction, two mating types of hyphae undergo plasmogamy to produce a **zygosporangium,** a thick-walled sexual structure in which karyogamy produces a zygote.

Asexual spores are also produced from sporangia on the ends of modified erect hyphae.

Rhizopus, the black bread mold, illustrates the life cycle of the zygomycete group (fig. 18.3). These organisms were a problem for bakers before mold inhibitors were used in commercial breads. Now sodium propionate and other materials are added to bread to prevent the growth of this mold and consequent spoilage. This fungus also grows well on fruits and vegetables and causes spoilage during storage.

3 ▶ Before this lab, two mating types of *Rhizopus* (+ and –) were inoculated onto a sterile potato dextrose agar in a petri plate. Observe a culture through your dissecting microscope. Identify the hyphae spreading from the points of inoculation to form white mycelial mats. All members of this division are coenocytic; hyphae lack cross walls (aseptate) and the single, large cytoplasm contains many haploid nuclei. Alternatively, you may view a prepared slide of *Rhizopus* containing both asexual and sexual reproductive structures.

You should be able to see asexual **sporangiophores,** erect hyphae, that bear tiny, black, spherical bodies, **sporangia,** at the tip. Asexual spores are produced by mitosis in these and, when released, will be wind-carried to new locations where they germinate to produce new mycelia.

In addition to asexual reproduction, *Rhizopus* can reproduce sexually. *Rhizopus* is **heterothallic,** meaning that there are two mating types. Where hyphae of different types meet, each hypha sends out a bulge of cytoplasm that eventually forms a bridge between the two hyphae (fig. 18.3). The outgrowths of each hypha, containing several haploid nuclei, are walled off from the parent hyphae to form **gametangia.** The sexual phase begins when the gametangia fuse, producing a single cell with several haploid nuclei, the **zygospore.** It can become dormant under adverse conditions and is resistant to desiccation for months. Under favorable conditions, the zygospore becomes active: its haploid nuclei pair and undergo karyogamy to produce diploid nuclei and the diploid nuclei rapidly divide by meiosis. Thus, genetic recombination occurs in this simple organism. The zygospore then germinates, producing a short **sporangium** that releases haploid spores. These spores will be air-carried to new locations and germinate to produce new hyphae, thus completing the life cycle.

4 ▶ Zygospore formation should be visible as a black zone at the region where the two mycelial mats overlap. Carefully remove some hyphae from the black line and make a wet-mount slide. If a black zone is not evident, prepared slides of *Rhizopus* zygospores may be substituted. Make a second slide, using hyphae from one edge of the mycelial mat. Look at both slides with the low-power objective on your compound microscope. Can you see zygospores?

Draw the heavy-walled zygospore below.

❓ What is the difference between an asexual spore and a sexual zygospore? How will the zygospore produce spores? How will these differ from asexually produced spores?

❖ Make a list of what you think are the distinguishing characteristics of the Zygomycota.

Phylum Ascomycota: The Sac Fungi

This phylum includes such diverse forms of fungi as the yeasts, powdery mildews, and the cup fungi found on decaying vegetation. About half the species associate with algae to form lichens. The blue, green and red molds on spoiled foods are ascomycete fungi as are the tree diseases, chestnut blight and Dutch elm disease, which have virtually made these tree species extinct in the United States. About 60,000 species have been described.

For the most part, the hyphae of these fungi are septate, but perforations in the cross walls allow cytoplasm to flow from one compartment to another. All fungi in this division share a common characteristic: the **ascus,** a saclike reproductive structure that produces haploid spores following sexual union. In lab topic 10, you looked at the ascus in *Sordaria* and observed the results of meiosis. Review your notes on that lab topic. In this lab topic, you will look at an example of a mildew and a cup fungus.

Figure 18.4 Life cycle of the parasitic powdery mildew of lilac.

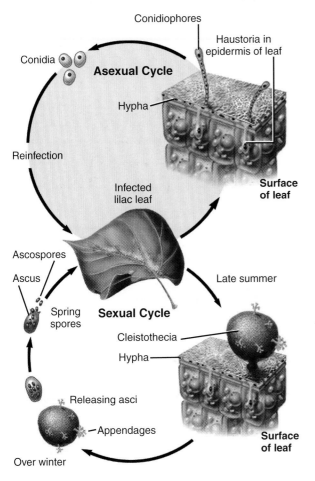

Conidiophores

Conidia

Haustoria in epidermis of leaf

Asexual Cycle

Hypha

Reinfection

Infected lilac leaf

Surface of leaf

Ascospores

Ascus

Late summer

Spring spores

Sexual Cycle

Cleistothecia

Hypha

Releasing asci

Appendages

Surface of leaf

Over winter

Mildews

The general term mildew is used to describe discolorations and odors caused by fungi growing in moist areas. Gardeners use the term to describe over a 1,000 species of parasitic fungi that grow on the leaves of ornamental plants, often giving them a white powdery appearance. Lilac leaves often have this appearance in late summer, but many other plants also are afflicted by powdery mildews. Losses from powdery mildew are due to reduction of aesthetic value of ornamental plants and weakened plants that might be susceptible to invasion by pathogens that would kill them.

5 ▷   From the supply area, obtain a lilac leaf that has powdery mildew. Look at it with a dissecting microscope. You may be able to see hyphae on the leaf surface and "powdery" **conidiophores** projecting up from the surface (Fig. 18.4). Cells in these structures go through mitosis to produce asexual spores called **conidia** that are carried by air currents. When these spores land on suitable host plants (many mildews infect only certain plants), the spores germinate. Specialized regions of the hyphae called **haustoriae** form pegs that penetrate the epidermal cells of the leaf and absorb nutrients. Hyphae grow over the leaf surface to produce

more conidiophores, and so the infection spreads. If a slide of a cross section of an infected leaf is available, look at it to see if you can find any haustoria. They are difficult to see, but you should see hyphae on the leaf surface.

In the late summer, sexual reproduction occurs. Because the same leaf may be infected by hyphae originating from different conidia, both + and − mating types may be present. The hyphae of different mating types will undergo plasmogamy and form a fruiting body called a **cleistothecia.** In nature these appear as dark pin-head-size flecks on leaf surfaces. Karyogany will occur inside the cleistothecia.

If a slide with cleistothecia is available, look at it with your compound microscope, or make a slide by scraping leaves in the lab. Note the ornate appendages on the cleistothecia. Inside one of these you would find several asci, each containing eight spores. These eight cells form when the zygote divides by meiosis to produce four cells and each of these divides once by mitosis. The cleistothecia is an overwintering body. It will fall to the ground when the leaf does. In the spring, it will release its spores, called **ascospores,** and reinfect new leaves. It is interesting to note that these fungi follow only a parasitic lifestyle and do not grow in decaying leaf litter or on other materials.

Cup Fungi

The fruiting bodies of some ascomycete fungi grow to a large size and are known as **ascocarps.** The edible morel or "sponge" mushroom is an ascocarp as are the cup fungi of the genus *Peziza.*

Examine preserved or fresh ascocarps of *Peziza.* It should be obvious why the name cup fungus is used. The inside of the cup is lined with many asci in different stages of development. As the asci dry, ascospores are explosively released into the air.

6 ▷   Obtain a slide of a section of *Peziza* and examine its structure, comparing it to the artist's conception in figure 18.5. Can you see individual hyphae on the slide? Are they septate or coenocytic? *septate*

Cup fungi are heterothallic, having two mating types. The cup-shaped fruiting body contains three types of hyphae. Hyphae containing haploid nuclei grow from germinating spores of the + and − types to form the bulk of the cup. In the center of the mass, a third hyphal type called an **ascogenous hypha** forms from the plasmogamy of + and − hyphae. Since one of the strains often donates its nuclei into the cytoplasm of the other, the hyphae are given names: an antheridium is a nuclear donor and an asgonium is a nuclear recipient. Each cell of the ascogenous hypha has two nuclei, one of the + type and the other of the − type. These cells are known as **dikaryons.**

The ascogenous hyphae grow upward to the surface of the cup. Here the + and − nuclei undergo karyogamy to produce a diploid zygote. It then divides by meiosis to produce four nuclei, which each divide by mitosis to form eight ascospores inside a saclike **ascus.** The inside of the cup is lined with thousands of these asci (pl.).

Observing Fungal Diversity and Symbiotic Relationships

Figure 18.5 Ascomycete cup fungus structure. (a) Hyphae of opposite mating types fuse to produce dikaryotic hyphae. These grow to produce an ascocarp fruiting body in the shape of a cup. Nuclear fusion occurs in certain cells lining the cup. The diploid zygote divides by meiosis and then by mitosis to produce eight spores held in a saclike structure, the ascus. Spores are released and germinate to produce haploid hyphae. (b) Photomicrograph of ascospores in ascus sacs. (c) Living cup fungus *Sarcoscypha coccinea*. (d) Morel mushrooms are also ascomycetes.

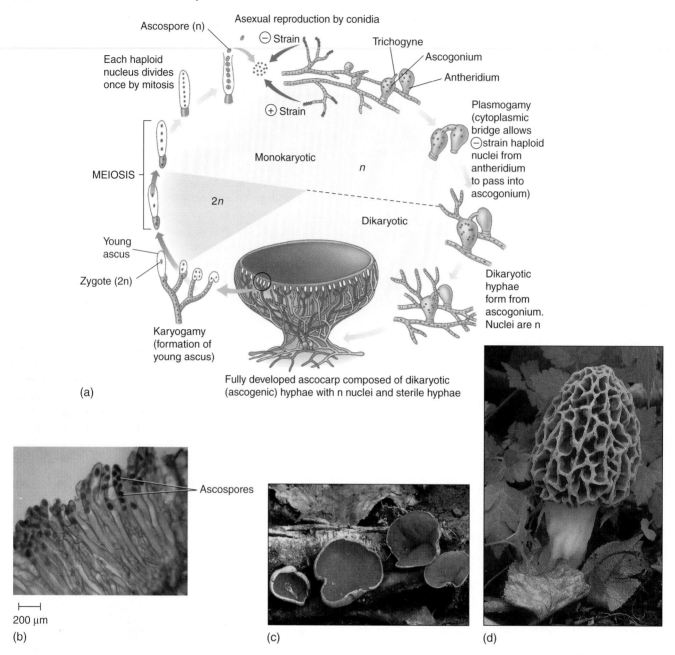

(a)

Fully developed ascocarp composed of dikaryotic (ascogenic) hyphae with n nuclei and sterile hyphae

200 μm
(b)

(c)

(d)

What is the maximum number of spores in an ascus? Are the spores haploid or diploid? What do spores produce when they germinate?

Figure 18.6   Three examples of basidiomycete fungi: (*a*) poisonous mushroom *Amanita muscaria;* (*b*) bracket fungus growing on side of dead tree; (*c*) a puffball releasing spores.

(a)   (b)   (c)

Make a list of what you think are the distinguishing features of the Ascomycota.

Smuts and rust of various grain crops and vegetables are also included in this group. About 25,000 species have been described.

7⊳   If mushrooms are growing in the lab or if you can find some growing on campus, carefully clear away the soil around the base of one to see if you can uncover the mycelial mat. Use a gentle stream of water from a squeeze bottle to separate the soil particles from the hyphae. Mushrooms grow quickly because materials can be rapidly transported through the perforated septa of the mycelium to the growing fruiting body. Sketch the relationship of the mycelium to the base of the mushroom.

Phylum Basidiomycota: The Club Fungi

This is a varied phylum of saprophytic and parasitic fungi as well as mutalistic mycorrhizae. All are characterized by **basidia,** club-shaped structures that each produce four spores (basidiospores). Puffballs, mushrooms, toadstools, and bracket fungi are the familiar fruiting bodies (basidiocarps) of this group (fig. 18.6). Beneath the fruiting bodies are extensive, dispersed mycelial mats, which "feed" on decaying vegetation. The hyphae are always septate. At some stage of development, the septae are perforated, allowing cytoplasmic and nuclear exchange between adjacent cells.

8⊳   Examine the structure of a common edible mushroom. Note the **cap** (pileus) with the **gills** on its undersurface and the supporting **stalk** (stipe). This fruiting body, a **basidiocarp,** is an aggregate of hyphae as is the ascocarp in the cup fungi (fig. 18.7). One important difference is that all of the hyphae in a basidiocarp are dikaryotic. Fungi in this group are heterothallic, having hyphae of the + or – type. When + and – hyphae meet, they fuse and give rise to dikaryotic hyphae that continue to grow. Aggregated growth of such hypha results in a mushroom.

Figure 18.7 Life cycle of a basidiomycete mushroom. (*a*) Haploid hyphae from two mating types fuse to produce a dikaryotic mycelium; each has two nuclei, one derived from each parent. This dikaryotic stage will form a fruiting body, and nuclear fusion occurs in the basidia located on the gills. The zygote thus formed divides by meiosis, producing basidiospores which germinate and form haploid hyphae. (*b*) Scanning electron micrograph of a basidium, showing basidiospores forming.

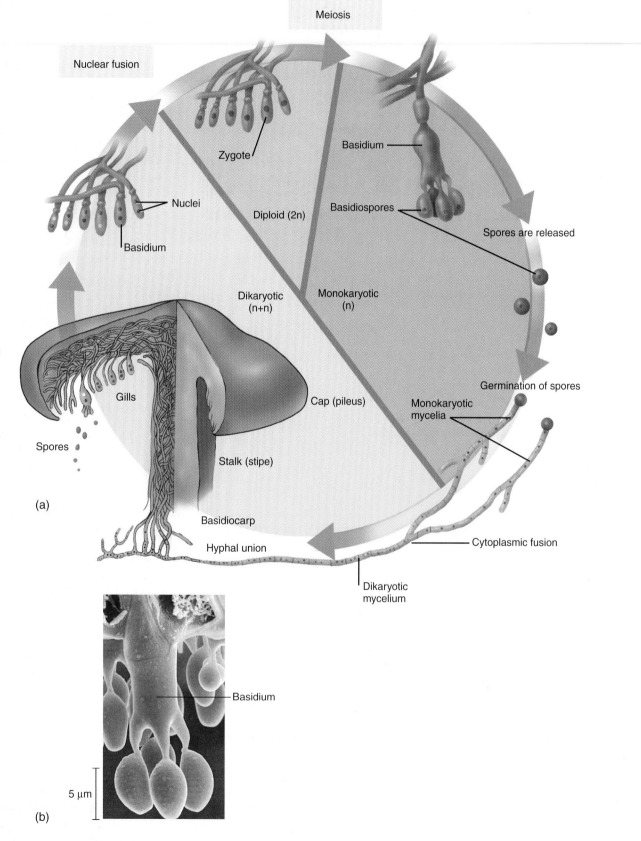

9 Split your mushroom longitudinally through the cap and stalk. Look at the cut surface with a dissecting microscope. Can you see the hyphae?

10 Obtain a prepared slide of a mushroom's gills and use your compound microscope to locate a **basidium.** Karyogamy of the + and – nuclei occur in the basidia, and then the zygote nucleus divides by meiosis to produce four haploid nuclei. The nuclei migrate to the lower surface of the cell where each is partitioned into a **basidiospore.** When released, these spores are wind-borne to new locations where, if conditions are favorable, they germinate to produce either + or – hyphae. A single mushroom can produce billions of basidiospores. Below, draw a cell bearing basidiospores.

Many mushrooms are edible, while others are among the most poisonous organisms known. There is no easy method for separating the edible and poisonous species. The *Amanita* mushrooms, for example, contain a chemical that blocks the function of RNA polymerase in human cell nuclei, meaning that no RNA is made. One bite of an *Amanita* mushroom (fig. 18.6*a*) contains enough poison to kill an adult human being. There are no known antidotes for fungal poisons.

Make a list of what you think are the distinguishing features of the Basidiomycota.

Fungal Associations

Lichens

These composite organisms consist of a fungus growing in close association with a photosynthetic green alga or, in some cases, with cyanobacteria (fig. 18.8). The fungus and alga together produce a unique superorganism, the lichen, that can colonize environments that neither could inhabit alone. The fungus contributes to this mutualistic association by absorbing minerals and moisture from the environment. The algae are nestled among the fungal hyphae, where they benefit from the absorptive processes, and, in turn, produce carbohydrates and other organic molecules, which are absorbed by the fungus. About 5 to 10% of the lichen's dry weight is due to algal cells.

Lichens are found from the arctic to the tropics, growing on rocks, trees, and soils. About 25,000 kinds of lichens have been described. They range in color from black and white to delicate shades of green, yellow, brown, and red. Lichens produce acids that gradually break down the rocks they grow on, contributing minerals to the buildup of soil. Lichens that contain cyanobacteria are able to fix nitrogen, contributing nitrates to developing soils.

11 Examine the examples of whole lichens available in the lab. Note the three general forms (fig. 18.8). (1) **Crustose** lichens grow as a crust on surfaces. (2) **Foliose** lichens are more or less leafy in appearance. (3) **Fruticose** lichens are shrublike with branching and intertwined fibrous parts.

Ask your instructor if you should dissect a lichen. Place a small piece in a drop of water on a microscope slide, tease it apart with dissecting needles, while looking through your dissecting microscope.

12 Obtain a slide of a foliose lichen. Look at it with your compound microscope. Compare the general organization to figure 18.8*d*. Are the algae scattered or concentrated in layers? _Seem concentrated in layers_

Lichens are able to live in harsh environments because they can survive long periods of desiccation. When it rains or fog rolls in, a lichen can absorb 3 to 35 times its own weight in water. As water is absorbed, the algae become photosynthetically active and remain so until a dry period begins. Because lichens are dry most of the time, they have very low growth rates, increasing in diameter from less than 1 to 10 millimeters per year. On this basis, some large lichens may be thousands of years old.

Lichens do not reproduce sexually, although the fungus or algal members may do so individually. Fragmentation of the lichen asexually propagates the association. Some lichens form special structures called **soredia,** which are minute fragments of fungus and algae that may be wind-carried to new locales.

Mycorrhizae

Many fungi live in a beneficial symbiotic relationship with the roots of living plants and contribute to their growth. If pine trees are planted in sterile soil, their growth is slow. If a small amount of soil from a natural pine forest is added,

Observing Fungal Diversity and Symbiotic Relationships   **231**

Figure 18.8 Lichens: (*a*) a crustose lichen; (*b*) a foliose lichen; (*c*) a fruticose lichen; (*d*) structure of a lichen in cross section; with algal cell; (*e*) scanning electron micrograph of fungal association.

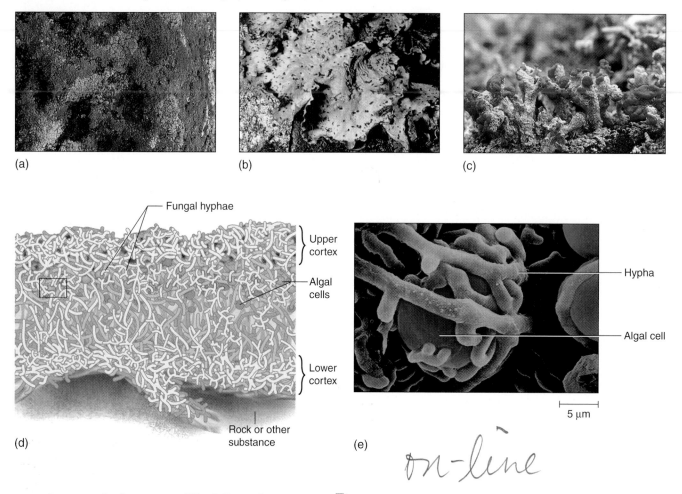

(a)　　　　(b)　　　　(c)

(d)　　　　(e)

on-line

growth promptly increases. Why? Fungal spores or mycelia in the added soil started hyphal growth and formed associations with the pine tree roots. The hyphae spread out from the roots and aided in mineral and water absorption, greatly increasing root efficiency (fig. 18.9). In turn, the fungi obtain carbon compounds from the roots of the plants. Only a few families of vascular plants characteristically lack mycorrhizae. Fossil roots over 400 million years old contain mycorrhizal fungi, attesting to a long and stable coevolution.

There are two types of mycorrhizal relationships found in nature: **endomycorrhizal** and **ectomycorrhizal.** In an endomycorrhizal relationship, the hyphae of a zygomycetous fungus penetrate the root cells, whereas in an ectomycorrhizal relationship, the hyphae of a basidiomycete fungus surround the root cells and extend into the soil. Few ascomycota fungi form mycorrhizae. The associations are usually species specific.

13▶ Study the demonstration slides of mycorrhizae in association with roots. Sketch the association in the circle.

Figure 18.9   Ectomycorrhizal hyphae form a mantle surrounding the root and penetrate between, but do not enter, cortical root cells.

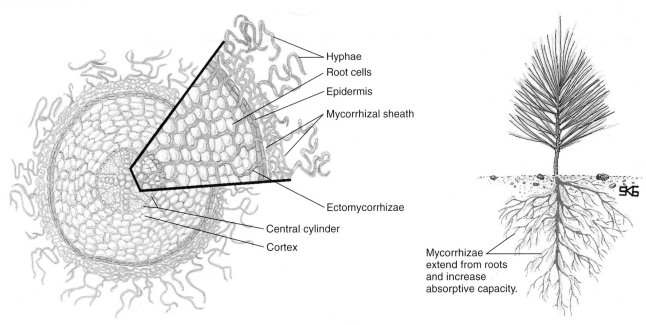

Hyphae
Root cells
Epidermis
Mycorrhizal sheath
Ectomycorrhizae
Central cylinder
Cortex

Mycorrhizae extend from roots and increase absorptive capacity.

Fun with Fungi (optional)

Below are two activities that your instructor may choose to assign. Each gives a chance to look at fungi more carefully.

Killer Fungi

We tend to think of fungi as rather passive organisms, growing on moist surfaces that contain nutrients. Some fungi, however, are not passive but are quite active in trapping food. Fungi in the genus *Arthrobotrys* grow in the soil and produce hyphae that form little constricting and adhesive loops which function like sticky lassos. Small nematode worms that also inhabit the soil crawl into these loops which constrict, trapping the worm. The force can be considerable and almost cut the worm in half. When a nematode enters a mycelium, it is immobilized. Enzymes secreted by the fungus on the immobilized worms kill and digest them and the fungus absorbs the digestion products. Because soil nematodes can damage the roots of crop plants, there is interest in using the fungus as a biological control agent for the worms rather than chemically treating the soil.

It is possible to buy strains of the *Arthrobotrys* fungus and to culture them on cornmeal agar. If microscopic nematodes in the genus *Rhabditis* are added to the culture, many will be captured by the fungus and will be digested.

14▶   Your instructor will have set up cultures of *Arthrobotrys* and, before lab, will have added nematodes to the culture.

Follow the instructor's directions to see if you can find nematodes that have been snared.

Fungus Amon-gus

Fungal spores are everywhere, floating in the air. When they land on suitable substrates, they germinate and produce hyphae, eventually a mycelial mass, and finally reproductive structures. Chances are that you never observed a complete life cycle. This will be your opportunity.

15▶   Your instructor will have some small sealable sandwich baggies and loaves of bread in the lab. You will use the bread to trap and grow some of these fungi. Take a slice of bread and go outside and wave it through the air several times. Place it in the baggie along with ten drops of distilled water. Seal the bag and take it back to your room. Poke a few small holes in the bag to allow for air exchange. Place the bag in a dark warm place.

Each night look at the bag and keep a journal of the changes that occur. Note such things as when you first see mycelial growth, what percent of the slice is covered by fungi (at each observation period), and when fruiting bodies appear. Bring your bread mold and journal to class next week.

Summary

At the end of this lab topic, find the summary table for fungal diversity and fill in the required information.

Learning Biology by Writing

Write a short, 200-word, essay describing the importance of fungi in ecosystems as decomposers, parasites, and mutualistic organisms.

Your instructor may ask you to turn in answers to the Lab Summary or Critical Thinking Questions that follow.

Internet Sources

Considerable information is available on the World Wide Web on the topic of lichens. A starting place for locating information is the Internet Resource Page for Bryologists and Lichenologists at http://www.unomaha.edu/~abls/resources.html Check out the Images and Information section. An interesting site about fungi in the news is Dr. Tom Volk's site at http://botit.botany.wisc.edu/toms_fungi

---

Lab Summary Questions

1. What are the distinguishing characteristics of a fungus?
2. List the distinguishing characteristics of the four phyla of fungi studied in this lab.
3. If you were given a sample to identify, what characteristics would you use to determine if it was a lichen or a pure fungus?
4. In general terms, describe a sexual life cycle in fungi. What stages are haploid and which ones are diploid? When is nuclear division by mitosis and when by meiosis?

5. Why do fungi produce both sexual and asexual spores? Does one have an advantage or advantages over the other?
6. Compare the life cycles of fungi from the phylum Ascomycota with those from the phylum Basidiomycota. How are they similar? How do they differ?
7. Give examples of three different symbiotic relationships that fungi have with plants or protists. Indicate which are parasitic and which are mutualistic. Explain why.
8. What are the ecological roles of fungi in ecosystems?

Critical Thinking Questions

1. The *Suillus lakei* mushroom is commonly found growing around the base of conifers, especially the Douglas fir. Offer a possible explanation of this association of a fungus and plant.
2. DNA testing revealed that an *Armillaria bulbosa* fungus growing in the Upper Peninsula of Michigan covered 15.9 hectares (38 acres) (it may well be the largest organism in the world). Explain how this basidiomycete could get to be so large. Would you expect to find larger examples?
3. Many fungi produce and secrete antibiotics that inhibit bacterial growth. What is the advantage to the fungus in doing this?
4. Crustose lichens growing on rock surfaces may increase in diameter only a millimeter or so each year. If a lichen started growing on the day you were born, how many millimeters in diameter would it be now? How big is that in inches?

Summary table for fungal diversity.

Characteristic	Chytridiomycota	Zygomycota	Ascomycota	Basidiomycota
Common names?				
Asexual or sexual or both?				
Ploidy of spores?				
Septate or aseptate hyphae?				
Describe plasmogamy.				
Instant or delayed karyogamy?				
Names of unique structures?				
Fruiting body name?				
Meiosis produces what?				
Forms mychorrhizae?				
Dikaryotic or coenocytic hyphae?				
Habitat?				
Ecological role?				

Observing Fungal Diversity and Symbiotic Relationships

LAB TOPIC 19

Investigating Early Events in Animal Development

Supplies

Preparator's guide available on WWW at
http://www.mhhe.com/dolphin

Equipment

Incubators at 20° C and 37° C
Dissecting microscopes
Compound microscopes
Refrigerator

Materials

Prepared slides

Sea star development from egg to larval stage
Early cleavage stages of *Cerebratulus* sp.
16 hr chick blastoderm (wm)
18 hr chick (cs)
33 hr chick (wm)
72 hr chick (wm)
Live sea urchins in reproductive condition
Photographs, models, or plastic whole mounts of 24-
    and 48-hour stages in the development of a chick
Depression slides
Slides and coverslips
Thermometers
Dropper bottles
Pasteur pipettes
Glass rods
Syringes
Petroleum jelly
Miscellaneous beakers

Solutions

Seawater
0.5 M KCl
3.5% NaCl
50% ethanol as fixative

Student Prelab Preparation

Before doing this lab, you should read the introduction
and sections of the lab topic that have been scheduled
by the instructor.

You should use your textbook to review the
definitions of the following terms:

archenteron	gastrula
blastodisc	mesoderm
blastomere	primitive streak
blastula	radial cleavage
cleavage	spiral cleavage
ectoderm	totipotent
endoderm	zygote

You should be able to describe in your own words
the following concepts:

Cleavage of the zygote to produce a blastula
Significance of gastrula formation
Gastrula formation in bird's egg

As a result of this review, you most likely have
questions about terms, concepts, or how you will do
the experiments included in this lab. Write these
questions in the space below or in the margins of the
pages of this lab topic. The lab experiments should
help you answer these questions, or you can ask your
instructor during the lab.

Objectives

1. To identify the stages in the early development of
   sea stars
2. To compare radial to spiral cleavage
3. To observe fertilization and early cleavage in sea
   urchins
4. To identify and describe the early stages of chick
   development

Background

Development is one of the truly amazing processes in both plants and animals. Cells from different parents fuse to produce a deceptively simple-looking zygote that, in turn, undergoes a series of cell divisions to produce a multicellular adult. The adult is made up of hundreds of different kinds of cells that perform myriad highly integrated and coordinated functions. For most organisms, although the stages of development have been outlined, the underlying mechanisms that control development remain an enigma. It is known that the developmental program is encoded in the genetic material, but no one yet understands exactly how certain genes are activated and others are repressed at just the right times during development.

An animal's embryonic development can be divided into four major stages: (1) **fertilization,** when the sperm pertrates the egg and is soon followed by syngamy, the fusion of the egg and sperm nuclei; (2) **cleavage,** the mitotic divisions that partition the large cytoplasm of the fertilized egg into smaller cells; (3) **gastrulation,** a morphogenetic cellular movement that produces an embryo with three layers of cells; and (4) **organogenesis,** the process whereby specific organs develop from the primary germ layers.

The specifics of development are different for different animals, but there are many similarities between species in general development. As development progresses, all embryos become more complex as previously nonexistent tissues and organs appear and the embryo becomes capable of performing new functions. This increase in complexity is called **differentiation.**

However, the developing organism is not simply a random assortment of new cell types and organs; these new features always have specific spatial relationships with existing cells and with those that will form later. This development of form is called **morphogenesis** and comes about as a result of cell movement and growth as the embryo uses yolk materials for a source of energy and for chemical building units. During morphogenesis, the cells must "communicate" to inform each other of their location during migration movements and to trigger the differential use of genetic information.

The physical characteristics of eggs are related to the environment in which an animal lives, the place where the embryo develops, and the stages in the life cycle of the species. Eggs of animals that live in aquatic environments generally have gelatinous coats, which protect the egg from physical and bacterial injury, and moderate amounts of yolk, which supply the embryo with sufficient energy to achieve the developmental stage of self-feeding.

Land animals that release their eggs for development outside of the mother have eggs covered by hard shells that provide protection against physical injury and desiccation. However, such a protective device is not without its problems. Fertilization must occur before the shell is formed because the sperm cannot penetrate the hard structure. This means that fertilization must be internal in the female because the shell is produced by cells in her reproductive tract, not the zygote. The embryo inside a shelled egg must exchange O_2 and CO_2 with the environment, must have sufficient energy and raw materials to develop to an advanced stage, and must have a means of disposing of nitrogenous wastes. Inside such eggs, four extraembryonic membranes grow out from the embryo, surround it, and function in gas exchange, waste storage, and nutrient procurement.

In mammals, since the developing embryo is carried inside the mother's uterus, no outer shell or jelly coats are necessary. When the egg implants in the uterus, extraembryonic membranes surround the embryo and also form the **placenta,** a highly vascularized organ that brings extraembryonic capillaries close to the mother's capillaries where nutrients and wastes are exchanged.

In animals, the way in which an egg undergoes early development is strongly influenced by the amount and distribution of yolk in the egg. Eggs of some species have little yolk and it is uniformly distributed in the egg's cytoplasm, as in the starfish and sea urchin you will study in this lab. Others have a large amount of yolk that displaces the active cytoplasm to one pole of the spherical egg. Such eggs are characteristic of fish, amphibians, reptiles, and birds.

LAB INSTRUCTIONS

You will observe developmental patterns in an echinoderm, a marine worm, and a bird, which have very different kinds of eggs. You will study the early developmental patterns of each organism to see the effects of egg type on development. Because these comparisons are detailed and time consuming, the lab topic may take two laboratory periods if all parts are done.

Sea Star Development

The sea star is used to illustrate early development because the events of cleavage and gastrulation are especially easy to see. More complex development patterns in other organisms may be compared to this simple pattern.

1⃞ Obtain a microscope slide containing developmental stages of the sea star from the unfertilized egg through the gastrula stages. Scan the slide under low power with the compound microscope and locate the stages described in this section and shown in figure 19.1. These slides are usually *thicker* than most slides you have studied. *Do not use high power,* or you may push the objective through the coverslip while focusing.

The **unfertilized egg** contains a single egg nucleus and a large amount of rather equally distributed yolk surrounded by a cell membrane. When the sperm penetrates the egg, a **fertilization membrane** forms as materials stored beneath the cell membrane are released. This new

Figure 19.1 Stages in echinoderm (sea star) development: (*a*) unfertilized egg; (*b*) zygote surrounded by fertilization membrane; (*c*) two-cell stage (blastomeres); (*d*) four-cell stage; (*e*) morula; (*f*) blastula; (*g*) early gastrula; (*h*) late gastrula.

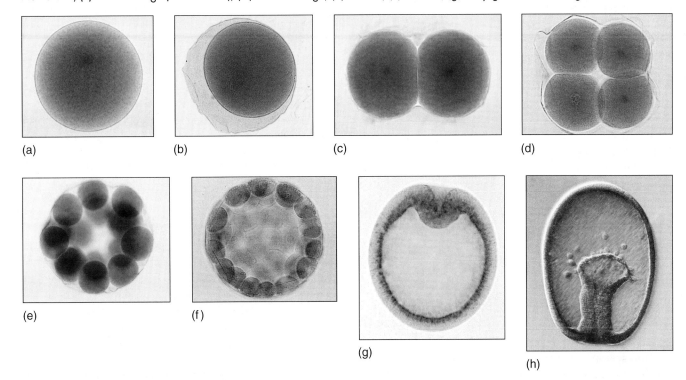

(a)   (b)   (c)   (d)

(e)   (f)   (g)   (h)

membrane may be visible as a "halo" above the surface of fertilized eggs on your slide. This membrane acts as a barrier that prevents the penetration of additional sperm. Think of the genetic consequences of polyspermy: the fusion of more than one sperm with an egg. Once inside the egg, the haploid sperm nucleus migrates through the cytoplasm. It fuses with the haploid egg nucleus, in an event called **syngamy** that forms the diploid nucleus of the **zygote.**

For several hours following fertilization, the zygote undergoes a series of rapid mitotic cell divisions without any intervening periods of growth. This **cleavage** process involves the duplication of chromosomes followed immediately by mitosis and cytokinesis, followed again by chromosome duplication, followed by mitosis, and so on. The result is that the large uninucleate fertilized egg is divided into smaller and smaller cells, each with a single nucleus. If the type of cell division is mitosis, do the nuclei of the developing embryo contain the same or different information than the nucleus formed by syngamy?

The cleavage divisions are usually synchronized in all cells, so that the embryo goes through a series of stages in which it contains 1, 2, 4, 8, 16, 32, and 64 cells. Around the 32-cell stage, the number of cells becomes difficult to count, and the developing embryo is referred to as a **morula,** meaning mulberrylike.

Find examples of 2-, 4-, 8-, and 16-cell stages on your slide. Note that the size of the developing embryo does not increase as cell number increases. The existing material of the embryo is merely partitioned by cleavage into smaller cells.

2▶ If ocular micrometers are available on your microscopes, you can test this hypothesis.

The cells in the 1- through 16-cell stages are spherical. The volume of a sphere equals $4/3 \, \pi r^3$. If you measure the diameter of cells in each stage, you can calculate the volume of cytoplasm in each stage by:

$$\text{Zygote volume} = 4/3 \, \pi r^3 = \underline{\hspace{1.5cm}}$$

$$\text{16-cell volume} = 16(4/3 \, \pi r^3) = \underline{\hspace{1.5cm}}$$

◆ Within an acceptable margin of error are the volumes equivalent? Do you accept or reject this hypothesis?

The developmental process requires energy. Where does this come from? Would you predict any effect on cytoplasmic volume?

In the transition from blastula to gastrula, a lot has happened. The basic shape of the animal has appeared. Most animals have a digestive tube within a body tube plan of organization, if you neglect the appendages. Think of the bodies minus appendages for an earthworm, insect, and human to test this idea. In addition, the gastrula stage is the first time we see distinct tissues, and these will further develop to yield the many tissues found in an adult.

Beyond the gastrula stage, differentiation continues. In the larva stage, the embryo is free swimming or drifting with a mouth and anus. At this stage, it is capable of feeding itself and continues to grow, using external energy sources. At a later stage, a metamorphosis occurs, and the larva changes to a miniature version of the adult sea star.

As cleavage continues, the cells eventually form a hollow ball of cells that is ciliated on its outer surface. The beating of the cilia allows the embryo to swim out of the enveloping fertilization membrane. This stage is called a **blastula,** the cells are **blastomeres,** and the central cavity is the **blastocoel.** Find a blastula on your slide and identify these structures.

Several hours after blastula formation, a second major morphogenic event occurs. Cells at one end of the blastula undergo rapid growth and move into the blastocoel. This infolding is called **invagination** or **gastrulation** and results in a two-layered embryonic stage called the **gastrula.** The gastrula then elongates, forming a tube within a tube cylindrical body.

The two layers of cells in the gastrula are called **primary germ layers.** The inner layer is the **endoderm** and will form the lining of the digestive system and digestive glands. The outer layer is the **ectoderm** and will form the skin and the nervous system of the adult. The midgastrula and late gastrula stages can be identified by the elongation of the gastrula and changes in the shape of the endoderm tube. The elongated tube is called the **archenteron,** or primitive gut, and its opening to the outside is called the **blastopore.** It will become the anus of the sea star. The inner end of the archenteron eventually develops two lateral pouches that will grow outward and pinch off, forming a third germ layer called the **mesoderm.** The start of this process is shown in figure 19.1*h*. Muscles, connective tissues, and gonads will develop from the cells in this layer. Find a late gastrula stage on your slide and sketch it in the circle in the next column. Label all the structures indicated by boldface terms in this paragraph.

Cleavage Patterns in Invertebrate Animals

Echinoderms, such as the sea star you just studied, have what is called **holoblastic** cleavage. Holoblastic refers to the fact that when the cells (blastomeres) divide, the daughter cells are usually of equal size and completely separate from one another. Such division processes are characteristic of eggs in which the yolk is evenly distributed. In contrast to holoblastic cleavage, animals that have eggs with large amounts of yolk often have **meroblastic** cleavage. In this case, the daughter cells resulting from a division are often unequal in size with the larger cells containing a greater amount of yolk. You will see an example of meroblastic cleavage when you study avian development.

Echinoderms also have **radial** cleavage (fig. 19.2). Each successive mitotic spindle and cleavage plane forms at right angles to the previous one so that the cells are arranged on radii extending from the center of the cell mass. In contrast to radial cleavage, many invertebrates have **spiral** cleavage. In fact, this basic difference correlates with the two clades of higher animals, the Protostomes (spiral cleavage) and Deuterostomes (radial cleavage). (See lab topics 21–23.) In spiral cleavage, after the four-cell stage, the mitotic spindles form at oblique angles

Figure 19.2 Comparison of radial and spiral cleavage. (a) In radial cleavage in the deuterostome clade, the spindles form either parallel or at right angles to the axis running through the poles of the cells. Consequently, cells sit directly on top of cells underneath. (b) In spiral cleavage in the protostome clade after the four-cell stage, the mitotic spindles form at oblique angles to the polar axis of the cells. Because cytokinesis always occurs across the center of the spindle, the oblique spindles result in displacement of the newly forming cells. Arrows indicate the direction of displacement.

(a) Radial cleavage

| Polar view | Side view | Side view |
| 8-cell stage | 8-cell stage | 16-cell stage |

(b) Spiral cleavage

| Polar view | Side view | Side view |
| 8-cell stage | 8-cell stage | 16-cell stage |

to the previous one so that the cleavage planes are also displaced to oblique angles.

3 ▷ To observe evidence of spiral cleavage, look at a slide of the early developmental stages of *Cerebratulus* sp., a marine ribbon worm. Find an eight-cell stage where you are looking down on the organism from the top, a polar view. Note the positions of the top four blastomeres relative to the positions of the lower four. The top blastomeres lie over the grooves between the lower blastomeres. Sketch the stage.

4 ▷ Change slides and look again at the slide of sea star developmental stages. Find an eight-cell stage in polar view and note the relative positions of the upper and lower four blastomeres. They lie directly on top of one another with the cells and grooves in register. Sketch the stage below.

Look again at the eight-cell stages of both the sea star and *Cerebratulus*. Are each of the cells in each embryo the same size? Which of the organisms has holoblastic cleavage?

Experimental Embryology with Sea Urchins (Alternative Activity)

Sea urchins, as well as sea stars, are members of the phylum Echinodermata and show the same early developmental patterns. Sea urchins are found in both the Atlantic and Pacific oceans. The breeding cycles are such that the West Coast species are fertile in the fall and winter, whereas the East Coast species are fertile from April to September. This makes sea urchins ideal for teaching laboratories because animals in breeding condition are available throughout the year from biological suppliers.

Sea urchins in breeding condition can be induced to release their gametes by a variety of methods. Mild electrical shock, temperature shock, injection of potassium chloride, or surgical removal of the gonads have all been used successfully to obtain eggs and sperm. The separately collected eggs and sperm can be mixed on a microscope slide so that fertilization and development can be observed.

Obtaining Gametes

Because it is difficult to tell the difference between the sexes in sea urchins, several sea urchins may have to be used to obtain both eggs and sperm. One male and one female, however, will provide enough gametes for a whole lab section.

5 ▷ To induce shedding, inject three to four animals with 2 ml of 0.5 M KCl, as shown in figure 19.3. Do it one at a

Investigating Early Events in Animal Development **241**

Figure 19.3 Technique for collecting sea urchin gametes. (*a*) Inject 0.5 M KCl into soft tissues near mouth to stimulate gamete release. (*b*) Place animals oral side down in dish and watch for gamete release from aboral surface. (*c*) Place female over beaker of cold seawater with aboral surface down in water. Eggs will sink to bottom. (*d*) Sperm should be removed from male's aboral surface and stored undiluted in a dropper bottle.

(a)

(b)

(c)

(d)

time, through the soft tissue surrounding the mouth. Place the animals in a dry bowl with the **oral** (mouth) surface down and watch the **aboral** surface now on top for the release of gametes through the genital pores. After a few minutes, gametes will be emitted. Males release a white suspension of sperm, and the females a yellowish-brown suspension of eggs (one species has red/purple eggs).

Collect the eggs and sperm as follows (see also fig. 19.3):

Females should be placed over a small beaker filled with cold seawater. Put the aboral surface down and immersed in the water. When shed, the eggs will sink to the bottom of the beaker. To facilitate union with sperm later, these eggs should be washed three times with seawater. Swirl the water in the beaker, allow the eggs to settle and decant the supernatant. The eggs will remain viable for two to three days if stored in the refrigerator in seawater.

Sperm should be collected by holding a male over a sterile petri plate with the oral surface up so that the sperm drip into the plate. They can be transferred to a cold, clean

dropper bottle for storage up to 24 hours. No seawater should be added until the sperm are to be used. At that time, add three to four drops of sperm to 25 ml of cold seawater to create a suspension that will remain active for 20 to 30 minutes.

Observing Fertilization

Fertilization events are temperature sensitive and rarely occur above 23° C. If your lab room is hot, cool all slides and suspensions in a refrigerator before starting the experiment.

To observe fertilization, take a depression slide and add one to two drops of seawater containing 10 to 30 eggs. If depression slides are not available, add a few washed grains of sand to a drop of water containing the embryo on a normal slide. This will keep the coverslip elevated. Examine these eggs under the 10× objective on the compound microscope and try to locate the egg nucleus. The yolk is concentrated in one hemisphere of the egg called the **vegetal hemisphere.** It is usually on the bottom as the egg floats. The other hemisphere is called the **animal hemisphere.**

While watching through your microscope, add a drop of the diluted sperm suspension to the slide. Carefully observe the gametes. You will probably not see the single sperm that fertilizes the egg. When a single sperm fuses with the egg, it will release a **fertilization membrane,** which you may be able to observe. Your textbook should provide some of the details on the development of the fertilization membrane. In brief, sperm penetration causes changes in the sodium and calcium ion balance in the egg. This causes release of material from cortical granules just beneath the egg's surface membrane. This release pushes the jelly coating of the egg away from the egg membrane which, in turn, prevents any other sperm from fusing with the egg. This is a block to what is called polyspermy, the fusion of multiple sperm with eggs. Can you think of genetic reasons why this is important?

If petroleum jelly is placed around the edge of the coverslip, the slide can be kept and later developmental stages observed during the next 24 hours. (Be sure the temperature does not exceed 23° C.) Alternatively, about an hour before lab your instructor may have added sperm to a suspension of eggs. You could make a slide of these fertilized eggs to see some of the later events. Refer to the following schedule of developmental events to see the approximate timing for different stages.

Sperm contacts egg	0 minutes
Fertilization membrane forms	2 minutes
1st cleavage (2-cell stage)	1 hour
2nd cleavage (4-cell stage)	1 hour, 30 minutes
4th cleavage (16-cell stage)	2 hours, 30 minutes
Blastula (about 1000 cells)	7 to 8 hours
Gastrula	12 to 15 hours
Pluteus larva	2 days

7▷ While waiting for the first cleavage to occur, obtain a regular microscope slide and add a drop of diluted sperm

suspension. Add a coverslip and observe the slide under high magnification (or oil immersion) with your compound microscope. Sketch what you see.

Separation of Blastomeres (Optional)

It is possible to separate the blastomeres of a two- or four-cell embryo from an isolecithal egg and to have each cell develop into a normal adult (though the new embryos will be smaller than normal because each cell starts with less yolk material than a normal zygote). The first two divisions in an isolecithal egg pass through both the animal and vegetal poles of the egg and produce four equivalent cells. In other egg types, the cells produced are not equivalent, and separation of the cells does not result in viable embryos. Cells capable of developing into complete embryos are said to be **totipotent.** Since humans have isolecithal eggs and the blastomeres remain totipotent for the first few cleavages, this explains how identical twins, triplets, or quadruplets may be born.

Animals whose blastomeres are totipotent during the early stages of development are said to have **indeterminate** cleavage. Some animals have determinant cleavage: their blastomeres are not totipotent and the fate of the cells is determined or committed early in the cleavage process. The blastomeres cannot develop into whole organisms because their genetic program has committed them to developing into a specific part of the organism.

Hypothesize what will happen if you separate totipotent blastomeres which had indeterminate cleavage.

8▷ You can test this hypothesis using 2- and 4-cell stages of sea urchins. To separate the blastomeres, the fertilization membrane must first be removed from the cleaving egg.

Figure 19.4 Egg development in chicken. Albumin (egg white) and shell develop as egg passes through oviduct. Embryo is at gastrula stage when egg is laid.

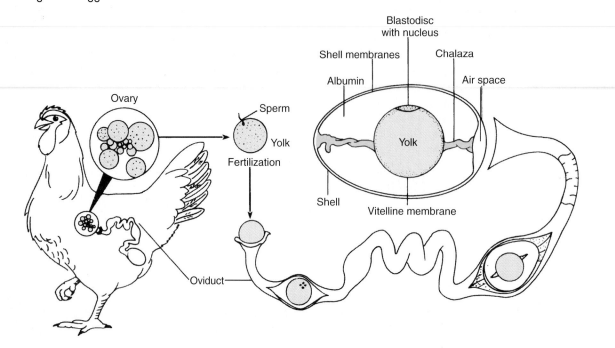

Fertilize about 100 eggs in a 20-ml beaker or test tube. Wait for approximately 30 minutes while they sit at 23° C or less. Take the eggs and draw them rapidly into a Pasteur pipette; then rapidly expel them. Repeat this aspiration process 15 to 20 times. Now examine a few eggs under your microscope to be sure the fertilization membrane has been removed.

Remove the seawater over the eggs and replace it with calcium-free 3.5% saline. (In the absence of Ca++, the blastomeres will not stick together tightly.) Add and withdraw fresh, calcium-free saline several times to ensure that all the old calcium ions are washed away.

After 30 minutes, aspirate the two- or four-cell embryos back and forth in a Pasteur pipette. Observe the cells under the microscope. If the blastomeres have separated, dilute the suspension of blastomeres with seawater. Add about ten blastomeres to a depression slide. Add a coverslip, ring with petroleum jelly, and observe the slide over the next 48 hours. Alternatively, the stages can be incubated in a test tube and samples can be periodically withdrawn for observation. Your instructor can add alcohol to fix them 24 and 48 hours from now.

Does the data that you collected lead you to support or reject the hypothesis made earlier?

Compare the rate of development of the separated blastomere embryos to that of the normal embryos in the first part of this exercise. Compare the sizes of the blastula and gastrula stages in these two experiments. Describe the differences.

Chick Development (Alternative Activity)

Because of time limitations, it will not be possible to observe all of the developmental stages of the chick. Three stages have been chosen to illustrate chick development: 16-, 18- and 24-hour embryos. You will look at prepared slides.

Bird Gametes and Mating

Copulation in birds is rather straightforward. Eggs are fertilized before the shell is deposited by accessory glands in the female's uterus. Therefore, fertilization is internal. Male and female birds lack external genitalia. Their excretory digestive, and reproductive systems all empty into a single chamber called a **cloaca.** Most birds mate when the male briefly mounts on the back of the female while she elevates her tail feathers exposing her cloaca opening. Known as "cloacal kissing," the male presses the lips of his cloaca to hers and passes sperm to her. Sperm are produced in two internal testes in the male and pass via a duct to a storage area prior to mating. A single ejaculation contains millions of flagellated sperm. The sperm enter the female's reproductive tract from her cloaca and are stored. As eggs are produced and released from the ovary, sperm are released from the storage area and fertilize the egg. As eggs descend the oviduct, the egg white (albumen) is added and finally the shell. Some birds, especially ducks and geese, do have penises. In 2001, scientists found that the Argentine Lake Duck has a penis over 42 cm long (almost the same as body length), leading to speculation on how they copulate and

Figure 19.5 Side view of developing chick embryo. (a) Blastodisc confined to top of yolk. (b) Several cell layers accumulate. (c) A cavity appears in cell mass when blastula forms.

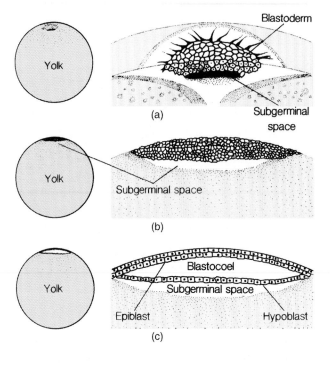

(a)

(b)

(c)

Figure 19.6 Formation of primary germ layers in avian egg. (a) The blastodisc is a mass of cells found on top of the yolk; (b) The cell mass splits to form a cavity yielding a blastula-like stage where the upper layer is the epiblast and the lower is the hypoblast; (c) Cell migrate down into the cavity from the epiblast at the primitive streak and along the edges of the disc, forming the three primary germ layers as indicated.

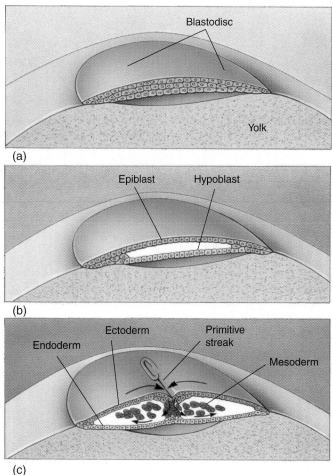

(a)

(b)

(c)

what evolutionary selective pressures ever led to such a disproportionate organ.

Since the eggs of birds (also fish and reptiles) contain large amounts of yolk, the nucleus and cytoplasm are confined to a small disc-shaped area called the **blastodisc,** which sits on top of the yolk. Fertilization occurs at the blastodisc, and cleavage divisions are confined to this area. The large yolk does not divide as in the starfish; the developing chick embryo "rides" atop the yolk during development (fig. 19.4). When slides are prepared, the yolk is removed and only the blastodisc area is on the slide.

Chick Cleavage and Gastulation

9▶ Obtain a whole mount slide of a 16-hour old chick embryo. You will be looking at the entire blastodisc after it has undergone several cell divisions. As the cleavage divisions occur in the blastodisc, a flat layer of cells called the **blastoderm** is produced. As the cleavage continues, the blastoderm becomes several cell layers thick with a fluid-filled space, the **subgerminal space,** developing between it and the yolk. The blastula is formed when the cell layers of the blastoderm separate into an upper **epiblast** layer and a lower **hypoblast** layer (fig. 19.5). The space between them is the avian equivalent of the **blastocoel.** Only the epiblast cells will form the tissues of embryo.

As development continues, cells in the center of the epiblast migrate down from the surface and away from the cen-

ter area in a process that is equivalent to gastrulation. These movements result in the recognizable **primitive streak** stage (fig. 19.6). Find the primitive streak on your slide. Epiblast cells migrating into the primitive streak delaminate and enter the blastocoel. Some remain there to become mesoderm. Others displace hypoblast cells to become endoderm. Cells remaining in the epiblast become ectoderm.

10▶ Now get a slide of a cross section through the blastoderm for an 18-hour chick. Compare what you see to figure 19.7. By this time most of the cellular migrations are complete and new structures are beginning to appear from the primary germ layers. These are: ectoderm on the outside will produce the skin and feathers as well as the nervous system; mesodermal layers in the blastocoel will form the muscles and bones and other connective tissues; endoderm will form a digestive tube. A **neural tube** may

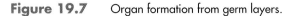

Figure 19.7   Organ formation from germ layers.

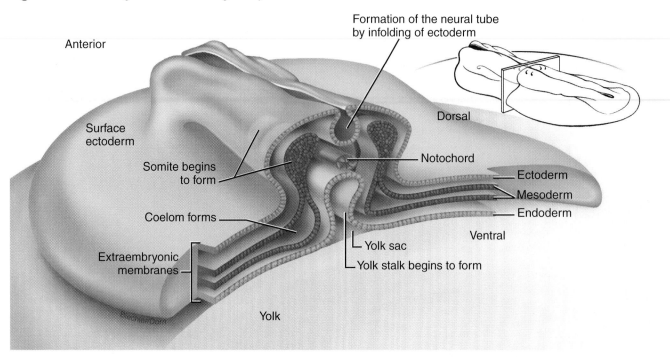

be developing from the infolding of the ectoderm. It will become the nerve cord. A notochord should be visible below it. It will be replaced by the vertebral column.

Later Developmental Stages

11▶Study photographs or whole mount microscope slides of 33-hour and 72-hour chick embryos. Note the developing nervous system and the anterior-posterior axis.

By 33 hours, the relatively unorganized structure of the blastoderm is gone, replaced by an obviously animal-like embryo (fig. 19.8). The **neural tube** runs the length of the embryo and its anterior portion is beginning to differentiate into regions of the brain. A crude **heart** has

formed. **Somites,** segments of body muscle, are beginning to appear.

By 72 hours, note that the anterior half of the chick has rotated and lies on its left side, while the posterior half remains dorsal side up (fig. 19.9). Considerable differences may also be noted in the nervous and circulatory systems. The neural tube has differentiated into a five-part brain and a nerve cord. **Optic cups** are developing from the optic vesicle, and the optic lens is starting to form. An auditory vesicle should be visible. The heart has changed from a muscular tube to a two-chambered organ, and aortic arches with associated gill slits have developed. Note the vitelline arteries and veins. What are their functions?

Figure 19.8   Dorsal view of 33-hour chick.

Figure 19.9   Whole mount of a 72-hour chick embryo.

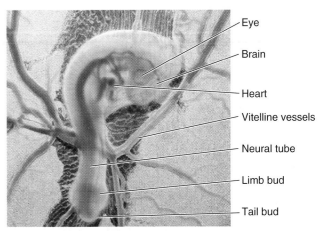

- Eye
- Brain
- Heart
- Vitelline vessels
- Neural tube
- Limb bud
- Tail bud

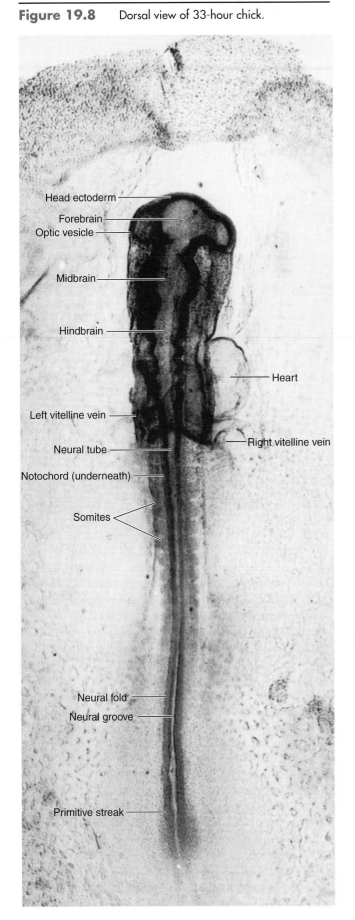

- Head ectoderm
- Forebrain
- Optic vesicle
- Midbrain
- Hindbrain
- Heart
- Left vitelline vein
- Right vitelline vein
- Neural tube
- Notochord (underneath)
- Somites
- Neural fold
- Neural groove
- Primitive streak

Learning Biology by Writing

Compare and contrast radial cleavage with spiral cleavage. Describe how the gastrula stage forms from a blastula and discuss the significance of the gastrula stage, *i.e.,* why is it important. Discuss how events in the gastrula and following stages differ in protostomes and deuterostomes.

As an alternative assignment, your instructor may ask you to answer the Lab Summary and Critical Thinking Questions that follow.

Internet Sources

Check the WWW for current research on early developmental stages in animals. Use the search engine Google at www.google.com and enter the phrase research on blastula or research on gastrula. Scan the sites returned and write a summary that is of most interest to you.

Lab Summary Questions

1. Describe how a fertilized sea star egg develops into a blastula.
2. Why is the gastrula considered an important stage in development? How does the gastrula stage reflect the general adult body plan of most animals?
3. Describe the events that occur when a sea urchin's egg is fertilized.

4. Compare and contrast radial and spiral cleavage.
5. If a four-cell stage is separated into individual blastomeres that each develop into a complete embryo, what does it indicate?
6. In your own words, describe the role of the "yolk" in the development of an embryo in a bird's egg.
7. Describe the fomation of a blastula and gastrula in a bird egg.

Critical Thinking Questions

1. Describe the different ways that frog, chicken, and human embryos have for procuring nutrients, exchanging gases, and treating wastes.
2. Twins may be identical or fraternal. What is the difference? Are the cells in early human embryonic stages totipotent? Support your answer with reasons.
3. Reflect on the general life cycles of a bird and an invertebrate such as a starfish. Recognize that many adult birds are smaller than starfish. Why are bird eggs always larger than invertebrate eggs? Consider in your answer the energy budgets for development and the role of larval free-living stages.
4. Many invertebrates produce several thousand eggs during their lifetimes. Why is the world not over run by invertebrates?
5. The eggs of most terrestrial animals are surrounded by shells while those of aquatic species are not. Why do you think this is so? In species with shells would you expect that fertilization is internal or external? Why?

LAB TOPIC 20

Animal Phylogeny: Investigating Animal Body Plans

Supplies

Preparator's guide available on WWW at
http://www.mhhe.com/dolphin

Equipment

Compound microscopes
Dissecting microscopes
Saltwater aquarium with living demonstration
specimens of sponges and anemones

Materials

Preserved specimens
Leucosolenia
Gonionemus
Taenia (tapeworm—demonstration)
Ascaris
Prepared slides
Leucosolenia, longitudinal section
Hydra, longitudinal section
Obelia, whole mount
Cnidocyte, demonstration slide
Dugesia, cross section and whole mount
Clonorchis sinensis or other fluke, whole mount
Tapeworm, scolex and mature proglottids
Pinworm, demonstration
Trichinella, demonstration
Living *Dugesia*
Living *Hydra*
Living vinegar eels (nematodes)
Petri dishes
Raw meat or egg yolk
Dissecting pans and instruments
Watch glasses
Spring water

Student Prelab Preparation

Before doing this lab, you should read the introduction
and sections of the lab topic that have been scheduled
by the instructor.

You should use your textbook to review the
definitions of the following terms:

acoelomate	Nematoda
bilateral symmetry	Platyhelminthes
body plan	Porifera
Cnidaria	pseudocoelomate
coelom	radial symmetry

You should be able to describe in your own words
the following concepts:

The advantages of cellular specialization to form
tissues and organs
The advantages of bilateral symmetry and a body
cavity
How body plans become more complex as you
move from sponges to nematodes

As a result of this review, you most likely have
questions about terms, concepts, or how you will do
the experiments included in this lab. Write these
questions in the space below or in the margins of the
pages of this lab topic. The lab experiments should
help you answer these questions, or you can ask your
instructor during the lab.

Objectives

1. To study the functional anatomy and life cycles of
representatives from four simple animal phyla
2. To illustrate the organizational differences between
animals with and without tissues
3. To illustrate asymmetry, radial symmetry, and
bilateral symmetry
4. To illustrate the differences between animals with
and without body cavities
5. To collect evidence that tests the following
hypothesis: Animals in the four phyla (studied
during this lab) can be arranged in a sequence
from simple to complex.

Background

What makes animals different from other organisms? Animals are heterotrophic multicellular, eukayotic organisms.
They normally feed by ingesting food and digesting it internally. Their cells have no cell walls and are held together

Figure 20.1   A phylogenetic hypothesis showing possible evolutionary relationships among clades of animal phyla to be studied. Clade names are shown in green boxes. Animals can be assigned to clades on the basis of answering four questions in yellow diamonds: (1) Does animal have tissues? (2) What is animal's symmetry? (3) What are the embryological developmental patterns? (4) Are there similarities in the gene sequences?

Hypothetical Phylogenetic Relationships of Phyla to be Studied

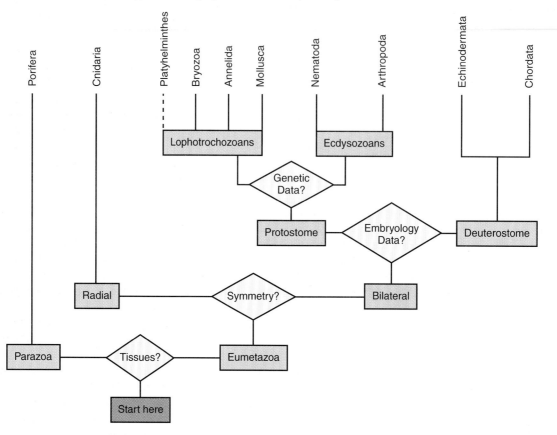

by external structural proteins such as collagen and by specialized cellular junctions. Most animals have specialized excitable tissues, nervous and muscle tissues, which allow coordinated movements. Most are sexually reproducing and the diploid phase of the life cycle is predominant. The zygote goes through a series of developmental stages starting with cleavage and progressing through blastula and gastrula stages. Many animals have independent larval stages, which are self-feeding, and then metamorphose into sexually mature adults.

The evolutionary sequence and phylogenetic relationships in the animal kingdom have become topics of debate in recent years as new data have become available. Most agree that animals originated from an ancestral stock of colonial flagellated protists, sequentially diversifying over time into the many forms seen today. About 35 phyla of animals are recognized. Vertebrates, animals with backbones, and the ones that usually attract our attention, are in only one of these phyla. All of the others are invertebrates and may be less familiar. In this book ten of the 35 phyla are discussed. These phyla are shown at the top of figure 20.1.

For the past 70 years or more, the 35 phyla have been arranged into what we now call clades, a grouping of phyla that share common characteristics. Until recently, these clades were based on answering four rather simple questions about an animal's anatomy and embryology. These are:

1. Does the animal have tissues?
2. What is the symmetry of the body: radial or bilateral?
3. Does the animal have a body cavity?
4. During embryonic development, how does the egg initially divide into many cells and how do the mouth and anus form?

In recent years the basis for forming clades has shifted emphasis from only morphological characteristics to include molecular similarities and differences. The techniques of molecular biology allow DNA to be isolated from an organism and the nucleotide sequence in the genes determined. Once the sequences are known, new techniques in bioinformatics can be used to analyze them. Using computers and statistical tests, it is possible to compare samples of DNA from two or more different organisms. The analyses allow scientists to say with a certain probability that two animals share a certain percentage of their nucleotide sequences with each other. The underlying reasoning in bioinformatics is that if two animals are evolutionary related, they will have similar genes. The closer the animals are in a evolutionary sense, the greater will be the similarity in their genes. Those with similar genes then are considered to be members of the same clade. Analysis of two types of genes have proven very useful: those for rRNA synthesis and those for a group of genes controlling embryological development of body plan. These are called *Hox* genes.

In the last ten years or so, molecular data on gene sequences have started to accumulate allowing genetic comparisons that were not previously possible. When the composition of clades based on genetic data are compared to clades based on the old morphological approach, many of the groupings remain the same, thus validating the hypotheses that certain animal groups are related. However, significant disparities have also been found where the morphological approach suggests one relationship but the genetic data suggest another. The earlier hypotheses based on morphological data are then rejected because they have been falsified by new evidence. What is interesting is that several of the clades based on morphological data are also supported by the molecular data. For example, the clades based on tissue organization and body symmetry are supported as are the clades based on embryological data. What has not been supported are the clades based on body cavity development. Apparently, body cavity is not something that evolved in a systematic fashion. It may have appeared in different lineages or disappeared, and it is clear that one type of cavity did not evolve into another over evolutionary time.

Figure 20.1 can be used to show the usefulness of these cladistic ideas. Let's assume that you were given data about an unknown organism that met the criteria for being an animal. You are asked to assign it to one of the ten phyla you will study in this course. Figure 20.1 tells you that you should first ask, Does it have tissues? If it does not, then it is a member of the clade Parazoa which contains one phylum, Porifera. On the other hand, if it has tissues, then it is a member of the clade Eumetazoa which contains most animals. If you establish that it is on the Eumetazoan branch, you must now determine its symmetry. If it is radially symmetrical, then it must be a member of the phylum Cnidaria. If it is bilaterally symmetrically, as most animals are, more information will be needed to narrow down the choices.

Assuming your unknown is a bilaterally symmetrical animal, you will now need information about its embryological development. Figure 19.1 in the previous lab topic shows the early developmental stages of a starfish. Most animals go through similar stages as they develop from zygotes. Look at the last photo in the sequence. This is called a gastrula stage and it is when the digestive system develops. In one clade, called the **protostomes,** the opening of this tube to the external environment will become the mouth. In the other clade, called the **deuterostomes,** the opening that you see in the figure will become the anus.

There are additional differences between the two clades but these will not be discussed until later lab topics. Let's assume that the data you have will allow you to determine if it is a protostome, as most animals are.

Note that the protostomes are divided into two clades based on data from gene analysis: lophotrochozoans and ecdysozoans. Access to genetic data would allow you to decide what clade your unknown animal belonged to, but that would take some insights to interpret. Alternatively, you could make use of certain traits that correlate with the genetic data. The animals in the lophotrochozoan group use subcellular structures called cilia to move water over their outer surfaces for feeding or for locomotion. The ecdysozoan clade lacks cilia. All ecdysozoans are covered by a cuticle, an outer skeleton which protects the body and is periodically shed as the animal grows. Assuming that your animal had ecdysozoan characteristics, you now know that it is a member of the phylum Nematoda or phylum Arthropoda. You would now use the phylum characteristics for the final assignment to a phylum. For example, if it was worm-like, had no appendages, and the exoskeleton was made of collagen, it would be a nematode.

The example you have just worked through shows the utilitarian value of grouping organisms into clades. With the appropriate information at five decision points, an animal can be classified into a phylym. However, there is more to it. Applying the underlying philosophy of systematics, those animals found in a clade are thought to be genetically similar, *i.e.* related through evolution.

The assumption is that animals evolved from a colonial flagellated protist nearly 700 million years ago (mya). We are not sure what this animal looked like as it is probably extinct today. However, we can hypothesize that it most likely was simple without tissues. From these animals evolved a group that had tissues, cells specialized for particular functions, and that formed organs. This primitive eumetazoan could have been radially or bilaterally symmetrical, although it is hypothesized that radial symmetry is the primitive condition and bilateral developed later.

The bilaterally symmetrical group is the more interesting because it gave rise to most of the types of animals we see today. Consequently, the evolutionary development of bilateral symmetry is considered a major event in animal evolution. Bilateral symmetry most likely affected nervous system development. A bilaterally symmetrical animal moves through its environment one end first. This may have caused selection for clustering of sensory receptors with associated nerves on the end we now recognize as the head with nerve tracts that led to other body regions. The general adaptability of this body plan is attested by the number of animals having it. During the Cambrian Explosion from about 565 to 525 mya, the three major clades with bilateral symmetry appeared in the fossil record. These are Deuterostomes. Protosotomes-Lophotrochozoans, and Protostomes-Ecdysozoans. The origins of most of the animals in the 35 phyla we see today are traceable to this time.

Animal Phylogeny: Investigating Animal Body Plans　　**251**

In this and subsequent lab topics on animal diversity, you will look at 10 of the 35 phyla. The intent is to give you some common experience with animals that you might never have seen before. As you study the animals, there will be many new anatomical and taxonomic terms to learn. Flashcards and memorization sessions will help you. However, do not neglect to comprehend the big picture. The big picture is gained by comparing and contrasting one animal group with another. For example, all animals must perform the following functions:

- Feeding and digestion
- Obtaining oxygen and losing carbon dioxide
- Getting nutrients and oxygen to remote cells
- Maintaining salt and water balance as well as eliminating metabolic wastes
- Locomotion and support
- Sensing and reacting to the environment
- Reproduction

As you work through the animal kingdom, you should be able to describe how each phylum studied has solved the problem of performing this list of functions. If you are able to do this, you will achieve a sense of diversity that is fascinating rather than bewildering.

LAB INSTRUCTIONS

In this lab topic, you will study representatives from four phyla: **Porifera, Cnidaria, Platyhelminthes,** and **Nematoda.** These animal phyla were chosen for the first lab topic on the animal kingdom because they nicely illustrate differences in tissue organization, symmetry, and development of body form in the bilaterally symmetrical animals.

Before starting this lab, I will propose a hypothesis for you to test, to falsify if you can. The hypothesis is: *"There is no evidence that animals in the four phyla studied here can be arranged in a sequence from simple to complex."* If you find evidence that you can, then you have falsified the hypothesis and must accept an alternative. Formulate an alternative hypothesis and state it below.

Phylum Porifera

The 9,000 or so species of sponges in the phylum Porifera are among the simplest multicellular animals because they lack distinct tissues and organs. Phylogenetically, the sponges are an early branch away from the main evolutionary patterns seen in the animal kingdom.

Body Plan Sponges have a simple body plan consisting of about two cell layers supported by fibers and secreted mineral elements around a central cavity called a **spongocoel.** The spongocoel is not a coelom or a digestive system; it is part of the water movement system of the sponge. The body walls of sponges are perforated by numerous openings that allow water to flow into the spongocoel. Unique cells with flagella, called **choanocytes** are found only in this phylum, and move water in through the pores on the body surface and out through a large opening, the **osculum** (see fig. 20.2).

The body walls of sponges are supported by a skeleton consisting of either (1) calcium carbonate crystals, (2) silica crystals, or (3) fibers of a protein called **spongin.** Some have both spongin and spicules. These three types of skeletal elements form the basis for dividing sponges into three taxonomic classes. Body shapes vary and do not have a regular symmetry, although they are often chimney-like.

1. ▷ Obtain a whole-mount slide of a simple sponge. Look at it only with your scanning objective. The slide is thick and if you use the other objectives, they will smash the coverslip. Note the base, the pores on the surface, and the **osculum.** The skeleton made of calcium carbonate should be visible as **spicules.** In the lab there should be other specimens of spicules, including the intricate glass skeletons. What do you think their skeletons are made of? _____

You may want to use the stereoscopic microscope to observe the specimen.

2. ▷ ***Functional Anatomy*** Now obtain a slide of a longitudinal section of *Leucosolenia* and look at it first with the dissecting microscope. Compare the section to figure 20.2. What type of body plan does *Leucosolenia* have? _____

Using your compound microscope, look at the cell organization. How many cell layers are there between the outside and the central spongocoel? _____ Look closely at the cells lining the spongocoel, using your high power objective. Although difficult to find, you may see choanocytes (fig. 20.2). The coordinated beating of their flagella drives water out the osculum and draws water in through the surface pores. This waterstream is essential. It brings in a constant stream of food particles as well as oxygen and removes carbon dioxide and ammonia produced in metabolism.

Figure 20.2 Longitudinal sections through different types of sponges showing the relationship of spongocoel to choanocytes and canals: (a) asconoid body type; (b) syconoid body type; (c) leuconoid body type.

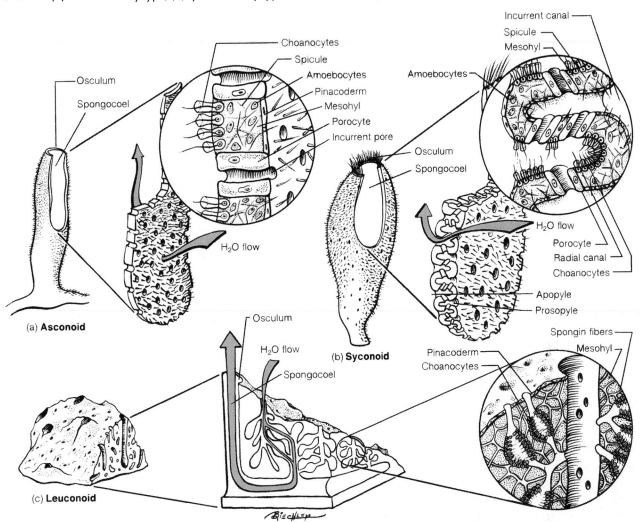

(a) **Asconoid**

(b) **Syconoid**

(c) **Leuconoid**

Sponges are filter feeders. They feed by removing small particles of organic matter from the water entering the spongocoel. The movement of the flagellum on each choanocyte causes a vortex to form in the collar region of the cell. Particles settle into the base of the collar where they are engulfed by extensions of the choanocyte's membrane to form food vacuoles. Do sponges have intracellular or extracellular digestion? _____

3 ▷ Other cell types should be visible on the slide. The external surface of the sponge is composed of **pinacoderm** cells. The pinacoderm layer is interrupted by **porocytes,** cells that form hollow cylinders through which water enters the spongocoel. Between the pinacoderm and choanocyte layers is a gelatinous matrix, the **mesohyl,** in which spicules are embedded and often extend through the pinacoderm. Amoebocytes in the mesohyl manufacture the spicules. These cells also can transform into gametes during the reproductive season. Sponges are sessile. They lack muscles and nerves and do not move.

Reproduction Sponges can reproduce asexually by fragmentation and budding. Small pieces of sponge that are broken off will grow into new individuals. Sexual reproduction is the usual way sponges reproduce. Most sponges are **hermaphroditic** and produce both eggs and sperm. Sperm, derived from amoebocytes in the mesohyl, are released by one individual and are drawn into a second in the feeding current. They are picked up by choanocytes that lose their collar after ingesting sperm and carry sperm to the eggs in the mesohyl. The resulting zygote develops into a flagellated larva.

Parts (b) and (c) of figure 20.2, show the basic body organization of more complex sponges. The **sycon** body type can be thought of as an accordionlike folding of the ascon body type. The **leucon** type is a more complex folding where individual chambers are created. Zoologists think that the location of choanocytes in distinct chambers makes them more efficient in capturing food particles. Examine the specimens of bath sponges in the lab. What type of body plan do they have? _____

Animal Phylogeny: Investigating Animal Body Plans

In your study of sponges did you find any evidence that they have organs or organ systems? Describe the reasons for your answer.

How do you think sponges perform the important physiological functions of respiratory gas exchange, circulation, and excretion?

Phylum Cnidaria

Animals in the phylum Cnidaria (previously known as the Coelenterata) have tissues and rudimentary organs. They are radially symmetrical and lack any definite head or tail. The phylum includes about 10,000 species commonly known as jellyfish, sea anemones, and corals. Most species are marine. The name *Cnidaria* comes from the term **cnidocytes,** which are unique stinging cells found in these animals.

Body Plan   Cnidarians have bodies consisting of two well-defined tissues, the outer **epidermis** and an inner **gastrodermis** that lines the digestive system. They are described as being diploblastic (two layers of cells). Between these cell layers is a layer of gelatinous material called **mesoglea.** The rudiments of a neuromuscular system are also seen for the first time in this group. Cnidarians have a hair-net-like organization to their nervous systems. There is no "brain" or major coordination center, nor are there nerve cords.

The cnidarians have three basic body forms: a planula larval stage and two adult forms—a sedentary **polyp** stage and a free-swimming **medusa** (jellyfish) stage. The life cycles of many species involve one or more of these stages. The phylum Cnidaria contains three taxonomic classes: **Hydrozoa,** which includes *Hydra, Obelia,* and the Portuguese man-

of-war; **Scyphozoa,** which consists mainly of marine jellyfish; and **Anthozoa,** which includes sea anemones and corals.

A Polyp: Hydra

Hydra is a genus of sessile hydrozoans that live attached to stones and vegetation in freshwater streams and ponds. Most are brown or cream colored but some species in the genus *Chlorohydra* are green because of symbiotic algae. *Hydra* has only a polyp stage in its life cycle; it does not have a medusa stage. Your lab work will show you that *Hydra* is more complex than the sponges just studied because it has well-defined tissues and rudimentary organs (gastrovascular cavity, nervous system, and gonads) and is capable of involved behaviors.

4▶ ***Functional Anatomy***   Get a living *Hydra* from the supply area and put it in a small watch glass with a small amount of water. Study the animal with your dissecting microscope. Depending on its condition, it may be asexually budding another *Hydra* or it may have a swelling of the body wall, which is a developing gonad.

Let the animal sit for several minutes while you watch it intermittently. It should extend its tentacles and body.

Identify the mouth, tentacles, and body column (fig. 20.3). Gently poke it with a probe. Describe the behavior you observed. From this behavior would you conclude that *Hydra* has a nervous system? Muscles?

Your instructor may ask you to feed your *Hydra*. Use an eye dropper to add several *Daphnia* next to the *Hydra*. If you are lucky, it will immobilize one and ingest it.

Return the *Hydra* to the stock container when finished observing.

Figure 20.3   Anatomy of *Hydra*.

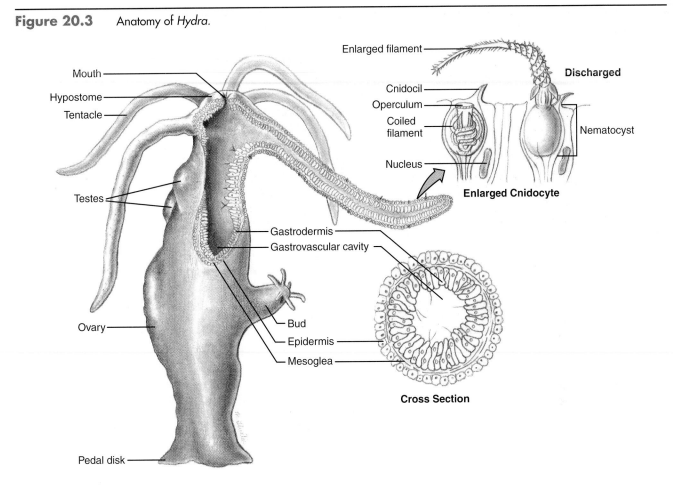

Cross Section

5 ▷   Look at a slide of a longitudinal section of *Hydra* (fig. 20.3). How many well-defined cell layers do you see in the body wall? _____ Identify the following structures: **gastrovascular cavity, gastrodermis, mesoglea, epidermis, mouth,** and **tentacles.** Did you see any evidence of specialized organs?

Examine the tentacles closely. You may be able to find a cnidocyte containing a **nematocyst,** an apical organelle containing an extensible thread that can entangle or penetrate small prey. Prey caught on the tentacles are moved to the mouth and stuffed into the gastrovascular cavity. Cells of the gastrodermis secrete digestive enzymes into the cavity digesting the food. Anything that is not digested is forcefully regurgitated by contractions of the body wall. *Hydra,* as all cnidarians, has a sacklike digestive system that lacks an anus. Anything entering or leaving must pass through the mouth. In contrast to sponges, digestion is extracellular rather than intracellular. Gastrodermal cells secrete digestive enzymes into the gastrovascular cavity.

If living *Hydra* are in the lab, try this method for seeing nematocysts. Add a living *Hydra* to a drop of 5% acetic acid on a slide. To conserve animals slice it in half longitudinally and share with someone. Add a coverslip and press methylene blue or fast green to the edge of the coverslip and wick the stain under the coverslip by touching a paper towel to the other side. Observe the edge of the tentacle with your high power objective and low light intensity. You should find threadlike structures extending from the tentacles. What are these? Alternatively, your instructor may give you a prepared slide of discharged cnidocytes.

How do you think that *Hydra* performs the important physiological functions of gas exchange, circulation, and excretion?

Reproduction   *Hydra,* like most polyps, can reproduce both asexually and sexually. Budding is the most common form of asexual reproduction. Buds form as an outgrowth of the parent's body wall that lengthens and forms tentacles. It eventually breaks off and lives independently.

Figure 20.4   The anatomy of the jellyfish *Gonionemus* sp.

Gastrovascular cavity
Mesoglea
Radial canal
Manubrium
Mouth
Velum
Circular canal
Suctorial pad
Cnidocyte batteries
Tentacle
Statocyst location

Subumbrellar Surface

Sexual reproduction is often triggered by changing environmental conditions to less than optimum. Gonads develop as clusters of gamete-producing cells in the body wall. These temporary gonads are seen as swellings. Those near the base are ovaries and will produce eggs. Those near the mouth will produce flagellated sperm.

Sperm swim to the eggs and fertilize them in position in the body wall. The resulting zygote may not immediately develop into a new individual. When it begins to develop, it divides to produce a mass of cells called a blastula that, in turn, develops into a larval stage. Cilia on the surface cells give the larva mobility. After swimming and drifting, the larva settles to develop into a new polyp.

A Medusa: Gonionemus

Commonly called jellyfish, the medusa is one of the alternative body forms found in cnidarians. Some cnidarians have no medusae in their life cycles and exist only as polyps. Hydra was an example. Others have only a medusa and no polyp. Yet others have both polyp and medusa in their life cycles. Members of the genus *Gonionemus,* found in shallow bays on both coasts of North America, have both polyp and medusa stages, but the polyp stage is quite small and will not be studied here.

Obtain a preserved specimen of *Gonionemus* in some fluid in a small dish from the supply area. You may want to look at this with your dissecting microscope to see some of the details of its anatomy.

Body Plan   The animal is obviously radially symmetrical. Any way you look at it, you cannot identify a head or a tail. There are obvious upper and lower surfaces, respectively called the **umbrellar** and **subumbrellar** surfaces. The body consists of an outer cellular covering and inner cellular lining of the digestive system. In between the two cell layers is a thickened **mesoglea** layer composed of polysaccharides and proteins that are hydrated, giving the body a jellylike consistency. About 95 to 98% of the jellyfish's weight is due to the water hydrating the mesoglea. This gives them buoyancy nearly equal to that of sea water; therefore, the animal needs to expend little energy in swimming to remain suspended as it hunts for food.

Functional Anatomy   As you study your specimen, compare it to figure 20.4. Observe the subumbrellar surface and identity the **manubrium** hanging from the center. At its end is the **mouth** which opens into the **gastrovascular cavity.** It connects to four **radial canals** that pass to the periphery of the bell-shaped body where they connect with a **circular canal** passing around the circumference. Tentacles hanging from the edge of the bell contain batteries of cnidocytes that can stun and capture prey. The tentacles move the prey into the mouth, hence to the gastrovascular cavity where extracellular digestion breaks it down. Digestion products are carried through the radial and circular canals to remotely located cells. Despite its complicated canal structure, the gastrovascular cavity is a sac with only one opening, the mouth. Anything that cannot be digested must be regurgitated.

Medusae can actively swim and respond to stimuli. Epitheliomuscular cells in the bell can contract and make the bell smaller, expelling any water beneath it. A thin shelf of tissue, the **velum,** passing around the rim narrows the cross-sectional area for the water to escape, thus increasing the velocity obtained from a single contraction. Although you probably will not be able to see them, small organs of equilibrium called statocysts are located at the base of some of the tentacles. They contain small calcareous concretions surrounded by nerves. If a medusa tilts in the water, the statocyst shifts and contact nerves on one side. Outputs from these nerves cause compensatory swimming movements.

The gonads are best seen from the subumbrellar view. They are attached to the radial canals. Sexes are separate. Eggs and sperm are released into the sea and fertilization is external. You will look at sexual reproduction in more detail in the next specimen.

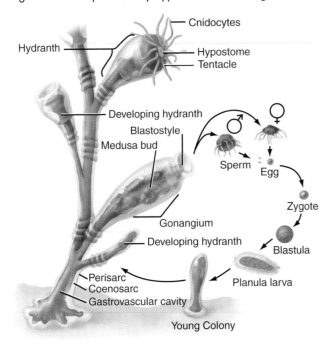

Cnidocytes
Hydranth
Hypostome
Tentacle
Developing hydranth
Blastostyle
Medusa bud
Sperm
Egg
Zygote
Gonangium
Developing hydranth
Blastula
Perisarc
Coenosarc
Gastrovascular cavity
Planula larva
Young Colony

How do you think an organism like *Gonionemus* performs the important functions of gas exchange, excretion, and circulation?

A Colonial Form: Obelia

Cnidarians in the genus *Obelia* are small marine animals that attach to seaweeds and pilings in shallow waters along the Atlantic and Pacific coasts of North America. *Obelia* is studied because it has a life cycle that contains both polyp and medusa stages.

7   Get a prepared microscope slide of *Obelia* from the supply area. Look at it with the scanning objective before switching to medium power. If available, also pick up a slide of the medusa stage of this animal.

Body Plan   At first this colony might look more like a plant to you than an animal. There is a central stalk with many branches. The colony is sessile and attaches by a basal holdfast to suitable substrates. At the tip of each branch there is a polyp (fig. 20.5). You should see two types of polyps: feeding polyps called **hydranths,** with tentacles, and reproductive polyps called **gonangia,** which lack tentacles. This is a colony of individuals, some of which are specialized for gathering food and others that are specialized for reproduction. The reproductive polyps are dependent on the feeding ones for nourishment. The similarity of the hydranth's anatomy to *Hydra* should be easy to recognize. The gonangia bud off small medusa which escape as individuals and sexually reproduce. If you have a slide of a medusa, note the similarities to *Gonionemus* as you look at it.

Functional Anatomy   Return to viewing the colonial form. Note how the body is surrounded by a translucent noncellular covering, the **perisarc.** It serves as an external skeleton supporting and protecting the living part that is collectively called the **coenosarc.**

The hydranths gather food in much the same way that *Hydra* does. The tentacles contain **cnidocytes** that immobilize prey that are brought into a **gastrovascular cavity** through the **mouth.** Digestion is extracellular. What is interesting is that the gastrovascular cavities of all individuals in the colony are interconnected. Scan the colony and note the continuous chamber in the coenosarc from one individual to the next. This means that if one hydranth captures prey, the digestion products will be shared by all in the colony. This common gastrovascular cavity is a big sac with many openings. Any undigested food is regurgitated through one of the mouth openings.

If you look carefully at his organism you will not see any organs for gathering oxygen or releasing carbon dioxide and ammonia. There is no circulatory system. How does a small animal like this perform those functions?

Reproduction   *Obelia* reproduces sexually. You should be able to see buds on the reproductive polyps. Small medusae will be formed in these buds and will break off and be released into the surrounding water. Both male and female medusae are produced. This is a form of asexual reproduction that allows a single sessile colony to produce many mobile reproductive individuals, increasing the chances of genetic outcrossing. Males produce sperm and females produce eggs. Fertilization is external and the zygote develops into a ciliated larval stage that will swim and drift before settling on substrate where it produces a new colony.

8 ▶ Stop your lab work for a moment. Pair off with another student and quiz each other about sponges and cnidarians. Explain how they feed and digest, obtain oxygen and get rid of carbon dioxide and ammonia, move and respond, and reproduce. Fill in the information required in table 20.1 at the end of this lab topic.

Phylum Platyhelminthes

Approximately 20,000 species are found in the phylum Platyhelminthes. It contains three classes: Turbellaria are called free living flatworms; Trematoda are flukes; and Cestoidea are tapeworms. Modern genetic analysis places the flatworms in the clade Lophotrochozoa. At one time flatworms were thought to be primitive animals because they lacked a body cavity and often had rudimentary organ systems. It is now thought that these are secondary characteristics. We are looking at them as our third group of animals because, like cnidarians, they have sac-like digestive systems while illustrating further organ development.

Body Plan These animals are bilaterally symmetrical with definite anterior and posterior ends and dorsal and ventral surfaces. Commonly called flatworms, the animals in this group are compressed dorso-ventrally and lack any appendages. They have an organ level of organization with functioning systems for digestion, excretion, movement, coordination, and reproduction. Flatworms are **acoelomate,** meaning that they do not have a body cavity as do vertebrates (and many others). Instead, the internal organs are surrounded and touched by tissues called parenchyma. Flatworms are **triploblastic,** meaning that their tissues are derived from three layers: a endoderm layer forms the lining of the gut; mesoderm forms most of the body tissues; and ectoderm gives rise to the outer covering of the animal. Most of the animals in this group are parasitic, living in the digestive systems of vertebrates. Consequently, you will see that many of the organ systems are reduced, while the reproductive system is greatly enhanced. For example, tapeworms lack a digestive system and absorb digestion products from the host's intestine. They have limited muscle development since they live anchored in the gut.

Class Turbellaria

Members of the genus *Dugesia* are common freshwater planarians in the United States. They live under rocks and sticks in streams and lakes.

Figure 20.6 Internal anatomy of planaria, a flatworm. Nervous and excretory systems are not shown and will not be visible on your slide. Diagram shows digestive system on left and reproductive system on right. Both systems are found on both sides.

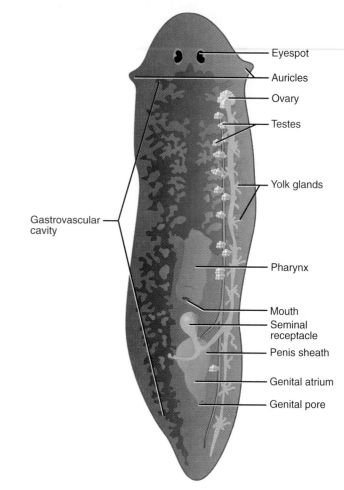

- Eyespot
- Auricles
- Ovary
- Testes
- Yolk glands
- Gastrovascular cavity
- Pharynx
- Mouth
- Seminal receptacle
- Penis sheath
- Genital atrium
- Genital pore

9 ▶ Functional Anatomy Obtain a living planarian, place it in a petri dish in some spring water, and observe its locomotion. Is there a definite "head" end? What type of symmetry do you see?

Observe what happens when you gently touch the planarian's head with a probe or when you turn the organism over. Normal locomotion occurs when cilia located in cells covering the ventral surface beat in a mucus trail secreted by the animal. Their movement is very smooth and steady. Muscles in the body wall allow the twisting, shortening, and extensions seen when the animal is disturbed.

10 ▶ Place a very small piece of raw meat in the dish and observe. What does the planarian's ability to move and behavior tell you about its muscle and nervous systems?

Note the two light-sensitive **eyespots** on the head end. Although not visible, there are clusters of nerve cell bodies adjacent to these, forming a primitive integrating center. Nerve cords run along the animal's length.

11▷ Obtain a stained whole mount of *Dugesia* and compare it to figure 20.6. The branched gastrovascular cavity should be clearly visible. How many lobes extend anteriorward? _____ Posteriorward? _____

How does a planarian, feed? They are carnivorous scavengers. The mouth is on the ventral surface about halfway along the body. A tubular **protrusable pharynx** connects to it. The tube can be extended. Enzymes released through the pharynx partially digest any food, and the resultant slurry is sucked into the gastrovascular cavity where digestion is completed.

Each of the lobes of the saclike gastrovascular cavity is highly branched. What is the advantage of this branching in an animal that lacks a circulatory system?

Flatworms lack an anus. Do they have a complete or an incomplete digestive system? _____

The excretory and nervous systems will not be apparent in your specimen. Special staining techniques must be used to see them. The reproductive system will not be studied in detail in this organism. No specialized respiratory gas exchange organs are present nor is there a distinct circulatory system. How do you think planarians carry out these important functions?

12▷ Now obtain a microscope slide of a cross section of a *Dugesia* and note the obvious organ and tissue organization. Compare the slide to figure 20.7.

Can you see well-defined layers of cells?

Are organs present?

Does the planarian have a body cavity distinct from its gastrovascular cavity?

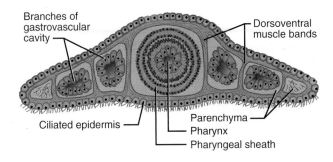

Figure 20.7   Cross section through anterior region of the planarian. Note absence of body cavity.

Branches of gastrovascular cavity — Dorsoventral muscle bands — Ciliated epidermis — Parenchyma — Pharynx — Pharyngeal sheath

Are the tissues and organs seen in this cross section more or less complex than those seen in sponges? Cnidarians?

The ventral surface of the worm can be identified by finding the **ciliated cells** on the surface: these allow the worms to move in a gliding fashion. **Gland cells** on the ventral surface secrete mucus, which aids in locomotion. Some cells on the dorsal surface contain darkly staining **rhabdites.** When provoked, planarians release the sticky contents of the rhabdites, producing a repellent slime.

Beneath the surface find the ends of the **longitudinal muscles.** What happens to the animal's shape when they contract? Also find the **circular muscles,** which pass around the animal's body. What happens when they contract? **Dorsoventral muscles** should also be visible passing from the dorsal to the ventral surfaces. When they contract, what happens to the shape of the worm?

Class Trematoda

All adult members of class Trematoda, commonly called **flukes,** are parasitic in vertebrate hosts. After the eggs are shed, they go through complex life cycles involving larval stages that infect intermediate hosts, usually snails.

Clonorchis is a fluke that lives in human bile ducts, releasing eggs into bile flowing to the intestine from where they are voided with the feces. Figure 20.8 shows the life cycle of this organism. Eggs develop into intermediate larval stages that first inhabit snails and then fish as intermediate hosts. Eating raw or improperly prepared fish containing *Clonorchis* larvae leads to infection in humans. The adults mature in the human body, producing eggs that start the cycle over again. It is a common parasite in Asia.

Animal Phylogeny: Investigating Animal Body Plans

Figure 20.8 Life cycle of the human liver fluke *Clonorchis*. Infected human passes eggs in feces. A miracidium larva, hatching from an egg, infects a snail. The miracidium asexually reproduces in snail, giving rise to cercaria larvae. These escape from snail and penetrate fish. In fish, cercaria encyst in muscle. Humans are infected when they eat raw fish containing encysted larvae.

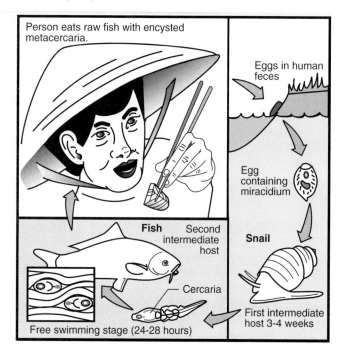

Figure 20.9 Anatomy of an adult human liver fluke.

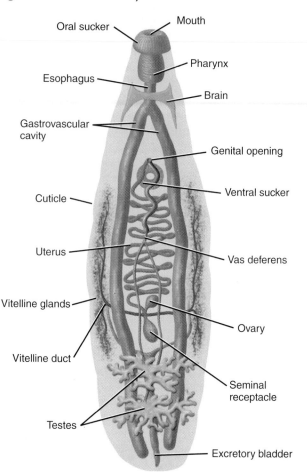

13 Obtain a slide of *Clonorchis sinensis*. Look at the slide under the low-power objective and find the structures shown in figure 20.9.

Functional Anatomy Like its free-living relative you just studied, *Clonorchis* has a dorsoventrally flattened body. It lacks eyespots and a protrusable pharynx. At the anterior end is an **oral sucker** which it uses to maintain its position in the bile duct. A second sucker is located on the ventral surface. The mouth is in the center of the sucker. It ingests cells lining the bile duct. A short **esophagus** leads to two branches of the **gastrovascular cavity.** Would you expect to find an anus in this animal? How are nutrients distributed throughout the organism?

Would you expect to find a body cavity in a flatworm like this?

Reproduction The reproductive system occupies a substantial part of the body. These animals are hermaphroditic. Consequently, they have both male and female reproductive organs, but self-fertilization does not occur.

Locate the paired **testes** in the posterior third of the worm. Small ducts lead from the testes to the **genital** opening located near the ventral sucker.

The **ovary** is anterior to the testes. It produces eggs that pass into the **uterus** where they are fertilized by sperm, held in the **seminal receptacle** from previous copulations. Yolk from the **vitelline glands** combines with the fertilized egg and a shell is formed before the eggs are released through the genital opening.

As is the case with most parasitic organisms, the digestive, nervous, and muscular systems of *Clonorchis* are reduced, and the reproductive system is enlarged. This ensures that large numbers of eggs are produced to compensate for the great odds against successfully completing a life cycle. Each egg ingested by a snail will produce several larvae, amplifying the reproductive potential even more. Further development depends on the larvae infecting a fish and ultimately being eaten by a human. Many larvae simply die, but the more larvae there are, the greater the chances of survival for the species.

Figure 20.10   Anatomy of a tapeworm: (a) external anatomy of whole worm; (b) internal anatomy of a mature proglottid; (c) scanning electron micrograph of hooks on scolex; (d) photo of mature proglottid.

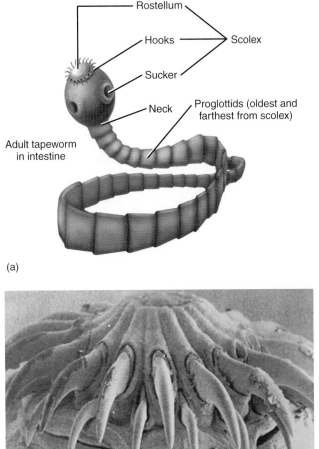

(a)

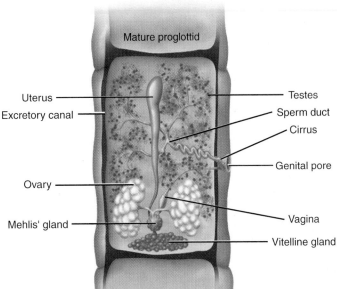

(b)

(c)

(d)

Class Cestoidea

There are about 1,000 species of tapeworms; all are parasites living in the intestines of vertebrate hosts. The general body form is similar to a long ribbon made up of small units called **proglottids.** Some tapeworms may be over fifteen meters long! They are highly adapted to a parasitic way of life. The anterior end has special holdfast structures (fig. 20.10*a*) which anchor the animal to the intestinal wall. They lack a digestive system and absorb nutrients from the intestinal fluids, sometimes causing malnutrition and weight loss in the host.

14   From the supply area, obtain a composite or separate slides showing a scolex, mature, and gravid proglottid. From these representative samples taken from different body regions, you should be able to piece together an un-derstanding of a tapeworm's anatomy. There may also be whole tapeworms on display so that you can see the body regions that the slides were made from.

The **scolex** is on the anterior end. You can see the hooks and suckers that anchor it in the intestine. Behind the scolex is the **neck** region. Here repeated cell divisions produce new very small proglottids by mitotic cell divisions and differentiation. As new ones are formed, older ones are displaced toward the posterior so that the oldest proglottids are at the other end.

Look at a representative mature proglottid. Long tapeworms may contain thousands of these units. As you study the mature proglottid, refer to figure 20.10*b* to identify the organs. Most of the internal structure of the proglottid is devoted to reproduction.

Animal Phylogeny: Investigating Animal Body Plans

Tapeworms are hermaphroditic and each proglottid will contain both male and female reproductive organs which both empty their gametes into a common genital atrium. Internal, self-fertilization in common. Fertilized eggs develop in the uterus as a desiccation resistant shell forms around the developing zygote. Proglottids that are filled with mature eggs are called **gravid** proglottids and are found near the end of the animal. A single gravid proglottid may contain up to 100,000 eggs. Gravid proglottids are shed with the host's feces and break open, releasing their eggs to contaminate an area. Thus, the bulk of the tapeworm's body is virtually a reproductive machine, producing a huge number of eggs. This is most likely related to the risk inherent in a parasite's life cycle which depends a lot on chance happenings.

Tapeworm life cycles depend on the eggs being eaten by an intermediate host. The dog and cat tapeworm depends on a flea eating some of the eggs. Fleas living near the anus become infected with a larval stage. If a grooming dog or cat ingests a flea by licking or biting the anal area, the larval stage develops into a tapeworm when it reaches the second host's intestine.

Phylum Nematoda

The 90,000+ species of animals in this phylum all have well-developed tissues and organs. They are also bilaterally symmetrical and have body cavities. There are several classes, but we will treat them as one group. Recent DNA sequence analyses place the nematodes in the ecdysozoan clade with the Arthropods, even though they are quite different in appearance. The nematodes are being covered here with the simpler phyla because they illustrate in a rather simple organism additional body features that are found in almost all other animals of greater complexity.

Body Plan Compared to the animals you have studied so far today in the lab, there are two major differences that you might notice immediately as you begin your dissection. The animal has a body cavity and a tube within a tube organization.

Functional Anatomy *Ascaris* will be used to demonstrate the functional anatomy of nematodes. It is a parasite living in the intestines of swine and humans. It is similar in appearance to most free-living nematodes, except it is much larger.

15 ▷ Obtain a preserved *Ascaris* and look at its external features. Identify the **mouth** and the **anus.** Males will have a hooked posterior end. What is the sex of your specimen? _____ The body is covered by a noncellular **cuticle** composed of the protein collagen secreted by underlying cells. As the cuticle is shed, a new one replaces it. Why must a growing animal with an exoskeleton shed it?

Place the specimen in a dissecting pan and pin the anterior and posterior ends. Cut the animal longitudinally along the middorsal line and pin the body wall down to expose the internal organs. Flood your dissecting tray with water, so that the organs float. This will make structures easier to see. Identify the structures shown in figure 20.11.

When you cut through the body wall, you will notice that all of the internal organs are exposed because they lie in a cavity. This cavity is called a **coelom.** Having a coelom offers many advantages to animals. In fact, the vast majority of animals have coeloms, attesting to its importance. What are these advantages? As fluids ebb and flow during body movements, the fluids act as a circulatory system, distributing nutrients and removing wastes from remote cells. Fluids can serve as a hydroskeleton by transmitting pressure changes from muscle contraction in one body region to another. A body cavity also allows internal organs to move independently of movements in other organs or in the body wall. Lastly, a body cavity allows organs such as the stomach to expand and store food or gonads to enlarge during breeding season. The coelom in nematodes, as well as in several other animals, is known as a pseudocoelom because its lining is not completely developed.

The tube-within-a-tube body plan should also be obvious with the digestive tube running from the mouth to the posterior anus and, in normal life, surrounded by the body wall tube. Is a tubular digestive system more or less efficient than the sac-like digestive system of *Clonorchis?* _____ Why?

What is noteworthy about a tube is that it has two ends, a mouth and an anus. With this organization, indigestible food no longer need be regurgitated. It simply passes along a tube where digestion sequentially occurs and any residue passes out the anus.

Two nerve cords run the length of the animal but are different to see. There is no major nerve center that would qualify as a "brain."

Reproductive System In male worms, the testes are not paired, A single thin tubular **testis** passes into a vas deferens that conveys sperm to an enlarged seminal vesicle where it is stored. During mating two spicules are inserted into the female genital pore as sperm are transformed.

The female reproductive system is Y-shaped. The single vagina branches into two uteri that gradually narrow to form oviducts and finally small tubular ovaries. See figure 10.3.

Fertilization is internal, usually in the ovary region, and then a shell is laid down around the zygote as it passes toward the vagina. Juveniles are usually already formed by the time the egg is released.

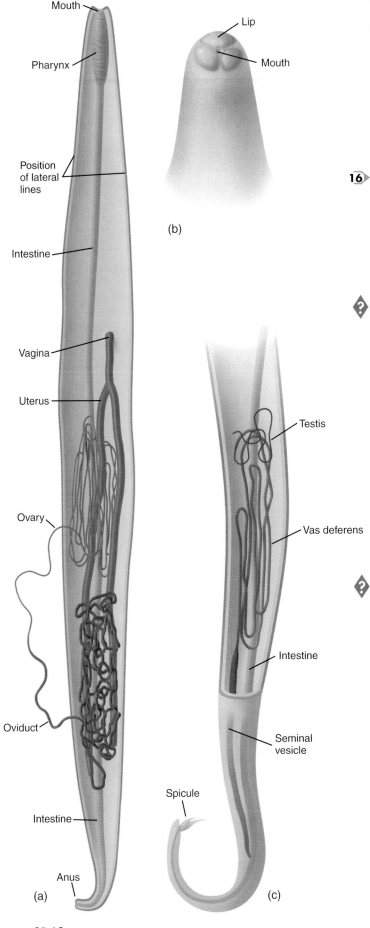

Figure 20.11 Anatomy of *Ascaris*: (*a*) internal anatomy of female; (*b*) oral view of female; (*c*) posterior end of male.

The eggs pass out of the host's digestive system with the feces. When ingested by a suitable host, the juveniles hatch and burrow out of its intestine to the liver and lungs where they mature. Mature worms burrow back to the intestine. A fascinating question is: How are the worms guided as they burrow through the host?

16 Figure 20.12 shows the worm in cross section. Note the open body cavity with the free-floating digestive and reproductive systems unattached to the body wall and surrounding tissues. The body wall consists of a noncellular **cuticle** covering the animal, a cellular **epidermis** that secretes the cuticle, and a **longitudinal muscle** layer. No circular muscles are found in the body wall. What types of movements would you predict that *Ascaris* can make? What are the structural and protective advantages of having a semirigid cuticle?

Nematodes lack specialized organs for gas exchange and circulation. How do you think they perform these important physiological functions?

Figure 20.12 Cross sections of *Ascaris:* (a) male and (b) female.

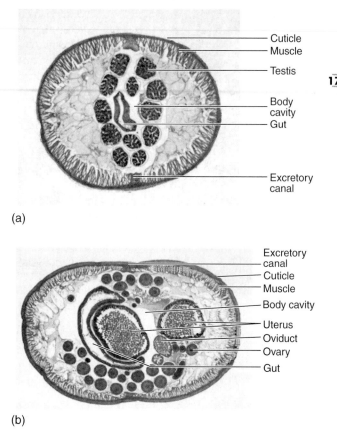

(a)

(b)

In your own words, describe the major differences in body plan between a nematode and planarian when viewed in cross section.

When finished with your dissection, dispose of the carcass as directed. Be sure to wash your hands and dissection instruments thoroughly. There is a possibility that all of the *Ascaris* eggs were not killed by the preservative; you would not want an infestation.

Vinegar eels

17▶ To finish your study of nematodes, you will observe living nematodes. Vinegar eels are small nematodes that live in rotting vegetation and can be cultured in vinegar. Make a slide by putting a drop of the culture on a slide. Remember the nematodes are at the bottom of the culture because they cannot swim.

Add a coverslip and look at the nematodes through your compound microscope. Note the thrashing movements. Because nematodes have a cuticle that prevents shortening, contraction of the longitudinal muscles causes them to bend side to side. You should be able to see the same organs inside as you saw in *Ascaris*.

Rogue's Gallery

While most nematodes are free living, many are parasites of animals and plants. In the lab are two demonstration slides that you should take a look at: pinworms and Trichinella. Hookworm and filarial worms are other nematodes that are human parasites as well.

Pinworm *Enterobius vermicularis* is a common parasite of children both here in the United States and abroad. It lives in the large intestine where mature females will migrate to the anal region to lay up to 16,000 eggs. Migration causes irritation and itching in the perianal area. A child scratching the anal area contaminates his/her hands. As children often put there hands in their mouths, the cycle is **18▶** completed as new eggs are ingested. Take a look at the demonstration slide of this nematode.

Trichinella *Trichinella spiralis* is a serious parasite of rats, pigs, and humans as well as other mammals. Infection is called trichinosis. Fortunately, public health practices have made it rather uncommon.

The life cycle is keyed to the eating habits of carnivores. Adults live embedded in the lining of the intestine where they mate. Females will bear up to 1,500 juveniles. The immature forms burrow out of the intestine and enter the circulatory system where they are carried throughout the body. They burrow into the muscles where they are encapsulated by reactions of the host's tissues. The life cycle ends there unless the host is eaten by another carnivore. If contaminated muscle is ingested, the juveniles leave the cyst and take up residence in the new host's intestine where they produce juveniles to continue the cycle.

Since this parasite can go through pigs, humans are potential hosts if they eat raw pork. Public health officials have broken the cycle in two ways. We all know that we should not eat raw pork, and trichinosis is the reason why. Second, it used to be the practice to feed garbage scraps to hogs. If the garbage contained pork muscle that had not been cooked, the parasite continued in pig herds. There are now laws that prohibit the feeding of uncooked garbage to **19**▸ hogs, and this parasite is the reason why. The chances of eating infected pork are very low.

The demonstration slide of this nematode is a section of infected muscle. What you will see in the muscle tissue is a cyst with a small worm coiled inside waiting to be eaten!

Learning Biology by Writing

Write an essay summarizing your observations that support or fail to support the null hypothesis that these four phyla can be arranged in a phylogenetic sequence from simple to complex. Be sure to recognize those characteristics related to parasitism as secondary developments and not as major phylogenetic trends.

Your laboratory instructor may ask you to answer the Lab Summary and Critical Thinking Questions that follow.

Internet Sources

Many of the species in phylum Platyhelminthes and phylum Nematoda are parasites. Scientists who study these animals belong to professional societies that promote research and dissemination of information. Use your browser to connect to one of the following: American Society of Parasitologists at http://www-museum.unl.edu/aspl Society of Nematologists at http://www.ianr.unl.edu/son/ What did you find most interesting at these sites?

Lab Summary Questions

1. Fill in table 20.1 on the next page to summarize your observations on the evolution of body organization.
2. Describe the changes in the digestive system as you move from the sponges through the cnidarians and flatworms to the nematodes.

3. What are the advantages of a tube-within-a-tube body plan compared to a body plan in which the gut is saclike?
4. Contrast radial and bilateral symmetry. Why is nervous system development coordinated with bilateral symmetry?
5. What is a body cavity, and why is it beneficial to an animal?
6. None of the animals studied in this lab had organs specialized for respiratory gas exchange or circulation. How do these animals perform these important physiological functions?
7. At the beginning of this lab, the following hypothesis was made: Animals in the four phyla studied here can be arranged in a sequence from simple to complex. What observations have you made today that falsify this hypothesis?

Critical Thinking Questions

1. Radially symmetrical animals lack a brain and nerve cord. Why?
2. Sponges are sessile and often green in color. Why aren't they considered plants?
3. To control parasitic infections by flukes, snails are often removed from ecosystems. Why?
4. If dogs are kept free of fleas, they will not have tapeworms. Why?
5. If you were offered a generous serving of wild boar that was cooked so that the center was red (or rare), why should you refuse it?
6. Brushing against some jellyfish can kill you. Why?

TABLE 20.1 Summary of characteristics in simple animals

	Porifera	Cnidaria	Platyhelminthes	Nematoda
Common names				
Tissues?				
Symmetry?				
Name organ systems present				
Coelom?				
Herbivore, predator, or parasite?				
Unique features				
Describe body plan				
Habitat				

LAB TOPIC 21

Protostomes I: Lophotrochozoans and Development of Complexity

Supplies

Preparator's guide available on WWW at
http://www.mhhe.com/dolphin

Equipment

Compound microscopes
Dissecting microscopes

Materials

Preserved specimens for demonstrations
    Polychaetes
    Chiton
    Snail
    Tusk shells
Preserved clams for class
Live earthworms (night crawlers) for class
Frozen (thaw) whole squid for class
Live leeches and polychaetes for demonstration
Live *Lumbriculus variegatus* culture kit
Prepared slides
    Earthworm, cross section
    Clam, cross section
    Bryozoan zooid, *Cristatella*
Dissecting trays and instruments
Capillary tubes

Solutions

10% ethanol

Student Prelab Preparation

Before doing this lab, you should read the introduction and sections of the lab topic that have been scheduled by the instructor.

You should use your textbook to review the definitions of the following terms:

Annelida	gastrula
blastopore	lophophore
Bryozoa	Lophotrochozoa
clade	mesoderm
coelom	Mollusca
ectoderm	protostome
endoderm	schizocoelom
enterocoelom	spiral cleavage
Ecdysozoa	trochophore larva

You should be able to describe in your own words the following concepts:

    What a clade is and why it is important
    What a gastrula is and why it is important
    The body plan of an annelid
    The body plan of a mollusc

As a result of this review, you most likely have questions about terms, concepts, or how you will do the experiments included in this lab. Write these questions in the space below or in the margins of the pages of this lab topic. The lab activities should help you answer these questions or you can ask your instructor during the lab.

Objectives

1. To illustrate a lophophore and trochopore
2. To illustrate the development of complexity in bilaterally symmetrical animals
3. To dissect an earthworm, clam, and squid
4. To collect data to test the hypothesis that annelids and molluscs are more complex than the animal phyla studied previously

Background

In the previous lab, you began your study of the animal kingdom. You saw how animals with tissues had a more complex body plan than those without. Typically, animals with tissues were more diverse in their appearance and had organ systems to perform many of the processes necessary for life. You saw how those with tissue organization, the Eumetazoa, could be divided into two groups, those with radial symmetry and those with bilateral symmetry. Radially symmetrical animals are not very diverse. Only two phyla out of 35 have that body symmetry. Bilateral symmetry, on the other hand, had tremendous potential and is found in 31 of the 35 recognized phyla.

Figure 21.1 Two embryological development patterns are found in bilaterally symmetrical animals. In the protostome pattern the blastopore becomes the mouth and mesoderm develops from a splitting of cell masses. In deuterostomes the blastopore becomes the anus and mesodern forms from cells splitting off the developing gut.

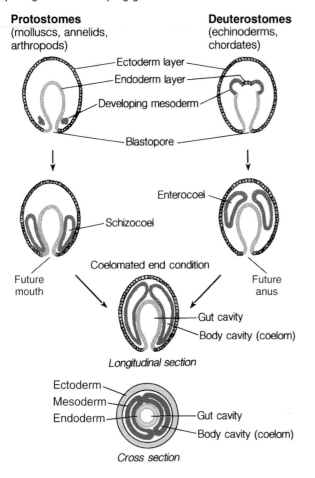

From these layers will come all other cells that are found in the adult animal, but the origins of the adult cells are not random. They follow very strict lines of development. The outer layer of the late gastrula is called the **ectoderm.** It will form the outer covering of the animal and the nervous system. The cells lining the digestive tube of the gastrula, called the **endoderm,** will develop into the lining of the digestive system and often digestive glands. The remainder of the adult animal's body (muscles, skeleton, connective tissues, reproductive tissues, *etc.*) will come from a third layer of cells, called the **mesoderm.**

In addition, most of the bilaterally symmetrical animals have a tube-within-a-tube body plan with a body cavity **(coelom).** You have seen a coelom, if you have ever cleaned a fish or a turkey before cooking. It is the cavity in which the organs are found. A coelom has many functions: (1) organs contained in it can enlarge or move independent of the body wall, such as the stomach extending while feeding and later emptying; and (2) fluids in the cavity can cushion internal organs from injury, serve as a primitive circulatory system, or act as a hydrostatic skeleton. The embryologica origin of the coelom differs between the protostomes and the deuterostomes.

In the **protostomes,** the mesoderm originates as a solid mass of cells that grows inward from the endoderm near the blastopore (fig. 21.1). This group of cells then splits to form a sac that grows out to line the body cavity. The term **schizocoelomate** describes how the body cavity develops in protostome animals by splitting of mesoderm buds to form sacs. In the **deuterostomes,** which includes the echinoderms and chordates (lab topic 23), the mesoderm arises as an outpocketing of the endoderm to form sacs at the end away from the blastopore (fig. 21.1). These sacs grow out to line the coelom. The term **enterocoelomate** describes the origin of the coelom from the digestive lining in these animals. While the protostome-deuterostome dichotomy seems very logical and easy to apply, it is often difficult to make the distinction in practice. The embryology of many animals is difficult to observe to make a determination of the fate of the blastopore, let alone to determine how the mesoderm and the coelom develop. In fact, the embryological developmental sequence of many animals is yet to be described. See lab topic 19 for a discussion of embryology. Over the years as new embryological evidence has been discovered, there have been debates over whether a phylum should be considered a protostome or deuterostome. Only recently, with the advent of molecular techniques and genetic analysis, has a new and easier method appeared.

Genetic analysis supports the protostome/deuterostome dichotomy (splitting into two). About four phyla are considered deuterostomes, the rest, some 27 phyla, are protostomes. The molecular data indicate that in the protostomes, there are two additional clades based on genetic differences, a group called the **lophotrochozoans** and another called the **ecdysozoans.** In this lab you will look at some representative lophotrochozoans and in the next lab you will study the most successful ecdysozoan phylum, Arthropoda.

In this lab topic we will begin to look at bilaterally symmetrical animals in detail and will continue to do so for two more labs. The bilaterally symmetrical animals are divided into two clades (groupings) based on embryological characteristics (see fig. 20.1). Understanding the gastrula stage of embryonic development and the development of the digestive tube in bilaterally symmetrical animals is the key to understanding the basis for these clades. In one clade, the **protostomes,** the initial opening of the digestive tube (blastopore) in the gastrula will become the mouth in the adult animal. In the other group, the **deuterostomes,** that blastopore does not become the mouth. Instead it becomes the anus (fig. 21.1).

Correlated with the fate of the blastopore in these two groups is a second embryological characteristic: development of the mesoderm. Bilaterally symmetrical animals are **triploblastic,** meaning that three distinct layers of cells are found early in development, usually in a late gastrula stage.

Although the lophotrochozoan clade is based on genetic similarities, it is a diverse group. The name, lophotrochozoan is a made-up word, new to the language of biology, which attempts to capture two morphological characteristics that are seen among members of the clade. Not all of the phyla have both characteristics. About four of the phyla have a feeding structure called a lophophore which consists of ciliated tentacles. This includes a phylum of small animals called bryozoans which you look at in this lab. Several, but not all, of the other phyla have a larval stage in their life cycle called a trochophore. Annelid worms and mollusks are included in this group. You will look at a trochophore larva and then study the functional anatomy of adult representatives from both phyla. You will not be looking at the 15+ other phyla that are included in this group.

Figure 21.2 (*a*) A colony of *Cristatella* sp; (*b*) an individual zooid with a prominent lophophore.

(a)

LAB INSTRUCTIONS

After looking at a lophophore and trochophore that give the unusual name to this clade of protostomes, you will study the functional anatomy of an earthworm, clam, and squid.

CLADE LOPHOTROCHOZOA

We will start our study of this clade by looking at some animals that demonstrate the distinguishing characteristics: the lophophore feeding organ and the trochophore larva.

Phylum Bryozoa (Ectoprocta)

To find animals with lophophores we must look among three or four phyla of lesser known animals, most of which are quite small in size. Bryozoans, also known as moss animals, ectoprocts, and lace corals, have been chosen to illustrate this group because study materials are readily available. There are about 5,000 species. Most are marine but about 50 species are found in freshwater where they are reasonably common as sessile-encrusting colonies.

1▷ Obtain a prepared whole mount slide of *Cristatella* sp., one of the freshwater species. Other species may be substituted. Look at it with your scanning objective. High magnification is not required to see the structures of interest.

Body Plan All but one bryozoan species are colonial. An individual is called a **zooid** (fig. 21.2). How many zooids are on your slide? _____ Colonies develop when zooids bud to form new individuals. The colonies of some species will have different kinds of zooids: some with jaw-like structures and others, with whip-like filaments. Both of these prevent other organisms from attaching to the colony surface. Zooids are bilaterally symmetrical and have a well-defined body cavity, tissues, and organs despite a microscopic size.

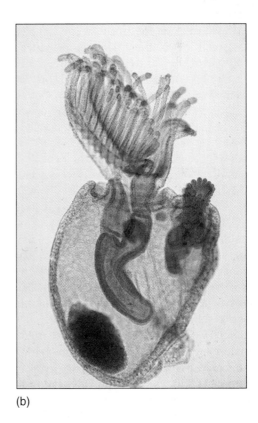

(b)

Colony shape can be erect and branching in some species so that the colony resembles a plant. The "stems" in these cases are actually hollow cylinders containing the zooids. Other times, the colonies will be flat encrusting mats on surfaces or a floating gelatinous mass. The animals live inside an exoskeleton, called a **zoecium,** made of the polymer chitin. It can be hardened by calcium carbonate deposition. *Cristatella* produces a jellylike colonial matrix with thousands of zooids embedded in the surface "facing" outward.

Figure 21.3 Representatives of phylum Annelida: (*a*) polychaete marine annelid; (*b*) the Christmas tree worm is a polychaete fan worm; (*c*) a freshwater leech.

(a) (b) (c)

Functional Anatomy Note the crown of ciliated tentacles extending from one end. This is the **lophophore** and is used in filter feeding. The beating of the cilia on the tentacles moves small protists towards the mouth located in the center of the crown. Particles are caught in mucus which passes into the mouth. The tentacles also function in gas exchange and excretion. Bryozoans have a complete digestive system which is U-shaped. The anus is located external to the lophophore so that fecal material does not foul the feeding apparatus. Bryozoans lack a circulatory and excretory system. As in arthropods and mollusks, muscles attach directly to the exoskeleton. The skeleton has openings in it that allow the animal to extend soft parts, but these can be quickly retracted by muscle contraction.

Freshwater reproduction is usually asexual, either by fragmentation of the parent colony or by internal budding to form a structure called a **statoblast.** It is composed of a group of cells surrounded by two chitinous valves. Statoblasts are able to overwinter and survive freezing while the parent colony dies. Growth can be rapid; a colony in Yugoslavia was observed to grow to 38,000 individuals over a five-month period.

Sexual reproduction also occurs. Individuals are hermaphroditic. Motile sperm are released at the tips of the tentacles and caught in the lophopores of other individuals. Eggs are brooded inside the exoskelton. The zygote develops into a motile larval stage that drifts in currents before settling to produce a new sessile colony.

Return your slide to the supply area. Fill in the information for phylum Bryozoa in table 21.1 at the end of this lab topic.

Phylum Annelida

Reflecting the obvious external and internal segmentation, "segmented worms" is the common name for the 15,000 species of animals in this phylum. Three taxonomic classes are in the phylum. The **Class Polychaeta** is the most diverse, containing about 11,000 species (fig. 21.3a and b). These are marine worms that burrow in the sea floor or coral

reefs, ingesting the substrate and digesting any organic detritus it contains. Some are carnivorous, feeding on mollusks. All have a tubular segmented body that bears lateral fleshy extensions of the body wall called parapodia which are used in locomotion and gas exchange. A group called the fan worms has feathery outgrowths from the anterior end which are used to filter protists from the water. Some make tubes of mucus and bits of sand. The life cycle of a polychaete worm includes a trochophore larva. The 1,000 or so species of leeches are in the **Class Hirudinea** (fig. 21.3). Most live in freshwater, although some are terrestrial forms living in moist rainforests. Leeches are carnivores and scavengers. Some feed on invertebrates, but others are bloodsucking parasites that feed by temporarily attaching to other animals. All leeches have at least one sucker and lack parapodia. Annelids in the **Class Oligochaeta** are commonly known as earthworms. Although there are several species that live in freshwater and marine habitats, the most familiar examples are terrestrial. They feed on organic debris and are important in turning over soil and as members of aquatic food chains. You will dissect an earthworm to develop an appreciation of annelid functional anatomy.

Annelids are bilaterally symmetrical animals with the protostome pattern of development and lophotrochozoan gene sequences. Body size ranges from less than 1 mm to 3,000 mm. Annelids clearly illustrate the tube-within-a-tube body plan. Annelids have a coelom surrounded by a muscular body wall with a complete digestive system running the length of the animal. They have well-developed organ systems for digestion, circulation, excretion, nervous coordination, movement, and reproduction.

Larval Anatomy

2 A trochophore larva (fig. 21.4) is found in the life cycles of some polychaetes and marine mollusks. It has become a morphological icon for the clade lophotrochozoa, even though it is not found in the life cycles of any freshwater or terrestrial species. So that you will have some idea of what a trochophore is, we use it here to start our study of the annelids.

Figure 21.4 (a) A trochophore larval stage of a marine polychaete; (b) artist's interpretation of larval anatomy.

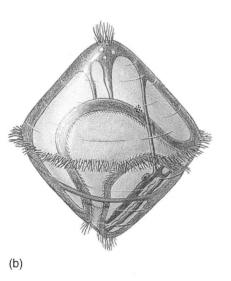

(a)

(b)

These free-living larvae hatch from fertilized eggs, usually within 24 hours or so following fertilization. The larva is fully functional except that it lacks a reproductive system. A band of cilia allow it to swim and to collect food. There is a fully functional gut. It has light sensory structures, touch receptors, and muscle bands. Following a period of drifting and feeding in the sea as plankton, the trochophore will metamorphose into an adult form in annelids and, in mollusks, into a second larval form called a veliger.

Annelid Diversity

3▷ Your instructor will have demonstration specimens of living leeches and polychaetes in the laboratory. Look at the specimens and note the obvious similarity to the earthworm.

◆ What characteristics do all annelids share?

◆ Based on external anatomy, how do leeches differ from earthworms? How are they similar?

◆ Based on external anatomy, how do polychaetes differ from earthworms? From leeches? How are they similar?

Earthworm Dissection

Earthworms can be obtained inexpensively at bait stores. Live worms are preferable for dissection. They can be anesthetized by submersion in tobacco water made with a crushed cigarette and then in 80% ethanol for a few minutes prior to dissection.

External Anatomy

4▷ Note the obvious segmentation of the earthworm. External structures in the earthworm are usually located by reference to the segments numbered from the anterior end. You can identify the anterior end by locating the **clitellum,** a swollen band covering several segments in the anterior third of the worm (see fig. 21.7). The clitellum secretes a cocoon around the eggs when they are released. Earthworms are hermaphroditic, having both male and female organs, so every worm has a clitellum.

Protostomes I: Lophotrochozoans and Development of Complexity

The mouth is located in the first segment, and over-hanging the mouth is a fleshy protuberance. The anus is in the last segment. There are no obvious external sense organs at the anterior end. The dorsal surface is identifiable by the **dorsal blood vessel,** which appears as a dark reddish line.

The surface of the worm has an iridescent sheen because light is refracted by the **cuticle,** a thin noncellular layer of collagen fibers. They are secreted by the cells of the underlying epidermis. Mucus secreting cells in the epidermis lubricate the body surface and keep it moist.

Run your fingers back and forth along the sides of the worm and feel the projecting chitinous bristles called **setae.** Separate sets of muscles allow these to be extended or retracted. Each segment has two pairs of setae on the ventral surface and two pairs on the side. How would the setae help the earthworm in locomotion?

Find the **excretory pores** located ventrolaterally in each segment, except for the first few and the last one. In the region of segments 9 to 15, the reproductive system openings are found on the ventral surface. Depending on the species, the **male pore** is usually in segment 15 surrounded by fleshy lips, and the **female pore** is in segment 14 just anterior to the male pore. In the grooves between the ninth and tenth and eleventh segments are the small openings of the seminal receptacles, part of the female reproductive system where sperm are stored following copulation. You may need to study this area with your dissecting microscope to see the openings indicated.

Internal Functional Anatomy

5 To examine the internal anatomy of the worm, lay it in a dissecting pan dorsal side up, pin it through the fleshy lip, stretch it slightly, and pin it through the last segment. Open the worm as in figure 21.5 by first cutting from clitellum toward anus.

Cut to one side of the dorsal blood vessel and keep points of scissors up against body wall to avoid cutting internal structures. In the area of segments 1 through 5 be very careful, because if you cut too deep you will destroy the "brain" and pharyngeal area. After the incision is made, pin the body open by putting pins at a 45° angle through every fifth segment of the body wall. This will provide a quick reference as you try to find various structures. When the specimen is opened and secure, add enough water to the pan to cover the open worm. This prevents the tissues from drying and floats organs and membranes for easier viewing.

Figure 21.5 Procedure for dissecting an earthworm.

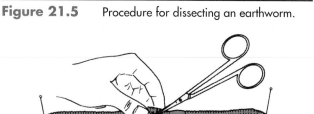

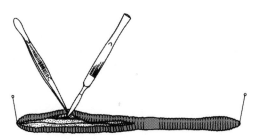

(1) Pinch skin with fingers. Cut through body wall (off center) from the clitellum to the anus. Do not damage internal organs by jabbing points downward.

(2) With scalpel, cut through septa on both sides. Pin body wall to tray.

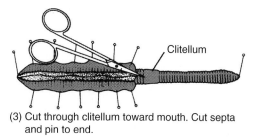

Clitellum

(3) Cut through clitellum toward mouth. Cut septa and pin to end.

Look at the internal organization of the worm and the obvious tube-within-a-tube body plan. Note the anterior concentration of structures (fig. 21.6). Also observe how the body cavity (coelom) is divided into compartments by cross walls called **septae.**

Each of the compartments is normally filled with fluid, which acts as a hydrostatic skeleton. Well-developed **circular** and **longitudinal muscles** in the body wall exert pressure on this fluid and allow a variety of movements.

If only the circular muscles contract, how will the shape of the worm change? How does the shape change when the longitudinal muscles contract?

Figure 21.6 Dorsal view of the internal anatomy of an earthworm: (*a*) appearance when first opened; (*b*) removal of reproductive organs reveals structures beneath.

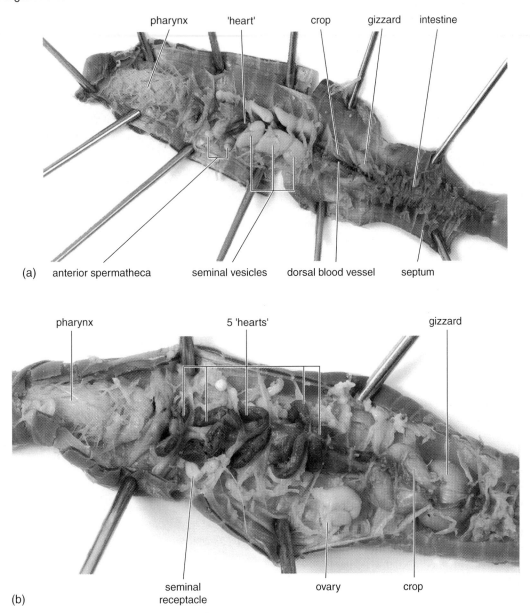

Digestive system Find the tubular digestive tract running from the mouth to the anus (fig. 21.7). Behind the mouth is the muscular **pharynx.** It appears to have a "fuzzy" external surface because several dilator muscles extend to the body wall. Contraction of these muscles expands the pharynx and sucks in particles of soil and detritus. Peristaltic waves of muscle contraction sweep ingested material down the **esophagus** into a thin-walled storage area, the **crop.** The material passes into the muscular **gizzard,** which grinds it into a fine pulp. Digestion and absorption of organic material take place in the **intestine** and undigestable soil particles are voided through the **anus** as castings. What is the longest part of the digestive system? How does its length relate to its function?

Figure 21.7 Lateral view of the internal anatomy of an earthworm.

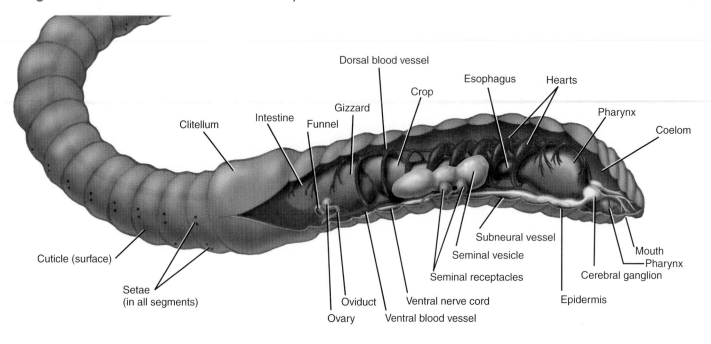

Does the earthworm have a complete or incomplete digestive system?

Circulatory system Earthworms have **closed circulatory** systems. Their blood is red because it contains the oxygen-carrying protein hemoglobin similar to the protein found in human red blood cells. Oxygen diffuses into the capillaries near the surface of the body wall and is carried throughout the body.

Find the **dorsal blood vessel** overlying the digestive tract. (fig. 21.7). It collects blood from the capillaries in each segment. Blood flows anteriorward in this vessel. Surrounding the esophagus are five pairs of pulsatile arteries, the **"hearts."** They pump blood from the dorsal vessel into the **ventral blood vessel.** Blood in the ventral vessel flows posteriorward, passing into capillaries in the body wall in each segment and into subneural vessels under the nerve cord.

Earthworms have no special respiratory structures. Respiratory gases are exchanged by diffusion across the body surface.

Excretory system Each segment except for the first few and the last contain a pair of **nephridia,** or excretory structures, that filter coelomic fluid and remove waste products. (See fig. 21.9 and lab topic 30.) To see a nephridium, try this. Put a drop of water on a slide. Use forceps to remove a septum from the side of the intestine and mount it in the drop. Add a coverslip and observe with your compound microscope. Adding a stain such as methylene blue or neutral red may improve viewing.

Nervous system Refer to figure 21.9 and note the position of the nerve cord and ganglia. Carefully remove the alimentary canal from the posterior third of the worm and look for the ventral nerve cord with its string of segmented ganglia. Remove a ganglion, mount it in a drop of water on a slide, and observe it with your compound microscope.

How many lateral nerves come from each ganglion? _____ Does the arrangement of the ganglia correspond to the segmentation? _____

Return to your worm and try to trace the nerve cord up into segments 1 through 5. This will be difficult. Locate the **subpharyngeal ganglion.** Trace it forward to the **circumpharyngeal connectives** that connect to a pair of **cerebral ganglia** lying above the pharynx. This complex is considered the worm's "brain."

Reproductive system Because earthworms are hermaphroditic, their reproductive systems are complex (fig. 21.8). Self-fertilization does not occur. The male system consists of three pairs of large **seminal vesicles,** which span body segments 9 to 13. They surround the two pairs of tiny **testes** located on the posterior surfaces of the septae separating segments 9 and 10, and 10 and 11. Sperm are released into the chambers of the seminal vesicles where they mature before being released via very small ducts, the **vas deferens.** Sperm exit through the **male gonopores** located in grooves on the ventral surface of segment 15.

The female system consists of a small pair of **ovaries** located on the septa between segments 12 and 13 (fig. 21.8). Eggs released from the ovaries enter the coelom before passing into the **oviducts,** which carry them to the **female gonopores** located in grooves on the ventral surface of segment 14. In segments 9 and 10 are two pairs of the **seminal receptacles,** which receive sperm from the partner during copulation.

Figure 21.8   Lateral view of reproductive organs in an earthworm.

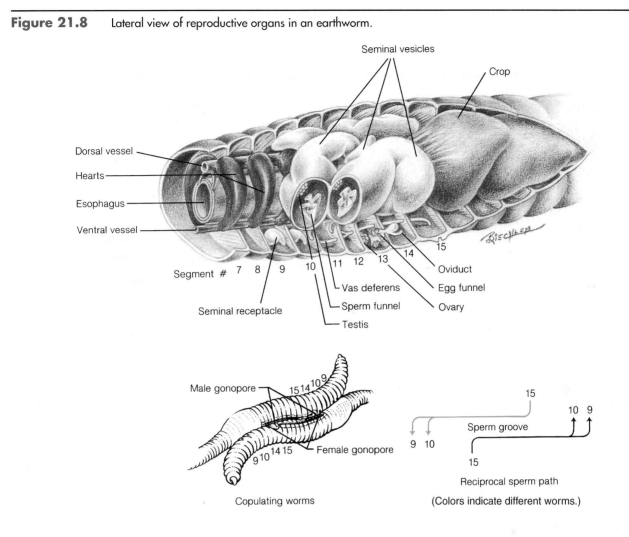

Male gonopore

Female gonopore

Copulating worms

Sperm groove

Reciprocal sperm path

(Colors indicate different worms.)

When two worms copulate, there is a reciprocal exchange of sperm (fig. 21.8), which are stored by the receiving partner. Observe the longitudinal **sperm grooves** on the external ventral surface, which facilitate this exchange.

Fertilization occurs externally sometime after copulation. Eggs are released through the female gonopore into a collar of mucus produced by the clitellum. The collar containing the eggs travels forward over the worm and sperm are released into it as it passes over the openings of the seminal receptacles, fertilizing the eggs. The collar slips off the anterior of the worm and becomes a cocoon in which the embryo develops.

Histology

6 Obtain a slide with stained cross section of the earthworm. Using the scanning objective of your microscope, compare the section to figure 21.9.

First note the obvious. The worm has open spaces surrounding the organs. This is the **coelom** or body cavity. The organs are free to move independent of body wall movements. Correlate the structures in the slide with those you saw in the dissection. Note the composition of the body wall, including the **cuticle,** the **epidermis,** the **circular muscles,** the **longitudinal muscle** layer, and the **peritoneum,** which lines the coelom. You may see some **setae**

and their associated musculature in the body wall. Observe the **typhlosole,** the dorsal infolding of the intestine, which greatly increases the area of absorption. Surrounding the intestine, find the **chlorogogue layer,** which functions like the vertebrate liver, storing glycogen and fat and metabolizing amino acids. Locate the three major longitudinal blood vessels: the **dorsal, ventral,** and **subneural vessels.** Finally, identify the **ventral nerve cord** and locate the three giant fibers plus the numerous smaller associated fibers. The giant fibers convey impulses quickly from one end of the organism to the other.

You are finished with your dissection of the earthworm. Dispose of the animal according to the directions of your lab instructor.

7 ***Live Blackworms***   Your instructor may have live California blackworms or mudworms *(Lumbriculus variegatus)* in the lab. These small oligochaetes are found in organic sediments in shallow water near the shores of lakes, ponds, and marshes throughout North America. They are easily raised for classroom use.

Use a disposable pipet to transfer a blackworm to a petri dish in a drop of water. Obtain a capillary pipet and coax the animal to enter it. This will require some trial and error.

Figure 21.9   Cross section of an earthworm.

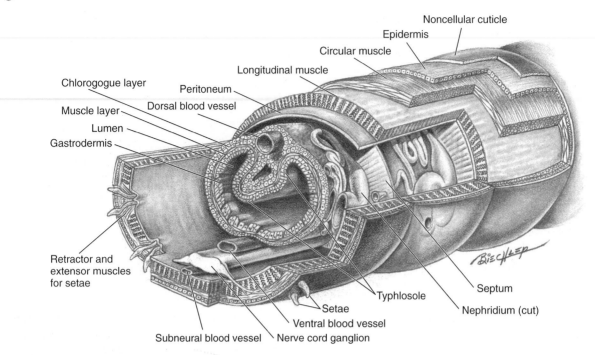

Noncellular cuticle
Epidermis
Circular muscle
Longitudinal muscle
Peritoneum
Dorsal blood vessel
Chlorogogue layer
Muscle layer
Lumen
Gastrodermis
Retractor and extensor muscles for setae
Setae
Ventral blood vessel
Subneural blood vessel
Nerve cord ganglion
Typhlosole
Septum
Nephridium (cut)

When the animal is in the capillary tube, place the tube on your compound microscope stage and observe it with the scanning objective. You should be able to see blood flowing in its vessels and to locate the "hearts" at the anterior end. Study the pattern of blood flow. Below, make a quick sketch of the worm and indicate the pattern of circulation from the hearts until the blood returns to the hearts.

How often does the heart beat in a minute? _____ Can you see waves of contraction sweeping along the dorsal blood vessel?

Return your blackworm to the supply area. Turn to table 21.1 at the end of this lab topic and fill in the information for phylum Annelida.

Phylum Mollusca

The 150,000 species of molluscs are bilaterally symmetrical soft-bodied animals with a protosome pattern of development and lophotrochozoan gene sequences. Marine species have a trochophore larva in their life cycle which is taken as evidence of their relation to the annelids, but mollusks lack segmentation and the coelom is not prominent, usually being confined to a cavity around the heart and gonads. Molluscs have well-developed organ systems for feeding, digestion, circulation, respiration, excretion, movement, nervous coordination, and reproduction. Common representatives are chitons, snails, slugs, clams, oysters, octopuses, and other less well-known animals. The body plan of a mollusk has four anatomical regions: head, visceral mass, foot, and mantle (fig. 21.10). In many, but certainly not all, the mantle secretes a calcareous shell that surrounds the animal. The molluscan body plan has allowed a great number of adaptations to different ways of life. This is reflected in the number of species and the eight taxonomic classes. Three of the classes are represented by few species and will not be discussed here.

Molluscan Diversity

8▷ Several specimens of different kinds of molluscs are on demonstration in the laboratory (fig. 21.11). They will be arranged by taxonomic class, and you should write brief descriptions of each under the appropriate following headings. As you look at these specimens, be sure to identify these six regions of the body: **head, mouth, visceral mass, foot, mantle,** and **shell.** Also note the type of symmetry exhibited.

1. Class *Polyplacophora.* These marine animals are commonly known as chitons. They live in intertidal

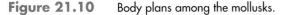

Figure 21.10   Body plans among the mollusks.

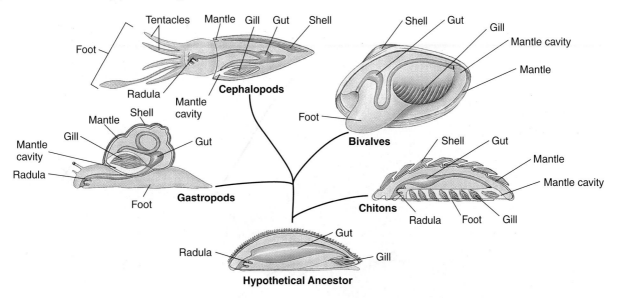

Tentacles Mantle Gill Gut Shell

Foot

Radula Mantle cavity

Cephalopods

Shell Gut

Gill

Mantle cavity

Mantle

Foot

Bivalves

Mantle Shell

Gill

Mantle cavity

Radula

Gut

Foot

Gastropods

Shell Gut

Mantle

Mantle cavity

Radula Foot Gill

Chitons

Radula

Gut

Gill

Hypothetical Ancestor

areas. They are subjected to wave action, and alternate drying and submerging caused by tides.

How does their anatomy equip them for such a harsh environment? How many plates are on the dorsal surface? _____ Where are the gills? _____ Write your description of the body regions below.

2. Class *Scaphopoda.* These marine animals are commonly known as tusk shells for obvious reasons. Look at the examples of shells on display in the lab.

3. Class *Gastropoda.* Commonly known as snails, slugs, limpets, conchs, and whelks, the gastropods are the largest class in the phylum. All gastropods undergo torsion and spiraling during embryological development. Torsion involves the 180° rotation of the visceral mass so that the digestive and nervous systems have roughly a U-shape. Spiraling involves the coiling of the visceral mass inside a shell. Write your descriptions of body regions below. What is the adaptive advantage of the shell?

4. Class *Cephalopoda.* Commonly known as nautiluses, squids, and octopuses, these animals are active marine predators. Nautiluses have an external shell, but in others the shell is internal or absent. Write your descriptions of body regions below. You will look at a squid later.

5. Class *Bivalvia.* This group includes clams, oysters, mussels, and scallops. They are sedentary, filter-feeders characterized by a laterally compressed body surrounded by a two-piece shell. The shell is lined by the mantle. The head is indistinct. A large muscular foot can be extended and is used in burrowing. You will dissect a clam in the lab and do not need to write a description.

Class Bivalvia

External Anatomy

9 ▷ Obtain a specimen of a freshwater clam. Note the two **valves** that make up the halves of the shell. The shells are hinged on the **dorsal** surface of the animal, near the swollen **umbo** area (fig. 21.12). The ventral surface is opposite the hinge, along the opening between the two valves.

In some preserved specimens, the muscular **foot** will be extending from the ventral gap between the valves. It also extends anteriorward. On the posterior surface, toward the dorsal surface, darkly pigmented **siphons** may also be

Figure 21.11 Representatives of phylum Mollusca: (*a*) a chiton; (*b*) a tusk shell; (*c*) a conch or whelk.

(a)

(b)

(c)

visible in the gap between the shells. Holding the animal dorsal up and anterior away from you, identify the right and left valves.

Now study the external surface of one of the valves. Use a scalpel to scrape the surface. The outer layer that you are removing is the **periostracum.** Composed of the insoluble protein conchiolin, it protects the calcium carbonate portion of the shell from chemical and physical erosion. Is $CaCO_3$ eroded by acids or bases? _____ The lines on the shell represent periods of growth, but are not annual growth lines. The shell and its periostracum are secreted by

cells at the edges of the mantle, a living tissue lining you will see when you open the clam. The umbo region is oldest and the regions at the edge have just been added to the shell.

Internal Functional Anatomy

10▶ To open the clam for study, you must reach in through the gap between the valves and cut the anterior and posterior **adductor muscles,** which pass from shell to shell. See figure 21.12 for their locations. Insert a scalpel through the gap, pass its point along the inside of the left valve to the muscle region, and slice.

CAUTION
Be careful: Sloppy technique may result in your damaging internal organs or, worse yet, cutting yourself.

After opening, free the soft mantle adhering to the left valve and cut the elastic ligament that forms the hinge. You should now be able to remove the left valve leaving the soft body parts in the right valve. Note how the soft tissue of the **mantle** forms an envelope around the visceral mass. The mantle performs three important functions: (1) it secretes the shell; (2) it has cilia on its internal surface, which help draw in water; (3) it is a major site of respiratory gas exchange along with the gills. Examine the left valve to see the **pallial line** about 1 cm from the edge. This is where the mantle attaches to the valve.

Examine the dorsal-posterior edge of the mantle and find the thickened, pigmented area. If you match up the left and right mantles, they form two tubes, a ventral **incurrent siphon** and a dorsal **excurrent siphon.** Cilia on the surfaces of the siphons, mantle, and gills (ctenidia) draw water into the mantle cavity over the gills and out the excurrent siphon. This water brings a stream of oxygen and suspended, microscopic food to the animal and carries away wastes. In some marine clams, the siphons may be a third of a meter long, allowing the clam to live buried in sand or mud.

Lay back the mantle to expose the **visceral mass** and muscular foot. When circular muscles in the foot contract, it extends in much the same way that a water-filled balloon elongates when squeezed at one end. Once extended into mud or sand, the tip of the foot expands to anchor itself and retractor muscles pull the body toward the anchored foot, providing a reasonable form of locomotion. Some marine clams can burrow a foot or more in wet sand in as short a time as one minute.

Feeding and Digestion Most of the visceral mass is covered by two flaps of tissue, the **ctenidia,** or gills. Although they also function in respiratory gas exchange, the primary function of these structures is in gathering food. Clams are filter-feeders, obtaining food by filtering algae, detritus, and bacteria from water entering the mantle cavity through the siphons.

Figure 21.12   Anatomy of a partially dissected freshwater mussel.

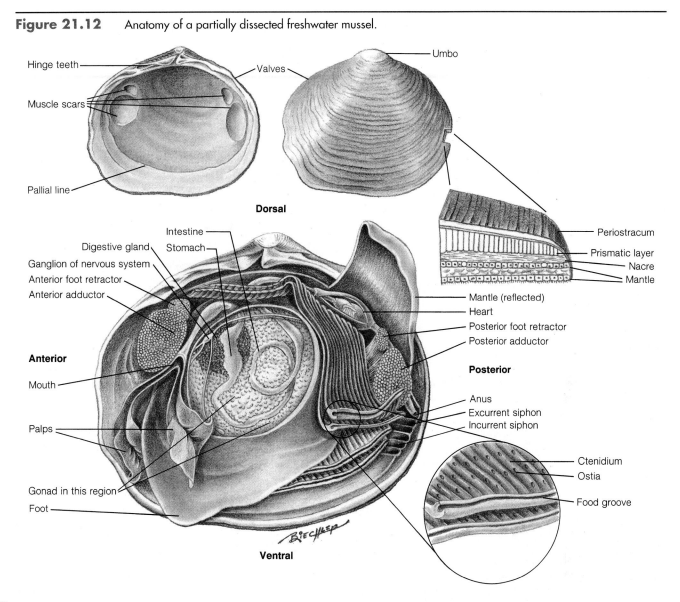

11 ▷   The structure of the ctenidia is best seen in slides of cross sections of bivalves viewed through a microscope (fig. 21.13). Get your microscope and a prepared slide. The shell was removed before the animal was sectioned to make the slide. Find the visceral mass hanging like a pendant with curtains of tissue surrounding it on both sides. The inner two curtains are the ctenidia and the outermost is the mantle.

Look carefully at one ctenidium on the microscope slide. Note that it is hollow with a **water tube** passing up the center. Close examination will reveal that the surface of the ctenidium has small porelike openings, the **ostia.** Water entering the mantle cavity passes through the ostia into water tubes and then upward to **suprabranchial chambers** (fig. 21.13). From there the water flows to the excurrent siphon.

Any particles contained in the water stream are filtered out at the ostia where they are trapped in mucus that coats the surface of the ctenidia. Cilia on the surface of the ctenidia move the mucus down to the food groove that passes along the edge of a ctenidium. Find the **food groove** on your slide.

Figure 21.13   Cross section of a freshwater mussel. Shell will not be included on microscope slide.

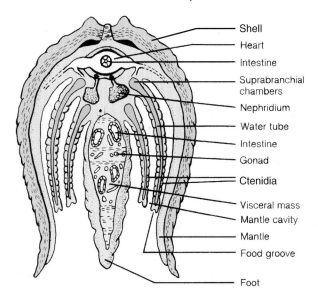

Now return to your dissection specimen. Look at the tip of the ctenidia where the food groove is located. Follow the food groove anteriorward to the region of the cut anterior adductor muscle. In this region between two flaps of tissue called **palps,** you should find the **mouth** of the clam. Food particles trapped in mucus enter the digestive system here and pass to the **stomach.**

To see the stomach, remove the ctenidia covering the visceral mass. Scrape and pick away the tissue of the visceral mass to expose the digestive tract. Tissue of a large, greenish-brown digestive gland surrounds the stomach and a lighter gonad surrounds the intestine. The intestine winds its way dorsally and then posteriorly to the anus located near the excurrent siphon in the region of the posterior adductor muscle. Does the clam have a complete or incomplete digestive system? _____

Circulatory System Clams have open circulatory systems. The circulating fluid is **hemolymph.** It enters the heart directly from the surrounding tissue spaces, the **hemocoel,** and is pumped through arteries to the tissues. At the tissues, the hemolymph does not enter capillaries. Instead, it is released directly into the tissue spaces from where it eventually percolates back to the heart.

To see the clam's heart, look at the dorsal surface of the visceral mass just anterior to the posterior adductor muscle. Remove the thin membrane in this area to reveal the **heart** sitting in the **pericardial cavity.**

The heart is three-chambered: one **ventricle** and two **atria.** As the intestine travels from the visceral mass to the anus, it passes through the ventricle. Two **aortas** leave the ventricle, one passing anteriorward and the other posterior.

Excretory System Beneath the heart, find the paired dark brown **nephridia** (kidneys). They remove nitrogenous wastes from the hemolymph. Urine passes from these to a pore in the suprabranchial chamber above the ctenidia. Water flowing through this chamber carries wastes out the excurrent siphon.

You will not study the nervous or reproductive systems of the clam. Dispose of your clam as instructed.

Class Cephalopoda

Squid Dissection

Squid have streamlined bodies and a jet propulsion mechanism that make them fast-moving predators. Large eyes, grasping arms and tentacles, and a beaklike structure for biting and tearing make them even more formidable. Depending on the species, squids can range in size from 2 cm up to 15 meters for giant squids.

Squid obtained from seafood suppliers make the best materials to dissect as the tissues look fresh and retain a natural color and consistency. Obtain a squid in a dissecting pan from the supply area along with dissecting instruments.

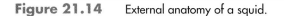

Figure 21.14 External anatomy of a squid.

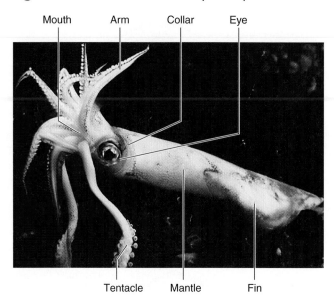

Mouth Arm Collar Eye

Tentacle Mantle Fin

External Anatomy Squids offer an excellent lesson in the adaptability of the basic four-part molluscan body plan: head, visceral mass, foot, and mantle.

As you look at your squid, note the enclosing muscular **mantle** which is set off from the rest of the body by a **collar** (fig. 21.14). The mantle contains specialized pigmented cells called **chromatophores.** Each of these cells has tiny muscle fibers attached to it, and the animal can expand or contract its chromatophores to change colors in seconds. The mantle tapers to a point at one end with two muscular fins. These are like diving planes on a submarine and, when the animal is swimming, can be used to steer up or down as well as to the side.

At the opposite end, note the tentacles and arms, 10 in all. These are modifications of the foot region in the basic molluscan body plan. Therefore, this would be the theoretical ventral surface, making the apex of the mantle the dorsal surface. Look at the suckers on the arms and note the small teeth around the periphery of each. They help hold prey. Spread the arms and look into the circle made by their bases to see the **mouth.** A horny **beak** should be seen protruding from the mouth.

Study the collar area and find the **siphon** (sometimes called the funnel) which is very important in locomotion. If you hold the animal up with the arms ventral, then the surface on which the siphon is found is considered the anatomical posterior surface. Squids swim at speeds up to 20 mph. They do so by water jet propulsion. Note the interlocking cartilages on the mantle and siphon. They support the funnel and mantle in this area. When muscles in the mantle are relaxed, water flows into a chamber beneath the mantle (mantle chamber) through the collar. Circular muscles then contract, first in the collar area to prevent water escape, and then in the rest of the mantle. As pressure builds, the water

Figure 21.15   Internal anatomy of squid; (*a*) male; (*b*) female.

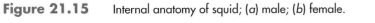

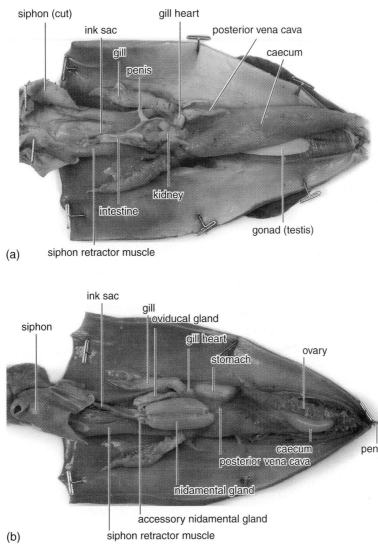

siphon (cut)
gill heart
ink sac
posterior vena cava
caecum
gill
penis
kidney
intestine
gonad (testis)
(a)   siphon retractor muscle

ink sac
gill
oviducal gland
siphon
gill heart
stomach
ovary
caecum
posterior vena cava
pen
nidamental gland
accessory nidamental gland
(b)   siphon retractor muscle

jets out the siphon. Slit open the funnel to see the tongue-like valve which regulates water flow through the funnel. Through a rapid cycling of relaxation and contraction phases, a continuous jet of water is produced. Muscles attached to the siphon can turn it in different directions which, along with the fins and trailing arms, helps the squid steer very complex paths as it pursues prey such as small fish.

The **eyes** are well developed and are image forming. There is an iris that regulates light entering the eye and behind it a spherical lens which focuses the light. The lens is focused by muscles that move it back and forth. It does not bend as do the lenses of vertebrates. Remove an eye and cut the lens free to see it.

13 ▸ *Functional Internal Anatomy*   To see the visceral mass, the mantle has to be cut to reveal the **mantle cavity.** Do so by laying the animal in the pan with its posterior side up. How do you tell which side is posterior?

Cut from the collar on one side of the siphon to the apex and pin the tissue folds back. The mantle cavity is not a coelom. It is a cavity that is continuous with the external environment. It contains the gills and visceral mass. Figure 21.15 shows the internal anatomy of a male and female squid.

General Anatomy   In the region of the siphon, note the large retractor muscles running through the mantle cavity. Some of these adjust the position of the siphon and others can retract the squid's head. The **anus** at the end of the intestine will be situated near the inner opening of the funnel. Its location assures that fecal material will pass out of the body and not foul the gills.

Near the midline, below the base of the siphon find the dark **ink sac.** It actually empties into the rectum and out through the anus. When disturbed, the pigment melanin is released from the sac into the mantle cavity and is expelled.

The cloudy suspension leaving the siphon confuses would-be predators. If the sac has ink in it, remove it by snipping it at both ends and put it in a dish. Take a dissecting needle and dip it in the ink and write something on paper. This is what artists call sepia ink.

Respiratory System   In the mantle cavity you should see two long feathery structures which are the gills. They function in gas exchange. They are constantly bathed by water entering the mantle cavity. Their delicate structure is protected by the surrounding mantle.

The organs of the visceral mass are covered by a thin transparent tissue (membrane). If you carefully remove it, you will be able to find the organs. If your specimen is a female, two large **nidamental glands** may be present covering the other organs. They produce a jelly which is laid down around the eggs as they are shed. Remove the glands to reveal the structures beneath.

Circulatory System   Cephalopods have a closed circulatory system with blood traveling from the heart to capillary beds and draining back to the heart through veins. The blood contains the respiratory pigment hemocyanin. Look at the base of a gill on each side and identify small yellowish **branchial hearts.** They receive blood returning from the mantle and head and pump it into the gills through an afferent branchial vein. Essentially, these are auxiliary pumps that assure blood flow through the gill where respiratory gas exchange takes place. Blood returning from the gills drains into efferent branchial veins which carries it to a larger **systemic heart (true heart)** located on the midline between the two branchial hearts. It is three chambered with two atria and a ventricle. It pumps the oxygenated blood to the head through the cephalic artery and to the mantle through a posterior arota. See if you can locate these vessels.

Excretory System   A pair of triangular shaped **kidneys** in a single nephridial sac can be found cephalic to the branchial hearts towards the midline. The vena cava flow enters the renal sac. Nitrogenous waste exchange takes place in the kidneys. Urine is voided through **nephridiophores,** opening into the mantle cavity toward the head. Wastes are expelled from the cavity through the siphon.

Digestive System   Most of the digestive system is not readily visible as it passes in a long U-shape from the mouth to the anus. However, some careful tracing will allow you to see most parts. Remove the siphon and make a median incision in the mass below it to expose the muscular **buccal blub** which bears the two horny **beaks.** Open the beaks and look into the buccal cavity to see **radula.** The beaks tear off chunks of flesh from the prey and the radula further pulverizes the food. Remove a portion of the radula and look at it with your microscope to see the minute teeth.

The thin walled esophagus passes through the triangular, yellowish liver which sits on the midline. The esophagus emerges from the other end of the **liver** and passes through the **pancreas** before opening into the thicker-walled **stomach.** Extending from the stomach to the apex is a **caecum.** Food passes into the caecum for storage and flows back into the stomach as digestion proceeds. It can be small or large depending on whether your animal fed before being caught. The **intestine** leaves the stomach near the point where the esophagus entered and travels toward the collar ending in the **anus** identified earlier.

Nervous System   Cephalopods have a highly developed nervous system. Push the buccal bulb and its retractor muscles aside to reveal two white **stellate ganglia** on the back wall of the mantle. Giant nerve fibers should be seen radiating from them. These control the contractions of the mantle in swimming. There are other ganglia in this area, but you need not look for them.

Reproductive System   The sexes are separate in cephalopods and the gonads are not paired. The gonads extend from the apex of mantle cavity to about midway to the head.

If you had a female, you removed the **nidamental glands** as you looked at the internal organs. The single **ovary** extends from mid body to the apex. It sheds eggs into the mantle cavity and they enter the **oviduct.** Push the ovary and caecum aside to find the oviduct underneath them. Near the left branchial heart, the oviduct enlarges to form the **oviducal gland** that secretes a shell around the egg. The oviduct continues cephalically and flares to form the **ostium** that opens into the mantle cavity. Eggs pass out of the siphon in strings. Find a student who dissected a male. Explain to each other the anatomy of the reproductive systems in your specimens.

In males, the single **testis** is in a similar location to the ovary in females so that the same organs will have to be moved to reveal it. Sperm are shed into the mantle cavity from an opening in the testis. They enter a **sperm duct** that conveys them to the region of the left branchial heart. As they pass they are formed into long **spermatophores.** They are stored in a spermatophore sac and are released from the **penis,** which is simply the cephalic end of the duct. During mating, the male inserts one of his arms into his mantle cavity to remove a spermatophore that was released from the sperm duct and transfers it to the mantle cavity of a female. Find a student who dissected a female and explain to each other the reproductive anatomy of your specimens.

Skeletal System   Remove the visceral mass and feel the back mantle wall. The hard structure is the squid's **pen,** a longitudinal stiffening element that is analogous to a shell although it is made of chitin. Cut into the wall and dissect it out.

You are now finished with the dissection of your squid. Dispose of the remains as directed by your instructor.

Turn to table 21.1 at the end of this lab topic. Fill in the information for phylum Mollusca.

Learning Biology by Writing

You have had the opportunity to study the general anatomy of an earthworm, clam, and squid as well as animals in lab topic 19. As you consider the representatives of the six animal phyla that you have studied, what evidence do you have to accept or refute the hypothesis that annelids and molluscs are more complex than the sponges, cnidarians, flatworms, and nematodes? Summarize your arguments in a one-page essay.

Your instructor may ask you to turn in answers to the Lab Summary and Critical Thinking Questions that follow.

Internet Sources

Use the WWW to discover what kind of discussions are occurring among scientists about Lophotrochozoa. Type the term into GOOGLE. You will get lots of "hits." Many will be class notes from other colleges. Scan the list looking for articles written by scientists who are arguing a point or reporting research. Read a few of these written by people working at reputable institutions or published in journals. Write a couple of paragraphs summarizing what they are saying.

Lab Summary Questions

1. What are the characteristics of a protostome animal? Which phyla are considered protostomes?

2. Describe in your own words what a lophophore is. Do the same for a trochophore larva. Why are animals with these structures grouped in the same clade?

3. Describe the body plan of an annelid. Briefly describe how an earthworm feeds/digests, exchanges respiratory gasses, circulates fluids, excretes wastes/salts, and moves.

4. What are the diagnostic features of the phylum Mollusca?

5. Explain how a clam is a filter feeder. Draw a diagram that shows the pathway of food particles from water into the mouth of a clam.

6. Turn to the next page and be sure to complete table 21.1, Summary of Lophotrochozoan Characteristics.

7. Describe what anatomical and physiological adaptations you saw in squid to a predatory way of life.

8. What evidence can you present from your observations to indicate that annelids and molluscs are more complex than cnidarians or nematodes?

Critical Thinking Questions

1. Based on your knowledge of an earthworm's anatomy and physiological systems, explain why large earthworms come to the surface after a heavy rain.

2. Provide arguments from your study of molluscs to support this statement: "The molluscan body plan allowed the development of adaptations to many ways of life."

TABLE 21.1 Summary of lophotrochozoan characteristics

	Phylum Bryozoa	Phylum Annelide	Phylum Mollusca
Common name(s)?			
Tissues?			
Symmetry?			
Lophophore?			
Trochophore?			
Name organ systems present			
Type digestive system?			
Type circulatory system?			
Type skeletal system?			
Describe body plan			

LAB TOPIC 22

Protostomes II: Ecdysozoa and Great Diversity

Supplies

Preparator's guide available on WWW at
http://www.mhhe.com/dolphin

Equipment

Compound microscopes
Dissecting microscopes
Berlese funnel (see fig. 22.10)

Materials

Horseshoe crab
Spiders
Acorn and gooseneck barnacles
Crayfish
Anesthetized live crickets
Miscellaneous insects for demonstration
Living *Daphnia*
Dissecting pans and instruments
Microscope slides
    Mosquito mouthparts (demo)
    Fly mouthparts (demo)

Solutions

Petroleum jelly
Congo red dye
Yeast
Test tubes

Student Prelab Preparation

Before doing this lab, you should read the introduction and sections of the lab topic that have been scheduled by the instructor.

You should use your textbook to review the definitions of the following terms:

Arachnida	cuticle
Chelicerata	exoskeleton
chitin	Insecta
coelom	open circulatory system
Crustacea	Uniramia

You should be able to describe in your own words the following concepts:

The general body plan of an arthropod
The diversity of the Phylum Arthropoda

As a result of this review, you most likely have questions about terms, concepts, or how you will do the experiments included in this lab. Write these questions in the space below or in the margins of the pages of this lab topic. The lab experiments should help you answer these questions, or you can ask your instructor during the lab.

Objectives

1. To illustrate the diversity of the arthropods by studying the external anatomy of a horseshoe crab, a spider, *Daphnia*, and a barnacle
2. To study the internal anatomy of a crayfish
3. To study the internal anatomy of a cricket
4. To note the evolutionary significance of the arthropod anatomy
5. To use a taxonomic key to identify insects to the level of order (optional)
6. To collect evidence to accept or refute the hypothesis that all Arthropods have an exoskeleton, a segmented body, and jointed appendages

Background

One of the great surprises coming from the genetic analysis of animal relationships conducted over the past 10 years is discovery of a clade of protostomes called the Ecdysozoa which are quite different from the lohotrophozoans. This new clade of seven phyla includes animals as diverse as nematodes, which you studied in lab topic 20, and arthropods (barnacles, crayfish, and insects, *etc.*) which you will study today, as well as many small wormlike creatures. The molecular analyses establishing this clade have been confirmed by investigation of two different sets of genes, those for rRNA and those for *Hox* genes that control body patterning during development. See figure 20.1 for the phylogenetic relationships based on this data.

Given the molecular similarities, biologists began to look for morphological similarities. The first to be noticed, and the one from which the name of the new clade was derived, was that all of the included animals have an exoskeleton that is shed as the animal grows. Another word for shedding is **ecdysis.** A second characteristic is that all of the animals in the ecdysozoan clade lack surface cilia both in their larval and adult stages. Ecdysozoans do not use cilia for locomotion nor do they use cilia for generating feeding water currents as many lophotochozoans and deuterostomes do. The ecdysozoans are a significant group of animals. Something like 75% of all of the known animal species are ecdysozoans, primarily because of the inclusion of the arthropods, the most diverse animal phylum.

The ancestor of the ecdysozoans is visualized as a small wormlike animal that was a burrower. The presence of an exoskeleton would have protected the animal's tissues from abrasion. The development of an exoskelton would have precluded the animal from having surface cilia because cilia are extensions of living cells and exoskeletons are nonliving layers of materials covering the cells. Surface ridges and spines would have developed to anchor the animal. This would have allowed the animal to force its way into soft mud or dig down with serpentine motions, as you observed in nematodes.

In this lab, you will study members of the phylum Arthropoda. Nearly a million species of arthropods have been described and the world population level of arthropods is estimated at a billion billion (10^{18}) individuals. Crustaceans, insects, millipedes, spiders, and ticks are but a few examples of this diverse phylum that has representatives in virtually every environmental habitat from the ocean depths to mountain tops, including aerial environments.

As in the annelids, the bodies of arthropods are segmented. This similarity led many to believe that the annelids and arthropods were closely related. However, the molecular evidence does not support such a relationship and the phyla are placed in separate clades. In the arthropods, the segments are often fused to make distinct body regions, such as the **head, thorax,** and **abdomen.** Some arthropods have a **cephalothorax** in which the head and thorax regions are fused.

Arthropods have a non-cellular **cuticle** composed of chitin, a complex polysaccharide stiffened by calcium salts and cross-linked proteins. Acting as an exoskeleton, the cuticle protects but is jointed to allow movement. For terrestrial species, the cuticle is an effective barrier preventing desiccation and may explain why the arthropods have been so successful on land. In order for individual arthropods to grow, they molt their exoskeletons, and most species have several developmental stages in their life cycles. Muscles are in distinct bundles rather than being part of the body wall and allow a variety of movements, especially in combination with the jointed appendages.

The digestive system of arthropods is more or less a straight tube leading from the mouth to the anus. Arthropods have an open circulatory system with a dorsal artery and heart that pumps hemolymph anteriorward. Because the exoskeleton limits diffusion, gas exchange in large arthropods is through special structures, such as gills, book lungs, or small tubules called tracheae. The nervous system consists of two ventral cords with ganglia serving as integrating centers, allowing complex behaviors and locomotion. Arthropods have distinct excretory organs that remove nitrogenous wastes from body fluids and which function in electrolyte balance. The sexes are usually separate.

Four evolutionary lineages can be seen in the arthropods. These are:

Trilobitamorpha: about 4,000 species of extinct trilobites known only from fossils, disappearing about 250 million years ago during the great Permian extinctions. Bodies were segmented with similar appendages on most segments. Modern arthropods tend to have specialized appendages on different segments.

Chelicerata: about 65,000 species of horseshoe crabs, mites, spiders, scorpions, and ticks; name reflects the clawlike feeding appendage called a chelicerae. They lack the antennae, compound eyes, and jaw-like mandibles found in the other lineages. The body has an anterior cephalothorax and a posterior abdomen. You will look at a horseshoe crab and spider as representatives of this clade.

Crustacea: about 40,000 species of crabs, crayfish, barnacles, and many others; primarily aquatic although there are terrestrial species (pill bugs); have two pairs of antennae, compound eyes, and appendages that branch. You will do a major dissection of crayfish, observe the anatomy of a microcrustacean (*Daphnia*), and observe the external anatomy of barnacles as representatives of this clade.

Uniramia: centipedes, millipedes, and insects; primarily land-dwelling animals although there are aquatic stages in some life cycles; have one pair of antennae, compound eyes, and non-branching appendages. Insects are the most diverse arthropods with over 900,000 named species. You will look at the anatomy of a cricket as a representative of this clade.

There is some debate among biologists about how these lineages are related. Some consider each lineage a subphylum of the Arthropoda. Others would elevate each lineage to phylum status and do away with the name Arthropoda, although it might still be used as a non-taxonomic clade name. Others suggest that the Crustacea and Uniramia be lumped together based on the fact both have jaw-like mandibles. They would place them in a clade called Mandibulata. Others would place only the insects and crustaceans together in Mandibulata and retain the Uniramia designation for millipedes and chelicerates. All of this is confusing for the student who simply wants to know: what should I learn? Unfortunately, you need to know a little about all of this and at the moment there is not an agreed upon right answer to the classification within the arthropods. On the other hand, this controversy makes it an exciting time to become a biologist. The debate indicates there is much work to be done before we will know the relationships within the group. Will you be among those who solve the problem?

Figure 22.1 Ventral view of horseshoe crab.

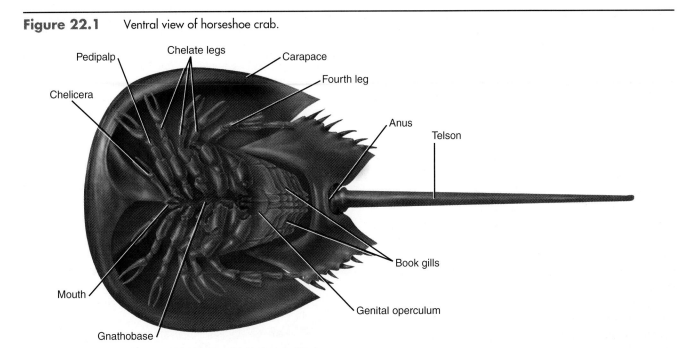

LAB INSTRUCTIONS

The intent of this lab is to introduce you to the diversity of arthropods. You will look at several very different animals that reflect the three living evolutionary lineages.

Subphylum Chelicerata

The 65,000 or so species of chelicerates are primarily terrestrial animals, although at one time in geological history they rivaled the crustaceans for predominance in marine ecosystems. Chelicerates include horseshoe crabs, scorpions, spiders, and mites as well as many extinct forms.

The characteristics of the group include:

- Body usually consists of a distinct cephalothorax and an abdomen;

- Cephalothorax bears six pairs of appendages: chelicerae, pedipalps, and four pairs of walking legs;

- No antennae.

Horseshoe Crab

Horseshoe crabs are marine scavengers that consume molluscs, worms, and plant material as they slowly move over the sandy bottoms of shallow bays and oceans. Despite its name, it is not a true crab but belongs to subphylum Chelicerata, class Merostomata.

1 ▷ Look at the horseshoe crab on demonstration in the laboratory. From a dorsal view, the exoskeleton is fused into a **carapace** that covers the 7 segments of the cephalothorax and the 12 segments of the abdomen. A hinge joint marks the dividing line between the body regions and allows the animal to bend. A tail-like **telson** extends from the tip of the abdomen. A pair of **compound eyes** are prominent on the anterior of the cephalothorax and two simple eyes **(ocelli)** are found on either side of a small spine located on the midline anterior to the compound eyes.

Turn the animal over and look at its ventral surface (fig. 22.1). Six pairs of jointed appendages surround the medially located mouth. The small first pair are **chelicerae** used in manipulating food. All but the last pair end in **chelae,** pincerlike structures used to grasp food. The base of the appendages bear spiny **gnathobases,** which macerate food and force it into the mouth as the appendages are moved in a kneading fashion.

The ventral surface of the abdomen is covered by six pairs of flat plates. The first of these is the **genital operculum.** Two **genital pores,** the openings of the reproductive system, lie beneath the operculum. Under the remaining five plates are **book gills,** so-called because in life the gills fan back and forth as you would fan the pages of a book.

Spider

2 ▷ Obtain a preserved garden spider in a dish of water and examine its external anatomy with your dissecting microscope. Most specimens will be female because males are smaller and not routinely sold by biological supply houses.

Examine the external anatomy with your dissecting microscope. Note the two body regions, the **cephalothorax** (also known as the prosoma) and the **abdomen** (also known as the opisthosoma), which are connected by a slender **pedicel,** a modified first abdominal segment (fig. 22.2). Locate the eight **ocelli,** simple eyes, on the anterior dorsal surface of the cephalothorax.

Figure 22.2 Spider: (a) dorsal view; (b) ventral view; (c) sagittal section showing internal organs.

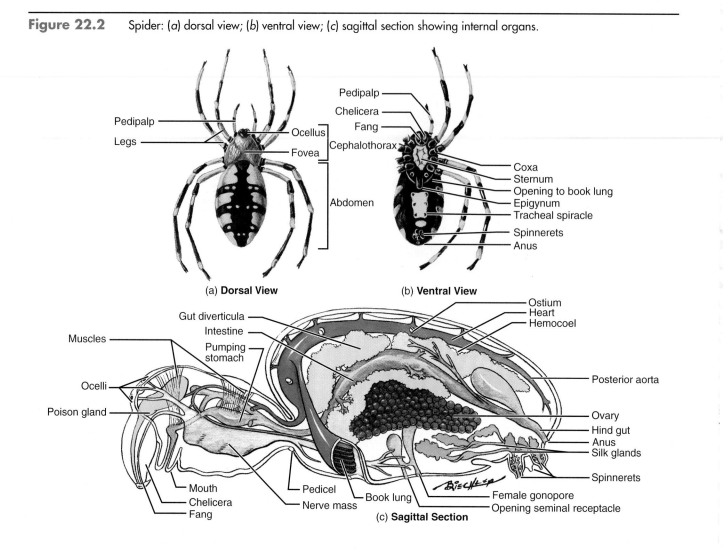

(a) **Dorsal View**

(b) **Ventral View**

(c) **Sagittal Section**

Turn the spider over and examine the ventral surface. Six pairs of appendages should be visible. The first are the **chelicerae,** which terminate in hollow, hardened **fangs** that inject a paralytic poison into prey. The second pair of appendages are the **pedipalps,** which serve a sensory function in feeding. Behind the pedipalps are four pairs of **walking legs,** each composed of several segments.

On the ventral abdominal surface, find the paired slits, **spiracles,** that open into the **book lungs.** Insert the tips of fine forceps into the opening and fold back the exoskeleton to reveal the plates of the book lungs. Hemolymph that contains the respiratory pigment hemocyanin circulates in the plates and exchanges respiratory gases with the air in the book lung compartment.

On the median line of the abdomen near the spiracle, find the platelike **epigynum,** which covers the female genital opening. The openings of three pairs of **spinnerets** should be visible on the ventral surface near the tip of the abdomen. Silk glands beneath the openings exude a viscous solution of silk protein that hardens when exposed to air.

Find the small paired **spiracles** on either side of the midline of the ventral surface of the abdomen. These open into a system of tubules, the **tracheae,** that supplement the gas exchange capabilities of the book lungs. How does this system of gas exchange differ from that of the earthworm?

The internal anatomy of the spider will not be studied, but it is shown in outline in figure 22.2. Trace the digestive system from mouth to anus. Note the poison gland and the ducts into chelicerae. This poison immobilizes prey, and the pumping stomach removes liquified tissue from the prey. Note the location of the ovary and the well-developed nervous system. The circulatory system consists of a pulsatile dorsal vessel that moves hemolymph from the posterior to anterior. This system is open and fluids percolate from front to back before being pumped forward again.

Figure 22.3 Internal anatomy of an acorn barnacle.

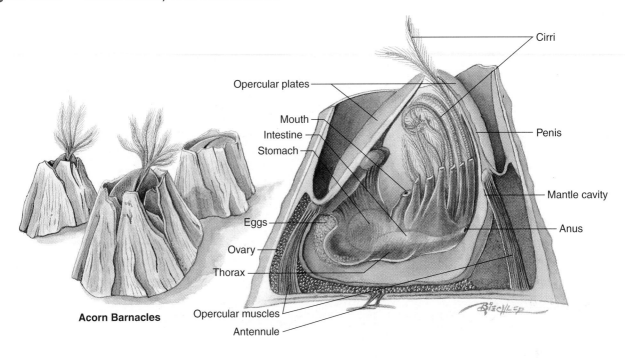

Acorn Barnacles

Subphylum Crustacea

Most of the 40,000 species of crustaceans are marine where they are the predominant types of animals. No other phylum has as many marine species. Many other crustacean species live in freshwater and there are even a few terrestrial species such as the pill bugs that live in damp terrestrial habitats. They range in size from a millimeter or so long to lobsters that can approach 20 pounds. They are a very diverse group as your study of barnacles, freshwater *Daphnia,* and crayfish will demonstrate.

What distinguishes the animals in subphylum Crustacea from the rest of the arthropods?

- Head develops from the fusion of five body segments each bearing a pair of appendages: 2 pairs of antennae, 1 pair of mandibles, 2 pairs of maxillae.
- Head can be fused with the exoskeleton of the thorax to form a carapace. Some species will be encased in lateral valves.
- Abdominal segments are usually distinct and the terminal segment bears a telson.
- Appendages are biramous, meaning they branch into two segments.

Barnacles

Sessile members of the class Crustacea, barnacles live attached to rocks, ships, whales, and seaweeds. Some barnacles have a long stalk, the gooseneck barnacles; and others have a compact body, the acorn barnacles (fig. 22.3). Examples of both are on demonstration in the lab.

3▶ Obtain a specimen of a preserved acorn barnacle in a dish of water. Although it superficially resembles a mollusc because of its calcified exoskeleton, dissection will show that it has a segmented body and jointed appendages, which are diagnostic of the phylum Arthropoda.

The barnacle is surrounded by a **carapace** composed of calcified lateral and upper plates. The four upper plates comprise the **operculum,** which is open when the animal is feeding. While looking through your dissecting microscope, carefully pry open the opercular plates to expose the animal.

The anatomy of the barnacle can be confusing. Picture an arthropod lying on its back with its legs projecting upward inside the surrounding plates. Each of the thoracic segments bears branched appendages called **cirri.** To feed, the animal opens the opercular plates, extends its thorax, and rapidly sweeps the cirri through the water, netting small particles, which are conveyed to the mouth. Find the mouth, and opposite it, the anus. A single, long penis should also be visible. Barnacles are hermaphroditic, and possess ovaries as well as testes. Cross fertilization is the rule, but being sessile, barnacles can only mate by extending the penis to nearby barnacles.

Clean up your barnacle dissection as directed by your instructor.

Daphnia

4▶ Commonly known as the water flea, this microcrustacean inhabits ponds and lakes in North America and Europe.

Make a slide of living *Daphnia* by catching one in a pipette from the lab culture and transferring it with some

Figure 22.4   *Daphnia:* Internal anatomy.

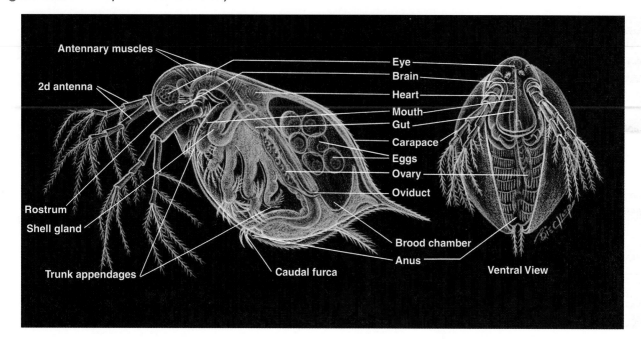

Antennary muscles

2d antenna

Rostrum

Shell gland

Trunk appendages

Eye
Brain
Heart
Mouth
Gut
Carapace
Eggs
Ovary
Oviduct

Brood chamber
Anus

Caudal furca

Ventral View

water to a slide. Use your dissecting microscope to watch it swim. Add a coverslip and observe it through your compound microscope (fig. 22.4). Ranging in size from 0.5 to 5 mm, *Daphnia* are covered by a transparent **carapace,** an exoskeleton that wraps around the body in a bivalve fashion. The head bears a conspicuous, single compound **eye,** and two large **antennae.**

Water fleas live about 100 days under optimum conditions. Because the exoskeleton constrains growth, it is periodically molted: once every 24 hours in juvenile forms and once in two or three days in adults. Identify the structures labeled in figure 22.4.

Reproduction is usually parthenogenic (unfertilized eggs develop) and results in female offspring. Only under stressful conditions are males produced and sexual reproduction follows their appearance.

Water fleas swim in a jerky fashion by stroking the antennae back and forth. These animals are filter-feeders. Legs inside the carapace terminate in comblike **setae,** which sieve bacteria, algae, and small particles from the water. Food balls are transferred to the **mouth** and take about one-half to three hours to pass through the digestive system. Often the gut appears green because it contains algae sieved from the culture.

5▸ Feeding behavior can easily be observed. Prepare a suspension of stained yeast by adding 3 g of wet yeast and 30 mg of Congo red stain to a test tube containing 10 ml of water. Boil gently. A *Daphnia* can be immobilized by using a dissecting needle to place a very small dab of petroleum jelly on a microscope slide. Cover it with a drop of water. Add a *Daphnia* and carefully begin to remove the water. The animal should fall on its side and become stuck to the jelly. Add back enough water to cover the animal.

Dip a dissecting needle into the yeast suspension and remove a small drop. Transfer this to the *Daphnia* slide while watching through your dissecting microscope. You should be able to see the sieving action of the leg combs and red yeast should be taken into the digestive tract.

Return the *Daphnia* to the supply area and clean your slide.

Crayfish

You will use a crayfish for your demonstration dissection of a crustacean. Obtain a crayfish, dissecting instruments, and a dissecting pan filled with wax from the supply area. Fresh crayfish can be obtained from restaurant suppliers and are much better to dissect because the tissues retain natural colors.

6▸ *External Anatomy*   Examine the external anatomy of the crayfish (fig. 22.5). The body is divided into two major regions, the posterior **abdomen** and the anterior **cephalothorax** covered dorsally by the **carapace.** The abdomen is composed of six segments and ends in a flaplike projection, the **telson.** Thirteen fused segments make up the cephalothorax. The **cervical groove,** a transverse (cross body) groove in the carapace, marks the separation of the head segments from the thorax. The head ends anteriorly in the pointed **rostrum.**

Lay the animal on its back and examine the appendages. On the last abdominal segment are two flattened lateral appendages, the **uropods.** Contraction of the ventral abdominal muscles draws the telson and the extended uropods under the body, allowing the crayfish to swim rapidly backwards. On the other abdominal segments are small **swimmerets.** If your specimen is a male, the swimmerets on the first abdominal appendages will be pressed closely together and pointed forward. They serve as sperm transfer

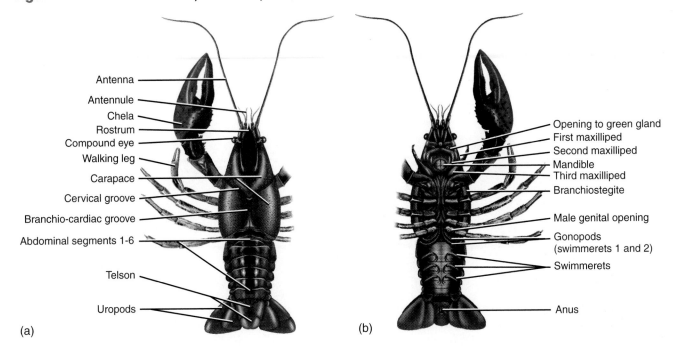

(a)

Antenna
Antennule
Chela
Rostrum
Compound eye
Walking leg
Carapace
Cervical groove
Branchio-cardiac groove
Abdominal segments 1-6
Telson
Uropods

(b)

Opening to green gland
First maxilliped
Second maxilliped
Mandible
Third maxilliped
Branchiostegite
Male genital opening
Gonopods
(swimmerets 1 and 2)
Swimmerets
Anus

organs (gonopods), receiving sperm from openings at the base of the last pair of walking legs. If your specimen is a female, these swimmerets are small. Be sure you observe both males and females in the lab.

The large appendages of the cephalothorax are the **walking legs.** The first pair of legs are called **chelipeds** because they bear **chelae** (pincers), which are used in grasping and tearing food as well as in defense.

Several appendages surround the mouth. The heavy **mandibles** are the jaws that grind food. There are also two pairs of large **maxillipeds** and two pairs of **maxilla** that handle food and carry sensory organs for taste and touch. Probe the mouthparts to find the second maxilla, which bears a large elongated plate, the **gill bailer.** The sculling action of the bailer draws water over the gills that are located under the sides of the carapace. The **antennae** and smaller **antennules** are the last appendages on the head. They have sensory functions.

The gills are contained in branchial chambers, located laterally in the cephalothorax. The lateral carapace covering the gills is called the **branchiostegite.** What are the advantages and disadvantages of having the gills enclosed by the exoskeleton?

Cut away the branchiostegite on one side of the crayfish. Be careful not to damage the underlying gills. Gradually clear away the material to reveal the feathery gills. Water is drawn in through an opening at the posterior end of the branchial chamber, passes over the gills, and exits anteriorly as a result of the action of the gill bailer. Cut off a gill and mount it in a drop of water. Add a coverslip and look at it with your compound microscope. Sketch what you see below.

Figure 22.6   Dorsal view showing internal organs in a partially dissected crayfish.

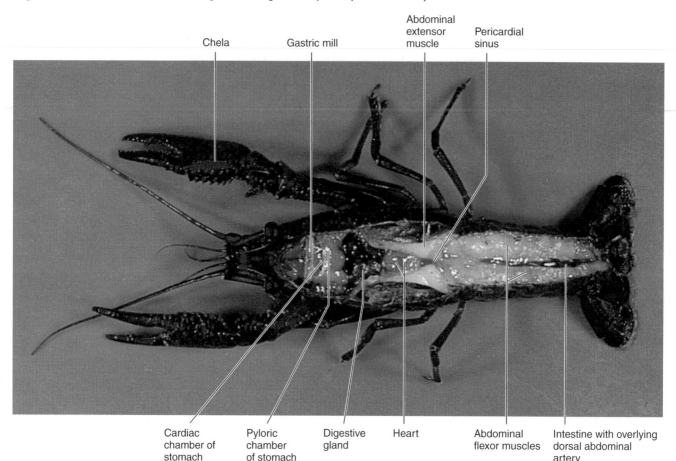

Chela   Gastric mill   Abdominal extensor muscle   Pericardial sinus

Cardiac chamber of stomach   Pyloric chamber of stomach   Digestive gland   Heart   Abdominal flexor muscles   Intestine with overlying dorsal abdominal artery

Hemolymph, crustacean blood, circulates through the gills and comes in close proximity to water passing over the gills, where carbon dioxide and oxygen exchanges occur. How is the respiratory system similar to that of the molluscs? How is it different?

7 ▸ *Internal Anatomy*   You will now study the internal organs of the crayfish. With scissors, cut very carefully to one side of the dorsal midline of the carapace from where it joins the abdomen forward to a region just behind the eyes. Use a scalpel or a probe to carefully separate the carapace from the underlying **hypodermis,** the tissue that secretes the carapace. Completely remove the carapace. Now cut through the exoskeleton covering the abdomen from the first to last segments. Carefully remove the exoskeleton to expose the underlying structures. Remove the gills and the membrane separating them from the internal organs. Flood the tray with water so that the organs float.

Compare the opened crayfish to figures 22.6 and 22.7. Starting in the abdomen, note the two longitudinal bands of **abdominal extensor muscles** that extend forward into the thorax. Contraction of these muscles straightens the abdomen from a curled position. Beneath the extensors, find the segmented **abdominal flexor muscles.** Contraction of these muscles bends the abdomen ventrally and allows the crayfish to swim rapidly backward in escape reflexes. Passing along the dorsal midline of these muscles is the **intestine** with the overlying **dorsal abdominal artery.** If you are dissecting an injected specimen, this artery should contain colored latex. Carefully remove the long bands of the extensor muscles. How has segmentation led to increased mobility?

Figure 22.7   Sagittal section of crayfish showing positions of internal organs.

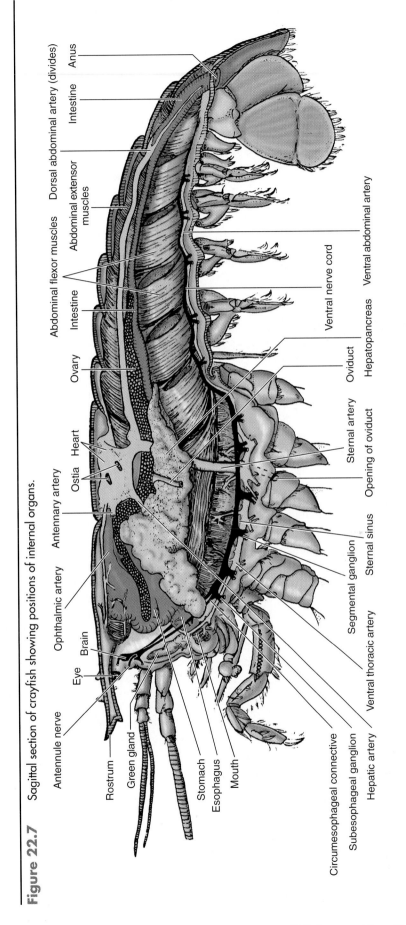

Antennule nerve

Eye   Brain

Rostrum

Green gland

Stomach
Esophagus
Mouth

Ophthalmic artery

Antennary artery

Ostia   Heart

Ovary

Intestine

Abdominal flexor muscles

Abdominal extensor
muscles

Dorsal abdominal artery (divides)

Intestine   Anus

Ventral nerve cord   Ventral abdominal artery

Hepatopancreas

Oviduct

Sternal artery   Opening of oviduct

Segmental ganglion   Sternal sinus

Circumesophageal connective

Subesophageal ganglion

Hepatic artery

Ventral thoracic artery

Circulatory System Trace the dorsal abdominal artery anteriorward to where it joins the delicate membranous **heart** located in the posterior third of the cephalothorax. Hemolymph surrounding the heart enters it through small openings, the **ostia,** which should be visible. When the heart contracts, small flaps of tissue seal the ostia from inside the heart, and hemolymph in the heart is forced out into the arteries. These arteries are often difficult to locate, but should be visible after careful study. As figure 22.7 indicates, these include the single **ophthalmic artery,** the **paired antennary arteries,** paired **hepatic arteries,** and a single **sternal artery.**

Arthropods have open circulatory systems. When hemolymph is pumped away from the heart in arteries, it leaves the arteries and enters the spaces in tissues at the periphery of the animal. Hemolymph in the tissues is displaced by new hemolymph being pumped in and percolates back to the pericardial sinus where it enters the heart through the ostia and is pumped to the periphery again. Hemolymph contains the respiratory pigment hemocyanin. When it is pumped through the sternal artery it enters the **sternal sinus** from which it flows into the gills before returning to the heart. This circuit allows respiratory gas exchange to occur.

Reproductive System From your examination of your specimen's external anatomy, you should know whether you have a male or female. In females, the paired ovaries lie ventral and lateral to the pericardial sinus. If collected in breeding season, the ovaries may be enlarged and filled with eggs. In males, the paired testes should be visible on either side of the midline beneath the pericardial cavity.

Digestive System Trace the digestive system from the mouth through the **esophagus** to the **stomach** and into the **intestine.** Find the anus ventrally at the base of the telson. Locate the large **hepatopancreas,** a large gland that secretes digestive enzymes into the stomach. Cut the esophagus and intestine to remove the stomach. Open the stomach and observe the hardened areas of the wall, the **gastric mill** that grinds food. At the posterior end of the stomach, fine bristles prevent large pieces of ingested food from entering the intestine.

Nervous System Look on the floor of the body cavity and find the ventral nerve cord. **Segmental ganglia** are collections of neurons that function as integration centers. Lateral nerves leave the ganglia and innervate the appendages. If you follow the nerve cord anteriorly, you will find ganglia beneath and above the stub of the esophagus. These function as the crayfish's brain.

Many arthropods have **compound eyes.** Such eyes are made up of hundreds of individual light-sensitive units called **ommatidia.**

Cut the surface off one of the crayfish's eyes and mount it in a drop of water on a slide. Look at the slide with your compound microscope. Sketch what you see in the circle. Compound eyes give an animal mosaic vision. What does that mean?

Excretory System Remove the organs from the cephalothorax. Find the **antennary glands** at the base of the antennae. These are the crayfish's excretory organs and also are known as green glands, though they are not green in color.

Dispose of your crayfish as directed by your instructor. Pair off with another student in the lab and explain to each other how a crayfish gets oxygen to its tissues and the pathways of circulation of hemolymph.

Subphylum Uniramia

There are two taxonomic classes in this subphylum: Class Insecta, which you will study in detail, and Class Myriapoda (centipedes and millipedes) which will not be studied. The uniramians are fundamentally a terrestrial arthropod group although a few species have invaded freshwater habits for part of their life cycle.

In the past, the uniramians have been classified along with the crustaceans in a clade called the Mandibulata. Many question the implication that these animals stem from a similar ancestor and the current thinking treats them as separate subphyla.

The characteristics of the uniramians include:

- Body with two or three regions; all have head and abdomen but insects have a thorax in addition.
- All appendages uniramous and do not branch at ends as in crustaceans.
- Head appendages include 1 pair of antennae, 2 pairs of maxillae, 1 pair of mandibles.
- Have a unique tracheal gas respiratory system.
- Have unique malpighian tubule excretory system.

Class Insecta

There are about 28 orders of insects and well over 900,000 species, making them the most diverse group of animals. The adaptability of the insect body form has developed over 400 million years and allowed insects to be successful in virtually every ecosystem, except in marine environments where the crustaceans predominate. Insects are the most common animals in terrestrial environments, attesting to the exoskeleton's ability to prevent water loss as well as to support the animal. Insects are among the relatively few animals that fly and this may contribute to their success by allowing them to escape predators and to disperse widely. Ecologically, they are important members of food chains, are extremely important in the pollination of dicot flowers with which they have coevolved, and can be transmitters of diseases caused by viruses, bacteria, and protists. The organ systems of insects are highly developed and some are unique as your dissection work will demonstrate.

8▷ Body Plan Obtain a cricket, cockroach, or grasshopper from the supply area.

Note your specimen's noncellular exoskeleton composed of chitin and proteins which are secreted by underlying cells. Crickets and grasshoppers have what is called incomplete metamorphosis. As these animals mature, they undergo a series of molts (ecdysis) increasing in size and changing slightly in morphology; *e.g.,* wings develop. Many other insects undergo a complete metamorphosis, radically changing body form from larval to adult stages as in the caterpillar-to-moth transition. Your instructor may have on display some living examples of the stages in insect development. Be sure to look at them.

For your insect, note the three distinct body regions: head, thorax, and abdomen. The **head** bears a single pair of **antennae** which function in chemoreception (odors) and as touch receptors. Compound and simple (ocelli) eyes also occur on the head. Grasshoppers and crickets have chewing mouthparts. Use your dissecting microscope to observe the mouthparts of your specimen. Find the heavy **mandibles** with the hardened edges. Which way do the mandibles move to chew food, vertically or horizontally? Note the upper and lower "lips" which cover the mouth. Small **palps** attached to **maxilla** help maneuver food to the cutting surfaces of the mandibles.

Other insects have mouthparts specialized for biting or lapping. Microscopes in the demonstration area have slides of mosquito and fly heads to demonstrate their mouthpart specializations. Mosquitos have biting-sucking mouthparts and the house fly has lapping mouthparts. Be sure to look at them.

The **thorax,** which consists of three body segments (pro-, meso-, and metathorax), bears three pairs of **legs** and usually two pairs of **wings.** One pair of legs originates on each thoracic segment. The legs are jointed, allowing agile movements. The leg segments starting next to the body are named coxa, trochanter, large femus, spiny tibia, and a tarsus bearing a terminal claw. The metathoracic legs are specialized for jumping in grasshoppers and crickets, especially the femur. In insects that only walk, such as ants, the legs will be more alike. In male grasshoppers and crickets the last leg is also used to produce sound. Teeth on the femur are rubbed against the edge of the front heavy wing to produce the familiar stridulation or chirping sounds of these species. The wings originate from the meso- and metathoracic segments. The forewings are heavy and the hindwings membranous. Some insects, such as fruit and house flies, have only one pair of wings. The second pair is reduced to small knob-like projections called halteres.

The **abdomen** consists of 11 body segments and sometimes fewer in other species. Count the number in your specimen. Look along the side of the abdomen and note the 10 respiratory spiracles. These open into a system of tubes that pass throughout the body (see fig. 28.6) and are part of the tracheal respiratory gas exchange system. The last three segments are modified in the different sexes, either for copulation or egg laying. A pair of sensory **cerci** can be found on the eleventh segment. On the first abdominal segment, find the **tympanum** that functions in sound reception.

As you have studied the external anatomy of your insect, you should have noticed how the exoskeleton is arranged in plates called **sclerites,** some with rigid junctions, called **sutures,** others with flexible junctions. Pigments in and under the exoskeleton give many insects their distinctive colors. The sclerite covering the dorsal surface is the **tergum** and that covering the ventral surface is the **sternum.** Lateral plates are called **pleurons.**

Internal Anatomy Preserved insects are usually not good materials to dissect because the exoskeleton prevents rapid penetration of the preservative so that the organs are often digested by enzymes escaping from the digestive system. For this section, you will use a freshly killed cricket or cockroach. They can be killed by exposing them to carbon dioxide or placing them in a freezer.

Insects are compact animals with many organs and tissues crammed into a small space. Given the relatively small size of the animals, this makes dissection a challenge. In this section, you will have an opportunity to develop your microsurgery techniques as the dissection should be done while viewing through a dissecting microscope.

9▷ Obtain a freshly killed cricket or roach, a petri plate filled with wax, and four small pins. Remove the insect's

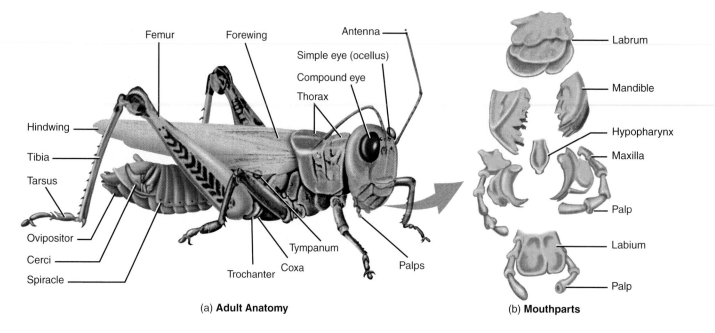

(a) **Adult Anatomy**   (b) **Mouthparts**

wings by cutting them off at the base. Fasten the animal to the wax, dorsal side up, by putting a pin through the head and the last abdominal segment. Now comes the crucial step—opening the animal.

Using fine, sharp scissors and forceps, loosen a tergite near the tip of the abdomen and insert the tips of the scissors. Cut along the lateral edge of the tergum. Cut forward to the thorax. Raise the dorsal flap and use a dissecting needle to work it free of underlying tissues. Be careful not to damage the dorsal aorta which lies underneath. Now continue your cut forward through the thorax. When at the front of the thorax, make a new cut across the body to the other side. Make a similar cut in the last abdominal segment. Now work the dorsal flap of the exoskeleton free and lay it over and pin it down. Flood the tray with saline to float the organs and keep them from drying. You will now systematically work your way down through the animal, exposing different organ systems. Use figure 22.9 as a guide to identification.

Circulatory System Insects have **open circulatory systems,** as do all arthropods. Vessels move a circulating fluid called **hemolymph,** from one body region to another where the fluid enters the spaces between tissues and gradually percolates back.

If your removal of the exoskeleton was done carefully, you should see the **tubular heart** on top of the body mass in a small depression. Hemolymph enters the heart from the surrounding spaces through minute lateral openings called **ostia** and is pumped forward by peristaltic waves through a **dorsal aorta** to the head. There it leaves the aorta and enters the tissue spaces, collectively called the **hemocoel.** Hemolymph gradually flows back to the pericardial space and enters the ostia to complete the circuit.

As you look at the organs, a chalky white **fat body** may fill the hemocoel. In females, the fat body can be re-

placed by swollen oviducts holding up to 600 eggs. Remove this material carefully to reveal the respiratory and digestive systems.

Respiratory System Insects have a unique respiratory system, the **tracheal system** which conveys oxygen directly to the tissues.

As you work though the fat body, you should occasionally see glistening white tubes. These are **tracheoles** of the tracheal system. They branch from main **tracheae** coming from the spiracles that you saw on the external lateral surface. This system of tubules branches into fine tubes that directly take away carbon dioxide and bring oxygen to every tissue in the body. The hemolymph does not serve as an intermediate carrier as blood does in many other animals. The finest tubules have fluid in them so that gas exchange occurs via a fluid medium. In larger insects, air sacs associated with the tubes help to ventilate them (see fig. 28.6). When the insect uses muscles for movement, the sacs are compressed and air is forced out and drawn in as the muscles relax.

An interesting sidebar on respiratory systems is a problem beekeepers encounter. A parasitic mite (a chelicerate) parasitizes bees. It invades and lives just inside the spiracles in the tracheal system. This interferes with the respiratory system and can kill a whole bee colony.

Digestive System Insects have a complete digestive system showing typical tube-within-a-tube architecture.

As you reveal the digestive system, note the different regions along its length. The foregut has three regions: the **esophagus, crop** (a storage area), and the **gizzard** (proventriculus) where chitinized plates on the inner wall grind the food into a fine pulp. Food leaves the foregut and enters the midgut consisting of a **stomach** with six fingerlike **gastric caeca** which secrete digestive enzymes. The hindgut or **intestine** leaves the stomach and passes back to a **rectum** and

Figure 22.9   Internal anatomy of a cricket. Fat body is not shown for clarity.

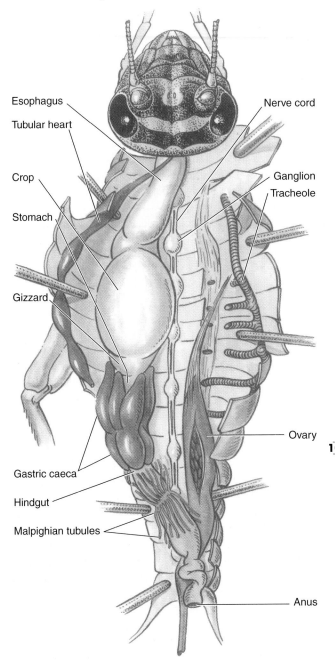

Esophagus
Tubular heart
Crop
Stomach
Gizzard
Gastric caeca
Hindgut
Malpighian tubules

Nerve cord
Ganglion
Tracheole
Ovary
Anus

anus. The rectum is a water-reclamation organ and produces a relatively dry fecal material. When insects molt, the lining of the foregut and hindgut are shed along with the exoskeleton.

Excretory System The excretory system of insects is unique and consists of structures called **malpighian tubules.** These can be seen radiating like threads from the middle of the hindgut. The cells in the walls of these tubules absorb nitrogenous waste materials from the hemolymph that bathes them, forming uric acid. Crystals of uric acid enter the hindgut and are compacted with fecal material before defecation.

Reproductive System The sexes are separate in insects and show sexual dimorphism (sexes look different). Based on your observation of external anatomy, you should know if your insect is male or female. If the gonads were not enlarged and filling the body cavity, remove the digestive tube to find them. You may be able to follow the ducts leaving the ovary or testes and proceeding to the genital opening.

Nervous System After removing the digestive and reproductive systems, you should be able to see the internal floor of the body cavity. Passing from anterior to posterior on the midline, you should be able to see two whitish **nerve cords** which form three paired ganglia in the thorax and five smaller ganglia in the abdomen. **Ganglia** are collections of nerve cell bodies and are where integration of nerve signals takes place. If this is not apparent, pour the saline out of the dish and add a few drops of methylene blue stain to the body cavity. Let it sit for a few minutes and then wash it out. The nerve cords should stain blue. Additional ganglia are found in the head but will not be observed in this dissection.

Muscle System The muscle system is too complex to work out in a short lab period. Insect muscles have a more complex arrangement than mammalian muscles because of the virtually unlimited surfaces for attachment on the inside of the exoskeleton.

You are now finished with your insect dissection. Dispose of the carcass and clean your dissecting instruments as directed.

Using Keys to Identify Insects (Optional)

10▶ If you are to do all of the dissections included in this lab, there probably will not be enough time to also do this portion of the lab. However, you instructor may choose to skip some of the dissections and use this activity to introduce you to the diversity of insects.

There are about 28 taxonomic orders of insects. For example, butterflies are in the order Lepidoptera while beetles are in the order Coleoptera. With close to a million species of insects in so many orders, experts turn to taxonomic keys to help them identify newly collected specimens.

A taxonomic key asks you questions which must be answered by looking at the unknown insect. For example it may ask, Does the specimen have wings? If yes, then you are directed to another set of questions such as, Does the specimen have one pair of wings or two? If it has one pair, then you are told the insect is in the order Diptera, the order of flies. Of course, other answers lead you off in other directions and to other questions that eventually allow you to assign an insect to one of the 28 orders. Similar keys have been made within orders to allow you to identify species. Can you imagine the number of questions you would be asked in order to identify a species of beetle, when there are over 500,000 species known?

If you are to do this activity, your instructor will have a collection of unknown insects in the lab. They will be numbered so that you can check your identification later against a master list. Keys to taxonomic order will also be available and you will be given directions in their use.

Protostomes II: Ecdysozoa and Great Diversity

Figure 22.10   Berlese funnel set up to collect leaf litter arthropods.

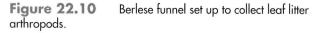

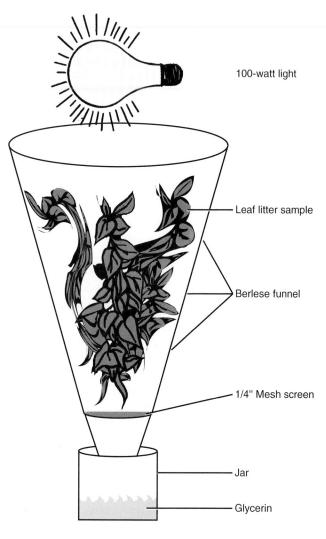

100-watt light

Leaf litter sample

Berlese funnel

1/4" Mesh screen

Jar

Glycerin

Field Collecting Terrestrial Arthropods (Optional)

11▶ Many insects as well as other arthropods live in leaf litter on forest floors. In this section of the exercise, you will use a classical technique called a Berlese funnel to collect small arthropods (as well as animals in other phyla) from a leaf litter sample (fig. 22.10).

Take a sample of a forest floor by digging with a spade down into the humus layer and collecting about a cubic foot of loose leaves and some top soil. Put it in a plastic bag for transport. Back in the lab, empty the contents into a large funnel called a Berlese funnel which has some 1/4" mesh screen across the bottom to prevent debris from falling out. Tape a jar of glycerin to the bottom of the funnel and set it up in a ring stand. Place a 100-watt light above the funnel. The heat from the light will dry the sample, forcing any animals in it to go deeper and deeper. At some point they will pass through the screen and fall into the glycerin and be killed. This may take several days and your instructor may set this up before lab.

Your job is to see what kinds of animals the sample contains. Expect to find all kinds of arthropods, such as insects of many different kinds, spiders, mites, pill bugs, centipedes, millipedes, and pseudoscorpions. In addition, expect to find other ecdysozoans such as nematodes. You may also find an occasional annelid and mollusk (snail).

When you are ready to examine the animals from the sample, dilute the glycerin in the trap jar with some alcohol and stir. After the animals fall to the bottom, pour off most of the fluid without losing any animals. Use an eyedropper to transfer some of the animals to a small watch glass. Identify them and tally the number present by using table 22.1. Enter the location where the leaf litter sample was collected.

TABLE 22.1	Types of animals from leaf litter.		
Sample from _____			
Animal name	**Tally marks**		**Class Totals**
Arthropod			
Insect			
Mite			
Spider			
Centipede			
Millipede			
Pill bug			
Pseudoscorpion			
Other			
Nematode			
Annelid			
Mollusc			

When you are finished with your identifications, go to the blackboard and write your totals in a facsimile of the table. Sum the class data. Based on the class data, what are the most common types of animals in the leaf litter from this location?

Let's assume that your leaf litter sample represents about 0.3 square meters of forest floor. How many of each type of animal would you expect to find in a hectare (= 10,000 square meters or 2.2 acres)?

Learning Biology by Writing

Write an essay that summarizes the evidence gathered in this lab that supports or refutes the hypothesis that all Arthropods have an exoskeleton, segmented body, and jointed appendages. Indicate how the diverse species studied are similar and how they differ from each other.

As an alternative, your instructor may ask you to turn in lab summary 22 at the end of this lab topic.

Internet Sources

Check the WWW for information about arthropods as vectors of diseases. Use your Internet browser to connect to Google at www.google.com. Type in Center for Disease Control. When connected to CDC find the search function. Enter the name of an arthropod-borne disease such as: Chagas; filariasis; Lyme disease; malaria; sleeping sickness; or West Nile virus. Read about the disease and answer these questions: (1) Name of disease? (2) Causative agent? (3) Arthropod vector? (4) World region(s) where common? (5) Estimated number of people affected?

Lab Summary Questions

1. What are the distinguishing characteristics of the clade Ecdysozoa? Phylum Arthropoda?
2. What are the general differences between the three living subphyla of arthropods?
3. Describe the circulatory and respiratory systems of a crayfish. Compare them to the systems found in an insect.
4. What are the similarities between annelids and arthropods? What are the differences?
5. What is a taxonomic key and how is it used?

Critical Thinking Questions

1. Why are there no really large arthropods? What could be engineering-type factors limiting body size?
2. How does the segmentation of an arthropod compare with the segmentation of an annelid?
3. What correlation may be made between body symmetry and locomotion?
4. Arthropods are the most diverse (greatest number of species) and the most common (greatest number of individuals) of all the animals. List several reasons why you think they have been so successful.

TABLE 22.2 Summary of phylum Arthropoda

	Subphylum Chelicerata	Subphylum Crustacea	Subphylum Uniramia
Common name(s)?			
Tissues?			
Symmetry?			
Name organ systems present			
Type digestive system?			
Type circulatory system?			
Type skeletal system?			
Unique features?			
Describe body plan			

LAB TOPIC 23

Deuterostomes and the Origins of Vertebrates

Supplies

Preparator's guide available on WWW at
http://www.mhhe.com/dolphin

Equipment

Compound microscopes
Dissecting microscopes

Materials

Preserved specimens
    Demonstration specimens of the five classes of
        echinoderms
    Sea star
    Branchiostoma (amphioxus)
    Adult sea squirt (*Ciona* or *Molgula*)
    Perca flavescens (perch); fresh fish can be substituted
Prepared slides
    Sea star arm, cross section
    Ascidian tadpole larva
    Branchiostoma, whole mount
    Branchiostoma, cross section
Dissecting pans and instruments
Photographs of various echinoderms

Student Prelab Preparation

Before doing this lab, you should read the introduction
and sections of the lab topic that have been scheduled
by the instructor.

You should use your textbook to review the
definitions of the following terms:

bilateral symmetry	Echinodermata
cephalization	radial symmetry
Cephalochordata	Urochordata
Chordata	Vertebrata
deuterostome	Water vascular system

You should be able to describe in your own words the
following concepts:

The general body plan of an echinoderm
The general body plan of a chordate
The organ systems you would expect to find in the
    animals you will study

As a result of this review, you most likely have
questions about terms, concepts, or how you will do
the experiments included in this lab. Write these

questions in the space below or in the margins of the
pages of this lab topic. The lab experiments should
help you answer these questions, or you can ask your
instructor during the lab.

Objectives

1. To illustrate the anatomy of echinoderms
2. To study the anatomy of the primitive chordates
3. To study the anatomy of a bony fish
4. To relate the anatomy of echinoderms, primitive
   chordates, and bony fish to evolutionary trends
5. To collect evidence to test the hypothesis that all
   chordates have certain common characteristics that
   suggest an evolutionary linkage among the groups

Background

With this lab we end our investigation of diversity and phylogeny within the Kingdom Animalia. We will do so by
looking at animals that are grouped in the clade Deuterosto-
mia. This is not a large clade and contains only 3 or 4 phyla
with approximately 59,000 species, only about 4% of the
species in the animal kingdom. During the lab you will
look at animals from the phylum Echinodermata that includes such animals as sea stars, brittle stars, sea urchins,
sea cucumbers, and feather dusters—and at animals from
the phylum Chordata that includes fairly simple animals
like sea squirts and lancelets, as well as those animals with
vertebral columns, fish through mammals.

Animals in the deuterostome phyla differ from the protostome animals in the following ways:

- Radial rather than spiral cleavage in early embryology;

- Enterocoelous body cavity development versus
  schizocoelous;

- Anus, rather than the mouth, developing from the
  blastopore in gastrula;

- Genetic similarities indicating a closer relationship among themselves than to other phyla grouped in the protostomes.

It is interesting to note that the new molecular data does not refute the relationships based on morphological data. In the strict logic of science, the hypothesis that Echinoderms and chordates as well as a few other phyla are related is not falsified. The hypothesis is understood to be true until someone can devise another independent test (experiment) to try to falsify it. This is what makes science so interesting: the constant suggesting of new ideas and then the logical testing of those ideas to determine if the idea has validity. The molecular data, however, have raised questions about how many phyla should be included in the deuterostomes and what their evolutionary relationships are to each other. Most of this debate is about a group known as the hemichordates which we will not be discussing.

LAB INSTRUCTIONS

You will observe the anatomy of four deuterostome animals: an echinoderm (a sea star) and three chordates (a tunicate, an amphioxus, and a fish).

Phylum Echinodermata

The approximately 7,000 species of echinoderms are spiny-skinned, slow moving or sessile marine animals that are voracious feeders. Adults have an unusual symmetry not found in any other animal. They have a pentamerous radial symmetry, meaning that the body is arranged into five (or multiples of five) repeating units radiating from a central area. However, their larval stages have bilateral symmetry which changes to pentamerous during development. Embryological development follows the typical deuterostome pattern (see lab topic 19 for a thorough discussion).

All adults have a calcareous endoskeleton composed of separate plates or ossicles forming an open meshwork with living tissues in the spaces. The bases of spines protruding from the surface articulate with the ossicles. Ossicles are often fossilized, allowing the identification of nearly 13,000 extinct species dating back to the Cambrian.

Echinoderms have a unique physiological system called a **water vascular system** that serves a variety of functions, the most prominent being locomotion. They have a complete tubular digestive system. The body surface, or pouches in some species, serves as a site for exchange of respiratory gases and nitrogenous wastes, and no other specialized organs are found to perform these functions. No specialized circulatory system structures are found other than ciliated cells lining the coelom which move fluids from one area to another. The nervous system is diffuse, usually consisting of a nerve ring, nerve net, and radial nerves. The sexes are sep-

arate individuals. Asexual reproduction can occur by regeneration of missing body parts from fragments.

Echinoderm Diversity

Six classes are recognized. Five of these will be available in lab for you to see. Go to the area where the display specimens are laid out and find the representatives of each class (fig 23.1).

Crinoidea (sea lilies and feather stars) have the following characteristics:
- featherlike arms branching into pinnules radiate from the body in multiples of five;
- body looks like a cup with mouth upward;
- sea lilies have body at end of stalk; feather stars can be free swimming;
- open ambulacral grooves on arms bear tube feet that filter food from water.

Concentricycloidea (sea daisies)
- disc-shaped body with ring of marginal spines and radiating arms;
- discovered in 1986 in deep water off New Zealand; about 1 cm in diameter.

Ophiuroidea (brittle stars and basket stars)
- body stellate with 5 thin arms joining a central disk;
- mouth opening on under-surface;
- ambulacral grooves closed with tube feet lacking suckers.

Echinoidea (sea urchins and sand dollars)
- globe-shaped or disc-shaped bodies with plates fused into solid test;
- no arms;
- mouth opening on under-surface;
- spines movable;
- ambulacral grooves closed;
- feed by scraping encrusting algae from surfaces.

Holothuroidea (sea cucumbers)
- soft-bodied and sausage-shaped;
- no arms;
- mouth opening at one end and anus at other;
- circle of feeding tentacles around mouth;
- skeleton reduced to widespread ossicles;
- tube feet reduced and sometimes absent;
- suspension and deposit feeders.

Asteroidea (sea stars)
- You will dissect a sea star and observe its anatomy in detail.

Figure 23.1   Diversity of echinoderms: (a) sea cucumber; (b) sea lily; (c) feather star; (d) sea urchin; (e) brittle star; (f) sea daisy.

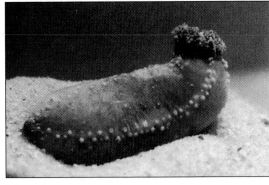

(a)

(d)

(b)

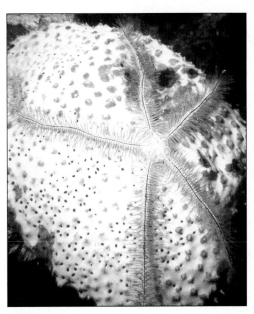

(e)

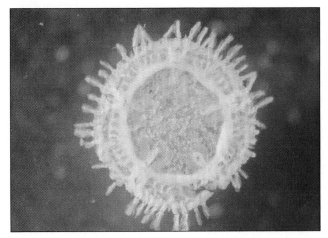

(c)

(f)

Figure 23.2   Anatomical features of a sea star. Animal is partially dissected to show different internal organ systems.

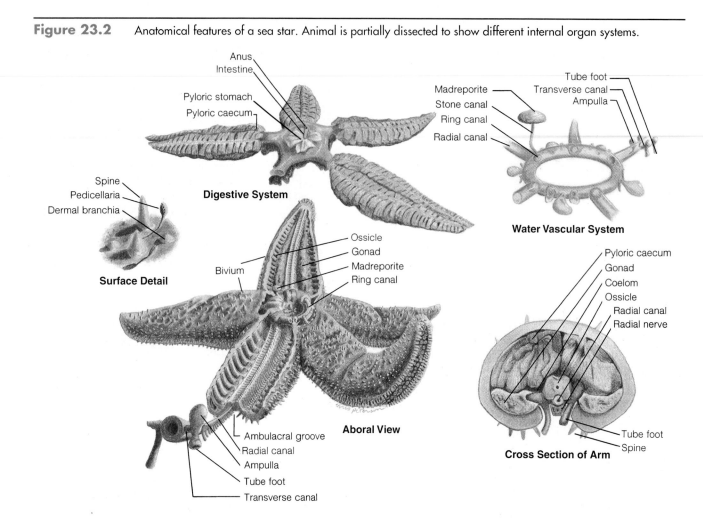

External Anatomy of Sea Star

1 ▷ Obtain a sea star and observe its pentamerous radial symmetry. The ventral surface, the underside, is called the **oral** surface, and the dorsal is called the **aboral** surface. Examine the central disc aboral surface and find the **madreporite,** a small light-colored calcareous bump or spot, which acts as a pressure-equalizing valve between the water vascular system and seawater (fig. 23.2). The difficult to see **anus** is located more or less in the center of the aboral surface of the central disc.

Observe the many spines protruding from the surface of the organism. These spines give the echinoderm ("spiny-skinned") phylum its name. If a sea star is submerged in water and its surface examined with a dissecting microscope, many small, pincerlike **pedicellariae** may be seen projecting from the surface of the animal. The pedicellariae remove encrusting organisms from the skin, keeping it extremely clean in comparison to many other invertebrates. Some echinoderms, such as sea urchins, use pedicellariae to hold objects for camouflage.

2 ▷ Scrape away some of the tissue at the surface and mount the material in a drop of water on a microscope slide. After adding a coverslip, observe the slide under the compound microscope and sketch some pedicellariae in the circle. Some of the scrapings may contain **dermal branchiae,**

which are hollow, bubblelike extensions of the body wall that aid in gas exchange. Alternatively, your instructor may have a prepared slide of pedicellariae for you to examine.

The body wall contains the skeleton of the sea star. It is made of a large number of interconnecting **ossicles.** These can best be observed in a dried specimen or one that has been soaked in NaOH to remove the flesh.

Turn the animal over and observe its oral side. The mouth is surrounded by a soft membrane called the **peristome** and is guarded by oral spines. Five **ambulacral grooves** radiate from the central disc down the center of each arm. Each groove is filled with **tube feet,** extensions of the water vascular system. Though they are almost impossible to see, a sensory tentacle and a light-sensitive eyespot are at the end of each arm.

Internal Anatomy

3 ▶ Cut 1 to 2 cm off the tip of an arm and then cut along each side of the arm to where it joins the central disc. Join the cuts by cutting across the top of the arm at the edge of the disc, and remove the body wall to expose the coelom within the arm (fig. 23.2). Note how the organs are free in the coelom. Ciliated cells line the coelom and move fluid through the cavity. Coelomic fluid carries oxygen, food, and waste products to and from the organs.

Now remove the aboral body wall covering the central disc without removing the madreporite. Do so by cutting carefully around the periphery of the disc and then cutting to and around the madreporite, freeing it from the body wall. Lift and remove the body wall carefully, noting where the intestine joins the anus in the piece you are removing.

Digestive System   In each arm are pairs of **pyloric caecae,** or digestive glands, that empty into the pyloric portion of the stomach. The **pyloric stomach** connects with the **intestine** on the aboral side; on the oral side, it connects with the **cardiac stomach** and **esophagus.** When the sea star feeds, the cardiac stomach and the esophagus are everted through the mouth and surround the material to be digested. Enzymes are secreted by the cardiac stomach so that partial digestion occurs outside the body. Food particles then are ingested and digestion is completed inside the body. Carefully remove the digestive glands to expose the next layer of organs in the arm. Is this a complete or incomplete digestive system? _____

Reproductive System   Under the digestive glands in each arm lie two **gonads.** Depending on the stage of the breeding cycle at the time the sea star was caught, the gonads may either fill the arm or be quite small. **Gonoducts** lead from each gonad to very small **genital pores** located at the periphery of the central disc on the aboral surface between the arms. Though the sexes are separate, they are difficult to distinguish except by microscopic observation of the gonad contents.

Water Vascular System   The water vascular system of the echinoderms is a unique internal hydraulic system associated with the functioning of the tube feet. Water enters the madreporite and travels via the **stone canal** to a **ring canal** that encircles the mouth (fig 23.2). From the ring canal, the water passes out into **radial canals** in each arm, and from each radial canal, several **transverse canals** lead to each pair of **tube feet.** Remove the gonad from the dissected arm to reveal the radial canal and the ampullae of the tube feet.

Figure 23.2 shows a cross section of a sea star and the relation of the radial canals to the tube feet and to the **ampullae** on the inside of the animal. Contraction of the ampullae extends the tube feet, whereas relaxation retracts them, providing a means of slow locomotion and a means of adhering to substrates. Cut across one of the arms and find the radial and transverse canals. Carefully try to reveal how the transverse canal joins the tube foot.

Nervous System   The nervous system of the sea star lacks cephalization, as is characteristic of radially symmetrical animals. Though you will not be able to see them, a **nerve ring** surrounds the mouth with **radial nerve cords** passing into each arm. At the junction of the ring and radial nerves are ganglia. The only differentiated sense organ is the eyespot at the tip of each arm, though there are other sensory cells located throughout the epidermis.

Note that in your almost complete dissection of this animal that you did not see a heart or blood vessels, nor were there any excretory organs. How do you think the animal performs the functions of circulation and excretion?

Clean up your area and dispose of your starfish as directed.

4 ▶ *Microscopic Anatomy*   Obtain a prepared microscope slide of a cross section of a sea star arm and compare what you see to figure 23.2. There are several things to note in this cross section. The pyloric caecum is the tissue filling most of the cavity and the gonad is beneath it. Both of these are in the body cavity, a coelom. You should be able to see the radial canal and the structure of one of the tube feet. Note the complex musculature.

Return your microscope slide to the supply area.

Phylum Chordata

The approximately 50,000 species of chordates range from simple invertebrate animals, such as sea squirts, to the relatively complex mammals and birds. All share eight characteristics:

1. All have **bilateral symmetry**.
2. All have a **deuterostome** pattern of development.
3. All have a **notochord** at some stage in their development.
4. All have a **dorsal hollow nerve cord** at some stage.
5. **Pharyngeal gill slits** appear during development or in adults.

6. All have a **post anal tail** at some stage.

7. Digestive system is complete with regional specializations.

8. All have a ventral contractile blood vessel or heart.

The phylum is split into three subphyla: Urochordata (about 3,000 species of tunicates, salps, and sea squirts), Cephalochordata (23 species of lancelets), and Vertebrata (about 47,000 species of fish, amphibians, reptiles, birds, and mammals). The first two subphyla contain animals that lack a backbone although they have notochords. They are called invertebrates along with all of the animals studied so far. The notochord in the vertebrates is replaced by a segmented column of bones, the vertebral column, which the clade name reflects. In this part of the lab, you will look at representatives of all three subphyla.

Subphylum Urochordata

Depending on the species, these marine invertebrates are sessile filter feeders (sea squirts) or drifting pelagic filter feeders (salps and larvaceans) that may form colonies several meters long. The distinguishing notochord is found only in the larval stages and disappears in the adult. If only adults were studied, the relationship to the chordates would not be apparent.

Larval Anatomy

5 We will start our study of the urochordates by looking at a slide of the larval stage, for it is only in this stage that we see the notochord. Obtain a slide of a **tunicate tadpole larva** and look at it only with your scanning objective.

Note the general resemblance to the tadpole stage of a frog, hence the common name (fig. 23.3c). The tadpole larva develops from the zygote and is free swimming for several days, aiding in the dispersal of the species. It will settle on a substrate and in a matter of minutes begin a metamorphosis. The adhesive glands can be seen on the anterior and aid in attaching the animal to a substrate.

Look at the tail, for it is here that you find the **notochord.** Made from cartilage like material, it stiffens the tail so that when tail muscles contract the tail bends rather than shortens to provide swimming movements. Above the notochord you may be able to make out a shadow of the nerve cord.

The main part of the body consists of a outer covering, the **tunica,** surrounding a large pharyngeal basket and various internal organs. The slits in the basket wall should be visible as a repeating pattern. We will talk about how the basket is used in feeding when we discuss adult anatomy.

When a larva settles on a suitable substrate, it undergoes a change in body form. The tail is reabsorbed with its notochord and nerve cord. The larva begins to filter feed and grows into an adult.

Adult Anatomy

6 Some adult tunicates will be available in the lab for you to observe. Look at them with your naked eye and through the dissecting microscope (fig. 23.3a and b). The outer covering of the animal is called the **tunica** and is the basis for the common name tunicate. It contains muscles that can contract and change the body shape. When they do, water often shoots out of an opening, hence their second common name, sea squirts.

Digestive System Opposite the basal end, find the two **siphons.** One opens into the **pharynx** and the other into a surrounding chamber called the **atrium.** In some species the tunica is translucent and allows you to observe the pharynx. As in the larval stage, it has slits in its walls, forming what is called a **pharyngeal basket.** The beating of cilia on cells lining the pharynx draws water through one of the siphons into the pharynx. The water passes through the slits into the atrium and then out through the other siphon. Suspended materials (such as algae, protists, and small larvae) cannot pass through the slits and are retained in the basket. Curtains of mucus produced by the pharynx flow downward and trap the food particles. Entrapped particles pass on to an **esophagus** at the base of the basket. Food enters the **stomach** where it is digested and into an **intestine.** The digestive tube is U-shaped and curves around so that the **anus** is located near the siphon through which the water stream exits so that fecal material is carried away.

Other Organ Systems Tunicates have no specialized respiratory system. Gas exchange occurs by diffusion across the body surfaces that are exposed to the water stream passing through the animal. Tunicates have an open circulatory system. A tubular heart is located below the stomach.

Tunicates are hermaphroditic, each possessing an ovary and a testis lying in very close proximity to each other and located near the loop of the digestive system. These closely placed gonads are shown as one structure in figure 23.3a. Gonadal ducts pass from the gonads to the region of the excurrent siphon where they open near the anus. When eggs or sperm are released, they are carried by the water stream into the surrounding water. Fertilization is external. Development is rapid and tadpole larvae settle within a few days to metamorphose into the adult body form. What is the advantage of having the gonoduct open near the excurrent siphon?

Subphylum Cephalochordata

The subphylum Cephalochordata contains a few small (5 cm long) fishlike species commonly known as lancelets. Lancelets live in shallow marine environments throughout the world. In some places, lancelets occur in sufficient numbers to be used as a human food source. The common

Figure 23.3 Stages of the life cycle of a tunicate: (*a*) internal anatomy of an adult; (*b*) adult sea squirt; (*c*) anatomy of a tadpole larva.

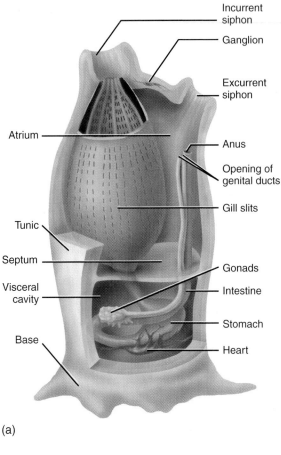

(a)

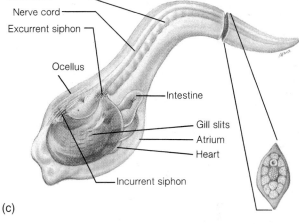

(b)

(c)

American species is in the genus *Branchiostoma,* though it is often called amphioxus, the former genus for the group.

External Anatomy

7▷ Immature lancelets are small enough to be mounted whole on microscope slides. Adults can be chemically treated so that the tissues are rendered transparent. They are then mounted in plastic. Obtain a slide or plastic mount and look at it through a dissecting microscope.

Note its general form and the absence of paired appendages (fig. 23.4). The **rostrum** is the anteriormost part of the lancelet. Beneath it is the **oral opening** surrounded by a fringe of **oral cirri.** Dorsal and ventral fins occur caudally. On the ventral surface, there are two pores: the **anus** is the most posterior opening, and the **atriopore** is anterior to the anus and is a water escape route for this filter feeder.

Internal Anatomy

8▷ Study a whole mount of a young lancelet with the dissecting microscope on high power with transmitted light and then

with the compound microscope on low power (fig. 23.4*a*). Do not use high power.

Digestive System Behind the oral cirri is the **vestibule,** a chamber lined by bands of ciliated cells collectively called the **"wheel organ."** The beating of the cilia draws a water stream containing oxygen and food particles through the oral opening into the vestibule. Mucus secreted into the vestibule traps the food particles. The mucus passes into the mouth at the rear of the vestibule. Trace the digestive tract from the mouth into the **pharynx,** a filterlike basket consisting of **gill bars** and **gill slits.** Water containing suspended food material passes from the vestibule through the mouth into the pharynx and is filtered through the slits and enters the **atrium,** the space surrounding the gill basket. Water then flows from the atrium to the outside through the atriopore. Food particles cannot pass through the gill slits. Food thus concentrated passes as a mucous string posteriorly into the esophagus, stomach, and intestine, and eventually out the anus (fig. 23.4*a*). The gill bars probably do not

Figure 23.4 Lancelet anatomy: (*a*) internal organs; (*b*) cross section in region of pharyngeal basket; (*c*) cross section in posterior half; (*d*) diagram of circulatory system.

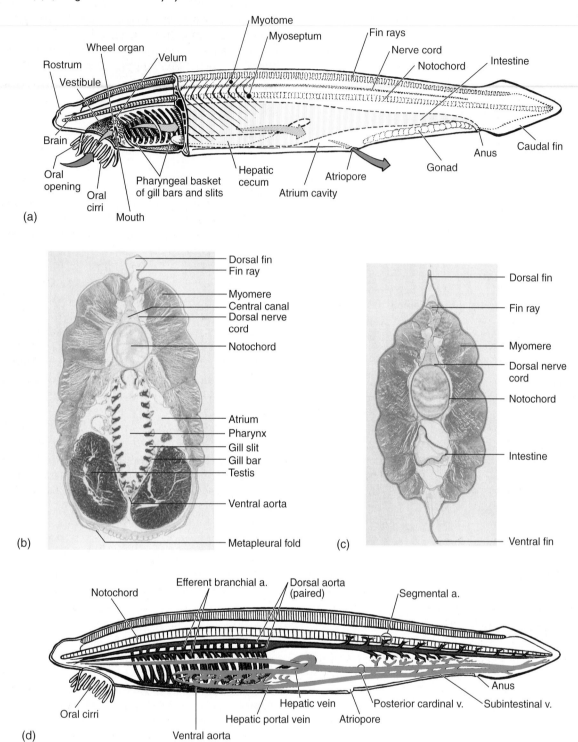

function in respiratory gas exchange. Most of this occurs across the body surface.

Musculature The muscles are arranged as V-shaped **myomeres.** When contracted, these muscles bend the body to allow the animal to swim and burrow. Lancelets commonly

burrow into sediments tail first with the oral opening remaining at the surface. The dorsal **notochord,** a cartilaginous rod, runs the length of the organism just dorsal to the pharynx and intestine. It stiffens the animal along its length so that contraction of the myomeres does not shorten the body. Instead, myomere contraction causes side to side movement that allows swimming locomotion. Dorsal to the notochord is the **nerve cord** with a small enlargement at the anterior end.

Circulatory System The circulatory system is difficult to observe (fig. 23.4*d*). The system is closed but lacks a definite heart. Blood is collected from the digestive system tissues by a **subintestinal vein.** It flows into the **hepatic portal vein,** which drains through the **hepatic vein** into the **ventral aorta** below the pharynx. **Afferent branchial arteries,** each with a contractile bulb at its base, leave the aorta and pass up through the gill bars. Pumping occurs by contraction of the branchial bulbs and the ventral aorta. Blood from the gills is collected by the **efferent branchial arteries** and flows into paired (right and left) **dorsal aortas** that join to form a median dorsal aorta posterior to the pharynx. This vessel supplies blood to the posterior tissues and to the muscles through the **segmental arteries.**

Reproductive System Locate the gonad located beneath the intestine in the caudal region (fig. 23.4*a*). The sexes are separate in lancelets and the gonad may be either an ovary or a testis. In sexually mature individuals it enlarges and may fill the atrium. Eggs or sperm are released into the atrium, and pass out through the atriopore with feeding water currents. Fertilization is external.

9 ▶ ***Microscopic Anatomy*** Look at the cross sections of *Branchiostoma,* in figure 23.4*b* and *c*. If slides are available compare them to the figure. Locate the notochord (roughly just above the center in the section). Dorsal to it is the nerve cord. Observe the nerve cord closely and note the hollow center, or **central canal,** within the nerve cord (a characteristic of chordates). Beneath the notochord will be the tube of the digestive system. In the section through the gill area, note the arrangement of the gill bars, slits, and atrium.

If you reflect on the general anatomical plan of amphioxus, its very general similarities to a fish are obvious. Except for the segmented backbone, jaws, and appendages, its general body plan is that of a fish. The similarity to the tunicate tadpole larva also should be apparent. This similarity probably means that both amphioxus and cartilaginous fish are related ancestrally through some organism much like a tadpole larva. Each group has evolved since then to the forms we see today. During the long periods of time (500×10^6 years ago) since the Cambrian, when fish appeared as the first vertebrates, other evolutionary changes have occurred to give rise to the other classes of vertebrates, including amphibians, reptiles, birds, and mammals.

Return your specimens to the supply area.

Subphylum Vertebrata

In the subphylum Vertebrata, all four chordate characteristics are present at some stage of the life cycle. However, during development in vertebrates, the notochord is replaced by the **vertebral column** (backbone) surrounding the nerve cord. Vertebrates are highly cephalized with well-developed sense organs and a distinct brain located at the anterior end of the dorsal nerve cord. The brain is encased in a skull, which together with the vertebral column, makes up the axial skeleton. Most vertebrates have two pairs of appendages supported by an appendicular skeleton. In contrast to the exoskeleton of many invertebrates, the vertebrate skeleton is a living endoskeleton capable of growth and self-repair. What are the advantages of this type of skeleton for a terrestrial existence? Describe why this type of skeleton places limits on the size of a vertebrate.

The circulatory system of vertebrates is a closed system consisting of a ventral heart and a closed vessel system of arteries, veins, and capillaries, with the respiratory pigment, hemoglobin, contained within red blood cells. The sexes are separate and reproduction is sexual.

This subphylum contains the largest number of chordate species. It includes seven classes: (1) Agnatha (jawless fishes), (2) Chondrichthyes (cartilaginous fishes), (3) Osteichthyes (bony fishes), (4) Amphibia, (5) Reptilia, (6) Aves, and (7) Mammalia. You will dissect a perch or trout, a representative of the 30,000 species of bony fish. In later labs, you will study vertebrate anatomy in a mammal, the fetal pig.

External Anatomy of a Fish

10 ▶ Preserved perch are often used for dissection but fresh fish (trout) from a fish farm can be used. They have the advantage of fresh tissues and natural color and may actually cost less than preserved perch. The anatomical drawings can be used with most species. Obtain a perch and wash it in tap water to remove the preservative.

CAUTION

Be careful: the dorsal fin of a perch contains spines that can give you a painful puncture wound.

Note the main regions of the body: head, trunk, and tail. Find the **dorsal fins** and the ventral **anal fin** (fig. 23.5). These fins help the fish maintain an erect position. Note the paired **pectoral** and **pelvic fins.** Used in a sculling motion, they allow the animal to move slowly. The caudal fin or tail is the main propulsive fin in fast swimming. The thin membranes of the fins are supported by skeletal elements, the **fin rays.** Cut the tips of the dorsal fin rays off with scissors to protect your hands during the subsequent dissection. Trout have a fatty adipose fin located posterior to the dorsal fin.

The body is covered with scales, larger in the perch than in the trout. Remove a few and mount them in a drop of water on a slide. Add a coverslip and observe. Scales provide fish with a thin, flexible armor. The rings that you see in the scales correspond to growth stages and can be

Figure 23.5 Internal organs of a perch.

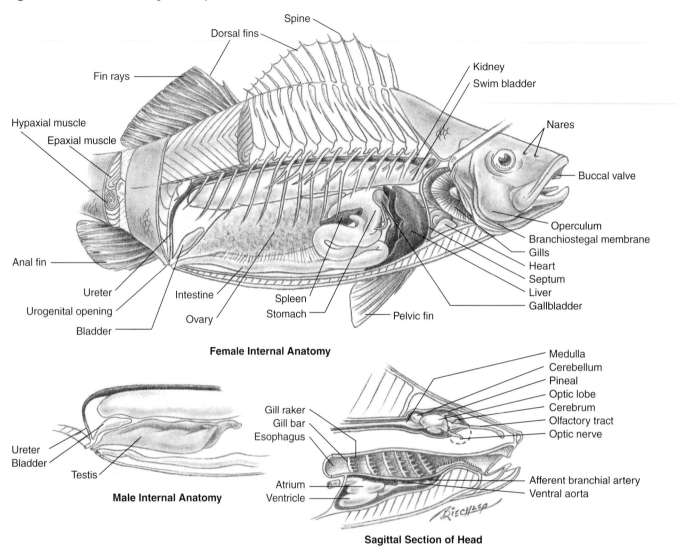

Female Internal Anatomy

Male Internal Anatomy

Sagittal Section of Head

correlated with age. Lipsticks and nail polishes containing "pearl essence" are made from silvery materials found in fish scales, primarily herring.

Respiratory System Pry open the mouth and examine the internal edge of the upper surface to see the **buccal valve,** a small flap of tissue that seals the mouth opening when the jaws are closed.

On each side of the head are the **opercula,** large plates that protect the gills and function as a pump. Along the posterior and ventral edges of the opercula, find the **branchiostegal membranes.** Fish pump water over their gills by the following mechanism. When the opercula are raised with the mouth open, the branchiostegal membranes close off the posterior gill openings so that water flows through the mouth into the pharynx. The mouth is then closed and sealed by the buccal valve. The opercula are then lowered, forcing water out the gill openings.

Lift up the operculum and reach in with scissors and cut one of the gill bars loose with its gill filaments. Remove it and

place it in a small dish with enough water to float the gills. Look at this with your dissecting microscope. The feathery gills have a tremendous surface area. Why is this important?

Examine the side of the body to find the **lateral line.** Sense receptors located here are sensitive to sound waves, water pressure, and weak electric currents.

Internal Anatomy of a Fish

11▷ To expose the internal organs, lay the fish on its left side and insert the tips of your scissors just anterior to the origin of the anal fin. Cut forward along the midventral line to the pectoral fins. Make a second cut parallel to the first but above the lateral line on the same side. Now connect the two cuts with a third at the anterior edge. Lay the right body wall back, free it from underlying tissues, and remove it.

Figure 23.6   Circulatory system of perch. Major arteries are in red and veins are in blue.

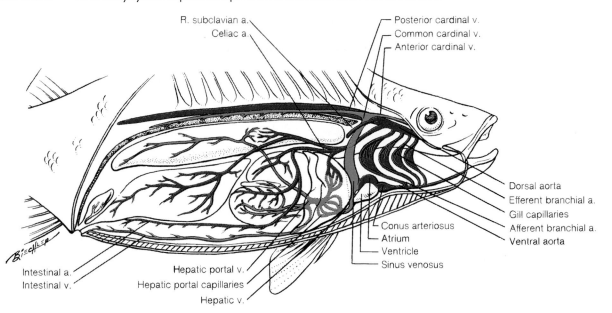

R. subclavian a.
Celiac a.
Posterior cardinal v.
Common cardinal v.
Anterior cardinal v.
Dorsal aorta
Efferent branchial a.
Gill capillaries
Afferent branchial a.
Ventral aorta
Conus arteriosus
Atrium
Ventricle
Sinus venosus
Intestinal a.
Intestinal v.
Hepatic portal v.
Hepatic portal capillaries
Hepatic v.

Note the various organs suspended in the coelom by **mesenteries,** thin, transparent projections of the shiny **peritoneum,** a tissue that lines the body cavity. Dorsal to the other organs is the long **swim bladder** (fig. 23.5). The air content of the bladder is actively regulated to maintain the buoyancy of the fish, allowing it to remain motionless in the water without settling to the bottom.

Digestive System   Remove the right operculum to expose the gills attached to the **gill arches** of the **pharyngeal wall.** Insert the tips of your scissors into the mouth and cut through the angle of the jaw back through the gills to expose the inside of the pharynx. Remove the structures you have cut through.

Examine the inside of the pharynx and visualize how water entering the mouth flows through the slits in the pharyngeal wall, over the gills on the gill bars, and out through the external gill openings. On the inside surface of the gill bars, note the **gill rakers,** which act as crude sieves in this species and direct ingested food toward the opening of the esophagus. In species that feed on suspended algae and detritus, the gill rakers may be highly developed and resemble a fine-tooth comb, which effectively removes small particles from water passing through the pharynx. In a predatory species, the gill rakers are widely spaced.

Pass a blunt probe through the mouth and down the **esophagus** into the **stomach** located dorsal to the liver. At the juncture of the stomach and the small intestine, note the **pyloric caecae.** These finger-like structures are outpocketings of the intestine and are thought to serve as temporary storage areas and possibly in absorption. Observe how the intestine leaves the stomach toward the anterior, producing a pouch. This is typical of predatory fish. This arrangement allows the stomach to expand to accommodate large prey. Trace the **intestine** from the stomach to the **anus.** The

spleen (not part of the digestive system) is the triangular-shaped organ located in the region between the stomach and the small intestine. Lift the lobes of the **liver** to find the globular, dark green **gallbladder.** What is its function?

Cut open the stomach to determine what your fish had eaten. List the contents below.

Circulatory System   Fish have a closed circulatory system. Find the **two-chambered heart** anterior to the **septum** in front of the liver, and ventral to the pharynx (fig. 23.6). The anterior chamber is the muscular **ventricle,** which pumps blood to the gills via the **ventral aorta** and **afferent branchial arteries.** Blood rich in oxygen and low in carbon dioxide leaves the gills via the **efferent branchial arteries.** Find the **dorsal aorta** dorsal to the gills. Blood collected from the gills flows via this major vessel to the rest of the body.

Blood is collected from the tissues via the **cardinal vein,** which runs just ventral to the vertebrae and from the digestive system via the **hepatic portal vein.** The latter flows into capillaries in the liver and is collected by the **hepatic vein.** The cardinal vein and hepatic vein flow into the **sinus venosus,** which returns blood to the **atrium** of the heart.

Deuterostomes and the Origins of Vertebrates

TABLE 23.1   Comparison of the characteristics of various phyla

Phylum	Symmetry	Coelom	Digestive System	Circulatory System	Respiratory System	Nervous System	Skeletal System
Porifera	None		Intracellular			Absent	Endoskeleton
Cnidaria	Radial	None	Gastrovascular cavity (sac) (incomplete)	Absent	Body surface	Ladder-like	Hydrostatic skeleton
Platyhelminthes		Acoelomate					
Nematoda		Pseudocoelom				Ventral cord	Exoskeleton
Bryozoa	Bilateral	Schizocoelom	Complete; (tube) mouth develops from blastopore		Lophophore	Ventral cord	
Annelida				Closed	Body surface or gills		
Mollusca						Cord	Exo- and endo-skeletons
Arthropoda				Open	Body surface or gills or book lungs or trachea	Ventral cord	Exoskeleton
Echinodermata	Bilateral, but radial in adult	Enterocoelom	Complete; (tube) anus develops from blastopore	None	Body surface	Not easily seen	Endoskeleton
Chordata	Bilateral			Closed	Lungs or gills	Dorsal cord with cephalization	

Excretory System   The excretory system consists of the paired **kidneys,** which are located dorsal to the swim bladder behind the peritoneum. Remove the swim bladder and find the dark brown organs firmly attached to the dorsal wall of the body cavity and extending almost its entire length. Along the median border of the kidneys are the **urinary ducts.** As they pass posteriorly, they unite into a single duct, which passes ventrally to the urinary opening. A urinary **bladder** may be seen branching from the duct's distal end.

Reproductive System   Perch have distinct males and females. Fish-farm-raised trout are usually all females. Females have a single large **ovary** lying between the intestine and swim bladder. If your fish is a female and was reproductively "ripe" when collected, the ovary will be filled with yellowish eggs. Eggs leave the posterior of the ovary via an oviduct that leads to a urogenital opening posterior to the anus.

If your fish is a male, paired **testes** will be visible caudal to the stomach. A **vas deferens** passes caudally in a longitudinal fold in each testis. These fuse to form a single duct that passes to a genital pore posterior to the anus.

Nervous System   To observe the brain of the perch, shave the skull bones off the front of the head to expose the brain.

The major divisions of the brain are: **telencephalon, optic lobes, cerebellum,** and **medulla oblongata** (fig. 23.5). Cut your fish in half by a transverse cut halfway down the body. Look at the vertebral column and find the nerve cord surrounded by bone.

What adaptations have you observed in this dissection that would explain why bony fish, in general, are larger in size than lancelets?

Table 23.1 summarizes the major evolutionary themes that you have observed in your brief survey of the animal kingdom in lab topics 20, 21, 22, and 23.

Learning Biology by Writing

All general biology textbooks indicate that chordates have a set of common characteristics: (1) a notochord; (2) a dorsal hollow nerve cord; (3) pharyngeal slits at some stage of the life cycle; and (4) a post-anal tail. In a short essay, describe the evidence you have from your lab work that supports or refutes this generalization.

Your instructor may also ask you to answer the following Lab Summary and Critical Thinking questions.

Internet Sources

Use the WWW to locate information about current research being done on cephalochordates. Use your browser to connect to www.google.com. Type in the search terms, *Branchiostoma* and research. Scan several of the sites that are listed and write a few paragraphs summarizing what you think are the most interesting sites. List the URLs at the end of your summary.

Lab Summary Questions

1. Why are echinoderms and chordates grouped together?
2. Describe the water vascular system of a sea star.
3. Briefly describe how a sea star performs the physiological functions of: (1) digestion; (2) gas exchange; (3) circulation; (4) excretion; and (5) body support.
4. Why are tunicates considered to be chordates?
5. Why is *Branchiostoma* considered an important evolutionary link among the chordates?
6. Have you observed any evidence that would allow you to refute the hypothesis that all chordates have a notochord, postanal tail, gill slits, and a dorsal nerve cord? Explain.
7. Explain how a fish ventilates its gills.
8. Describe how a large predatory fish swallows its prey and trace the pathway the food follows from mouth to anus.
9. Trace a drop of blood as it flows from a fish's ventricle to the intestinal wall and back to the ventricle.

Critical Thinking Questions

1. What traits suggest an evolutionary linkage between tunicates and lancelets?
2. Why are animals as different as a sea squirt and an eagle classified into the same phylum?

LAB TOPIC 24

Investigating Plant Cells, Tissues, and Primary Growth

Supplies

Preparator's guide available at
http://www.mhhe.com/dolphin

Equipment

Compound microscopes
Dissecting microscopes

Materials

African violet, geranium, or sycamore leaves
Apples, unpolished
Carrot
Celery
Elodea plants
Natural fibers (sisal rope, linen, or burlap)
Pear
Peppers, green and red
Potato
Red onion
Specimen house plants with waxy, unpolished leaves
Tomato
Young sunflower plants
Zebrina leaves
Razor blades and glass plates as cutting surfaces
Prepared slides
    Cross and longitudinal sections of *Cucurbit* stem
    Endosperm of persimmon
    Pine wood macerate
    Pumpkin or grape macerate
    Longitudinal section of root tip with root hairs
Solutions
    Blue food coloring
    IKI Stain
    Phloroglucinol stain
    Methylene blue stain
    Water in dropper bottles

Student Prelab Preparation

Before doing this lab, you should read the introduction and sections of the lab topic that have been scheduled by the instructor.

You should use your textbook to review the definitions of the following terms:

cambium	plastid
cell wall	periderm
collenchyma	phloem
companion cell	sclerenchyma
dermal tissue	sieve tube member
epidermis	tracheid
fiber	vacuole
ground tissue	vascular tissue
meristem	vessel
parenchyma	xylem
plasmodesmata	

You should be able to describe in your own words the following concepts:

Primary and secondary cell walls
Plant tissue systems
Primary and secondary growth

As a result of this review, you most likely have questions about terms, concepts, or how you will do the activities included in this lab. Write these questions in the space below or in the margins of the pages of this lab topic. The activities should help you answer these questions, or you can ask your instructor during the lab.

Objectives

1. To study the unique structures found in plant cells
2. To learn the characteristics of the basic types of cells found in plants and how the structure of those cells reflects their functions
3. To identify and compare the structure and function of plant tissues resulting from primary growth

Figure 24.1   Transmission electron micrograph of a sunflower (*Helianthus*) leaf primordial cell showing the cellular structures that make plant cells unique: cell walls, plasmodesmata, chloroplasts and a large vacuole.

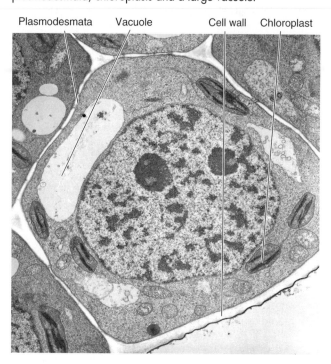

Plasmodesmata   Vacuole   Cell wall   Chloroplast

Figure 24.2   Transmission electron micrograph of cell wall showing middle lamella, primary wall, and secondary cell wall.

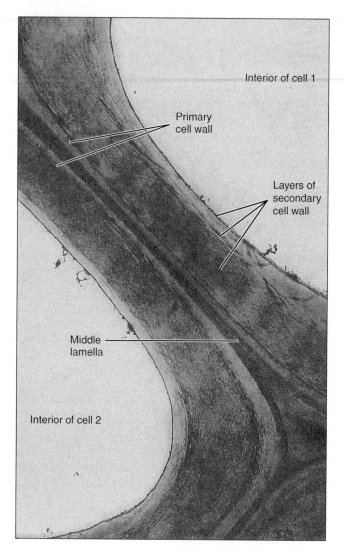

Interior of cell 1

Primary cell wall

Layers of secondary cell wall

Middle lamella

Interior of cell 2

Background

Approximately 400 million years ago, filamentous green algae were able to make the transition to land. The plants we see today have descended with modification from that ancestral stock. By dominating most terrestrial ecosystems, plants have influenced the evolution of bacteria, fungi, and animals that depend on them either directly or indirectly for food and shelter. Plants also influenced the physical environment, stabilizing it and contributing to the oxygen and carbon dioxide geochemical cycles of the global ecosystem.

Plant cells exhibit the eukaryotic cell plan. Therefore, they have chromosomes in a nucleus that divides by mitosis. In the cytoplasm, the typical eukaryotic organelles are found: mitochondria, ribosomes, endoplasmic reticulum, and Golgi apparatus, along with thousands of different kinds of enzymes required for the complex metabolism of these cells. In addition, there are other structures that are unique to plant cells. These include cellulosic cell walls, intercellular cytoplasmic connections called plasmodesmata, large central vacuoles, and plastids (fig. 24.1). In this lab, you will observe these unique structures in a variety of plant cells.

Collectively, the cytoplasm and nucleus of a plant cell are called the **protoplast,** and constitute the living part of the cell. A protoplast produces a cell wall and together these constitute what we would call a plant cell. During the life of a plant, protoplasts in certain tissues often make cell walls and then die, leaving the cell walls to function either

as supporting members of the plant body or as conduits for the transport of water and minerals. Thus, when looking at mature plant material, you will see that some cells may have living protoplasts with cell walls and others will be only a cell wall left by a protoplast that has died.

Cell walls are composed of cellulose microfibrils and other polysaccharides that are made by the protoplast. Following cell division, protoplasts produce a **primary cell wall.** Some cells, especially those that will be structural support cells, produce a **secondary cell wall** on the inside of the primary cell wall. It is much thicker and is impregnated with **lignin,** a phenolic polymer that stiffens the wall (fig. 24.2). This forms a very strong, but flexible, composite material. After cellulose, lignin is the second most common molecule found in plants.

Many of the living protoplasts in plant bodies are not separate cells, although it may appear so through a light microscope. Many, maybe even most, of the living cells in

Figure 24.3 Transmission electron micrograph showing plasmodesmata passing through middle lamella and primary cell wall of adjacent cells. These passageways allow materials to pass readily from cell to cell.

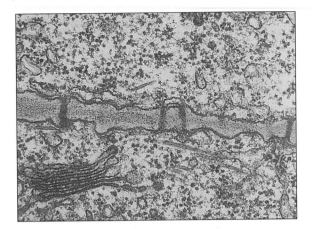

Figure 24.4 Three basic types of primary tissues differentiate from cells produced by apical meristems.

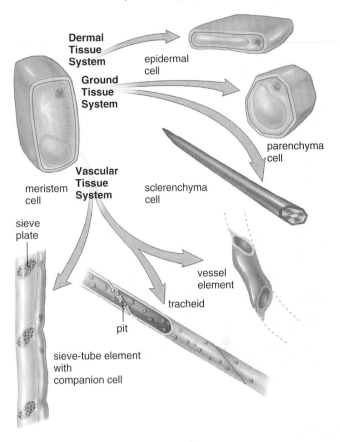

a plant are interconnected to adjacent cells by small tubes of cytoplasm that form numerous bridges. Called **plasmodesmata,** these bridges pass through openings in the cell walls that allow direct exchange of materials between cells (fig. 24.3). In cells that deposit secondary cell walls, no secondary cell wall material is deposited where there are plasmodesmata. These regions are called **pits.**

Mature plant cells often contain a large central **vacuole** (fig. 24.1). Besides storing water and minerals, the vacuole plays an important role in the growth strategy of plant cells. As primary cell walls are laid down, they are flexible but eventually limit a protoplast's ability to expand as it grows. To forestall this limitation on growth, newly formed protoplasts accumulate water in their central vacuole, enlarging the space surrounded by the primary cell wall so that when they later grow, the water can be replaced by cytoplasm or secondary cell wall material.

Plant cells contain a unique organelle called the **plastid.** Involved in photosynthesis and storage, these organelles have a variety of shapes and sizes. They are described in terms of the types of pigments they contain. Chloroplasts contain chlorophyll and are the organelles in which photosynthesis occurs. Chromoplasts contain yellow and orange-red pigments (carotenoids) and are responsible for colors in many flowers and fruits. Leucoplasts lack pigments and are organelles that store starches, proteins, and sometimes lipids. Plastids contain DNA and ribosomes (in addition to that found in the chromosomes and mitochondria). This has led to the concept that plastids originated from free living prokaryotes that were taken up by primitive eukaryotic cells at the time of the origin of plants.

Plant bodies have three regions: roots, stems, and leaves. Although these regions differ substantially in their form and function, they all contain similar basic cell types and tissues. The main differences are in the relative proportions of different kinds of cells and tissues as well as how they are organized into functional groupings. Four tissue systems are recognized in flowering plants: the dermal system, which comprises the covering of the plant body; the vascular system, which conducts materials through the plant; the ground system, which represents the metabolically active portion; and the meristematic systems, which are responsible for growth.

The **meristematic tissues** are the active regions of cell division in plants. They are responsible for plant growth, and all other cell types arise by differentiation from the daughter cells of these divisions. **Apical meristems** are regions at the tips of roots and stems where cell divisions lengthen the plant body. This is called **primary growth.** Division of an apical meristem cell produces two cells. One, called an initial, will continue to be a meristem cell and divide again. The other, called a derivative, will differentiate into promeristem, a transitional stage that commits the cell's progeny to develop into only certain types of tissues (fig. 24.4).

Some cells of the protoderm develop into the cell types of **dermal tissue** that cover the surface of the root and shoot systems. These cells protect the plant from invasion by microorganisms, prevent desiccation, allow for moisture and mineral absorption, allow for photosynthetic and respiratory gas exchange, and may produce structures or secretions that discourage herbivores.

Investigating Plant Cells, Tissues, and Primary Growth **317**

TABLE 24.1   Basic tissue types found in plants

Tissue Types	Description of Cell Types
Dermal tissue	Includes living epidermal cells, trichomes, root hairs, and guard cells; protects internal cells and functions in water absorption and transpiration; arises from apical meristem; in secondary growth, epidermis is replaced by periderm arising from cork cambium.
Ground tissue	**Parenchyma.** Component of ground tissue; most common living cells in plant body and found in leaf mesophyll, herbaceous stems, fruits, and rays of woody plants; functions in storage, secretion, and wound healing; arises from apical meristem and retains capability to divide (typically no secondary cell wall) and differentiate into other tissue types.
	Collenchyma. Component of ground tissue; living cells forming distinct strands beneath epidermis in stems and petioles that support the primary plant body; arises from apical meristem.
	Sclerenchyma. Component of ground tissue; includes living and dead fibers and sclerids; thick secondary cell walls with lignin provide support and mechanical protection; found throughout the plant, often in association with xylem and phloem; arises from apical meristem.
Vascular Tissue	**Xylem.** Component of vascular tissue; includes tracheids and vessels; functions in transport of water and minerals from roots to shoots; primary xylem arises from apical meristem and secondary xylem from vascular cambium; protoplast produces thick secondary cell wall and then dies; lumen of dead cells provides pathway for water transport.
	Phloem. Component of vascular tissue; includes living sieve tube members and companion cells; functions in distributing photosynthetic products and storage compounds throughout plant; primary phloem arises from apical meristem and secondary phloem from vascular cambium.
	Meristems. Cells that continue to divide and grow, forming tissues described above; apical meristems at root and stem tips allow growth in length (called primary growth); lateral meristems (also called cambiums—vascular and cork) found in roots and stems allow growth in girth (called secondary growth).

Some of the transitional tissues located interiorly will become procambium cells that divide to produce vascular tissues. The **vascular tissue system** allows materials to be transported from one part of the plant body to another: water from the roots to the above-ground parts, and products from photosynthetic cells to non-photosynthetic cells elsewhere in the body. In addition, vascular tissues often provide support to the shoot, allowing it to reach considerable heights.

Promeristem cells also produce the **ground system tissues.** This name fails to convey the importance of these tissues in the biology of plants. Most of the living cells in plants are found in ground tissues. Far from being filler material in the plant body, ground tissues are the most metabolically active cells. They are where photosynthesis occurs and where starches are stored, and they are active in producing reproductive structures.

In woody plants, additional meristematic tissues are responsible for what is termed secondary growth, growth in thickness, or girth of stems and roots. There are two lateral meristems. One, located in the bark, called the **cork cambium,** produces cork cells that characteristically cover the mature parts of older woody plants. The bark replaces the epidermis which was formed during primary growth. Beneath the bark a second lateral meristem, the **vascular cambium,** is responsible for producing secondary vascular

tissues, xylem to the inside and phloem to the outside. The addition of these tissues allows woody stems and roots to increase in diameter (secondary growth) as the apical meristem increases the length (primary growth). As a result, some woody plants can reach heights of over 100 meters and diameters measured in tens of meters.

The specific types of cells found in these types of tissue are described in table 24.1.

LAB INSTRUCTIONS

In this lab, you will study the basic types of cells found in plants. Where possible, you will find the cell type in live materials. After learning to recognize many of the basic cell types, you will study how they are organized into the tissues of the living plant.

Investigating the Unique Structures of Plant Cells

In this section, you will view slides from a variety of plant materials to observe those structures that are unique to plant cells: cell walls, plasmodesmata, vacuoles, and plastids.

Figure 24.5 Patterns of cellulose deposition in primary cell wall determine how the cell can expand. (*a*) When cellulose is laid down transversely to long axis of cell, the cell can elongate to form a tubular cell. (*b*) When laid down randomly, the cell will expand equally in all directions until it contacts other cells that may alter its shape. (*c*) Cellulose is secondary cell walls is laid down in layers oriented at right angles to each other and is impregnated with lignin. This forms a rigid casing that prevents the cell from expanding.

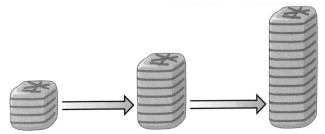

Transverse cellulose microfibrils: longitudinal expansion

(a)

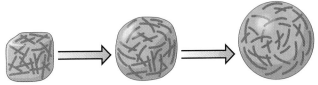

Random cellulose microfibrils: equal expansion in all directions

(b)

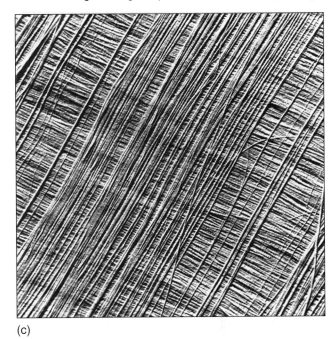

(c)

Cell Walls   All plant cells produce a primary cell wall following mitosis and cytokinesis. The newly formed protoplasts first produce a **middle lamella** composed of gel-like pectins, a polysaccharide material that holds plant cells together. Cellulose is then deposited between the middle lamella and protoplast to form the **primary cell wall.** Enzymes (cellulose synthase) for producing this cellulose are found in the plasma membrane of the protoplasts. They use thousands of activated glucose molecules from the cytoplasm to form the cellulose polymer. Forty to seventy cellulose molecules are laid down next to each other to form a cellulose microfibril. The microfibrils are laid down in patterns (fig. 24.5). Primary cell walls are dynamic structures. The walls are extensible allowing cellular expansion, they undergo reversible changes in thickness, and they usually are penetrated by plasmodesmata. Many plant cells produce only a primary cell wall and these cells have a living protoplast at maturity. Other cells, which usually lack a protoplast at maturity, add a secondary cell wall.

In **secondary cell walls,** cellulose is laid down in ordered patterns with one layer oriented almost at right angles to the layer beneath it. The layers are also impregnated with a phenolic polymer called **lignin.** Secondary cell walls are stiff compared to unlignified primary cell walls. Secondary cell walls do not allow cellular expansion, they do not undergo reversible changes in thickness, and they lack plasmodesmata. Sometimes the secondary cell wall can take up to 90% of the volume of a cell. The protoplasts' function is to make skeletal structures that are left behind to function as supporting members or conduits for water movement. Secondary cell walls are characteristic of the cells found in xylem and sclerenchyma tissues. All other tissues may have cells with only primary cell walls.

The flesh of pear fruit contains some cells with only primary cell walls (called parenchyma cells) and others (called stone cells or sclereids), with both primary and secondary cell walls. Plant anatomists use the stain phloroglucinol to differentiate between primary and secondary cell walls. Phloroglucinol stains lignin bright red. Since lignin is found only in secondary walls, the stain is used as a chemical test to distinguish secondary from primary cell walls that lack lignin.

To do this test, first put a drop of phloroglucinol stain (*caution:* this stain is made with HCL acid) on a slide. Use a razor blade to make a very thin slice of the pear's flesh and add it to the stain. After 5 minutes or so, add a coverslip and press it down to spread the material. Observe with your compound microscope.

Most of the cells on the slide will not be stained. These are parenchyma cells that have only thin primary cell walls. They store the carbohydrates that make the fruit sweet. Scan around the slide until you find groups of cells that are red. These cells, sclereids, have thick secondary cell walls containing lignin which reacts with the stain. You may have noticed hard, gritty lumps whenever you have eaten a pear. The grit is sclereids with their heavy secondary cell walls. Similar cells are found in the coating of seeds and in the walls of nuts.

Use high magnification to study some of the sclereids. Note the small cavity, called the **lumen,** where the protoplast

Figure 24.6   Endosperm cells from persimmon have very thick primary cell walls that are penetrated by pits containing plasmodesmata. The middle lamella is also visible as a faint line about halfway between adjacent cells.

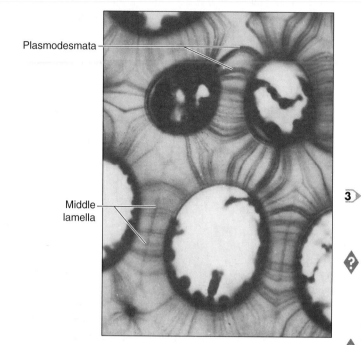

Plasmodesmata

Middle
lamella

lived. Observe the radiating lighter areas in the thick cell walls. These pit canals were once filled with plasmodesmata. Figure 24.3 is an electron micrograph of plasmodesmata passing through the cell walls of two adjacent cells.

2 ▶ *Plasmodesmata and Middle Lamella*   Obtain a prepared slide of endosperm from the seed of a persimmon *(Diospyros)*. This tissue supplies nutrients to the developing embryo in the seed.

As you look at this slide first with the medium power and then the high power objective, note the exceptionally thick primary cell walls. If you look at the wall midway between two adjacent cells, you should be able to see a faint line parallel to the surfaces of the cells. This is the **middle lamella,** a gel-like polysaccharide that holds the cells together. As you study the cell walls, you should also be able to see fine lines that seem to radiate from each cell. These are canals through which numerous plasmodesmata connect adjacent cells. Compare what you see to figure 24.6.

Can you think of any advantages there might be to having connections between the cytoplasms of adjacent cells?

Vacuoles   Most of a living plant cell's volume is taken up by one or more large membrane-bound vacuoles which often displace the cytoplasm (fig. 24.1). The membrane surrounding the vacuole is called the **tonoplast.** It regulates what passes into or out of the vacuole. Vacuoles are filled with water containing dissolved salts. They function as organelles of growth and as hydrostatic support structures. Following mitosis, many plant cells take up water into their vacuoles before the cell wall is laid down, creating space for later synthesis of cytoplasm, storage of materials, or synthesis of secondary cell wall material. Many structures in herbaceous plants are supported by hydrostatic "skeletons." When the vacuole takes on water, it presses against the surrounding cell wall, often just a flexible primary cell wall. The composite structure is stiffened, much the same as a tire stiffens when air pressure is increased in the inner tube.

3 ▶   Cut a red onion in fourths and remove one of the fleshy layers (leaves). Snap the layer backward with a twisting motion. Use your fingernail or forceps to peel a cellophane-thin layer from the ragged edge and mount it in a drop of water on a slide. Add a coverslip and observe. Can you see a large central vacuole? _____ What color is it? _____ These water-soluble pigments are called anthocyanins.

Add a drop of neutral red stain to the edge of the coverslip and wick it under the coverslip (see figure 3.4). Let the slide sit for 5 minutes and observe. Where is the cytoplasm of the onion cell? Draw a few onion cells below.

Now obtain a young leaf from the tip of a stem of the **4** ▶ aquatic plant *Elodea*. These leaves are only one or two cells thick. Make a wet mount slide and observe. Can you see the cell wall? _____ Is the plasma membrane visible? _____

Look for the following: vacuole, cytoplasm, nucleus, chloroplasts. What are the relative sizes of these structures in the cells? Make a list from largest to smallest.

Estimate the number of chloroplasts in an *Elodea* cell. Be sure to focus up and down as you view the cell since some of the chloroplasts may be in different planes of focus.

As you look at the *Elodea* leaf, you may see that the chloroplasts appear to move in some of the cells. This is called cytoplasmic streaming or cyclosis and is due to the action of cytoplasmic proteins similar to those in our muscles.

Plastids In *Elodea* you saw one type of plastid, a chloroplast. Chloroplasts are the sites of photosynthesis in plants. Non-photosynthetic cells lack chloroplasts, but have other types of plastids. In the lab there are green and red peppers, **5**▷ tomatoes, and carrots. Your assignment is to make very thin sections of these specimens, using a razor blade, and to determine whether the entire cell is colored or whether the color is localized in plastids inside the cells. This will take some practice and you should attempt to get sections that are as thin as cellophane. Only in thin sections will you be able to discern the structures. Make wet-mount slides of the sections as you get them. After you have finished viewing a slide, trade with someone else who has a slide of another tissue so that prep time is minimized.

❓ Describe what evidence you have that the color is localized in structures within the cells. Include sketches of a few of the cells you observed. Be sure to label the drawings.

6▷ Leucoplasts are colorless plastids that store starch. Potatoes contain substantial amounts of starch. Cut a very thin slice of a potato and make a wet-mount slide. Look at it to see if you can see leucoplasts in the cells. The stain IKI stains starch purple. Add a drop of the stain next to your coverslip and wick it under. Look at your specimen again. ❓ What color are the leucoplasts?

Investigating the Structure of Plant Tissues

Meristematic Tissue

Located at the tips of roots, stems, and branches, as well as inside the stems and roots of woody plants, meristematic tissues produce all of the other cells of the plant body.

There are three types of meristems. **Apical meristems** are found at the tips of roots and stems where they add cells to a plant's length during primary growth. **Lateral meristems,** found in the main shafts of stems and roots of only woody dicot plants, add the cells that increase the girth during secondary growth. In grasses a third type of meristem is found. Called **intercalary meristems,** these allow the rapid regrowth of leaves from dividing cells found at the leaf's base. This allows leaves to be lost to browsers (or lawn mowers) without killing the plant.

We will not look at meristematic tissues in this lab. The cells are small and rather uniform in appearance. You might review your notes from lab topic 9 where you stained the apical meristem cells in onion roots. In the next lab, you will observe apical and lateral meristems in roots and stems.

As cells produced by the apical meristems mature, they become specialized as tissues that perform particular functions in the plant. This process is called **differentiation.** The cells of the meristem produce three types of promeristem cells: protoderm, procambium, and ground promeristem. From these transitional meristems, the tissues differentiate to form the plant body. Although root and shoot systems vary considerably among angiosperms, all are made up of the same tissues and cell types. If you compared a root to a leaf or a monocot to a dicot, you would not find different tissues. Instead, you would discover that there are different arrangements and proportions of the basic tissues. These tissues are the dermal tissue system, the ground tissue system, and the vascular tissue system. In the next several sections of the lab, you will study the cells characteristic of these systems.

Epidermal Tissue

The **epidermis,** usually one cell thick, is a primary tissue originating from the protoderm produced by the apical meristem. It covers the plant body and mechanically protects the tissues beneath it, reduces desiccation by secreting waxes and oils, allows for gas exchange, and may be involved in water absorption. As you will see, some epidermal cells are specialized for particular functions. Some plants, called woody dicots, undergo secondary growth that has a devastating effect on the primary epidermis. Secondary growth involves the formation of new cells from the vascular cambium located several cell layers beneath the epidermis. Stop a moment and consider the geometry of this situation. What would you hypothesize would happen if cells were added beneath an epidermis that is already a tight covering? It seems logical to expect that the epidermis would come under tension and eventually split, exposing other tissues in the stem or root to attack by microorganisms and to the drying effects of air. Plants have evolved a developmental mechanism that prevents this from happening. As the plant expands, a secondary dermal system, called the **periderm,** develops. The outer layers of bark are examples of this system. When you are out and about campus, look carefully at the young twigs on a tree. You will see how the youngest twigs have a smooth epidermis, but the older ones have rough textured bark.

TABLE 24.2   Epidermal cell characteristics

Specimen			
Chloroplast present?			
Shape?			
Edge spaces?			

When you study roots and stems in the next lab topic, you will learn how the primary epidermis is replaced by the periderm. In this lab, we will concentrate on the primary epidermis and the types of cells found in it.

Basic Epidermal Cells   Basic epidermal cells are the types of cells that cover most surfaces of the primary plant body. They tend to be flattened cells and often will have cell wall patterns that form interlocking joints with adjacent cells. Lacking chloroplasts, their function is protection. They often secrete cutin, a wax that waterproofs their exposed surfaces.

7   In the lab are several plant specimens (fruits, stems, leaves) for you to use as study materials. Your challenge is to remove the surface layers from these plant specimens and make slides to see the structure of the epidermal cells. To illustrate basic epidermal cells, you should work only with materials that have a smooth (not hairy) outer surface. Before you make any slides of the materials, be sure to take a look at the specimens through a dissecting microscope so that you see the big picture.

To strip the epidermis from stems or leaves, fold the material over and then tear in such a way that the epidermis on the inside of the curvature is lifted off the stem or leaf. The piece of stem or leaf with the adhering epidermis should immediately be placed in a drop of water and the thin epidermis cut off. Add a coverslip and observe. If you coordinate with others in the lab, each can prepare a slide from a different kind of plant or a different region and then slides can be exchanged. The object is to develop a comparative knowledge of basic epidermal cell structure.

As you study the specimens, answer these questions:

Do these cells have chloroplasts?

What is their shape?

Are there spaces between the cells or do they butt against all adjacent cells?

Do you see any unusual structures in the cells?

Study at least four different examples of basic epidermal cell types and fill in the comparative data required in table 24.2.

How does the shape of epidermal cells reflect their function?

8   In the lab are some apples or some green house plants. The surfaces are not bright and shiny and seem to be covered by a dull whitish layer. Take a cloth and rub the surface of an apple or leaf. Why are they now shiny? How does this observation relate to the function of the epidermis?

Guard Cells   These specialized epidermal cells are found on leaves and stems where they surround a slit-like opening called a **stoma** (pl. **stomata**). Unlike other epidermal cells, guard cells contain chloroplasts. Guard cells change shape by taking on or losing water. Due to the orientation of the cellulose microfibrils in their cell walls, they bend when they take on water. This opens the stoma. When the guard cells lose water, they straighten and the stoma closes. Carbon dioxide used in photosynthesis, and oxygen that is produced, move into or out of the leaf through the stomata.

9   If you did not see guard cells when you were looking at epidermal cells earlier, prepare a leaf peel of a *Zebrina*

Figure 24.7 Trichomes on the surface of stinging nettle (*Urtica dioica*) have pointed tips hardened with silica. When the tips penetrate the skin of an animal, they break off and a toxin flows from the cell into the skin causing a painful reaction that discourages herbivory.

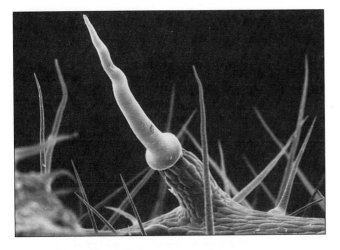

leaf as you did before. However, this time make a peel from the lower epidermis where guard cells are more common. After mounting the peel in a drop of water, look at it. Draw a pair of guard cells below.

Trichomes These hair-like specialized epidermal cells give a fuzzy appearance to stems, leaves, and flowers. Their functions are varied. They protect the plant from predators. Some trichomes, such as those of the North American stinging nettle *(Urtica dioica)*, have sharp hypodermic points (fig. 24.7) which can penetrate skin. The cells produce a toxin that both stings and irritates so that a "smart" animal soon learns to browse elsewhere. Trichomes also reduce airflow directly over the surface and can reduce drying. Their reflective surfaces also can keep the leaf from overheating.

10▷ Any leaf or stem with a fuzzy surface has trichomes. African violet, geranium, or sycamore leaves are common examples. Your job is to figure out a way to make a slide so that you can see trichomes through your compound microscope. You could try making a leaf tear or you might try scraping some of the cells from the surface with a razor blade. Before you make any slides of the materials, be sure to take a look at the specimens through a dissecting microscope so that you get the big picture about what you are studying.

Make a slide and draw some of the trichomes below. If there is more than one species with trichomes, be sure to look at several examples.

Root Hairs Epidermal cells of roots near the tip in the zone of differentiation can develop into root hairs. These are epidermal cells with tubular extensions that greatly increase the surface area for water and mineral absorption. Older root hairs are abraided by the soil but new ones keep forming as the root's apical meristem adds cells that differentiate into protoderm and eventually root hairs. Thus the absorptive zone of the root keeps advancing as the root elongates.

11▷ Take a radish seedling from a petri plate and mount it in a drop of water. Observe the root hairs through a dissecting microscope.

12▷ Get a prepared slide of a longitudinal sections of a root tip containing the root hair zone and look at it first with the scanning objective of your compound microscope. Note the single epidermal cell layer and how the "hair" is an extension of a single epidermal cell. Can you see any cross wall at the base of the hair? _____

❖ Given their function, would you expect root hairs to have waxy, cuticle-like epidermal cells found in leaves? Why?

Figure 24.8 Examples of three types of ground tissue cells. (*a*) Thin, primary-walled, round parenchyma cells are very common cells in stems, roots, and leaves where they are specialized for various metabolic functions. (*b*) Elongate collenchyma cells are characterized by unevenly thickened primary cell walls and provide support in growing parts of a plant. (*c*) Elongate sclerenchyma fibers are harvested to make rope and cloth. (*d*) Sclerenchyma fibers seen in cross section show how the lumen of the cell is almost completely filled with secondary cell wall material.

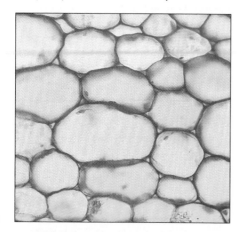
(a)

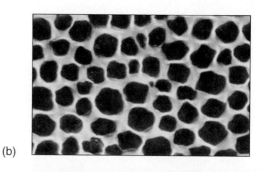
(b)

(c)

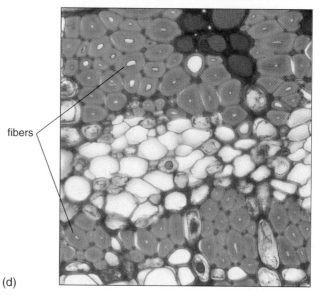
(d)

fibers

Ground Tissues

Arising from the cell divisions in the ground meristem of the primary plant body, ground tissues are the most common types of cells found in plants. They function in basic metabolism, storage, and support in plants. Three basic cell types are found: parenchyma, collenchyma, and sclerenchyma (fig. 24.8).

Parenchyma These are the most common cells in the primary plant body. Parenchyma cells may function in photosynthesis, storage, secretion, and a variety of other tasks in leaves, stems, roots, flowers, and fruits. They retain their protoplasts at maturity and are morphologically characterized by large vacuoles and thin primary cell walls, although they sometimes will develop a secondary cell wall. Although you cannot see them, plasmodesmata commonly interconnect the cells. Parenchyma cells can be almost any size or shape but are most often 14-sided polygons. Parenchyma is usually loosely organized with spaces between adjacent cells, forming fleshy structures as in the pith and cortex of some stems, roots, fruits, and flowers. Parenchyma cells retain the ability to divide even at maturity and can differentiate into other cell types.

When you looked at cells containing plastids at the beginning of this lab, you were looking at parenchyma cells found in leaves, roots, and fruits. Did the parenchymal cells stain with phloroglucinol? _____ What does this tell you about the composition of their cell walls?

13▶ Here you will look at parenchyma cells found in the stem of a non-woody herbaceous plant. Young sunflower (*Helianthus*) plants were cut from their roots before lab and the cut ends were immersed in a blue food color solution. This will preferentially stain the xylem, a vascular tissue, which you will study shortly. For now, you will look at the cells that make up the bulk of the stem.

Take the stem and cut a very thin cross section and mount it in a drop of water on a microscope slide. Add a coverslip and observe.

How would you describe the shape of the cells? Are the cells tightly packed or loosely organized? Draw a few of the cells below.

14▶ Now prepare a longitudinal section of the stem so that you can observe the parenchyma in the other dimension. Add to your notes about cell shape. Are the cells tubular or more compact?

Collenchyma Like parenchyma, collenchyma cells are alive at maturity. The cells tend to be longer than wide and are characterized by primary cell walls of uneven thickness so that the cell walls appear thick at the corners. In stems, petioles, and flowers, collenchyma occur as filaments or bands of cells just beneath the epidermis. Because they have only primary cell walls that are not lignified, they are flexible, yet they provide support by a hydrostatic mechanism. In stems, the tendency of parenchyma cells to take on water forces them against the collenchyma bands making the stem semi-rigid. Collenchyma is found only in the primary plant body.

The celery stalk (*Apium*) that we eat is a leaf petiole containing collenchyma strands just beneath the epidermis. Also in the petiole to the inside of the collenchyma will be parenchyma cells and vascular bundles of xylem and phloem. Obtain a piece of celery and peel out one of the strands. Once started, you should be able to pull out an entire strand from the length of the petiole.

15▶ Try to break an isolated strand to get an idea of its tensile strength. Use a razor blade to cut a section of parenchyma from inside the petiole. How does its tensile strength compare to that of collenchyma? Record your impressions below.

16▶ Cut two thin cross sections of celery petiole. Add one to drop of water on a slide and the other to a drop of phloroglucinol stain on another slide. Let this one sit in the stain for five minutes while you examine the unstained section. Locate the collenchyma by looking for shiny cell walls of uneven thickness (fig. 24.8*b*). Note their location. Now examine the other cross section for signs of staining. Do not confuse the vascular bundles with collenchyma stands. Do the cell walls stain with phloroglucinol? Do collenchymal cells have secondary cell walls? Draw a few of the collenchyma cells below.

Sclerenchyma Characterized by rigid, thick cell walls composed of both primary and lignified secondary cell wall material, sclerenchyma cells usually lack protoplasts at maturity. They are usually found in mature regions of a plant's body that have stopped elongating and that have achieved proper size and shape. Two types of sclerenchyma cells are found in plants: fibers and sclereids. Fibers are long slender cells forming clumped strands often in association with vascular tissues. Their hard cell walls support the plant and

Investigating Plant Cells, Tissues, and Primary Growth

can protect the phloem from the probing mouthparts of insects. Fiber strands of some species are economically important. Linen, burlap, jute, and sisal are examples of commercially important sclerenchyma fibers used to make cloths and ropes. Sclereids have very thick brittle cell walls and tend to be cuboidal (fig. 24.8d). They are found in layers in seed coats, such as walnut shell, where they protect the embryo. Fibers in stems support the aerial parts of the plants and when bent tend to snap back to the original shape. Sclerenchyma-rich stems do not droop when wilting but remain erect.

At the beginning of this lab, you observed sclereids when you stained for stone cells in pears. Here you will look at fibers. In the lab are some pieces of sisal rope, **17** burlap, or linen. Take a few fibers from these samples and mount them on a microscope slide in a drop of phloroglucinol stain. Let sit for a few minutes and then add a coverslip and observe with your compound microscope.

Can you see the individual fibers joined end to end? Sketch a few cells below.

18 Now obtain a prepared slide of a cross section of a grape stem. Find the heavy walled fibers in the stem. Note the very small lumen of the fibers. The protoplast essentially filled the lumen with secondary cell wall material and in so doing sealed its fate. The thick, lignified cell wall prevented materials diffusing to the protoplast and literally starved it. Did you see any pits in the cell walls of the fiber samples from the rope or cloth? _____ Pits are remnants left behind as a result of what?

Given the small size of the lumens in fibers, do you think they would make good conduits for water transport? _____ If not, what could their function be?

Vascular Tissue

Arising from cell divisions in the procambium during primary growth, and later from the vascular cambium if the plant is a species that undergoes secondary growth, vascular tissues include xylem and phloem. These tissues commonly occur in bundles that form conduction systems throughout the plant body. The tubular shape of their cells reflects their function as conduits. Through the dead tubular cells of the xylem, water and minerals move from roots to other parts of the plant body. Through the living tubular cells of the phloem, photosynthetic products are transported from leaves to other living non-photosynthetic cells or sometimes from areas of carbohydrate storage to areas where carbohydrates are needed. In vascular bundles, xylem is usually located toward the center of the stem or root and the phloem toward the periphery. In species that undergo secondary growth, xylem can occupy most of the volume of the stem or root while the phloem becomes the innermost layer of the bark.

Xylem and phloem are what are called complex tissues. In addition to the cells that are unique to xylem and phloem, sclerenchyma fibers and parenchyma cells are often integrated into vascular tissues. We will not discuss these inclusions here but will concentrate on those cells that are unique.

Xylem   Xylem is the principal water-conducting tissue in plants. Tracheids and vessel members are the unique cell types (fig. 24.9). These cells lack protoplasts at maturity. The protoplast's function is to form a tubular cell with a rigid, lignified secondary cell wall, and then to die, leaving the lumen as a conduit for water conduction. The thick secondary cell walls, often with spiral ridges, resist collapsing when water is pulled through the xylem by the negative pressure generated in transpiration. **Tracheids** are long, narrow spindle-shaped cells with tapered ends. **Vessel elements** are large barrel-shaped cells that join end to end to form tubes with perforated end plates between adjacent cells. Vessel members transport water more efficiently than do tracheids and are thought to have evolved from tracheids. Both types of cells have thin areas in their side cell walls called **pits** where plasmodesmata passed between adjacent protoplasts when the tissue was forming. The pits in mature dead cells allow lateral movements of water between adjacent conduits, an important consideration should a conduit become plugged with salts, debris, or a fungal intrusion. In woody species, wood is composed of secondary xylem that originated from cell divisions in the vascular cambium.

19 Obtain a small pine twig that was soaked in 45% acetic acid. Tease it apart in a drop of methylene blue on a slide. Add a coverslip and observe. Alternatively, a prepared slide of pine can be substituted for this preparation.

Gymnosperms lack vessel elements and tracheids are the principle conducting cells in their xylem. Observe the spindle-shaped tracheids with high magnification. Look at the cell walls carefully for evidence of pits. Draw a few of the tracheids below.

Vessels are considered more evolutionarily advanced than tracheids and are found in flowering plants (angiosperms). They are composed of short cell wall sections from different vessel elements that are joined end to end. Often cross walls are absent at the ends and water can often pass up to three meters before it encounters a pitted cross wall. This allows for rapid water movement and may be one of the reasons that angiosperms have become the dominant terrestrial plants.

20▷ Obtain a macerate of a pumpkin or grape stem. Look for the vessel elements among the other cell types on the slide. Sketch some of the vessel elements below.

21▷ Earlier in the lab, sunflower stems were placed in blue food coloring. You used these stems as a source of parenchymal cells. It is now time to use them again, but this time you will study the xylem. Since xylem is the water-conducting tissue, the blue coloring will have preferentially stained these cells. Take a 1-cm section of stem and tease it apart while looking at it through a dissecting microscope. Can you see the spiral thickenings in the vessel element cell walls? _____ What was their function?

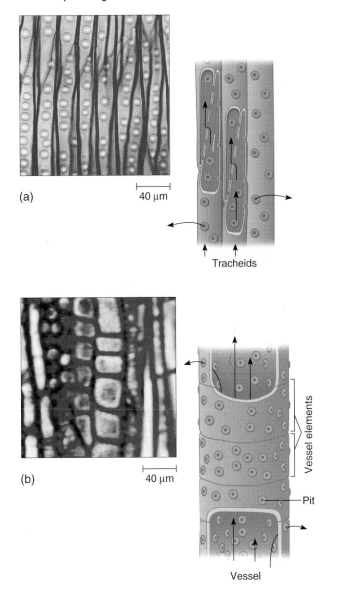

Figure 24.9 The unique cells found in xylem have stiff secondary cell walls penetrated by pits and lack protoplasts at maturity when the cell serves as a conduit for water transfer from the roots to the leaves. (a) Tracheids are long narrow cells with tapered ends that overlap. Numerous pits allow water to pass both through the ends and laterally to other cells. (b) Vessel elements are short and wide barrel-shaped cells usually lacking end walls that conduct water.

(a) 40 μm

Tracheids

(b) 40 μm

Vessel elements

Pit

Vessel

22▷ When you see an interesting group of cells, move them to a drop of water on a microscope slide and observe it with your compound microscope.

Phloem The cells of this tissues are the food-conducting tissues in plants. Phloem includes two unique cell types: sieve tube members and companion cells (fig. 24.10). **Sieve tube members** are cylindrical cells with thin primary cell walls that contain a living protoplast at maturity

Figure 24.10 The unique cells found in phloem have thin primary cell walls with plasmodesmata connecting the living protoplasts of adjacent cells. This arrangement promotes the transfer of organic materials among the living, non-photosynthetic cells of a plant. (a) Sieve tube member and companion cell in longitudinal section. (b) Similar cells in cross section with a sieve plate visible in a sieve tube member. (c) Artist's interpretation of the anatomical relationship.

Phloem cell types

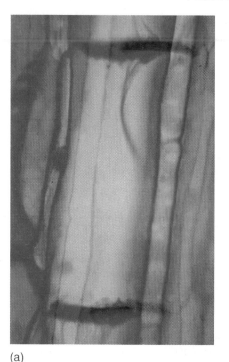

(a)

Companion cell
Sieve plates

100 μm

(b)

Sieve-tube member

Companion cell

Nucleus

Sieve plate

(c)

although they lack a nucleus. Sieve tube members join end to end to form long conduits. The end walls between adjacent cells, called **sieve plates,** are perforated by many pores through which large plasmodesmata pass. Organic materials flow through these cytoplasmic bridges from one cell to the next. Each sieve tube member has lateral plasmodesmata that link its cytoplasm with a living parenchymal cell called a **companion cell.** Companion cells load and unload organic materials being transported through the sieve tube members. The lives of the two cells are tightly coupled, and when one dies, the other dies as well.

Phloem has a trauma-control mechanism that prevents leakage of precious organic materials should a phloem conduit be broken by a grazing animal or accident. When a sieve tube member is injured, a slimy protein, called p-protein, surges through the adjacent cell and blocks the sieve plate pores so that no more phloem "sap" enters the injured cell,

thus conserving the precious sugars and other organics transported by the system.

23 Pumpkin stems (*Cucurbita*) contain large sieve tube elements. What is unusual is that they occur in bundles on both the inside and outside of the xylem (bicollateral bundles). Obtain a prepared slide of a longitudinal section of pumpkin stem. Look at it first under medium power with your compound microscope. Look for tubular cells that are joined end to end. Globs of p-protein may be found stuck to the sieve plates. These are sieve tube members. If the angle is correct, you may be able to see perforations in the plates between adjacent cells. Associated with the sieve tube elements are small companion cells.

If you could not see sieve plates and companion cells on your slide, look at the cross section of a pumpkin stem. Locate the phloem and look for cells that have a honeycombed cross wall.

Use the cross section of the pumpkin stem as a summary slide to see what you have learned in this lab. Since all flowering plants are composed of the cell types you have studied in this lab, you should be able to identify them in just about any slide that you look at. On this slide, try to identify the following:

Epidermal cells

Parenchyma

Collenchyma

Sclerenchyma

Xylem vessels

Phloem sieve tube members

Companion cells

If you are having trouble, see if your classmates or instructor can help you with the ones that you find difficult to remember.

Learning Biology by Writing

Because this was a descriptive rather than experimental lab, a report cannot be written. Instead, a descriptive essay is more appropriate. In about 200 words, describe how different cells arise from apical meristems. Describe the different types of cells found in plants.

As an alternative assignment, your instructor may ask you to turn in answers to the Critical Thinking and Lab Summary questions that follow.

Internet Sources

Understanding plant cells and their requirements has led to a whole new field called plant tissue culture. It is now possible to isolate parenchymal cells from a plant and to culture them in a test tube. When a mass of cells accumulates, they can be treated with plant hormones and they will differentiate into the three types of tissues that you have studied in this lab. By manipulating hormones, it is possible to get a group of cells to actually organize themselves into a miniature plant inside a test tube. If this plant is removed and placed in soil, it grows into a mature plant. Check out these www sites to see how these techniques are being used commercially:

Plant tissue culture information exchange
http://aggie-horticulture.tamu.edu/tisscult/tcintro.html

Caissons Laboratories Plant Tissue Culture Links
http://www.caissonlabs.com/links.htm

Alternatively, use GOOGLE and type plant tissue culture to find other sites.

Lab Summary Questions

1. What four unique structures do plant cells have that make them different from other eukaryotic cells?
2. Fill in the information required in table 24.3. Use *yes* or *no* except for last two columns.
3. Distinguish between these three basic types of plant cells: parenchyma, collenchyma, and sclerenchyma.
4. Describe how the four types of epidermal cells studied are adapted to their functions: basic, guard, trichome, and hair cells.
5. How does the structure of a vessel member differ from that of a sieve tube member? How does the overall structure of a vessel member and sieve tube member reflect the function of vascular tissues?

Critical Thinking Questions

1. A student wants to culture plant tissue so that he can asexually reproduce a fast-growing poplar tree that could be grown in plantations as a source of fibers for making wood. He decides that since most of the poplar tree is secondary xylem, he will culture isolate xylem cells and grow them in tissue culture. Why is this project doomed to failure from the start?
2. If plant roots are exposed to air and sunlight for a few minutes, and then the plant is placed back in the soil, it continues to wilt and then starts to recover. What is happening at the cellular/tissue level in the roots?
3. Would you expect chlorenchyma cells from a leaf to stain with phloroglucinol? Why? What about tracheids from an oak? Epidermal cells from a petunia?
4. When animals eat plant tissues, from what types of cells are they gaining energy and nutrients?

TABLE 24.3 Summary of plant cell types and characteristics

Cell Type	Found in tissue type	Primary wall	Secondary wall	Lignin	Cells retain cytoplasm?	Plastids	Shape	Function
Basic epidermal								
Guard cells								
Trichomes								
Hair cells								
Parenchyma								
Chlorenchyma								
Collenchyma								
Sclereids								
Fibers								
Tracheids								
Vessel elements								
Sieve tube members								
Companion cells								

LAB TOPIC 25

Investigating Primary and Secondary Growth in Roots and Stems

Supplies

Preparator's guide available on WWW at
   http://www.mhhe.com/dolphin

Equipment

Compound microscopes
Dissecting microscopes

Materials

Slides and coverslips
Prepared slides
   Basswood (*Tilia*), cross section
   Coleus shoot tip, longitudinal section
   Corn root, cross section
   Corn stem, cross section
   Medicago stem, cross section
   Oak cross and radial sections
   Ranunculus root tip, longitudinal section
   Ranunculus root, cross section
Twigs of hickory or buckeye with apical bud
100-ml beaker
Razor blades and forceps
Shallow pans
Live Materials
   Germinating radish seeds
   Germinating corn
   Potted, growing *Coleus* plants
   Sunflower plants about 12" tall; one well watered
      and other water deprived
   Wisconsin Fast Plants planted 3 weeks earlier
   Fresh celery
   Root vegetables such as carrots, turnips, ginger,
      radishes, or others as available

Solutions

0.5% methylene blue
Phloroglucinol stain

Student Prelab Preparation

Before doing this lab, you should read the introduction
and sections of the lab topic that have been scheduled
by the instructor.
   You should use your textbook to review the
definitions of the following terms:

apical meristem	Casparian strip
bark	cork

cork cambium	root hair
dicot	stele
endodermis	stomata
herbaceous	transpiration
monocot	vascular cambium
pericycle	vascular tissue
phloem	woody
root cap	xylem

You should be able to describe in your own words
the following concepts:

   Primary growth in stems
   Secondary growth in stems
   The transpiration-tension-cohesion theory of water
      movement in xylem
   How guard cells regulate stomatal opening

   As a result of this review, you most likely have
questions about terms, concepts, or how you will do
the experiments included in this lab. Write these
questions in the space below or in the margins of the
pages of this lab topic. The lab experiments should
help you answer these questions, or you can ask your
instructor during the lab.

Objectives

1. To recognize and compare tissue organization in
   primary roots of monocots and dicots
2. To compare the basis for primary and secondary
   growth of roots
3. To recognize the key role of the endodermis in root
   absorption
4. To observe the differences in stem anatomy
   between monocots and dicots, including examples
   of herbaceous and woody dicots
5. To be able to describe the basis for primary and
   secondary growth in stems
6. To demonstrate that water rises through the xylem
   tissues because of transpiration

25-1   **331**

Background

In previous labs you looked at how invasion of the terrestrial environment influenced the life cycles of plants. Last week you studied the types of cells found in plants. In this lab you will study how these types of cells are organized into functional units called tissue systems that allow vascular plants to live and grow in terrestrial environments. These tissue systems are organized into what we might call the architecture of the plant body. Just as buildings are made of basic types of building materials such as bricks, lumber, pipes, *etc.*, so plants are made from the basic types of cells you have studied. To continue the analogy, you know that more than one building can be made from the same list of basic materials. It is how they are put together that determines the final shape and function. So it is also with plants and their cellular building blocks.

The basic parts of the flowering plant body are the root system, stem, leaves, and reproductive structures, which include flowers and fruits. Each has a unique arrangement of the basic cell types which you will learn to distinguish and to correlate with function. Among flowering plants, there are two broad architectural types; monocots and dicots. At a minimum, you need to be able to distinguish the differences between them based on their anatomy.

If we think of architecture in evolutionary terms, the following is a reasonable, although hypothetical, architectural analysis of why plant bodies look the way they do today. Plants evolved from green algae. Think of the algae as resembling the chlorenchymal cells previously studied. These are the photosynthetic cells of the plant. Green algae require a very moist environment or they will die. How could the drying problem be solved architecturally? Encase the chlorenchymal cells in a casing of other cells that have specializations that prevent drying. These cells need not be photosynthetic since they can get "food" from their photosynthetic neighbors. Such cooperation would allow an association of cells to function when no longer in an aqueous environment.

However, photosynthesis requires carbon dioxide gas from the atmosphere and produces oxygen, so the surface must have some openings to allow gas exchange. This would also allow some water loss, but solutions to problems always require trade-offs. These simple architectural changes would have yielded a new type of encrusting organism, maybe a few cells thick. If it were thicker, then light energy would not be available to the cells on the bottom of the stack. While forms like this might have had some momentary success, the available surface area on land is limited and low encrusting plants would compete with each other for areas to grow.

How can architecture solve the available area problem? Go three dimensional and build high-rises! Any plants that were able to grow taller than their neighbors would intercept the sunlight first, and thus another architectural change could solve a problem. However, tall structures create other problems. A structure of any height requires support. Cellu-

lar adaptations that thickened cell walls would be one response. A second would be hydrostatic support, similar to the stiffening of a fire hose when pressure builds.

Once a stiff, upright structure was achieved, architecture presented an opportunity to increase efficiency. Development of flat projecting structures to house the photosynthetic cells would allow plants to intercept light that was passing through space beyond the stem. However, as our hypothetical plant got taller, other problems would develop. A thin, tall, upright structure with weight at the top is unstable and will topple. Vertical supports must be anchored in the ground. Additionally, as plants got taller there was a point at which water could not diffuse the distance from the moist soil to the elevated chlorenchymal cells faster than it was lost by evaporation from the leaves. What were the architectural solutions?

Anchorage could be accomplished by having some cells specialize to form underground structures, but becoming dependent on above ground parts. If these structures also became highly absorptive, they would facilitate water absorption. However, anything underground could not be photosynthetic, yet energy and building materials would be required to grow and function. If a plumbing system were added to the architectural plan, then two problems would be solved. The plumbing could conduct water to the elevated photosynthetic surfaces and could transport photosynthetic products to the underground parts supplying their needs. If the plumbing structures were reinforced, then they would add to structural strength, allowing the plant to grow even taller. Thus we can create a plausible scenario that explains the problems that were solved in the origin of plants with the development of a few mutually interdependent cell types. In today's lab, the architecture of stems and roots in modern plants will be investigated, and you will examine another important adaptation called secondary growth which accounts for most of the mass, and strength, of those plants known as dicots.

Stems As you saw in the past topic, stems elongate by primary growth resulting from the addition of cells at the growing stem tip. The cells produced by the apical meristem undergo a process of differentiation to produce different primary tissue systems: dermal, vascular, and ground tissues that are architecturally arranged into the stem. Much of the support function and just about all of the transport functions are performed by the vascular tissues, xylem and phloem. In simple terms, the protoplasts of xylem tissues make elongate cells with reasonable heavy secondary cell walls and then die, leaving a hollow conduit that conducts water and minerals. Phloem tissues are composed of living cells that function in distributing photosynthetic products throughout the plant and in distributing carbohydrates from storage areas to the rest of the plant.

The support functions are legendary, the type of stuff you see in the Guiness Book of World Records. Redwood trees in California and eucalyptus trees in Australia can reach heights of a 100 meters. In the other dimension, a sin-

Figure 25.1   Continuous columns of water in each xylem tube extend from the roots to the mesophyll of leaves. Water is drawn upward by transpiration.

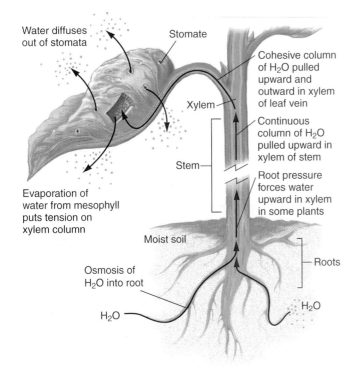

Water diffuses out of stomata

Stomate

Cohesive column of H$_2$O pulled upward and outward in xylem of leaf vein

Xylem

Continuous column of H$_2$O pulled upward in xylem of stem

Stem

Root pressure forces water upward in xylem in some plants

Evaporation of water from mesophyll puts tension on xylem column

Moist soil

Roots

Osmosis of H$_2$O into root

H$_2$O

H$_2$O

tension on the water column in the xylem. Water, along with dissolved minerals, is pulled from roots to leaves. This mechanism depends on the cohesiveness of water due to hydrogen bonding between adjacent water molecules.

Transport of organic molecules to other tissues occurs through the sieve tube members of the phloem. Carbohydrates are actively transported from the photosynthetic leaf cells into the phloem. In roots and regions of growth, these materials are removed from the phloem and used as energy sources or for making other kinds of molecules. Thus a solute concentration difference exists between the two "ends" of the phloem: high concentrations of the solute are present in the leaf, and low concentrations are in the root. Water diffuses from the xylem and enters the phloem by osmosis at the leaf end. This creates a pressure that drives the phloem fluid toward the roots where carbohydrates are removed and metabolized, and water reenters the xylem system. This mass flow results in the **translocation** of photosynthetic products. When plants break their dormancy after winter, flow of materials is reversed. Sugars from stored starches in the roots are transported to buds and cambiums where they fuel cell growth and division.

Roots   Roots have three functions. They absorb water and minerals from the soil. Roots serve as storage depots for photosynthetic products and these reserves fuel metabolism over winter in the living cells in the rest of the plant and provide the energy for leafing out in the spring. The tapping of maple trees in the spring takes advantage of the maple's capacity to store carbohydrates in its roots one season and mobilize them as sugars in another. Last, and certainly not least, roots anchor and stabilize the stem in the soil, an impressive task in the giants discussed earlier.

Compared to stems, we know much less about roots because they are more difficult to study. Their role in ecosystems where they unlock the soil's store of phosphate, sulfate, and calcium cannot be overstated. Herbivores not only gain energy, essential amino acids, and vitamins from eating plants, they also gain ions necessary for their metabolism. Some scientists have observed that because of roots, animals, including humans, do not have to eat dirt to gain minerals!

Root systems are often highly developed and account for 80% of the biomass of the plant. For example, a researcher found that a mature winter rye plant had a root system with a total length of 387 miles and a surface area of 2,554 square feet, containing 14 billion root hairs with a combined length estimated to be 6,603 miles. The most amazing fact, however, is that all of this was contained in less than two cubic feet of soil. Obviously, intimate contact occurs between soil and plant.

Mere absorption of water and minerals is not sufficient for plant survival. These essential materials must find their way from the roots to the actively metabolizing cells of the plant. As in stems, xylem and phloem are present in roots and perform this conduction function. Xylem tissues form a water-conducting conduit from near the tips of the roots to

gle banyan tree (stem) in India has a canopy (branches and leaves) that spreads over an area equal to about five football fields. It is difficult to imagine the stresses that the stems of such giants must encounter in high winds, and it speaks to the strength of the supporting xylem tissues. (This also says something about their root systems.)

Xylem also transports water from the roots to the leaves. In a lawn grass, this is not an impressive task, but for this to happen in one of those tall eucalyptus trees, over 300 pounds of pressure per square inch must be exerted to get the water from the soil to the leaves. How does the plant accomplish this? Two mechanisms are involved: root pressure, and transpiration-cohesion.

Root pressure is due to the osmotic uptake of water by a plant's root. Since root cells contain dissolved organic solutes, such as sugars and proteins, the *osmotic gradient* in wet soil goes from the soil into the root, literally forcing water into the xylem elements. It can account for raising the water column several feet.

The second and more important mechanism for transporting water involves the combined effects of **transpiration** and water cohesion (fig 25.1). Leaves have small pore-like openings called **stomata** (singular: stoma), usually on their undersurfaces. (See lab topic 26). Stomata are involved in gas exchange during photosynthesis but also allow water vapor to escape by diffusion. Lost water is replaced by vapor diffusion from the xylem into the leaf, resulting in a the tip of the stem. How tall was that species of eucalyptus mentioned earlier? Phloem must transport photosynthetic products from the leaves to the living cells at the root tips.

Roots elongate by adding cells at the growing root tip. The cells produced by the root apical meristem undergo a process of differentiation to produce the different primary tissue systems of the root: dermal system, vascular tissue system, and ground tissue system that are architecturally arranged into the root.

Secondary Growth Most dicots exhibit some secondary growth, and in woody plants it is a substantial contributor to body mass and strength of stems and roots. **Primary growth** increases the length of the plant body by adding cells at the tips of the stem, branches, and roots. **Secondary growth** is an increase in the girth of the components of the plant body by adding cells to the primary plant body. It is necessary because a plant that grows only in length (primary growth) will soon reach a point where a thin stem will not support it or its root will not be strong enough to anchor it. Those plants that lack the capacity for secondary growth are usually not very large or are aquatic where water supports them. By increasing girth, strength is added and plants can grow even larger, outcompeting neighbors for sunlight.

Secondary growth can occur because woody plants have two other meristems in addition to the apical meristems in their roots and stems. Called **cambiums,** one of these is located beneath the bark and is referred to as the **vascular cambium.** Division of it produces secondary xylem and phloem vascular tissues. The other is located in the bark and is called the **cork cambium.** Division of it produces cork cells that are components of the bark or periderm. The records for secondary growth are quite impressive. A chestnut tree in Sicily has a diameter of 18+ meters (58 feet!), yet that tree started from a single chestnut and once had a primary stem only a few millimeters across. Approximately 90% of the mass of this tree would be secondary xylem, the name given to xylem produced by the vascular cambium.

If you consider the geometry of secondary growth, a potential limitation is uncovered. The organization of stems and roots of woody plants can be thought of as cylinders within cylinders. If an internal cylinder expands as a result of adding cells by division of the vascular cambium, then the outer cylinders will be stressed and eventually will split longitudinally. Have you ever seen bark flaking off a tree and wonder why this happens? The explanation is secondary growth. Adding wood inside the bark stretches the bark, which gives way. You can see this as the deep longitudinal fissures in the bark of most large trees. Bark is replaced when the cork cambium produces cells that fill the voids as well as build up a surface layer. As secondary growth continues, the outermost layers of bark are stressed to the point that they no longer stay attached to the expanding new layers of bark and the old bark falls off.

LAB INSTRUCTIONS

In this lab, you will study the anatomy of roots and stems, noting the arrangements of basic tissues. You will learn the differences between monocot and dicot anatomy. In dicots you will see the effects of secondary growth. At the end you will do some experiments demonstrating water movement through xylem.

Whole Roots

Your lab instructor will set up demonstrations of the types of root systems found in plants.

1 ▷ Examine the specimen showing a **taproot** system consisting of a simple long conical root that penetrates the soil with many lateral branches from the main axis. Taproots are characteristic of gymnosperms and many dicots. Side branches bear root hairs at their tips. The parenchyma cells of taproots store carbohydrates; some taproots are used by humans for food. Can you name three?

2 ▷ Now look at the **fibrous root** system of a grass. Note how the root system repeatedly branches to form a complex network. Fibrous roots are characteristic of monocots.

In addition to their primary root system, many plants have **adventitious** roots, which arise as lateral extensions

3 ▷ from the base of the stem. Examine the root system of a mature corn plant to see these.

4 ▷ Obtain some germinating radish or corn seedlings in a petri dish and compare to figure 25.2. Look at it carefully through a dissecting microscope. Can you see the root cap at the apex? Note the root hair zone and how it begins only a few millimeters back from the tip. This occurs because the cells produced by the apical meristem require time to differentiate. Root hairs are part of what tissue system in plants? In older roots the root hairs cease to function, and die.

Root Histology

In this section, you will observe both longitudinal and transverse sections of roots from dicots and monocots.

Longitudinal Section of a Young Root

5 ▷ Examine a prepared slide of a longitudinal section of the dicot buttercup (*Ranunculus*) under low-power magnification. Compare your specimen to the diagram in figure 25.3. Note how the cell shapes change as you scan from the root tip toward the base of the root.

Figure 25.2   Primary root of a corn seedling. Note root hair zone.

Shoot

Primary root
with root hairs

Figure 25.3   Tissue organization in a young root.

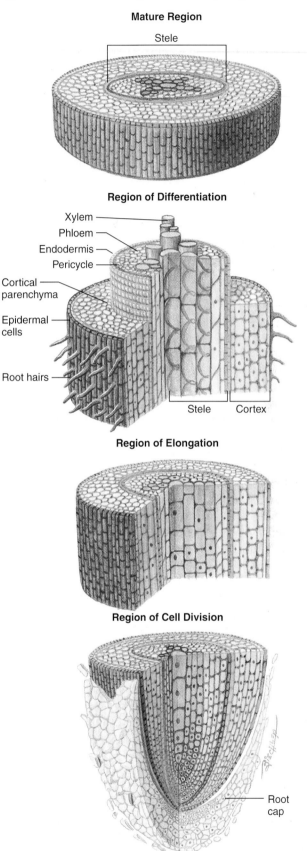

Mature Region

Stele

Region of Differentiation

Xylem
Phloem
Endodermis
Pericycle
Cortical
parenchyma
Epidermal
cells
Root hairs

Stele   Cortex

Region of Elongation

Region of Cell Division

Root
cap

Identify the **root cap,** a thimblelike covering of cells that protects the tip of the root as it pushes through the soil (fig. 25.4). The cells of the root cap are constantly renewed. It has been estimated that an elongating corn root can shed 10,000 root cap cells in a day. Once formed, root cap cells live four to nine days before they die, secreting a slimy material, called mucigel, that also lubricates the passage of the root through the soil. It is estimated that 20% of the carbohydrate made in photosynthesis gets used to make mucigel. Root growth, obviously, must contribute substantial amounts of carbohydrate to the soil where it supplies energy for the growth of soil fungi and bacteria after serving its lubrication function.

In the root tip behind the root cap is the **apical meristem,** the layer of cells that divides during root **primary growth.** You studied mitosis in the cells from this region in onion roots in lab topic 9. After cells are formed by division, they increase in size in the **region of elongation** behind the meristematic region. Cells in this region can increase in size 150-fold by water uptake. The elongation of these cells pushes the root forward up to 4 cm per day through the soil. Root growth is indeterminate, stopping only when the plant dies. It is this constant pushing forward that allows the roots to enter new soil areas and to tap the minerals dissolved in soil water.

Figure 25.4   Primary root tip. The root cap is separated, showing it is a distinct structure.

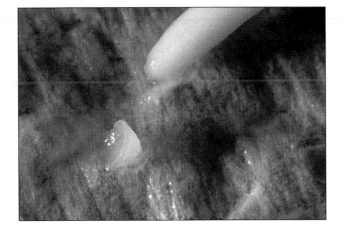

Further up the root, you may be able to see cellular differentiation. Several types of cells should be visible. The **epidermis** is the outer cell layer of the root. As the tip of the root penetrates new soil and epidermal cells mature, **root hairs** grow out from the epidermis and capture previously untapped water and mineral resources of the soil. The region of **absorption** is marked by the root hairs. Root hairs are epidermal tissues (trichomes) that are specialized for absorbing water and dissolved minerals. How many cells are in a root hair?_____ (Look for cross walls.) Plants may die when transplanted because the delicate root tips, with their root hairs, are torn off or dry in the moving process.

The roots of pines, birches, willows, and oaks often do not have well-developed root hairs. Instead, they have mycorrhizal roots in which filaments of a symbiotic fungus carry out the functions of root hairs. The fungal hyphae extend into the soil and penetrate between cells of the root cortex. They convey water and minerals into the root from the soil. (See fig. 18.9.)

In the zone of cellular differentiation, you may see a region called the **cortex** consisting of **parenchyma** cells. It extends from beneath the epidermis to a central vascular cylinder, the **stele.** The parenchyma cells of the cortex store starch and other materials in many plants. Note the loose arrangement of cells in the cortex. Water and dissolved minerals can travel through these spaces. Are the parenchyma cells normally living (with a protoplast) or dead?_____. The inner boundary of the cortex is a single layer of cells, the **endodermis.** As you will see, it is very important in regulating water and mineral absorption.

The stele is composed of all the cells inside the endodermis. The **xylem** and **phloem,** which conduct materials from and to the roots, are prominent here. Parenchymal cells, also found in stele, are living cells with relatively thin walls. They can divide to repair wounds and are physiologically active in storage. In young primary growth roots such as this, the xylem and phloem are primary tissues derived from the apical meristem.

As the root ages and thickens, secondary xylem and phloem will develop from a vascular cambium. When this happens, the epidermis is shed and replaced by the **periderm,** a bark-like covering. Because the root hairs are lost at this time, the older portions of roots with periderm do not absorb water and minerals.

Cross Section of Roots

While the age sequence of cells and differentiation is best seen in longitudinal sections as you just studied, tissue arrangements are easier to understand in cross sections. In this activity you will study cross sections of roots from a dicot and from a monocot. Be sure that you learn the differences in the arrangement of tissues in these two groups of flowering plants.

Dicot Roots

6 ▸ Obtain a prepared slide of a cross section of a buttercup root (*Ranunculus*), a dicot, and look at it with low power. Identify the epidermis, cortex, and stele as you did in the longitudinal section. Much of the bulk of young primary roots is contributed by the living parenchyma cells of the cortex.

The cylindrical stele should be readily visible at the center of the root (fig. 25.3, top section). Examine it with the high-power objective. At the center of the stele the large, heavy-walled **xylem** tissue may be seen in an x-like configuration, although it will sometimes have a symmetry based on three or five radiating poles. Between the radiating poles of the xylem locate the phloem.

In dicots, a **vascular cambium** develops between the primary xylem and phloem. During secondary growth of dicot roots, this cambium produces new **secondary xylem** to the inside and **secondary phloem** to the outside. Eventually, the addition of vascular tissue forms expanding concentric rings of secondary xylem and phloem. The expanding central cylinder of xylem causes tension in the surrounding primary phloem, cortex, and epidermis. These layers split longitudinally and slough off as secondary growth continues. The root also produces another cambium, the **cork cambium** (from the pericycle). Division of its cells produces a bark layer on the roots.

The outer boundary of the stele is the **pericycle** just beneath the endodermis. The pericycle is a narrow zone of parenchymal cells that maintain their ability to divide and form a meristematic tissue.

The endodermis is a physiologically important boundary that regulates water and mineral absorption. With a shape like a flattened box, the rectangular cells of the endodermis have a cell wall on each of their six faces; four sides have a band of waxy substance called **suberin,** but two are not suberinized (fig. 25.5). The waxless walls face the outside of the root on one side and the vascular tissues on the other. A good analogy to remember in understanding the structure of the endo-

Figure 25.5   Pathway of water and dissolved mineral absorption in a dicot root. Radial walls of endodermal cells are impregnated with the wax suberin, preventing water from passing between cells as it moves from the cortex to the stele. The only pathway for water is through the waxless tangential surface walls and the protoplasts of the endodermal cells, allowing them to regulate water and dissolved mineral uptake.

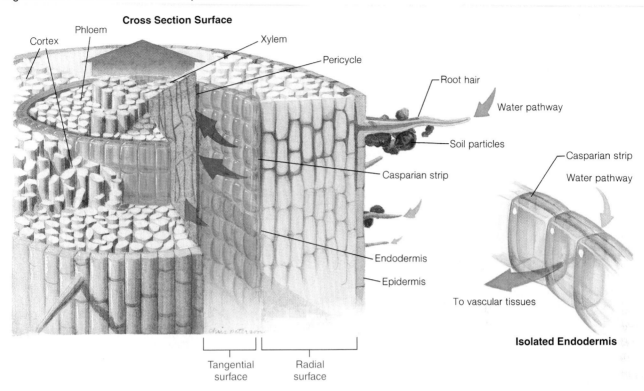

dermis is a brick wall: bricks are like cells with six faces, and in a wall, four of those faces are covered by mortar which is analogous to suberin. Because the wax-containing walls of adjacent endodermal cells abut one another, they form a water-impermeable barrier called the **Casparian strip.** It is not possible to see the Casparian strip on your slide.

When water and dissolved minerals enter the root, they travel the paths shown in figure 25.5. At the endodermis, suberin prevents water from diffusing around the cells, so water and dissolved minerals must pass through the plasma membranes into the protoplast of the endodermal cells, allowing them to regulate what passes to the xylem in the stele. Thus, these cells control absorption by the plant. Water and minerals that enter the stele and pass to the xylem are transported to other parts of the plant. As a root starts secondary growth in its aging regions and root hairs are lost from the surface, the endodermis also changes. Suberin will be secreted on all faces of the endodermal cells, essentially making it waterproof in the older parts of the root.

Lateral roots arise internally in the pericycle about one inch back from the tip. The growing tissues push their way to the surface through the endodermis, cortex, and epidermis (fig. 25.6). This process differs from lateral branching in stems where the branches arise from buds located at nodes on the stem surface. Nodes and associated buds are absent in roots.

Figure 25.6   Cross section of a water hyacinth root showing formation of a branch root from pericycle.

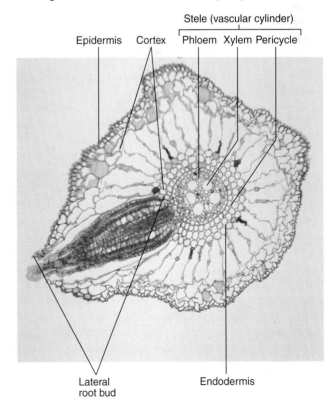

Investigating Primary and Secondary Growth in Roots and Stems

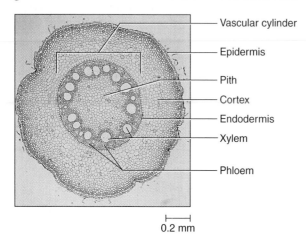

Figure 25.7   Cross section of a corn root, a monocot.

— Vascular cylinder

— Epidermis

— Pith

— Cortex

— Endodermis

— Xylem

— Phloem

|← 0.2 mm →|

Monocot Roots

7▷ Obtain a slide of a cross section of a corn (monocot) root (fig 25.7). Identify the outer epidermis and the parenchyma cells of the cortex. Note the intracellular spaces among the cortical parenchyma cells. Note the general similarity of the cortex to the dicot root but the obviously different organization to the stele. It is surrounded by an endodermis as in the dicot root, but the xylem and phloem cells are found in clumps around the periphery of the stele rather than at its center. The large cells to the inside of the pericycle are xylem. Phloem is located between the xylem vessels, just beneath the pericycle. No vascular cambium is found in monocot roots between the xylem and phloem. Consequently, they do not undergo secondary growth. The central area of the stele is filled with fairly uniform parenchymal cells and is called the **pith.**

Investigating Living Roots

8▷ If time permits, investigate the microscopic anatomy of roots by preparing freehand sections. Your instructor will have isolated roots from dicots and monocots in the lab, but will not tell you which is which. Your task is to examine the microscopic anatomy and to identify the unknowns.

Remove a root tip about 2 inches long from one of the unknowns. Take a two-inch square of Parafilm, a flexible plastic used to seal lab containers, and fold it in half. Place the root on the inside of the fold so that it is sandwiched between plastic. Place the preparation on a microscope slide and press on the Parafilm to hold the root in place. Cut a cross section of the root by slicing across it and the Parafilm. Discard the cutoff material. Add several drops of water to form a puddle at the newly cut edge. Now make a second cross section cut, making the section as thin as possible. Mount this thin section on a new slide in a drop of water and add a coverslip. Observe the cut face of the cross-sectioned root at first with the 10× objective of your compound microscope.

Identify the epidermis, cortex, stele, and xylem in your preparations. Which unknown has a tissue arrangement like a dicot? Which looked like a monocot?

Stem Structure

Stems support the flowers, fruits, and photosynthetic leaves in most plants. Conducting tissues in the stem bring water and minerals to the leaves and distribute photosynthetic products to the roots. Some stems are photosynthetic, as in cacti, and others are important in vegetative reproduction, food storage, and water storage. In this section of the lab, you will examine the basic organization of two types of dicot stems and a monocot stem. As with roots, you need to remember the differences in arrangement of the tissue types between monocots and dicots.

External Dicot Stem Structure

9▷ Examine a woody twig from a hickory, buckeye, or other tree. At the tip of the stem or its branches, find the **terminal** or **apical bud.** In the bud is the **apical meristem,** the source of cells for the primary growth that elongates the stem. In most plants, the apical bud produces a hormone that inhibits development of lateral buds, a phenomenon called apical dominance. Gardeners have long recognized this phenomenon. They obtain bushy shrubs and trees by clipping off the shoot tips, which removes the inhibition of the lateral buds, allowing branches to develop. Conversely, tall plants can be "forced" by trimming lateral branches, thus removing the lateral buds and directing the plant's energy reserves to the apex of the shoot.

Note the prominent **leaf scars** where leaves were attached (fig. 25.8). If you look carefully at a leaf scar, you can see the vascular bundle scars where strands of xylem and phloem entered the petiole of the leaf. The leaves grow out from locations on the stems called **nodes;** the stem segments between the nodes are **internodes.** Small pores called **lenticels** should be visible in the bark of the internodes. These loosely organized areas of bark allow metabolically active tissues in the twig to exchange respiratory gases with the atmosphere. The angle formed between a leaf and the internode above it is called the **axil.** Examine the area just above a leaf scar and you should see **lateral** or **axillary buds,** which can form branches of the stem.

Microscopic Herbaceous Dicot Stem Structure

10▷ Now obtain a prepared slide of a longitudinal section of a *Coleus* shoot tip. *Coleus* is a **herbaceous plant,** meaning that the stem shows limited secondary growth and the stem does not thicken very much. Annual plants such as these

Figure 25.8 Anatomy of a woody stem: (*a*) external features; (*b*) detail of leaf scar; (*c*) longitudinal section of terminal bud.

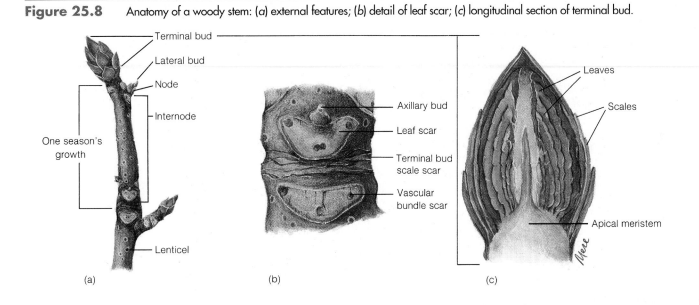

(a)   (b)   (c)

usually die at the end of the growing season. Using the scanning objective on your microscope, identify the lateral buds and terminal buds as you did for the woody stem. Note several **leaf primordia,** which surround and protect the terminal bud (fig. 25.9). What important group of cells necessary for continued growth is found in the terminal bud?

As the *Coleus* matures, flower buds may develop in the terminal bud. Note the **lateral bud** in the angle (axil) between the top surface of the leaf and stem. Lateral buds near the tip of a stem are in a state of dormancy, held in check by hormones produced by the terminal bud. As the stem elongates the distance between the terminal and lateral buds increases and the inhibitory effect is lessened allowing the lateral buds to develop into branches. In pines, firs and spruces this effect produces the pyramidal shape of the tree crown. If the terminal bud is damaged, the lateral buds quickly grow.

Examine the terminal bud under high power and find the **apical meristem** composed of darkly staining, rapidly dividing small cells. As cells are produced at the tip and subsequently elongate as they mature, the stem undergoes **primary growth** or growth in length. Examine the cells below the meristem and note their larger size and evidence of differentiation. Some will become vascular tissues; others will give rise to parenchyma cells, leaf primordia, and lateral buds. Trace the developing vascular bundles consisting of xylem and phloem passing into each of the leaf primordia.

Now obtain a cross section of a differentiated region of *Medicago,* a herbaceous dicot stem, and look at it under your 4× objective (fig. 25.10). Herbaceous stems undergo little secondary growth and are nonwoody. Note the general organization. In young stems resulting from primary growth, the

Figure 25.9 Longitudinal section of Coleus terminal bud.

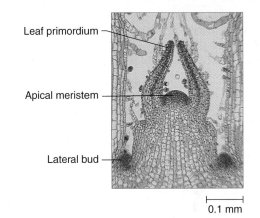

0.1 mm

outside of the stem is covered by a single layer of living cells, the **epidermis.** The epidermis protects the underlying tissues from drying, and often contains stomata openings. Epidermal cells on many stems form trichomes. Perhaps you have noticed the "hairy" stems of tomatoes which bear many trichomes. Waxes often are secreted by the epidermal cells to form a **cuticle.** Just beneath the epidermis is the **cortex,** a complex region. The bulk of the stem is **pith,** a soft matrix of living parenchyma cells. In the stems of some species the pith dies, producing a hollow stem.

The **vascular bundles** are arranged in a circle a short distance inside the cortex. Examine a single vascular bundle under high power. Xylem makes up the inward side of the vascular bundle. Most xylem is dead at maturity. The protoplast breaks down and the walls remain as water conduits. The larger of these are **vessels** and the smaller ones **tracheids** (fig. 25.11). The outer portion of the vascular bundle is phloem. Phloem is composed of larger **sieve tube**

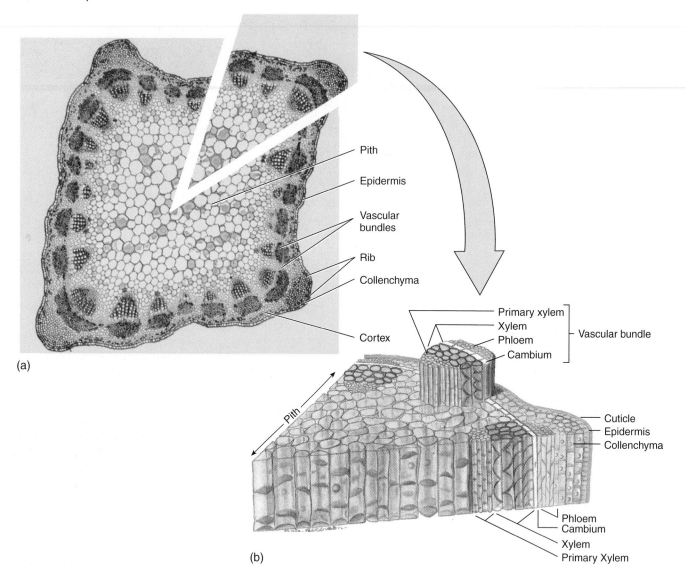

(a)

(b)

members and smaller **companion cells.** These cells must remain alive to function at maturity. Separating the xylem and phloem is a layer of **vascular cambium.** (see fig. 25.10). During stem secondary growth when the stem increases in diameter, division of the vascular cambium produces some secondary xylem and secondary phloem. The vascular tissues of the stem are continuous with those of the leaves and roots, forming an effective transport system throughout the plant. What functional reason can you suggest that would explain why most cells in the stem are elongated rather than being cuboidal?

Examine one of the "ribs" of the stem and identify the **collenchyma** cells with their unevenly thickened walls. These cells are living at maturity and provide support.

Internal Woody Dicot Stem Structure

Woody dicots, in contrast to herbaceous dicots, have stems that increase in girth year after year by secondary growth. As a plant increases in girth, its older xylem is buried in the added layers and cannot be supplied with oxygen. This physiological limitation has been solved by an interesting evolutionary adaptation. Only the cells near the surface of the trunk retain their protoplasts and are alive. Those near the center die, but their cell walls remain structurally intact and provide mechanical support for the crown of the plant.

A woody stem has three basic regions: bark, vascular cambium, and wood (fig. 25.12). In young twigs before secondary growth is initiated, the surface of the twig is cov-

Figure 25.11 Cells of the xylem and phloem: (a) scanning electron micrograph of cucumber xylem vessel; (b) scanning electron micrograph of phloem sieve plate in pumpkin and broken companion cell (cc).

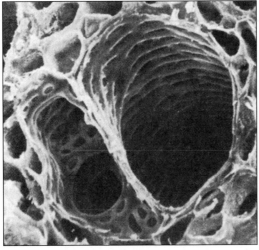

(a)

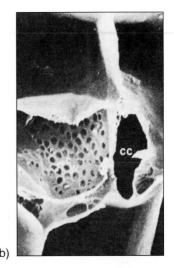

(b)

Figure 25.12 Tissue organization in a cross section of a two-year-old basswood stem, a woody dicot.

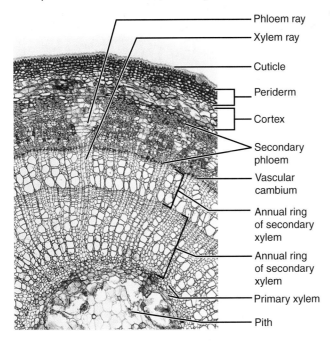

— Phloem ray
— Xylem ray
— Cuticle
— Periderm
— Cortex
— Secondary phloem
— Vascular cambium
— Annual ring of secondary xylem
— Annual ring of secondary xylem
— Primary xylem
— Pith

ered by epidermis. With the onset of secondary growth, bark develops as the epidermis splits and sloughs off. **Bark** is divided into an outer region, the periderm, and an inner region, the secondary phloem. The **periderm** consists of an outer **cork** layer, which prevents evaporation and protects the underlying tissues, and a **cork cambium,** which gives rise to the outer cork cells. **Secondary phloem** consists of sieve tubes, companion cells, sclerenchyma fibers, and storage parenchyma cells. These cells function in food conduction, storage, and support.

Beneath the phloem is the **vascular cambium.** Division of these cells produces secondary phloem to the outside and secondary xylem to the inside. Wood, the innermost region, consists entirely of **secondary xylem.** Its functions are water conduction and support.

Obtain a prepared slide of a cross section of a two-year-old basswood *(Tilia)* twig. Find the pith at the center of the section (fig. 25.12). The life span of the parenchyma cells forming the pith varies by species. As the pith ages, the protoplasts die and the cells accumulate tannins and crystals. Surrounding the pith is a narrow layer of primary xylem; both pith and primary xylem originated from the apical meristem and are the result of primary growth during the first year of life. To the outside is a relatively wide layer of secondary xylem laid down as an annual growth ring. Each annual ring consists of larger springwood cells and smaller summerwood cells. At one time the vascular cambium was next to the primary xylem, surrounding it as a cylinder of cells. As the vascular cambium cells divided, the daughter cell to the inside became secondary xylem, and the cell to the outside remained vascular cambium to divide again and again. Thus the cylinder of vascular cambium always enlarges and is located just outside of the xylem regardless of how many xylem cells are added in secondary growth. The vascular cambium separates the xylem from the outer phloem.

The phloem is composed of triangular sections, some of which point inward and others outward. Those that point inward, ending in sharp points, are **phloem rays** and are continuous with radial lines of cells, the **xylem rays,** which cross the xylem toward the pith. These cells conduct materials radially in the stem and its branches. In between the phloem rays are other triangular sections, ending in blunt points, that point outward. These consist of heavily stained **phloem fibers** and large, thin-walled **sieve tube members.**

Figure 25.13   Tissue organization in a mature oak stem.

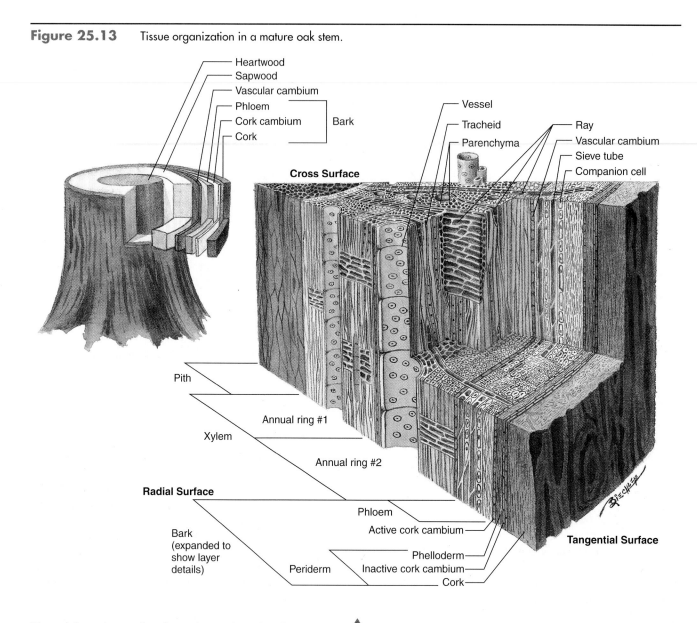

These join end to end to form **sieve tubes.** Small **companion cells** are associated with the sieve tube.

In young dicot woody stems, just outside the phloem is the **cortex,** a zone of loosely arranged, large parenchyma cells. The cortex is replaced by the **periderm** as the stem grows in diameter (fig. 25.13). The long axis of the cells in this zone follows the circumference of the stem rather than its long axis as do the vascular tissues.

The periderm is divided into three layers. Adjacent to the phloem are four to six layers of thick-walled cells, the **phelloderm.** External to the phelloderm are the thin-walled cells of the **cork cambium,** which gives rise to phelloderm and the outermost **cork cells,** which cover the mature stem. Cork cells secrete the waterproofing compound, suberin. It is a lipid. The bark of a woody plant consists of all the layers external to vascular cambium and thus includes the phloem and the components of the periderm.

As a tree grows in girth, what must happen to the existing bark as secondary xylem is added on the inside of the vascular cambium?

Structure of Wood

Wood is composed exclusively of years of accumulation of secondary xylem to the inside of the vascular cambium. If you look at a cross section of a tree trunk or branch two types of wood are visible (fig. 25.13). In the center the

Figure 25.14 Monocot stem: (a) cross section of a corn, *Zea mays*, shows vascular bundles are scattered throughout parenchyma; (b) enlargement showing a vascular bundle.

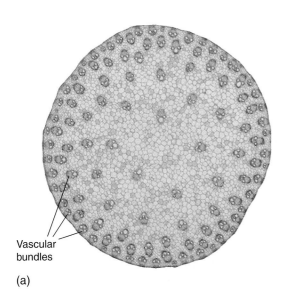

Vascular bundles

(a)

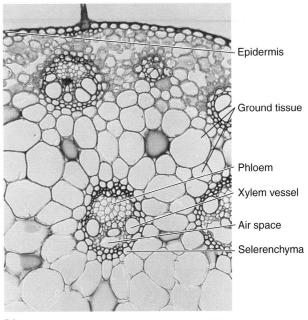

— Epidermis

— Ground tissue

— Phloem

— Xylem vessel

— Air space

— Selerenchyma

(b)

heartwood is darker in color due to the accumulation of resins, tannins and other metabolic wastes. It is the older wood. Just outside the heartwood, but not including the bark, is the **sapwood.** It is lighter in color and is younger, not having accumulated the metabolic products. It will gradually darken and become heartwood as new sapwood accumulates from division of the vascular cambium.

Commercial lumber is classified as softwood or hardwood. Softwoods are coniferous woods taken from pine, spruce, fir, or hemlock. Hardwoods are taken from dicotyledonous angiosperms, such as oak, cherry, ash, and several other species.

Both types of wood are prepared by sawing the secondary xylem layers into boards. The xylem of softwood has **tracheids** as the main structural cell type and lacks the **wood fibers** and **vessels** characteristic of hardwood. In this section you will look at the structure of oak, studying the wood in cross and radial sections (fig. 25.13).

Cross Section

13▶ Obtain a prepared slide of a cross section of oak wood and look at it first with low power. Bark is probably not included in this section. Compare it to the top of figure 25.13. Vessels and wood fibers make up the bulk of the xylem, although some tracheids are also present. The **pores** in the wood are the ends of the vessels. Identify the **annual growth rings** composed of larger cells in springwood and smaller in summerwood. Note the **wood rays,** which pass between rings.

Radial Section

A radial section is taken from a cylindrical stem by cutting a longitudinal section from the outer surface through the center of the cylinder. Such a section cuts the wood fibers and vessels in longitudinal section and shows the rays passing at right angles to the wood fibers and vessels.

14▶ Obtain a slide of a radial section of oak wood and look at it with low-power objective and then high-power. Identify the wood fibers, vessels, and rays. Look carefully at the walls of the vessels and note the **bordered pits,** which allow water to pass out of the vessels. Which cells are larger in diameter: vessels or wood fibers?_____

Internal Monocot Stem Structure

15▶ Examine a prepared slide of a cross section of a corn (maize) stem. Use the 4× objective. Refer to figure 25.14 and identify the cell types in your specimen.

Note how the monocot stem organization differs from the dicot stems studied earlier. The vascular bundles are not arranged in concentric circles, but are scattered, instead, throughout the ground tissue of the stem. Examine a **vascular bundle** under higher magnification. (Some people think these bundles resemble a monkey's face.) The bundle is surrounded by thick-walled sclerenchyma fibers. The xylem typically consists of four conspicuous vessels. The phloem is found in the area between the two larger vessels and extends from there toward the edge of the bundle. Look for sieve plates in

Figure 25.15   Anatomy of a grass plant. Stem is hollow in most species.

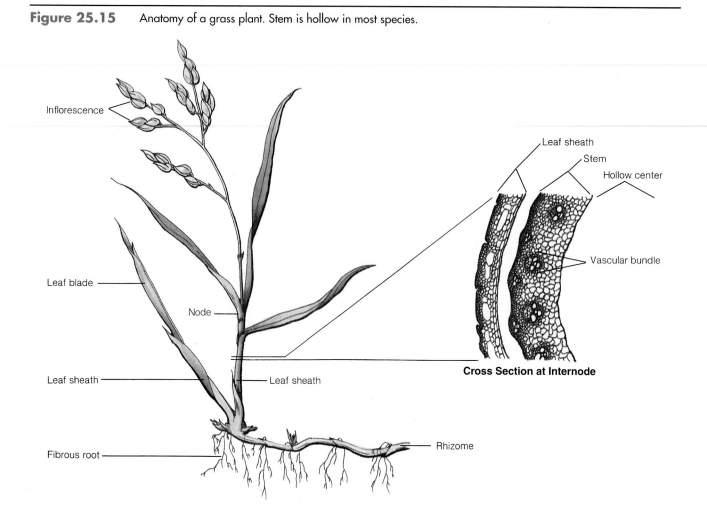

Inflorescence

Leaf blade

Node

Leaf sheath

Leaf sheath

Fibrous root

Rhizome

Leaf sheath
Stem
Hollow center
Vascular bundle

Cross Section at Internode

some of the sieve tubes. A vascular cambium does not occur in monocots. Consequently, monocots do not produce secondary xylem or secondary phloem; *i.e.,* they have no secondary growth.

The stem of corn is not representative of the stems of many grasses. Grass stems are often hollow, except at the nodes (fig. 25.15). In hollow stems, the vascular bundles are arranged as concentric circles in the walls of the stem, but vascular cambium is not present.

In grasses, a meristem is found at the base of each internode and at the base of the leaf sheath. These are located at the base of the grass plant. When we cut grass, the tip of the leaf and stem is cut off but the meristems close to the soil are not harmed. They produce new cells and the leaf and stem "shoot up" so that the lawn must be cut again, and again, and . . .

Branches arise from the lower portions of grass stems at the soil level. Usually, the branches grow more or less horizontally for a short time, often underground, to form **rhizomes** from which leaf and flower stems grow upward and fibrous roots grow downward. This type of growth is called tillering and allows a single plant to spread out from a single seed, crowding out competing plants.

Transpiration (Demonstration)

In the previous sections, you studied the anatomy of the transport system in plants. Now you will investigate the functions of the transport system.

During periods of active photosynthesis, the stomata of the leaves are open, allowing carbon dioxide to enter. At the same time, water escapes from the leaf and is replaced by conduction up the xylem.

16 Obtain a fresh piece of celery stalk with leaves. Cut off and discard the bottom two inches of the stalk while holding the stalk under water. Quickly place the cut end of the leafy stalk in a beaker containing a 0.5% solution of methylene blue dye. Place the celery in sunlight.

Let the stem sit in the dye for 30 to 60 minutes. Hypothesize how far the dye will move in an hour. At the end of an hour make cross sections at 1 cm intervals up the stem. How many centimeters did the dye travel in an hour? Does this value match your hypothesis?

Use a razor blade to make a thin section of a region that had a high dye concentration. Mount the section on a slide and look at it with your compound microscope. In what tissue is the dye located?

Describe the results below.

Testing the Water Tension Hypothesis

Water is thought to be pulled through the xylem system by tension created in the leaves when water is lost through evaporation. If this is the case, then it can be easily demonstrated. If sunflower plants are grown in pots until they are about one foot tall, they can then be subjected to different watering regimens. One pot will be well watered for a week before lab. The other will not be watered. If the theory is correct, the water in the xylem of the unwatered plant should be under greater tension than in the watered one.

17▷ To test this hypothesis, fill a tray with a methylene blue dye solution. Take the plants one at a time and bend the stems so that they are under the surface of the dye solution. Cut the stem while it is immersed and immediately remove it. The xylem which is under tension will take up the dye.

Students can then take the stems and cut them every 5 mm. By looking at the color of vascular bundles, they can see how far the dye has traveled.

How do the results of this experiment support or refute the water tension hypothesis for water movement in xylem?

Learning Biology by Writing

Because the activities in this lab were more descriptive than experimental, your instructor may want you to write a summary rather than a report. In about 200 words, describe how water enters a dicot plant and moves to the leaves. Include anatomical diagrams of the root, stem, and leaf showing the water pathway. Discuss the mechanism of movement.

As an alternative assignment, your instructor may ask you to turn in answers to the Critical Thinking and Lab Summary questions.

Internet Sources

Use the search engine GOOGLE to locate a research report on the World Wide Web dealing with how phloem transports organic molecules. Write an abstract of the report and include the URL.

Lab Summary Questions

1. Distinguish between primary and secondary growth in plant roots.
2. Draw diagrammatic cross sections of primary monocot and dicot roots on a piece of paper. Label the major tissues. Write a few sentences that describe how they differ.
3. Using both cross-section and longitudinal-section diagrams, describe the pathway for water and mineral absorption in plant roots.
4. Describe how the structure of the endodermis determines its function.
5. Distinguish between primary and secondary growth in stems.
6. As a stem undergoes secondary growth, what will happen to cell layers in a woody stem located to the outside of the vascular cambium and phloem?
7. Draw diagrammatic cross sections of monocot and herbaceous dicot stems. Indicate how they differ.

8. Draw cross sections of both a woody dicot stem and a herbaceous stem. Indicate how they are different.
9. Describe the process of transpiration and why it is an important process in plants. Include a brief discussion of how water and minerals move from the roots to the leaves.

Critical Thinking Questions

1. Pines depend on mycorrhizal relationships. A plant nursery was sterilizing soil before planting Christmas tree seedlings and found that their trees took on average two to three years longer to reach sellable size. Why?
2. Many trees will die after their roots are flooded with water or have cars parked on them, compacting the soil. Why?
3. People often spread fertilizer under large trees to make them grow even faster. Sometimes they put the fertilizer close to the trunk, thinking it will be absorbed faster. Why is this erroneous thinking?
4. What anatomical feature of the root would allow radishes to accumulate zinc in the parenchyma tissue of the root but not in stems and leaves?
5. The tallest trees in the world are around 100 m high. Do you think this is the maximum possible height or could trees grow even higher? Explain.
6. How can information about previous climate be obtained by studying the growth rings of trees? By observing stem structures?
7. If you cut the bark of a tree down to the vascular cambium, in a circle completely around the tree, the tree dies. Why?
8. A tree grows about one foot each year. You carve your initials in the trunk of a tree that is 15 feet tall. Your initials are four feet above the ground today. How far above the ground will they be 10 years from now?
9. What has happened to the epidermis, cortex, and primary phloem of a 100-year-old maple tree?

LAB TOPIC 26

Investigating Leaf Structure and Photosynthesis

Supplies

Preparator's guide available on WWW at
http://www.mhhe.com/dolphin

Equipment

Spectrophotometers
Kitchen blender
Compound microscopes
Desk lamps with outdoor flood lamps
Magnetic stirrer

Materials

Several different kinds of fresh leaves
Fresh *Elodea* cuttings
Frozen spinach
Prepared slides
    Dicot leaf, cross section
    C_4 plant, corn, leaf cross section
Flat-sided tanks at least 10 cm across
Ring stand and two clamps
Test tubes with one-hole stoppers
Syringe, 1 ml
1-ml pipettes with bend at the top end above 0 as in
    figure 26.10
800-ml beakers
Cheesecloth
Graduated cylinders
250-ml separatory funnel
125-ml Erlenmeyer flasks
Razor blades and glass plate

Solutions

Acetone
Petroleum ether
80% methanol
10% NaCl
Na_2SO_4 powder (anhydrous)
MgO powder
0.1 M $NaHCO_3$

Student Prelab Preparation

Before doing this lab, you should read the introduction
and sections of the lab topic that have been scheduled
by the instructor.

You should use your textbook to review the
definitions of the following terms:

chlorophyll   palmate
dicot   photosynthesis
epidermis   pinnate
guard cell   spectrophotometer
mesophyll   stomata
monocot   vascular bundle

You should be able to describe in your own words
the following concepts:

Light reactions
Dark reactions (Calvin cycle)
Absorption spectrum

As a result of this review, you most likely have
questions about terms, concepts, or how you will do
the experiments included in this lab. Write these
questions in the space below or in the margins of the
pages of this lab topic. The lab experiments should
help you answer these questions, or you can ask your
instructor during the lab.

Objectives

1. To observe the structure of dicot and monocot
   leaves
2. To measure the absorption spectrum of chlorophyll
3. To test the hypothesis that the rate of
   photosynthesis is proportional to light intensity
4. To determine by graphic analysis the light intensity
   at which a plant's oxygen production equals its
   oxygen consumption and to confirm this prediction
   by direct experimentation

Background

Photosynthesis supplies energy to virtually every ecosystem. Not only do plants, green algae, and cyanobacteria use this energy to grow and reproduce, but they also are eaten by animals or decomposed by bacteria and fungi, thus supplying these organisms' energy needs as well. In photosynthesis, light energy from the sun is captured through complex biochemical reactions and used to make new covalent chemical bonds. Carbon dioxide serves as a source of carbon, and water as a source of hydrogen in these reactions. It is estimated that 150 billion tons of sugar are produced annually by plants on a worldwide basis. An important by-product of this process is molecular oxygen, which enters the atmosphere replacing that which is consumed in respiration and various other oxidations. In addition, oxygen in the upper layers of the atmosphere is converted to ozone, which blocks the transmission of mutagenic ultraviolet light to the earth's surface.

Realize that energy, unlike materials such as carbon dioxide or water, does not cycle in ecosystems. Instead, energy flows in a direction toward dissipation. Plants are able to convert light energy to chemical bond energy where it is stored in the bonds of sugars. As plants use these sugars or are consumed by animals and these animals are in turn consumed by other animals, energy is lost as heat when the chemical bonds of the plant biomass are broken and reformed in animal biomass and so on in the food chain. As a general rule, only 10% of the energy available in ingested food is used to make new biomass; the remaining 90% dissipates as low-grade heat (entropy) throughout the universe. If photosynthesis suddenly ceased on this planet, life would eventually end because all energy stored in chemical bonds of biomass would be released as heat as one organism fed on another in ever shortening food chains.

Plants (along with algae and some bacteria) contain **chlorophyll,** a remarkable green-colored molecule that can absorb light energy and transfer it to its electrons. In plants, chlorophyll is located in the membranes of the **grana** found in those unique organelles called **chloroplasts** (fig. 26.1). These electrons can perform cellular work that ultimately results in using carbon dioxide to make organic molecules with energy stored in the chemical bonds. Work is done when the electrons pass through the **photosynthetic electron transport chain,** a collection of compounds located in the thylakoid membranes of the chloroplasts. The energy in the electrons is used to form **ATP** (adenosine triphosphate) by chemiosmosis, and the high-energy electrons may be transferred to the electron carrier **NADP⁺** (nicotinamide adenine dinucleotide phosphate), producing NADPH. These processes, collectively known as the **light reactions,** produce oxygen as a by-product. It escapes from the chloroplast into the atmosphere. Review the diagrams of the light reactions in your textbook and determine what is the source of oxygen released during photosynthesis. Does the oxygen come from CO_2 or H_2O?

Figure 26.1 The structure of a chloroplast: (*a*) transmission electron micrograph; (*b*) artist's three-dimensional reconstruction.

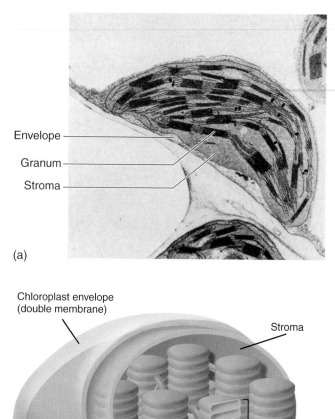

The ATP and NADPH produced during the light reactions are used as energy and hydrogen sources to build complex organic molecules. The set of reactions involved in building such molecules is called the **Calvin cycle,** or **dark reactions.** In these reactions, CO_2 molecules are bonded one at a time to already existing organic molecules, a process known as CO_2 fixation. The result is an increase in the size and number of organic molecules in the cell, using energy and hydrogen coming from the light reactions.

In most plants, glyceraldehyde-3-phosphate is the product of the dark reactions. It is then converted into hundreds of other compounds by other enzymes in the chloroplast or cytoplasm (fig. 26.2). Since glyceraldehyde-3-phosphate is a three-carbon compound, this type of plant is said to carry out **C₃ photosynthesis.** Some plants produce four carbon compounds in photosynthesis. These are called C₄ plants. Check your textbook if you would like details.

Figure 26.2 Relationship between photosynthetic CO_2 fixation and the synthesis of other biochemical molecules.

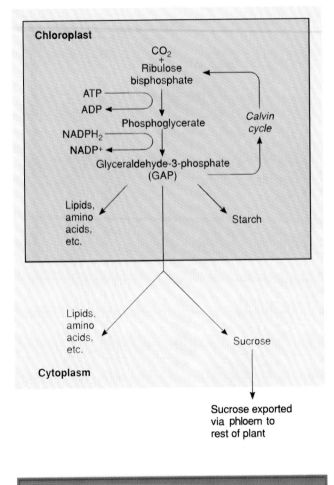

LAB INSTRUCTIONS

In this lab topic, you will observe the anatomy of leaves, determine the absorption spectrum of chlorophyll, and measure how light intensity influences photosynthetic rate.

Types of Leaves

1▷ Several types of dicot leaves will be on the demonstration table in the laboratory. Look at them and be sure that you understand the features indicated in boldface in the following paragraphs.

A dicot leaf consists of a flat **blade** and a stalk, or **petiole,** which attaches the leaf to the stem. Xylem and phloem pass from the leaf through the petiole and stem to the roots. At the point where a leaf attaches to a stem an axillary bud is found. The bud gives rise to branches of the stem. Leaves may be **simple,** consisting of a single blade and a petiole, or **compound,** consisting of several leaflets joined to a single petiole (fig. 26.3).

Vascular tissues (xylem and phloem) will be visible as **veins** in the leaves. Three venation patterns are found in leaves:

1. **Parallel**—veins pass from the petiole to the tip of the blade in a more or less parallel fashion. Monocots (grasses and their relatives) have leaves with parallel venation. If a vein branches into two equal parts, then it is called dichotomous venation. Ginkgo leaves (fig. 26.3e) have dichotomous venation; they are gymnosperms.

2. **Net venation**—veins are not parallel. Dicots (broadleaf plants) have leaves with net venation.

   a. **Pinnate**—a single main vein gives off smaller branch veins that run parallel to each other as in a feather.

   b. **Palmate**—several main veins radiate from where the petiole joins the blade as in fingers radiating from a palm.

Be sure to find examples of all types of venation among the demonstration materials. Record below the names of the species observed and whether they have simple or compound leaves and the type of venation.

Species	Simple or compound	venation
1.		
2.		
3.		
4.		
5.		
6.		
7.		
8.		

Many leaves display adaptations to specific habits. Leaves from dry areas are often thickened and have hairs that minimize water loss. Aquatic plants with floating leaves will have large air spaces in the leaf. There are often differences in the leaves of plants that are adapted to full sunlight or shade. Conifers have needles with reduced surface areas that reduce moisture loss in cold winters and dry summers.

Internal Leaf Anatomy

Although chloroplasts are found in all green parts of plants, the leaves are the primary solar energy collectors, whether the plants are ferns, pines, birches, or grasses. Leaves are remarkable organs having a high surface area per unit volume. Furthermore, many plants carry and position their leaves to capture sunlight effectively. Water evaporation through leaf

Investigating Leaf Structure and Photosynthesis   **349**

Figure 26.3   Types of leaves and venation: (*a*) palmately veined maple leaf; (*b*) simple but lobed leaf of tulip tree; (*c*) opposite, simple leaves of dogwood; (*d*) whorled leaves of bedstraw; (*e*) fan-shaped leaf of *Ginkgo,* with dichotomous venation; (*f*) palmately compound buckeye leaf; (*g*) pinnately compound black walnut leaf; (*h*) a grass leaf; (*i*) linear leaves of yew; (*j*) needles of pine.

surfaces creates a tension which draws water upward, through the xylem—an important transport function.

In this section, you will study the anatomy of leaves in both dicot and monocot flowering plants.

Structure of a Dicot Leaf

2 ▶ Hand cross sections of leaves can be easily made. The technique is outlined in figure 26.4. Once you have mastered the technique, a variety of leaves can be examined and compared. Try cutting sections from one of the sample plants in the lab. Mount the sections on edge in a drop of water on a slide. Add a coverslip and observe. Alternatively, obtain a slide of a cross section of a dicot plant. Look at it with the low-power objective of your compound microscope, switching to high power to see detail (fig. 26.5). How many cells thick is the leaf?_____

In nature, the surface of the leaf is usually covered by a **cuticle,** a layer of wax secreted by the underlying cells, which slows water loss through the large surface area of the leaf. The slide preparation process often extracts the wax from the leaf surface but you should see it on hand-sectioned material. Is a cuticle visible on your slide?_____

Find the **epidermis,** usually a single layer of cells at the surface of the leaf. These cells lack chloroplasts and have a thick cell wall on their outer surface. Closely examine the epi-

dermal cells on the lower surface of the leaf and find the specialized **guard cells** surrounding an opening called the **stoma** (plural: stomata). (See fig. 26.7.) Are stomata found more frequently on the upper or lower surface of a leaf?_____
Hundreds of stomata may be found in a square millimeter of leaf surface. Carbon dioxide enters a leaf through these openings and water and oxygen escape through them.

The inside of the leaf is composed of **mesophyll** tissue made of thin-walled parenchyma cells. When parenchyma cells contain chloroplasts they are called chlorenchyma. Mesophyll chlorenchyma accounts for most of the photosynthetic activity of a plant. Find the **palisade** mesophyll composed of one or more layers of closely packed cells. These are always near the upper surface of the leaf when it is on the plant. Beneath the palisade cells is **spongy** mesophyll composed of irregularly shaped cells separated by a labyrinth of air-filled, intercellular spaces which allow carbon dioxide, oxygen, and water vapor movement. Note how these spaces are continuous with the stomata. Which of the mesophyll layers has more chloroplasts?

Figure 26.4 Two-slide method for making cross section of a leaf.

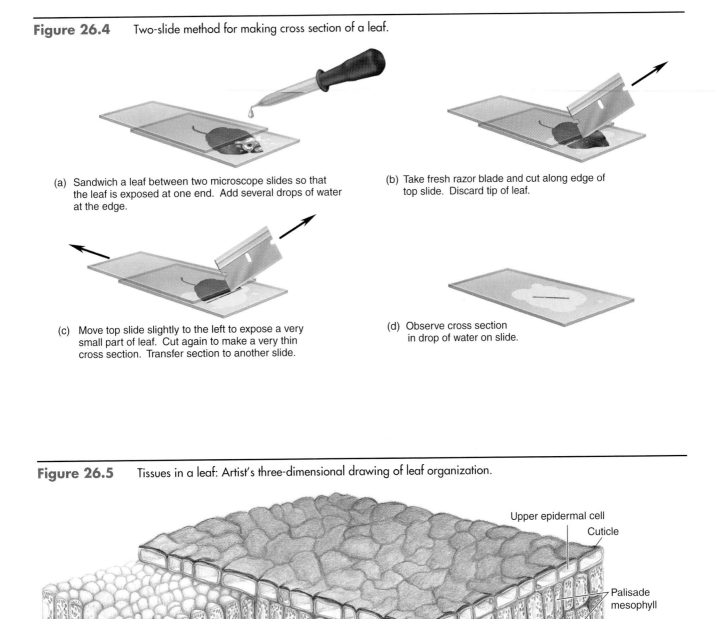

(a) Sandwich a leaf between two microscope slides so that the leaf is exposed at one end. Add several drops of water at the edge.

(b) Take fresh razor blade and cut along edge of top slide. Discard tip of leaf.

(c) Move top slide slightly to the left to expose a very small part of leaf. Cut again to make a very thin cross section. Transfer section to another slide.

(d) Observe cross section in drop of water on slide.

Figure 26.5 Tissues in a leaf: Artist's three-dimensional drawing of leaf organization.

Upper epidermal cell
Cuticle
Palisade mesophyll
Air space
Spongy mesophyll
Bundle sheath
Vascular bundle (vein)Xylem, up-permost; Phloem, lowermost
Lower epidermis
Guard cell
Stoma
Cuticle
Petiole
Blade

Figure 26.6 Scanning electron micrograph of a bluestem grass, *Andropogon*.

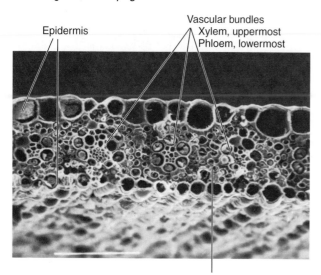

Epidermis

Vascular bundles
Xylem, uppermost
Phloem, lowermost

Mesophyll cells

You should also be able to see **vascular bundles** composed of xylem and phloem that transport materials to and from the mesophyll and the rest of the plant. What materials do they bring to the mesophyll?

What materials do they translocate from the mesophyll?

Note the tightly packed **bundle sheath** cells that surround the vascular bundles. They control the exchange of materials between the leaf and the vascular system.

3 Use different colored pencils to add arrows to the diagram in figure 26.5, showing the path of carbon and path of water through a leaf.

To determine if the structure of leaves is similar in a number of different plants, you will now hand-section leaves from several different species of plants. To do this, follow the procedure outlined in figure 26.4. Look at the edge of the sections first with the scanning objective, switching to the 10× objective of your compound microscope to identify the tissue layers in the leaf.

How can you distinguish between palisade mesophyll and the spongy mesophyll? Is the palisade layer toward the top or the bottom of the leaf? How many cells thick is the dermis in all leaves examined? Are all leaves about the same number of cells thick? Record your observations below.

Structure of a Monocot Leaf

About 65,000 of the 235,000 species of flowering plants are monocots. These include the bamboos, grasses, palms, lilies, tulips, orchids, and several other flowers. The leaves of these plants differ from the broadleaf you just studied (fig. 26.6). Monocot leaves are usually long and narrow with parallel venation rather than broad with a netlike vascular system.

4 Obtain a cross section of a corn leaf or alternatively cut hand sections from living monocots, using the technique in figure 26.4. Identify the following parts: cuticle; upper and lower epidermis; stomata; mesophyll, consisting of chlorenchyma cells with chloroplasts; and bundle sheath cells surrounding the xylem and phloem cells in a vein. Look closely at the chlorenchyma. Is it organized into palisade and spongy layers as in dicots?_____ Look closely at the upper and lower epidermis. Are stomata more common on one surface compared to the other?_____

Now carefully examine the upper epidermis and find the columnar **bulliform cells.** These thin-walled cells readily lose water on hot days and collapse. When they do, the corn leaf folds inward along its long axis. What purpose do you think this mechanism serves?

Figure 26.7   Artist's representation of stoma and guard cells from leaves.

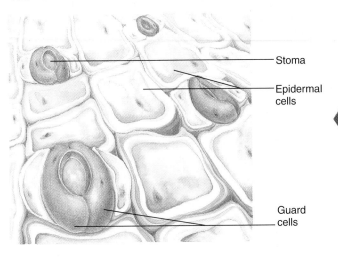

- Stoma
- Epidermal cells
- Guard cells

Guard-Cell Response to Osmotic Stress

The opening and closing stomata is related to the osmotic condition of the guard cells. When water is readily available in the leaf, the guard cells are turgid. Turgid guard cells, because of the orientation of cellulose fibers in the cell wall, have a crescent shape, which creates an opening next to adjacent cells (fig. 26.7). When water is not readily available, the guard cells become flaccid, losing their crescent shape and their edges become straight. When both guard cells straighten, they cover the opening of the stoma.

These conditions can be simulated by treating a strip of leaf epidermis containing guard cells with distilled water and a 5% salt solution.

5 ⊳ Place a drop of distilled water on a slide. Obtain a *Zebrina* leaf. Fold it in half with the underside inward (fig. 26.8). As it begins to break, pull the upper half away from the break. This should produce a cellophane-thin piece of the lower epidermis. Mount it in the water drop. Add a coverslip and observe. In the left half of the circle below, sketch your observations of the guard cells.

Figure 26.8   Technique for removing epidermis from underside of a leaf: Pinch leaf.

❖   What would you hypothesize would happen if you added several drops of a 5% salt solution next to and touching the coverslip?

Do so. Touch a piece of paper towel to the opposite side of the coverslip. This will wick the salt solution under the coverslip. Observe the guard cells again and sketch a few cells in the right half of the circle. How are the cells different? What caused this change in the guard cells? Must you accept or reject the hypothesis that you made?

It is sometimes fun to work with numbers to estimate the magnitude of an event or process. Some numbers are available regarding transpiration. A researcher counted the number of leaves on a 47-foot silver maple tree. He found that there were 177,000 leaves, with an average leaf area of 26 cm². If the transpiration rate per silver maple leaf is 0.01 ml/hour/ cm², how many liters of water would be lost from a mature tree in 12 hours?_____ How many gallons move through the vascular system and out the stomata of a tree this size in a day? (A gallon equals 3.9 liters.)_____ Another researcher working with tobacco estimated there were 12,000 stomata per cm². Assuming the same number for maple leaves, how many stomates did the silver maple have?_____

Photosynthetic Pigments

The light-absorbing properties of a chlorophyll solution will be studied using a spectrophotometer.

Extraction Procedure

6 ⊳ Chlorophyll can be extracted from plant tissue. You or your instructor will use a kitchen blender to homogenize 60 g of frozen spinach leaves in 150 ml of water containing 0.2 g of MgO powder. Pour the slurry into a 500 ml beaker, add 150 ml of acetone, cover with a watch glass or foil, and stir for 10 to 15 minutes on a magnetic stirrer. The resulting

Investigating Leaf Structure and Photosynthesis   **353**

Figure 26.9 A separatory funnel, used to separate a two-phase liquid mixture. The stopper must be removed before the stopcock is opened.

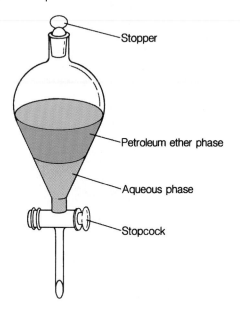

Stopper

Petroleum ether phase

Aqueous phase

Stopcock

TABLE 26.1 Light absorption characteristics of chlorophyll		
Wavelength	**Light Color**	**Absorbance**
420	_____	____
440		____
460	_____	____
480		____
500	_____	____
520		____
540	_____	____
560		____
580	_____	____
600		____
620	_____	____
640		____
660	_____	____
680		____

slurry should be filtered through six layers of cheesecloth into a beaker. This extract is a complex mixture of fats, sugars, orange carotenoid pigments, and chlorophyll.

CAUTION

Petroleum ether, methanol, and acetone are highly flammable. No flames are permitted in the laboratory during this procedure. Have absorbent material handy to soak up any accidental spills.

Contaminating compounds can be partially removed by a technique called solvent partitioning. This procedure is based on the solubility of nonpolar chlorophyll molecules only in nonpolar solvents, whereas more polar, contaminating compounds are soluble in the polar water and acetone solvents.

Add 50 ml of the filtrate to a separatory funnel. To this add 50 ml petroleum ether and 25 ml of 10% NaCl. The petroleum ether does not mix with the polar water-salt solution nor the acetone. Chlorophyll will be extracted from the filtrate as you shake the mixture. Cap and invert the separatory funnel, and open the stopcock to vent the system. Close the stopcock and shake vigorously. *Be careful!* Pressure will build up in the separatory funnel and should be released periodically by loosening the stopper. Place on ring stand to allow the two solvent systems to separate. The polar aqueous phase with contaminating compounds will settle to the bottom while the nonpolar chlorophyll solution will rise to the top. This may require 15 or more minutes for complete separation.

Drain and follow your instructor's directions to discard the lower water-acetone layer with any particulate interface (fig. 26.9). Now add 30 ml of 80% methanol to the flask and 15 ml of 10% NaCl. Shake and allow to separate. Drain and discard the lower aqueous methanol layer containing orange-yellow carotenoid and xyanthophyll pigments. Add 30 ml of 10% NaCl to the chlorophyll extract and shake again to remove residual acetone and methanol. Discard the aqueous phase. You should now have a reasonably bright green solution of chlorophyll.

Remove traces of water from the petroleum ether by adding about 2 grams of anhydrous Na_2SO_4 to the extract. Remove the Na_2SO_4 by filtering before proceeding to the spectrophotometer. You should now have several milliliters of clear green fluid containing the chlorophyll extracted from your sample. You will now measure its absorption spectrum.

The solvents used in this extraction are hazardous wastes and should be discarded according to directions given to you by your instructor.

Absorption Spectrum

7▶ Adjust a spectrophotometer to zero absorbance using pure petroleum ether as a blank. (Review fig. 5.5 for operating instructions.)

Add the chlorophyll extract to a spectrophotometer tube and read its absorbance across the visible spectrum at 20 nm wavelength intervals. If the solution is too concentrated (absorbance greater than 2 at 425 nm), dilute it with petroleum ether. If it is too weak, blow air through the extract to evaporate petroleum ether. This must be done in a hood. Remember to zero the instrument with petroleum ether at each new wavelength. Record your readings at each wavelength in table 26.1.

8▶ Observations of the color of light requested in table 26.1 can be obtained by placing a strip of paper in a clean, dry tube and inserting it in the spectrophotometer. If you look

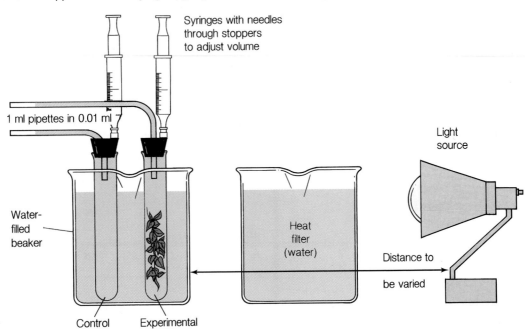

Syringes with needles through stoppers to adjust volume

1 ml pipettes in 0.01 ml

Light source

Water-filled beaker

Heat filter (water)

Distance to be varied

Control   Experimental

down into the tube with your hands cupped around the tube chamber opening, you should see the reflected light as it passes through the tube. Change the wavelength and record the colors in the table.

9   After the laboratory, when you are writing your lab report, you will graph this data. (See appendix B for graphing instructions.) Remember the dependent variable should be on the ordinate. Is wavelength or absorbance the dependent variable?_____ Label all axes and be sure to create a one-sentence figure legend that describes the sample.

Looking at your data in the graph that you just made, make a hypothesis that relates rate of photosynthesis to the wavelength of light striking a plant. Would you predict that photosynthesis would be greatest or least when the wavelength of light striking a plant corresponds to the wavelength of light maximally absorbed by chlorophyll? Test your hypothesis by consulting your text to determine what wavelengths of light are effective in photosynthesis (look up action spectrum in your textbook).

Light Intensity and Photosynthetic Rate

In this experiment you will test the hypothesis that, as the intensity of light increases, the rate of photosynthesis will also increase. The alternative hypothesis is that the rate will not change or may even decrease. Photosynthetic rate can be measured by determining the rate of oxygen production.

Some of the oxygen produced in photosynthesis is not released from the plant; instead it is consumed by aerobic respiration in the plant cells. The light intensity at which the rate of oxygen production equals the rate of oxygen consumption is called the **light compensation point.** Using the data collected to test the previous hypothesis, you will graphically determine the light compensation point for the aquatic plant *Elodea* and then experimentally test it.

Procedure

10   Obtain a piece of *Elodea* 5 to 6 inches long. Choose a healthy specimen with an actively growing leaf bud at the apex. Make a fresh cut across the basal end and place the sprig in a test tube with the cut end up. Fill the test tube with 0.1 M $NaHCO_3$ solution and add a rubber stopper containing a bent glass pipette and syringe, as shown in figure 26.10. The $NaHCO_3$ is a source of carbon dioxide for the plants. By making sure carbon dioxide is plentiful, photosynthesis will be limited only by light intensity.

The joint between the rubber stopper and the test tube must be dry to get a good seal and fluid must flow into the pipette as you push the stopper in. You must get a good seal between the tube and the stopper. If the stopper "creeps," it will give you erroneous readings. The position of the fluid in the pipette can be adjusted by raising or lowering the plunger on the syringe once the stopper is in place.

Prepare a second tube without *Elodea* to act as a control for temperature and atmospheric pressure fluctuations.

11   Place both tubes in a beaker of water at room temperature. Set up a heat filter and a lamp as shown in figure 26.10. Move the lamp so it is 75 cm away from the tubes with the heat filter in between. Dim the room lights and wait five minutes for equilibration.

TABLE 26.2   Oxygen production readings at 75 cm distance

	Time (Minutes)					
	0	2	4	6	8	10
Elodea						
Control						
Elodea—control						

TABLE 26.3   Oxygen production readings at 50 cm distance

	Time (Minutes)					
	0	2	4	6	8	10
Elodea						
Control						
Elodea—control						

TABLE 26.4   Oxygen production readings at 25 cm distance

	Time (Minutes)					
	0	2	4	6	8	10
Elodea						
Control						
Elodea—control						

What hypothesis are you testing in this experiment?

Adjust the fluids in the pipettes of both tubes so that they are at the 0.2 ml mark. Read the position of the fluid in the pipettes at two-minute intervals for ten minutes. Record your data in table 26.2. If you do not get a change of about 0.15 ml in ten minutes, you should extend the reading intervals to three to five minutes. If you change the time interval, be sure to change the times printed in table 26.2. At higher light intensities (shorter plant-to-light distances), much greater movement will be seen.

Oxygen is not very soluble in water. As *Elodea* produces oxygen in photosynthesis, the newly produced gas in the tube will force water to move into the pipette. There-

fore, fluid movement in the pipette is a measure of photosynthesis. If you measure the volume changes over a time period, you will be able to calculate the change in volume per minute which equals photosynthetic rate.

The changes you observed in the tube containing *Elodea* are the result of two simultaneous happenings: (1) the production of oxygen by *Elodea,* which always results in an increase in volume, and (2) fluctuations in temperature and pressure, which may result in either positive or negative changes in volume. To obtain a "true" reading of the volume change, you must subtract the control reading from the *Elodea* reading. Do this and enter the difference on the last line of table 26.2.

Move the lamp forward 25 cm to the 50 cm mark and let it sit for five minutes of equilibration. Adjust the fluid in the pipettes to 0.2 ml and repeat the experiment, recording the results in table 26.3. Calculate the difference between the experimental and control for this series.

Move the lamp forward to the 25 cm mark and let it sit for five minutes of equilibration. Adjust the fluid in the pipettes to 0.2 ml and repeat the experiment, recording the results in table 26.4. Calculate the difference between the experimental and control for this series.

Analysis

13▶ On one panel of the graph paper at the end of this lab topic, plot the corrected cumulative volumes as a function of time. Plot the data for each light intensity (as measured by distance between bulb and plant) on the same coordinates using different plotting symbols for each distance. Using a ruler, draw three straight lines that best fit the points for each treatment. The slopes of these lines are photosynthetic rates. Calculate the slopes at each of the light intensities. (Slope = ml oxygen produced per minute.)

14▶ On a second panel of the graph paper, plot these three slope values as a function of the distance of the plant from the light source. For point light sources, the intensity of the light at a given distance is inversely proportional to the square of the distance, that is $I = 1/d^2$. For an ordinary bulb at the distances in this experiment, the relationship is more nearly linear: $I = 1/d$. The place where the straight line joining the three points intersects the x-axis represents the predicted **light compensation point.** This is the light intensity (as measured by bulb to plant distance) at which oxygen release is zero. However, realize that photosynthesis is still producing oxygen but all of that oxygen is being consumed by aerobic respiration in the plant's cells.

Testing a Prediction

15▶ When the light is placed at the distance from the plants corresponding to the compensation point, the light intensity is such that there should be neither net consumption nor production of oxygen by the plant. Test this prediction (hypothesis) by placing the *Elodea* and control tubes at the compensation distance and measuring any changes that occur.

? Did you see any change? Does this mean that the plant is neither respiring nor photosynthesizing? Explain.

Demonstration of Starch Accumulation in Leaves (Optional)

As photosynthesis occurs and excess sugars are produced, starch accumulates in leaves. It is possible to see this starch accumulation by staining leaves with an I_2KI solution. Starch stains brownish purple.

Several days before the lab, *Nasturtium* or *Geranium* plants should be placed under constant illumination. On some of the plants, the leaves should be partially masked by tightly folding a 1″ strip of opaque construction paper around part of the leaf and securing it with a paper clip. A fun variation of this is to cut initials or letters in part of the paper so light can get through to the leaf. Other plants should be placed in a dark closet or cabinet for the same period of time. During the lab period, you or your instructor will obtain a leaf from each of the plants. If green and white variegated leaves are available, include them as a fourth factor. The edges of the leaves can be notched with one, two, three, or four V-shaped cuts for identification.

16▶ Place the leaves in boiling water for 5 minutes and then transfer them with forceps to hot 90% ethanol for a few minutes to remove the carotenoids and chlorophyll. Place each leaf in a dry petri dish. Spray the leaves with I_2KI solution. Describe the results.

Leaf from plant in *dark*

Leaf from plant in *light*

Leaf that was *masked* on plant in *light*

Variegated leaf from plant in *light*

Investigating Leaf Structure and Photosynthesis

Learning Biology by Writing

In this lab, you explored (1) the properties of chlorophyll and (2) the effects of light on the rate of photosynthesis. Your instructor will tell you which factors to include in your laboratory report.

To begin your report, outline the general purposes of your observations and experiments in a few sentences. Briefly describe the techniques you used. Include your results in the form of half-page graphs of any quantitative data that are to be included in the report. In the discussion, answer such questions as:

1. How does the chlorophyll absorbance spectrum relate to the expected action spectrum for photosynthesis?
2. What is the practical significance if the photosynthetic rate never rises above the compensation point?
3. How might you improve the experiments to gain more information?

As an alternative assignment, your instructor may ask you to answer the following Critical Thinking and Lab Summary Questions.

Internet Sources

As this lab demonstrates, chlorophyll is essential in photosynthesis. Despite its importance, there are many things that we do not know about it, and much research is being conducted to learn more. One type of research has to do with how chlorophyll is synthesized in plants, including what genes are involved. Use Google to search the WWW for information about chlorophyll synthesis and review the information available on a few of the sites. Write a short paragraph describing what type of research is being performed today on this topic.

Lab Summary Questions

1. Draw the anatomy of a leaf in cross section. Indicate the functions of the different cell layers found in a leaf.
2. Plot the data in table 26.1 and turn it in with this summary. Indicate with arrows on the x-axis the wavelengths of light most strongly absorbed by chlorophyll. See appendix B for directions on making graphs.
3. Plot the data on the last line of tables 26.2, 26.3, and 26.4 on one sheet of graph paper and turn it in with this summary.
4. Calculate the slopes of the three lines plotted in question 3 and graph the slopes as a function of the distance between the light source and the *Elodea*. Indicate the distance (really a light intensity) at which oxygen production equals consumption. Turn in the plot with this summary.
5. What happened when you placed *Elodea* at the light compensation distance from the light source?
6. Explain why a control tube was included in the light intensity experiments. What did it control for?

Critical Thinking Questions

1. During the chlorophyll extraction procedure, xanthophyll and carotene pigments were extracted and discarded. What are the advantages to the plant of having these accessory pigments?
2. Deep, clear tropical waters are dominated by Rhodophytes (red algae). Their color is derived from red pigments, phycobilins, that mask the chlorophyll. Why does this particular arrangement make them suited to their environment?
3. Discuss the adaptive differences as well as similarities you would expect to find in the structure of leaves taken from plants that lived in hot, relatively dry environments to those from wet, humid environments.
4. Ice and snow on lake surfaces in winter often reduces light intensity in the water below. If light intensity falls below the light compensation point, what is the implication for the ecosystem?
5. At night do green plants produce or consume oxygen? Why?

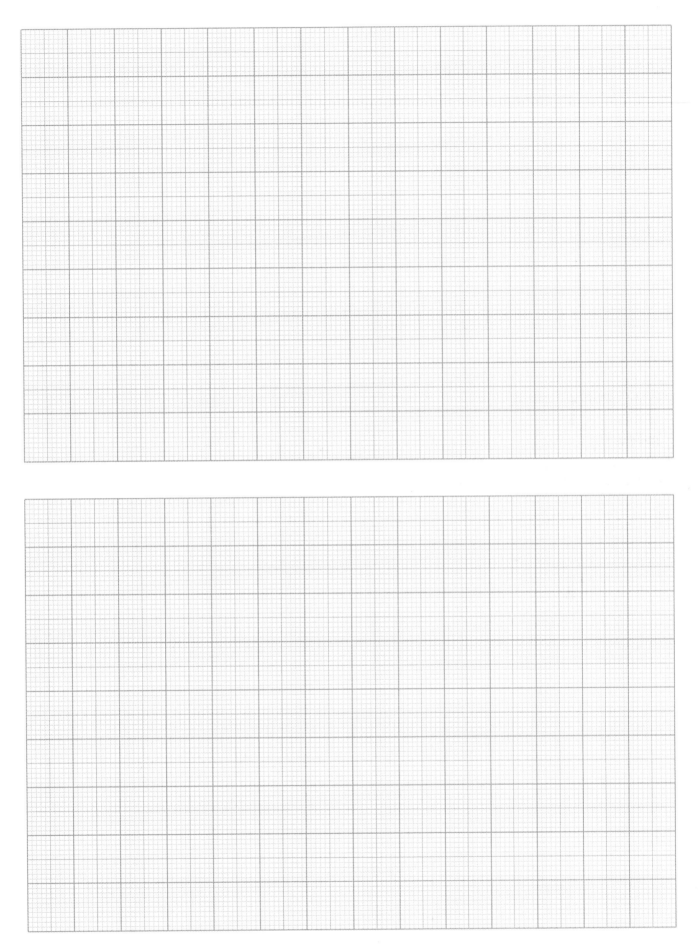

Investigating Leaf Structure and Photosynthesis

LAB TOPIC 27

Angiosperm Reproduction, Germination, and Development

Supplies

Preparator's guide available on WWW at
http://www.mhhe.com/dolphin

Equipment

Compound microscopes
Dissecting microscopes

Materials

Living flowers
  Flowers shedding pollen
  Six dicots varying in completeness and
    perfection
  Gladiolus or other large monocots
Tobacco hormone kit (optional)
Wisconsin Fast Plants (optional)
  Wild-type seeds
  Rosette seeds
  Growing system kit
Preserved flowers
  Oat inflorescences
  Corn tassel
Fresh fruits
  Apple, tomato, peach, bean, pepper, *etc.*
Soaked lima beans and corn kernels
Prepared slides
  Corn pistillate flower, longitudinal section
    (demonstration)
  Lily anther, cross section, developmental series
  Lily double fertilization, demonstration
  Lily eight-nucleate embryo sac, demonstration
  Lily ovule, megasporogenesis series
  Lily ovary, cross section
  Lily, stigma-style section
    showing growing pollen tubes
  Capsella embryo development
Miscellaneous
  Slides and coverslips
  Razor blades
  Dissecting needles
  10% raw sugar solution (do not use white or
    brown sugar)
  I_2KI stain in dropper bottles

Student Prelab Preparation

Before doing this lab, you should read the introduction and sections of the lab topic that have been scheduled by the instructor.

You should use your textbook to review the definitions of the following terms:

anther	meristem
antipodals	microspore
carpel	monocot
cotyledons	ovary
dicot	ovule
embryo sac	pistil
ethylene	polar nuclei
gibberellins	pollen
hypocotyl	radicle
integuments	seed coat
megaspore	synergids

You should be able to describe in your own words the following concepts:

How eggs are produced in plants
How sperm are produced in plants
How fertilization happens in plants
Double fertilization
Formation of embryo in seeds
Roles of plant hormones in regulating growth

As a result of this review, you most likely have questions about terms, concepts, or how you will do the experiments included in this lab. Write these questions in the space below or in the margins of the pages of this lab topic. The lab experiments should help you answer these questions, or you can ask your instructor during the lab.

Objectives

1. To observe how different flowers vary in their anatomy
2. To illustrate gamete formation and fertilization in flowering plants
3. To examine the development of a plant embryo in a seed and investigate fruit anatomy
4. To determine experimentally the role of hormones in regulating plant development and differentiation

Background

Flowering plants reproduce by two modes: asexually (vegetatively) by producing new individuals from an older individual's leaves, stems, or roots; and sexually with gametes, which give rise to an embryo contained in a seed. Both modes have their advantages.

Asexual, also called vegetative, reproduction produces robust progeny, which use energy reserves from their parents and thus may increase the chances for survival over those of delicate, sexually produced seedlings. Because asexually produced progeny are genetically identical to the parent, they also have the same genetic advantages that allowed the parent to reproduce asexually. Thus, a genetically superior individual may spread and multiply rapidly without risking genetic recombination inherent in sexual reproduction. Asexual reproduction can yield impressive results: a single aspen tree in the Rocky Mountains is estimated to have produced by repeated root sprouting 47,000 individual trees covering 200 acres. Cattails, prairie grasses, blueberries, blackberries, cherries, and strawberries (fig. 27.1) are examples of other species that are asexually active reproducers. Few annual plants reproduce asexually.

Many named cultivars of fruits and roses are propagated only vegetatively. For example, all McIntosh apple trees in the world today are traceable back to one tree that had superior fruit qualities. These offspring trees were first propagated by cutting branches from the original tree and grafting them onto roots of crab apple trees. Cuttings from subsequent generations have been used to produce the present number of trees. If you planted the seeds of McIntosh apples, would you expect the tree that grew to bear exactly the same kind of fruit as its parent? Why?

Figure 27.1 Asexual reproduction occurs by runner (stolon) formation in strawberries.

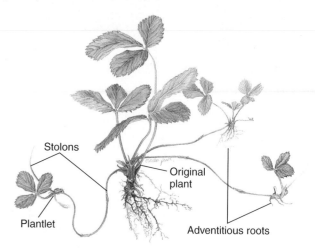

While the principal advantage of asexual reproduction is genetic constancy, the main advantage of sexual reproduction is genetic variability. It results from the meiotic production of sperm and egg with the built-in processes of crossing over and independent assortment, and the combination of genes from two parents at fertilization. No two individuals resulting from the fusion of gametes from the same parents are genetically alike. Each represents a unique combination of genes. These are tested by natural selection, thus increasing the probability of continued survival of the species over time during environmental changes. Many sexually propagating species also produce specialized seeds and fruits, which increases the possibility of dispersal to new locations.

The basic mechanisms in asexual reproduction are simple: cellular multiplication by mitosis and an ability of the new cells to differentiate into the tissues characteristic of roots, shoots, leaves, and other plant organs. Sexual reproduction is more complex. Spores are produced by meiosis. The spores develop into male or female gametophytes within the male or female parts of the flower. Pollen must be released to carry their sperm to the egg located in the ovule for fertilization to occur. Following fertilization, the development of the zygote into an embryo must be supported by the female parent. The seed containing the embryo must be dispersed and the seed must germinate. Moreover, all these stages of sexual reproduction must be coordinated with the seasons.

Sexual reproduction in flowering plants, in comparison to mosses and ferns (lab topic 16), is highly adapted to the terrestrial environment. Unlike the lower plants, water is not necessary for the sperm to reach the eggs. Following fertilization, zygotes develop surrounded by maternal tissue and are protected from drying. Mature embryos are encased in a seed coat and often remain viable for years in relatively dry environments. Some desert annuals, for example, grow only during intermittent wet periods. When it rains, the seeds germinate, grow, flower, and produce new seeds in the short period

TABLE 27.1 Differences between monocots and dicots

	Monocot	Dicot
Characteristic root	Fibrous	Tap
Arrangement of stem vascular tissue	Scattered bundles	Bundles in ring
Presence/absence of wood	Absent	Present or absent
Pattern of leaf veins	Parallel as in grass	Net as in oak or maple
Number of floral petals/sepals	Multiples of 3	Multiples of 4 or 5
Number of cotyledons in seed	One	Two
Pattern of growth	Primary only	Primary and secondary

of time when life-sustaining moisture is available. During the dry periods, no mature plants are present and the species survives as seeds that remain dormant until the next rain.

The morphological details of plant reproduction have been known for well over a hundred years. However, the mechanisms involved are still subjects of intense investigation, especially now that various molecular techniques are available. For example, many plants have both male and female organs; *i.e.,* they are hermaphrodites. A mechanistic question is what prevents them from self-fertilizing and becoming inbred? Modern work has shown several genes are involved in preventing this. Pollen falling on the female portions of flowers on the same plant do not germinate because there are genetic recognition factors that prevent it.

Another question has to do with pollen navigation. Pollen grains are microscopic, yet they often must produce a pollen tube over an inch long to deliver their sperm to the egg. This involves growth through female tissues. How does the pollen tube "know" where to go? Answering a question like this gets at the most fundamental of questions in biology, cell-to-cell communication.

Embryogenesis, the formation of the embryo from the zygote, accompanies the development of the seed. To achieve this multicellular stage, two developmental organizations must occur: a longitudinal apical-basal pattern and a radial pattern to produce concentrically arranged tissue systems. Over 50 genes have been identified that affect the development of these patterns, but there is much work to be done before we understand this most fundamental mechanism of plant development.

A last point of interest is the suspended state of animation (if you can say that for plants) that occurs in seeds. How is it that seeds can survive for such long periods of time and then begin an intense period of development? These are but a few questions that plant biologists are attempting to answer.

The intent of this lab is to allow you to look at several of the morphological aspects of plant reproduction and development. As you do so, think of these questions and others like them. It is wondering *why* that starts research projects and careers in science.

Sexual Reproduction

Flower Structure

Flowers contain reproductive gametophyte stages (pollen and embryo sac) of plants in the division Anthophyta. Traditionally, botanists divided the flowering plants into two taxonomic classes: the monocotyledons (65,000 species) and the dicotyledons (170,000 species). Table 27.1 summarizes the differences between these two groups. Recent genetic analysis indicates that the dicots should be subdivided into four clades based on genetic similarities. For simplicity, the monocot/dicot classification will be used here.

Review the life cycle of an anthophyte shown in figure 17.10 before you start your lab work.

Flowers are the reproductive organs of anthophyte sporophytes. Within the tissues of the flower, the microscopic male and female gametophyte stages are produced. You will first look at the sporophyte structures in flowers and then will use your microscope to study the tiny gametophyte stages.

A flower is essentially a short modified stem consisting of several nodes with very short internodes. Modified leaves at the nodes make up the floral parts.

Gladiolus

1 ▷ Obtain a fresh *Gladiolus* flower from the supply area. Using figure 27.2 as a guide, identify the following major parts:

1. **Calyx**—the outermost part of the flower made up of whorls of individual modified leaves, the **sepals.** May be green or colored as in *Gladiolus*.

Figure 27.2   Longitudinal section of a flower showing reproductive structures. Eggs are contained in ovules inside the ovary. Fertilization occurs when a pollen tube grows (from pollen landing on stigma) through style tissue to where it enters the ovule via the micropyle.

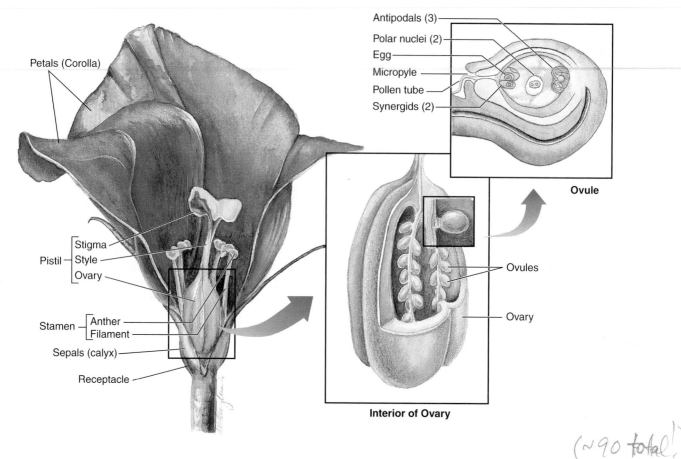

Interior of Ovary

2. **Corolla**—the usually colored part of the flower made up of whorls of modified leaves called **petals.** Sometimes the petals are fused as in the petunia. The calyx and corolla are collectively known as the **perianth.** In plants pollinated by animals (insects, birds, or mammals) the bright colors are an attractant and are a good example of coevolution.

3. **Stamen**—modified leaves composed of an **anther** on the end of a thin **filament.** Pollen develops in the anthers.

4. **Pistil**—sometimes called the carpel; a modified leaf or fused group of leaves that consist(s) of the terminal **stigma** where pollen attaches, the supporting **style,** and a basal **ovary** containing **ovules** in which eggs develop and are fertilized.

5. **Receptacle**—expanded tip of flower stalk that bears all four of the other floral structures.

Examine a stamen closely. Each anther consists of a **pollen sac** with a line of dehiscence (splitting) separating the lobes. Open a pollen sac by running the tip of a needle along this line and squeezing out the contents into a drop of water on a microscope slide. What does it contain? _pollen grains_

Take a razor blade and make a longitudinal section of the pistil. How many ovules are visible in the ovary? _~30 per carpel_ _(~90 total!)_

The flowers of dicots tend to have their parts arranged in multiples of fours or fives, such as five sepals, five petals, ten stamens, and five carpels. The flowers of monocots are based on multiples of three. Is a *Gladiolus* a monocot or dicot? _monocot_

Diversity of Flower Structure

This section is designed to introduce you to variations in flower structure.

A flower, such as the one you just studied, that has all four types of modified leaves (sepals, petals, stamen, and pistil) is known as a **complete** flower. Flowers that lack one or more of the four parts are **incomplete.**

Perfect flowers contain both stamens and pistils. Perfect flowers are bisexual. **Imperfect** flowers are a special case of incomplete flowers and lack either pistils or stamens. Flowers that contain only stamens are **staminate,** or male flowers; flowers that contain only pistils are **pistillate,** or female flowers. Imperfect flowers are unisexual.

Some plants will have individuals that bear only staminate flowers and others that bear only pistillate flowers.

TABLE 27.2   Diversity in dicot flowers

	Completeness	Perfection	Monecious or Dioecious	Name
1	Complete	Perfect	Monecious	Shamrock
2	Incomplete (no sepals)	Perfect	Monecious	Lily
3	Incomplete (no petals - bracts)	Pefect	Monecious	Anenome
4	Complete	Perfect	Monecious	Stock
5	Incomplete (no sepals?)	Perfect	Monecious	Tulip
6	Incomplete (no stamens)	Imperfect (no stamens)	Dioecious	Orchid

These species are said to be **dioecious** (of two households). You may have noticed that some holly or mulberry or ash trees do not produce fruits. The reason for this is that these species are dioecious, and only the female plants will produce fruits. Species having both staminate and pistillate flowers on one individual are said to be **monecious** (of one household).

2 ▶ Six flowers from different species are available in the lab. Each type is numbered. Examine these flowers and indicate in table 27.2 whether they are complete or incomplete and perfect or imperfect. Your instructor will give you the names of the flowers. (wheat is spike)

Grass Flowers

The flowers of grasses are highly specialized. Most grasses have flowers arranged in clusters called **inflorescences** (fig. 27.3). Two arrangements are found: the **spike**, in which the main axis is elongated and the flowers are directly attached; and the **panicle**, in which the main axis is branched with the branches bearing flower clusters. Because grasses are wind-pollinated, they lack the showy petals that attract pollinators. The grasses known as cereals and grains are a chief part of the human food supply and include corn, wheat, rye, and oats as well as sorghum and sugarcane.

3 ▶ Obtain an inflorescence of an oat from the supply area. What type of inflorescence do oats have? _panicle_

In oats, sepals and petals are missing and the flower is encased in modified leaves, generally known as **bracts.**

Individual flowers are called **florets** and are arranged in groups on a spikelet. Two bracts, called **glumes,** are found at the base of each oat spikelet and enclose several developing florets.

4 ▶ Look at a single floret with your dissecting microscope. It is enveloped by two other bracts, one smaller and the other larger with a long extension called the **awn.** At the base of the bracts are two protuberances, the **lodicules.** They swell and open the protecting bracts, as the floret matures.

5 ▶ Spread the bracts and find the three stamens. They are arranged around a central pistil that bears two feathery **stigmas.** Cut off the stigmas from a pistil and make a wet-mount slide. Examine the stigma with your microscope and draw it below. How is its form related to its function? The "hairs" can catch pollen grains. the

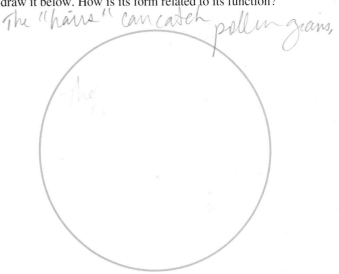

Pollen released by the stamens is air blown. The feathery stigmas entrap pollen, and a pollen tube grows down to the single **ovule** in the ovary at the base of the pistil. The seed develops in the ovary and is harvested as grain.

Corn differs from other grains because its flowers are imperfect. A single plant bears both male and female flowers. The staminate flowers are borne at the top of the plant as the **tassel** and release their pollen in the wind. The pistillate flowers are located lower on the plants at nodes and eventually develop into the **ears.** The **silk** found in ears is what remains of the styles that extended from every egg. Pollen that fell on the ends of these styles had to develop pollen tubes that were several centimeters long. Look at the

6 ▶ demonstration material of a corn tassel and a longitudinally sectioned pistillate inflorescence.

Figure 27.3 Structure of the inflorescence of oats: (*a*) panicle-type inflorescence; (*b*) side view of spikelet; (*c*) structure of a single floret.

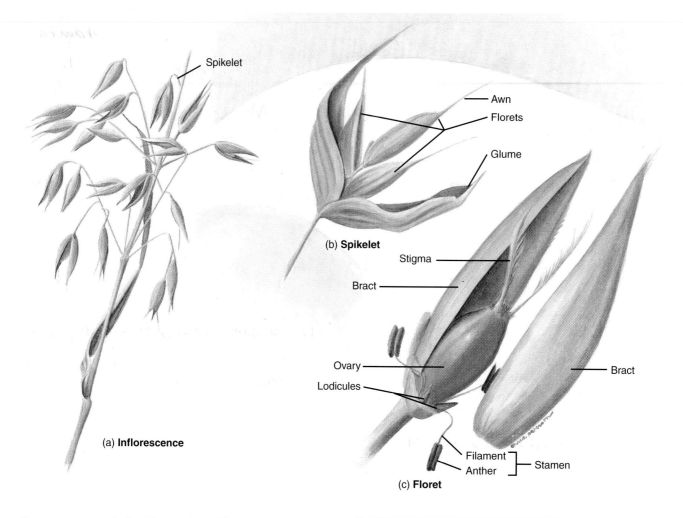

Spikelet

Awn

Florets

Glume

(b) Spikelet

Stigma

Bract

Ovary

Lodicules

Bract

Filament

Anther

Stamen

(c) Floret

(a) Inflorescence

Gametogenesis in Flowering Plants

As in all plants, flowering plants have an alternation of generations. The flower and plant that bears it represent the **sporophyte generation.** The gametophyte stage in flowering plants is microscopic and is embedded in the tissues of the sporophyte stage for protection and nutrition. Consequently, gamete formation does not immediately follow meiosis, as it does in animals. Instead, meiosis gives rise to haploid **microspores** in the anther or a haploid **megaspore** in the ovule. These haploid cells divide by mitosis to produce a small number of cells, which become either the male or female **gametophyte generation.** A pollen grain represents the male gametophyte. The female gametophyte is the embryo sac enclosed in the ovule of the parent.

Pollen Formation

7▷ Stamens are sometimes refered to as male organs in flowers although the anther and filament are sporophyte tissue and without sex. The male gametophyte develops inside the pollen grain. Remove a stamen from your *Gladiolus* flower and use a razor blade to make a cross section of the anther in a drop of water on a microscope slide. Observe

Figure 27.4 Photomicrograph of a cross section of microsporangia in an anther where pollen develops.

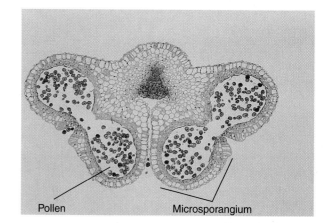

Pollen

Microsporangium

the section under low power with your compound microscope (fig. 27.4).

The four cavities you see are part of the four lengthwise chambers of the anther and are often called pollen sacs, or **microsporangia** (organs producing microspores).

8 Now get a prepared slide of a cross section of a lily anther. Other species can be substituted. Observe it with the high-power objective and note the pollen grains. Sketch the section and pollen.

Your instructor may set up a series of demonstration microscopes with sections of anthers at different stages of development. Look at these to see the sequence of events leading to mature pollen. In young microsporangia, **microspore mother cells** undergo meiosis to produce four haploid **microspores.** In the youngest anthers, the cavity will not be developed, although the outline of the microsporangium will be apparent. The larger cells lining the sporangium are called the tapetum and supply nutrients to the more centrally located microspore mother cells. In intermediate age anthers, the microspore mother cells will have completed meiosis I and some will have completed meiosis II so that clusters of two and four cells will be seen. Those in the four-cell clusters are microspores. *Lilium* has a diploid (2N) chromosome number of 24. How many chromosomes are in a microspore?_____

The microspores develop heavy walls as they mature into pollen. Maturation involves the synthesis of a heavy outer wall containing the polymer sporopollenin that protects the cells inside from drying. The haploid nucleus in each developing pollen grain divides by mitosis without cytokinesis to produce two nuclei, the **tube nucleus** and the **generative nucleus.** The generative nucleus divides again by mitosis to produce two male gametes or sperm nuclei. These nuclei represent a total of three cells and taken together represent the entire male gametophyte stage of the life cycle. These events may occur in the anther or after the anther splits releasing pollen.

Depending on the species, pollen grains are carried by either wind or animals to the stigma of a pistil. There a pollen grain germinates, producing a pollen tube, which grows through the tissues of the pistil to the ovule and fertilizes the egg (fig. 27.5).

Figure 27.5 Germinating pollen grain showing nuclei in emerging pollen tube.

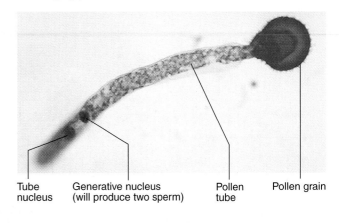

Tube nucleus · Generative nucleus (will produce two sperm) · Pollen tube · Pollen grain

Prepared slide

9 Obtain some pollen from fresh flowers and sprinkle the pollen onto a small drop of 10% raw sugar on a microscope slide. Add a coverslip and observe the specimen at intervals over the next 60 minutes. If pollen tubes do not appear, go to the supply table and get a prepared slide of pollen tubes. Two haploid nuclei will travel down this tube just before fertilization. They are the male gametes of the flower. Sketch a pollen grain and its tube below.

Egg Formation

10 The pistil is sometimes refered to as the female organ in a flower. This is not the case. The pistil is sporophyte tissue and is without sex. The female gametophyte stage develops within the ovules contained in the ovary. Return to the *Gladiolus* flower bud and cut a thin cross section of the pistil at the level of the ovary. Make a microscope slide by mounting the slice in a drop of water and adding a coverslip. Observe the slice using the scanning objective. Note that the specimen is a compound pistil made up of three

Figure 27.6 Structure of a lily's ovary: (a) photomicrograph of a cross section of an ovary; (b) structure of an ovule, containing the megasporangium in which the multicellular female gametophyte develops from the single megaspore mother cell; each gametophyte eventually produces one egg.

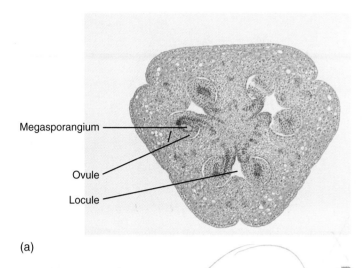

Megasporangium

Ovule

Locule

(a)

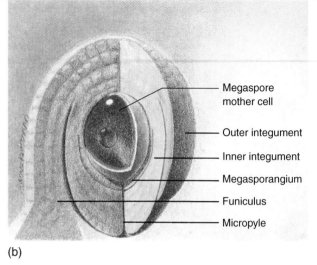

Megaspore mother cell

Outer integument

Inner integument

Megasporangium

Funiculus

Micropyle

(b)

fused **carpels** (see figure 17.11). Sketch the general structure of the pistil in cross section.

11

ovules

one carpel

Identify the three **locules,** or cavities of the ovary, and the two rows of ovules in each locule. The female gametophyte develops inside the ovule (fig. 27.6). During the evolution of flowering plants, natural selection favored plants in which the carpels folded inward to form the sealed ovarian structure now found in flowering plants. What is the advantage of an internally located ovule?

Obtain a prepared slide of a cross section of a lily ovary and look at it through your compound microscope. Identify an ovule and its embryo sac. Sketch them below.

Depending on the species, an ovary may contain one or more ovules. Each ovule contains a **megasporangium** (also called a nucellus) that will produce a single megaspore by meiosis. It develops into a microscopic, but multicellular, female gametophyte. Two integuments grow from the ovarian wall to surround the megasporangium except at a small region known as the **micropyle** (fig. 27.6b). The ovule is attached to the ovarian wall by a short stalk, the **funiculus.**

Within the megasporangium, the diploid **megaspore mother cell** divides by meiosis to produce four haploid megaspores (fig. 27.7). In most flowering plants, three of these die, and the fourth increases greatly in size. Its nucleus divides three times by mitosis to produce eight haploid nuclei. This large multinucleate cell is called the **embryo sac** and is the **female gametophtye** stage of the flowering plant.

Figure 27.7 Events in formation of a mature embryo sac, the female gametophyte: (a) through (e) show stages in development of female gametophyte; (f) events of fertilization in mature ovule; (g) embryo develops from zygote formed by fusion of egg and sperm nuclei and endosperm from fusion of two polar nuclei and second sperm nucleus.

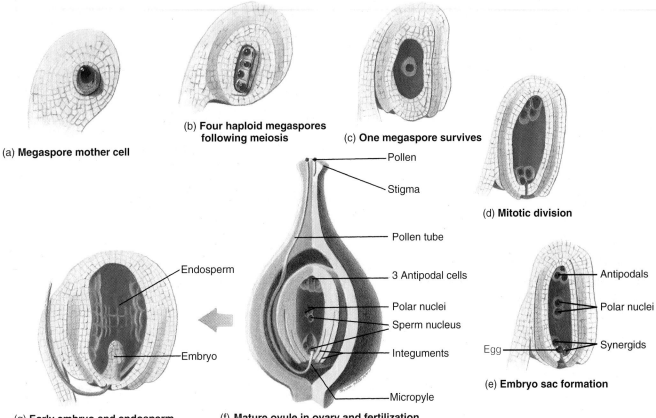

(a) **Megaspore mother cell**

(b) **Four haploid megaspores following meiosis**

(c) **One megaspore survives**

(d) **Mitotic division**

(e) **Embryo sac formation**

(f) **Mature ovule in ovary and fertilization**

(g) **Early embryo and endosperm**

12▶ Look at the demonstration slide of the eight-nucleate embryo sac. Be sure to count the nuclei. Draw the sac below.

While there is no one pattern of embryo sac (female gametophyte) maturation among all flowering plants, the following provides a reasonably general description of events in most flowers (fig. 27.7). The eight haploid nuclei in the embryo sac are arranged in two groups of four. One nucleus from each group of four migrates to the center of the cell. They are termed **polar nuclei.** Cell membranes now form around each of the three remaining nuclei in both groups, yielding a seven-cell stage. The **central cell** is a large binucleate cell, and the remaining six are small mononucleate cells at each end of the embryo sac. Those at one end, called **antipodals,** soon disintegrate. Of those at the micropylar end, one becomes the functional **egg** and the other two, called **synergids,** disintegrate.

Fertilization

13▶ Obtain a slide of a longitudinal section of a lily stigma and style with growing pollen tubes.

When pollen falls on the surface of the stigma, a sugary secretion stimulates it to germinate. The pollen tubes penetrate the loosely arranged tissue of the style and grow to the ovule. The micropyle provides the pollen tube with an entryway into the ovule. How the pollen tube navigates from the stigma to the micropyle is an interesting question. The loosely arranged tissue in the style allows this. What cannot be seen is what guides it. Are there chemical markers along the way, and if so how do they influence growth? It is questions like this that plant biologists are only beginning to work on.

Two haploid nuclei, termed **sperm,** pass down the pollen tube. One enters the egg and fuses with its nucleus to form a diploid zygote. The other enters the central cell where it eventually fuses with the two polar nuclei to form a triploid central cell. It develops into **endosperm,** a tissue that supplies nutrients to the developing embryo.

Flowering plants are unique in having **double fertilization;** both embryo and endosperm receive genetic material from two parents. The developing complex of embryo, endosperm, and embryo sac in the ovule become the seed of the plant. The ovarian wall becomes the flesh of the plant's fruit.

14▶ Slides showing fertilization in plants are expensive because they are difficult to prepare. Nonetheless, this is important to see and a demonstration slide will be available for you to view. Make a sketch of the embryo sac, micropyle, pollen tube, and male and female gametophyte nuclei.

Embryo Development

15▶ Obtain a slide of the fruit of the common weed shepherd's purse *(Capsella)* and look at it through your compound microscope. Several seeds are contained in the fruit. Within each seed, the developing embryo occupies most of the volume. Figure 27.8 shows some of the developmental stages of the embryo in the seed. Depending on its stage of development, your slide may contain only some of these stages.

Following fertilization, both zygote and endosperm grow and divide. The zygote divides by mitosis to produce a filament of cells called the **proembryo.** Cells at the base of the filament, located at the micropylar end, become **suspensor cells.** Their growth pushes the filament into the nourishing endosperm. Cells at the opposite (chalazal) end of the filament divide to produce at first a **globular** embryo stage.

Continued differential growth and division of the embryo lead to the appearance of characteristic regions. The globular mass of the embryo becomes **heart-shaped** as the seed leaves or **cotyledons** develop. The name *dicot* refers to the fact that these plants have two cotyledons. Monocots have only one. The cotyledons may accumulate food materials from the endosperm and become thick and fleshy as in peas or beans where the cotyledons represent the bulk of the mature embryo. In other species, such as wheat and corn, the cotyledons may be small and thin and the endosperm persists and provides starch, proteins, and lipids to the developing embryo. It is this biochemical content that makes them excellent foods.

In a mature dicot embryo, the **shoot apex** (= plumule) is seen as a small node of tissue between the cotyledons. The meristematic regions located here produce the shoot. Beneath the shoot apex and cotyledons is a short embryonic stem, the **hypocotyl.** In many plants, but not all, the hypocotyl elongates at germination. As it does, it forms a characteristic, inverted U-shape with the base in the seed and the tip attached to the cotyledons. When the curve of the U breaks the soil surface, it begins to straighten, pulling the cotyledons up. This mechanism avoids pushing the cotyledons and shoot apex through abrasive soil, which could injure the tissues. As the hypocotyl straightens, the new shoot begins to grow from the shoot apex. It connects with an embryonic root, the **radicle.** A root apical meristem located at the tip of the radicle will divide to produce new root cells at germination.

Coordinate your observations with others in the lab so that you see several developmental stages: filament; globular; heart-shaped; mature dicot.

Seeds

16▶ Obtain a lima bean that has been soaked overnight to soften the seed coat and hydrate the tissues. Find the **hilum** where the ovule attached to the funiculus (fig. 27.9). At one side of the hilum, the micropyle should be visible as a small distinct aperture. Peel away the seed coat to find the two **cotyledons** of the embryo. Separate the cotyledons and examine each half to see the other parts of the embryo. You should see the **shoot apex, radicle, and the hypocotyl.** Is the lima bean a monocot or dicot? _dicot_

17▶ Obtain a soaked corn kernel and cut it lengthwise at right angles to the broad side of the seed (fig. 27.10). Place a drop of I_2KI stain on each half. Starch will stain blue and the embryo will stand out against the background. Examine the halves to find the embryo in the base and on one side of the kernel. Note the large amount of endosperm in the kernel. A corn kernel is really a fruit consisting of a single seed to which the ovary wall has grown fast. The covering is called the **pericarp.**

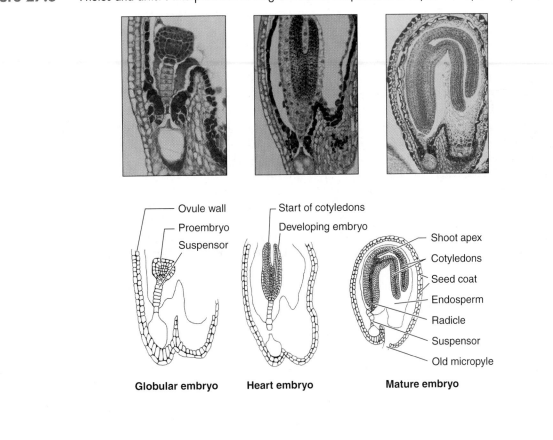

Globular embryo   Heart embryo   Mature embryo

Figure 27.9   Structure of a bean seed.

Figure 27.10   Structure of a corn seed.

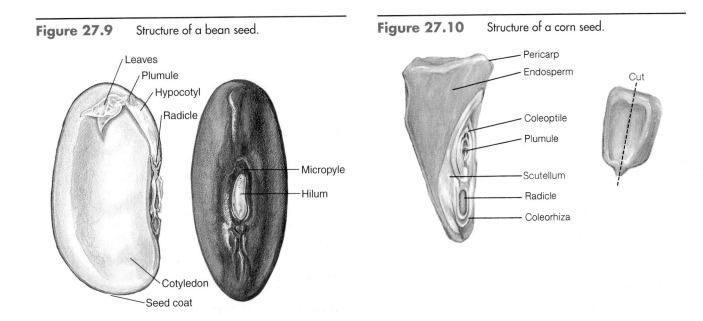

Figure 27.11 Development of a pear: (a) flowers; (b) early stage of fruit development showing ovarian wall enlargement; (c) ripe pears.

(a)

(b)

(c)

18▶ Draw the cut surface of the corn. Label the shoot apex, the surrounding sheath (coleoptile), the single cotyledon, which in grains is called the **scutellum,** the radicle and its enclosing sheath, the **coleorhiza,** the endosperm, and pericarp.

One of the factors that has allowed flowering plants to be the dominant, most diverse group of land plants is their capacity for wide dispersal. Fruits promote dispersal. Animals eat many fleshy fruits and seeds. The tough seed coat protects the embryo as it travels through the digestive tract. When the animal defecates, the seeds have been transported to a new location.

Some fruits, such as those of maples and ash, are winged and are transported by the wind. Others are spiny, such as cocklebur and beggar tick, and stick to the coats of mammals and birds. The variety of fruit types can be overwhelming (fig. 27.12). Botanists have devised a classification system that is outlined in table 27.3.

20▶ Look at several of the fruits available in the lab and assign them to the appropriate type. Make a list below of the fruits studied. *Cinquefoil – ? Achene.*

Apple – Pome
Plum – Drupe
Date – Berry
Sunflower – Achene
Almond
Hazelnut } *nuts*
Brazil nut
Kiwi – Aggregate
Blackberry – Aggregate
Orange – Hesperidium (Berry)
Ash – Samara
Eggplant – Pepo (Aggregate)

Fruits

Corn, pea pods, berries, watermelons, and nuts are all examples of fruits because they develop from the ovary of a flower and surround seeds. Vegetables, such as carrots, cabbages, and celery, differ from fruits in that they are not derived from the flower but develop from other parts of the plant.

The flesh of fruits develops from the ovary wall, the **pericarp,** which may thicken and differentiate into layers (fig. 27.11). Depending on the species, these layers may or may not be easy to distinguish.

19▶ Examine several fresh fruits, such as pears, plums, or apples, that have been cut in half. Do you see mature ovules (seeds)? The fleshy part is from the ovarian wall or the receptacle, depending on the species.

372 Angiosperm Reproduction, Germination, and Development 27-12

Green bean – legume

Figure 27.12 Examples of different types of fruits.

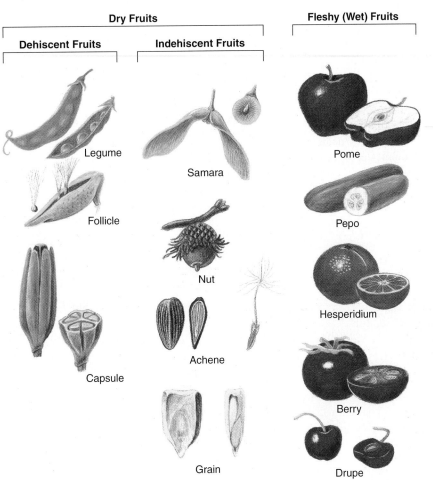

Dry Fruits

Dehiscent Fruits

Indehiscent Fruits

Fleshy (Wet) Fruits

Legume

Follicle

Capsule

Samara

Nut

Achene

Grain

Pome

Pepo

Hesperidium

Berry

Drupe

Kind of Fruit	Description	Examples
Simple	Develop from the ovary of a flower with single pistil	
Dry		
Achene	Fruit does not split open at maturity; seed remains attached to ovarian wall by stalk	Sunflower
Follicle	Fruit splits along one edge at maturity to release seeds	Milkweed, peony
Legume	Fruit splits along both edges at maturity to release seeds	Bean, carob, lentil, pea, peanut
Nut*	Hard ovarian wall encases a single seed and does not split open at maturity	Hazelnut, hickory nut
Grain	Seed coat is permanently united with ovarian wall; does not split open at maturity	Barley, oats, rice, wheat
Samara	Ovarian wall forms "wings" that aid in dispersal	Maple, elm, ash
Fleshy		
Berry	Entire ovarian wall is fleshy	Tomato, grape, date, avocado
Drupe	Outer zones of ovarian wall become fleshy, while inner zone is hard	Cherry, peach, plum, olive
Pome	Inner zone of ovarian wall is leathery and outer is fleshy incorporating accessory structures	Apple, pear
Aggregate	Develop from single flowers with multiple pistils	Raspberry, strawberry

TABLE 27.3 Classification of fruits by types

*Many so-called nuts are not true nuts; the peanut is a legume and walnuts, pecans, and cashews are drupes which lose the fleshy part during maturation.

Investigating the Effects of Hormones on Plant Development

Plant development is influenced by external factors such as water, mineral and light availability, and by internal factors, especially plant hormones. These relatively small, organic molecules play a major role in the regulation of plant growth by altering the expression of genes that lead to a change in the structure and function of the affected target cells. Five classes of plant hormones have been described: auxins, cytokinins, ethylene, abscisic acid, and gibberellins. **Differentiation,** the developmental process by which a relatively unspecialized cell undergoes a progressive change into a more specialized cell with specialized functions, is known to result from the influence of the ratio of the amounts of various hormones in the plant. In this section of the lab, you will experimentally determine some of the effects of hormones on plant growth.

Effects of Hormones on Seedling Growth Patterns

Wisconsin Fast Plants, developed by the plant breeder Paul Williams at the University of Wisconsin, go through their life cycle in about 40 days, from seed germination to production of plantable seeds, making them ideal for classroom use. These plants are in the genus *Brassica,* which is in the mustard family. Cauliflower, broccoli, rape seed, and turnips are also in this genus. You will use these fast plants to investigate the effects of hormones on plant development.

Brassica seeds with different genetic backgrounds and phenotypes can be ordered from biological supply houses. Wild-type plants have normal phenotypes and the plants grow several inches tall. A mutant strain of *Brassica* produces rosette plants that are very short because there is little elongation of the cells in the internodes. The nodes of the stem are adjacent to each other and the leaves that grow from the nodes give the plant a rosette appearance.

Plant breeders who first observed this hypothesized that the rosette plants were mutants that could not produce a class of plant hormones that cause stem elongation, the gibberellins. Your assignment is to conduct an experiment that tests this hypothesis. To do so, you should know that gibberellins can be absorbed by the leaves and the shoot apex. The gibberellins will be conducted by the vascular tissues throughout the plant.

21▷ The experiment may be started in one of two ways. One week before you are to administer the gibberellin you may be asked to plant seeds of wild-type and rosette *Brassica* in special quad pots so that plants will be available at the start of the hormone lab. Alternatively, your instructor may plant the seeds of wild-type and rosette *Brassica* eight days before the hormone lab.

When plants are available, carefully observe both the wild and mutant forms and describe them below. Explain why the mutant is called rosette.

Wild-type

Mutant type (rosette)

22▷ Your instructor will assign to you a quad of four plants which will include two wild-type and two rosette mutants. Measure the height of all plants in millimeters from the cotyledons to the very tip of the plant and record in table 27.4.

If this effect is due to a deficiency of gibberellic acid, what would you hypothesize would happen if you administered gibberellic acid to the plants?

Gibberellic acid can be administered by dissolving it in water and applying a drop of this solution to the leaves of the plant every other day. The gibberellic acid will be transported throughout the plant. Using your plants, create an experimental design with controls that investigates the effects of gibberellic acid on normal and mutant plants.

23▷ Your instructor will give you distilled water and a solution of gibberellic acid. Using a different pipette for each solution, apply single drops to all of the leaves on your plants. This procedure will have to be repeated every other day for the next week.

After at least seven days, measure the length of the stem between the cotyledons and shoot tip and record in table 27.4.

When all students have completed their individual measurements, the data should be put on the blackboard and an average calculated for each treatment. Record these

TABLE 27.4 — Height in millimeters of plants in hormone experiment

	Wild-Type Plants:		Rosette Plants:	
	Water Treated	**GA Treated**	**Water Treated**	**GA treated**
Your data				
Start				
End				
Class data summary				
Average height change				
% change (over time)				

in the bottom half of table 27.4. Now calculate the percent change in height during the week of the experiment for each of the treatments.

Restate your hypothesis and decide whether to accept or reject it based on your data.

will be identical except for the ratio of the concentrations of two types of hormones, auxins, and cytokinins. One medium will have more auxin than cytokinin, another more cytokinin than auxin, and the last will have equivalent amounts. These media will be in tubes labeled *A, B,* and *C,* but you will not know which tube contains which hormone. Read the section in your textbook that describes the effects of these hormones on cellular differentiation. Hypothesize what will happen if you place callus tissue on each of the three media. State your hypotheses below.

Differentiation in Plant Tissue Culture

This experiment is designed to demonstrate the totipotency of plant cells: the capability of plant cells to develop into any structure found in the plant. In testing for totipotency, you will determine that hormones play an important role. The experimental material will be tobacco cells that have been raised in culture producing a mass of undifferentiated cells called a **callus.** Callus cultures can be prepared in the lab or purchased from biological supply houses that grew the cells by taking a small sample of parenchyma cells from the stem of a tobacco plant and putting them into a growth medium containing vitamins, various salts, hormones, and sucrose as an energy source. The growth process takes weeks.

24▶ Your instructor will describe how to handle the tobacco callus. Strict sterile technique must be observed because fungal spores or bacteria can enter the culture and overgrow the callus destroying it. You will have to use sterile instruments, glassware, culture media, and work in a sterile transfer hood in order to be successful.

You will have available to you three variations of a growth media formula made up in agar. The three tubes

25▶ Your instructor will demonstrate how to use sterile techniques to set up your cultures. Use these techniques to divide the callus tissue you are given into thirds. Place one piece in each of three culture tubes containing different growth media formulations. These cultures will now be incubated for four to five weeks and you should observe them weekly during this time and make notes describing any differences that you see among them. At the end of the growth period, tell your instructor which tube had the high auxin concentration relative to cytokinin, which was low auxin, and which was equal. Support your answer with your experimental observations and cite your textbook indicating the known effects of these ratios of hormones.

Learning Biology by Writing

Because the activities in this lab were both observational and experimental, two alternative assignments are described. Your instructor will indicate which one to complete. To summarize your plant anatomy observations, write a brief essay describing the male and female gametophytes of a flowering plant. Indicate how double fertilization occurs and the advantages of seed as a dispersal mechanism. To summarize your experimental work with either tissue culture or whole plants, write a lab report in which you state the hypothesis that was investigated, describe your techniques, present your data, and come to a conclusion regarding your hypothesis.

Internet Sources

Plant tissue culture is becoming a standard technique in the propagation of plants because it is possible to clone individuals with desirable traits or, in some cases, it is possible to insert desirable genes into species of plants that never before had those genes. Search the World Wide Web for information on "plant tissue culture," using the quotes to limit your search to this exact phrase. Read the information available at a few of the sites found and summarize what is happening in this exciting field of research. Include the URLs for future reference.

As an alternative assignment, your instructor may ask you to turn in answers to the Lab Summary or Critical Thinking Questions that follow.

Lab Summary Questions

1. Describe the differences between monecious and dioecious species of flowering plants.
2. Describe how the male gametophyte stage of a flowering plant develops.
3. Describe how the female gametophyte stage of a flowering plant develops.
4. What is the difference between pollination and fertilization?
5. Describe fertilization in a flowering plant.
6. Outline the steps in endosperm formation.
7. A seed has been described as a mature ovule. How can this be true? Explain you answer.
8. Draw a four-stage developmental sequence of a flowering plant embryo showing a proembryo, globular, heart-shaped, and mature embryo inside a seed.
9. What evidence do you have that supports the hypothesis that plant hormones influence differentiation in plants?

Critical Thinking Questions

1. Make a list of several vegetables that are actually fruits. In making this list, what was your definition of a fruit?
2. If wind and insects carry pollen from several flower species at the same time, why doesn't the pollen of one species fertilize the eggs of a second species, thus forming hybrids?
3. The pistil is often described as the female part of the flower and the stamens the male part. Which structures actually produce the egg and sperm?
4. Why do you suppose flowers are fragrant and colorful?
5. Why are fruits often colorful and sweet?
6. Many fruits are harvested before they ripen and then are treated with ethylene gas just before they are shipped to stores. Why?

INTERCHAPTER

Investigating Animal Form and Function

In this section of the lab manual are a series of topics on the organ systems of animals. In each topic you are asked to dissect a fetal pig as a representative mammal. Some general instructions are given here to orient you to proper dissection technique.

Vertebrates

General Dissection Information

Fetal pigs are unborn fetuses taken from a sow's uterus when she is slaughtered. When the mother dies, the fetuses do as well. They are a by-product of meat preparation and are used in teaching basic mammalian anatomy. Often the circulatory system has been injected with latex so that the veins will appear blue and the arteries red.

Strong preservatives are used and can irritate your skin, eyes, and nose. Rinse your pig with tap water to remove some of the preservative to lessen irritation. If you wear contact lenses, you may want to remove them during dissection, since the preservative vapor can collect in the water behind the lens and be very irritating. If a lanolin-based hand cream is available, use it on your hands before and after dissection to prevent drying and cracking of the skin. Alternatively, your instructor may ask you to purchase rubber globes and use them during this and subsequent dissections.

You will use this same fetal pig in several future labs to study the circulatory, excretory, and muscular systems. This means that you must do a careful dissection each time so that as many structures as possible are left undamaged and in their natural positions. Two good rules to keep in mind as you dissect your animal are: *cut as little as possible and never remove an organ unless you are told to do so.* If you indiscriminately cut into the pig to find a single structure without regard to other organs, you will undoubtedly ruin your animal for use in future laboratories.

Many of the instruction for dissection in this lab and later ones use anatomical terms to indicate direction and spatial relationships when the animal is alive in normal orientation. You should know the meaning of such terms as:

Anterior—situated near head or, in animals without heads, the end that moves forward.

Caudal—extending toward or located near tail.

Cephalic—extending toward or located on or near head (also cranial).

Distal—located away from the center of the body, the origin, or the point of attachment.

Dorsal—pertaining to the back as opposed to **ventral,** which pertains to the belly or lower surface.

Median—a plane passing through a bilaterally symmetrical animal that divides it into right and left halves.

Posterior—toward the animal's hind end: opposite of anterior.

Proximal—opposite of distal.

Right-left—always in relation to the animal's right and left, not yours.

Sagittal—planes dividing an animal along the median line or parallel to the median.

1 ▷ Obtain a fetal pig and place it in a dissecting tray, ventral side up. Take two pieces of string and tie them tightly to the ankles of both right legs (the animal's right legs, not the legs to your right when the pig is lying on its back). Run the strings under the pan and tie each to the corresponding left leg. Stretch the legs to spread internal organs for easier dissection (fig. I.1). Do not pull so hard that you break internal blood vessels.

External Anatomy of Fetal Pig

Rows of **mammary glands** and the **umbilical cord** should be prominent on the ventral surface of your pig. Mammary glands and hair are two of the diagnostic characteristics of the class *Mammalia*. The umbilical cord attached the fetal pig to the placenta in the sow's uterus. Look at the cut end of the cord and note the blood vessels that carried nutrients, wastes, and dissolved gases from the fetus to the placenta, where they were exchanged by diffusion with the maternal circulatory system.

2 ▷ Determine the sex of your pig. Identify the **anus.** In females, there will be a second opening, the **urogenital opening,** ventral to the anus. In males, the urogenital opening is located just posterior to the umbilical cord. **Scrotal sacs** will be visible just ventral to the anus in males.

Figure I.1 shows the sequence of cuts that should be made to expose the internal organs. Make the cuts in the sequence indicated.

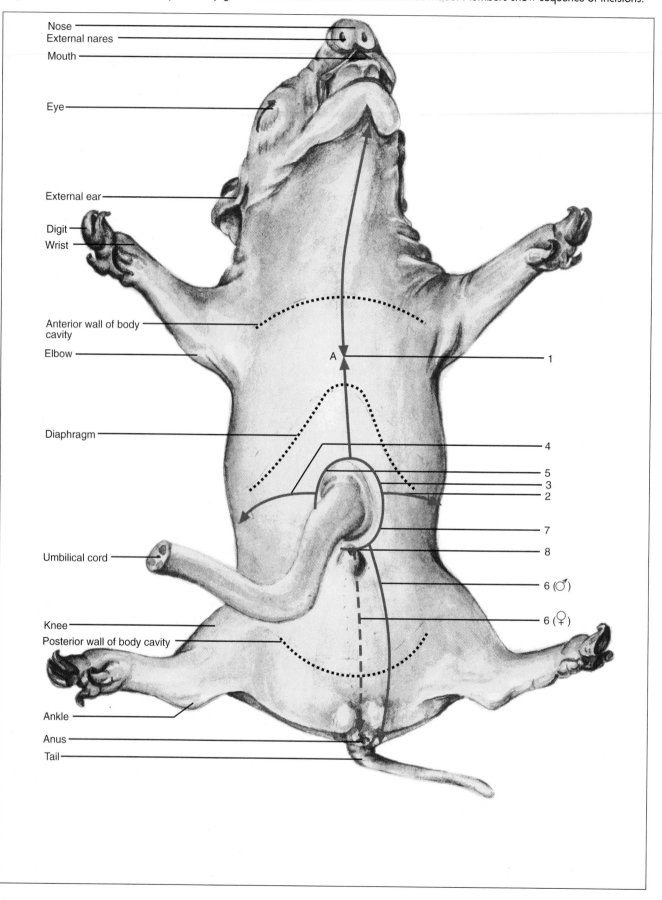

Nose
External nares
Mouth

Eye

External ear

Digit
Wrist

Anterior wall of body cavity

Elbow

Diaphragm

Umbilical cord

Knee
Posterior wall of body cavity

Ankle

Anus

Tail

A

1

4

5
3
2

7

8

6 (♂)

6 (♀)

LAB TOPIC 28

Investigating Digestive and Gas Exchange Systems

Supplies

Preparator's guide available on WWW at
  http://www.mhhe.com/dolphin

Equipment

Compound microscope
Dissecting microscopes

Materials

Fetal pigs
Dissecting trays
Dissecting instruments: scissors, forceps, blunt probe,
    razor blade or scalpel
Thread
Prepared slides
    Mammalian small intestine, cross section
    Mammalian lung section
Hydra culture
Daphnia cultures
Slides and coverslips
Pasteur pipettes with bulbs
Live crickets, grasshoppers, or roaches
Petri plates filled with wax
Demonstration dissection of crayfish gills and
    gill bailer
Live crayfish
India ink

Solutions

Methyl cellulose
70% ethanol
Ice

Student Prelab Preparation

Before doing this lab, you should read the introduction
and sections of the lab topic that have been scheduled
by the instructor.
  You should use your textbook to review the
definitions of the following terms:

alveolus	pharynx
bronchiole	smooth muscle
circular muscle	sphincter
esophagus	trachea
gill	tracheole
longitudinal muscle	villus
mucosa	

You should be able to describe in your own words
the following concepts:

  Incomplete and complete digestive tracts
  Pathway of food though mammalian digestive tract
  Anatomy of *Hydra* (see fig. 20.3)
  Pathway of air to mammalian lungs
  Mechanics of mammalian breathing

  As a result of this review, you most likely have
questions about terms, concepts, or how you will do
the experiments included in this lab. Write these
questions in the space below or in the margins of the
pages of this lab topic. The lab experiments should
help you answer these questions, or you can ask your
instructor during the lab.

Objectives

1. To investigate the feeding behavior of *Hydra*
2. To dissect the digestive and respiratory systems in
   a fetal pig
3. To observe the histology of small intestine and lung
4. To observe the anatomy of gill and tracheole
   respiratory systems
5. To measure human lung capacities

Background

Digestion involves the chemical breakdown of complex food
materials through the actions of enzymes. Chemically, it requires the hydrolysis of the covalent bonds holding together
large, polymeric molecules. Proteins are hydrolyzed into
amino acids and starches into sugars. The smaller molecules
resulting from digestion are easily absorbed. These small
molecules are used by the animal either as sources of energy
or as building units to make new molecules, such as proteins
and nucleic acids that are characteristic for the species. Thus,
when we digest bovine proteins in a hamburger, we reuse the
amino acids to make human proteins or convert the amino
acids to sugars that are used as an energy source.

Digestion can occur **intracellularly** or **extracellularly.** In simple organisms, such as protozoa and sponges, the individual cells of the organism ingest food materials by pinocytosis and phagocytosis and digestion occurs in food vacuoles inside of cells. Other organisms have extracellular digestion in special digestive organs that are either sac-like or tubular. Multicellular organisms, such as cnidarians and flatworms, have **incomplete** digestive tracts that are blind sacs; that is, food enters the mouth, passing into a chamber where enzymatic digestion and absorption occur, and nondigestible material must be expelled through the same opening. This seems to be a somewhat inefficient system because, when nondigestible material is expelled, recently ingested food may also be lost.

Higher animals have a **complete** digestive system, a continuous tube from the mouth to the anus in which food is sequentially broken down and absorbed. Different enzymes are secreted by glands at various points along the digestive tube, so that the digestion of different types of molecules occurs as food passes through the system.

Mere digestion, however, is not sufficient for processing food material. To be of value to the animal, absorption must also occur. In vertebrates, specialized regions of the digestive tube absorb nutrients, salts, and water from the digested material, or chyme. The undigested or nondigestible residues pass on to temporary storage areas before being defecated.

Arthropods and annelids have a digestive tube that passes more or less straight through the body from head to tail. The digestive tubes in such organisms include storage areas, grinding areas, digestive areas, and absorptive areas, but the tube is not longer than the organism. In vertebrates, on the other hand, the digestive tube is many times longer than the animal's length, with much of this length devoted to absorption. After absorption, the circulatory system carries food materials to cells throughout the body.

The cells of most animals are capable of aerobic metabolism. It requires an organism to have some mechanism for exchanging gases with its environment. Oxygen must move from the environment to every cell in the body, where it ultimately functions as a terminal electron (hydrogen) acceptor. Without oxygen, the mitochondrial cytochrome system would not operate, and cells could not carry out glycolysis, the Krebs cycle, and other oxidations of food materials. Furthermore, carbon dioxide, which is produced during the Krebs cycle and in other reactions, must pass from cells to the environment to maintain a consistent intracellular acid-base balance.

In small organisms with relatively low metabolic rates and large surface areas relative to body mass, free diffusion of these gases satisfies an animal's needs. Protozoa, sponges, cnidarians, flatworms, and roundworms have no anatomical specializations for respiration. Gases simply pass to and from the environment through their surface layers.

Larger aquatic animals, such as lobsters, clams, and fish, have developed specialized respiratory surfaces called gills. These featherlike surfaces allow the body fluids to circulate in a closed network that is separated by only a cell layer from the water in the environment. As the blood flows through the gills in one direction, water passes over the external gill surface in the opposite direction, allowing a very efficient exchange of respiratory gases.

Because gill surfaces must remain moist to function properly, they are not suited to gas exchange in terrestrial environments. Terrestrial organisms have **tracheal systems,** as are found in insects, or **lung systems,** as are found in many vertebrates. In these systems, oxygen-rich air enters the animal by a system of tubes and is brought in close proximity to the blood or the tissues where gas exchange occurs. There are exceptions to these generalizations, of course. Many adult amphibians have lungs, but also exchange gases directly through the skin and mouth surface, but as larvae they have gills. One family of salamanders commonly found in North America, the *Plethodontidae,* lacks lungs and depends entirely on cutaneous respiration.

The oxygen-carrying capacity of blood is increased by the presence of a respiratory pigment, such as **hemocyanin** (usually dissolved in hemolymph) or **hemoglobin** (usually in blood cells). These pigments bind loosely and reversibly with oxygen to facilitate oxygen transport from the gills to the tissues. Blood is oxygen deficient when it enters the gills or lungs. Oxygen diffuses into the aqueous portion of the blood and combines with the pigments. As the blood flows to areas where there is little oxygen, the diffusion gradient causes the oxygen to move into the tissues, where it is used by mitochondria. Conversely, a high concentration of carbon dioxide is usually present in active tissues. The CO_2 moves down the diffusion gradient into the blood where it simply may dissolve or may combine chemically with the pigments. The carbon dioxide is, in turn, carried back to the respiratory organs where it is exchanged with the environment.

LAB INSTRUCTIONS

You will study the digestive and respiratory systems of several animals.

Invertebrate Feeding Behavior in *Hydra*

Hydra is a small, freshwater cnidarian related to the jellyfish and sea anemones (see fig. 20.3). It lives attached to submerged rocks, leaves, and twigs. Hydra's body is organized simply, consisting of only two layers of cells surrounding a hollow cavity. However, the organism is highly specialized for food gathering; it uses tentacles to capture food and transfer it into the digestive (**gastrovascular**) cavity. The food is then digested by enzymes that are secreted by cells lining the cavity. Because of the organism's small size, the cells of *Hydra* obtain nutrients from the digestive cavity by simple diffusion. Each body cell also exchanges oxygen and carbon dioxide directly with the surrounding water and releases cellular metabolic wastes directly into the water as well.

1▷ Obtain a living *Hydra* from the supply table and place it in a small dish that has been thoroughly washed so it has no chemical residues from other uses. Observe the *Hydra* with a dissecting microscope against a dark background. Sometimes a *Hydra* contracts into a small ball after it is transferred, but it should relax, attach to the glass, and resume its normal functions before long. If the *Hydra* attaches to the surface water film, sink it by dropping water on it from a dropper.

Locate the **pedal disc** (the animal's point of attachment), the body **column,** the **tentacles,** and the **hypostome,** a small raised region surrounded by the bases of the tentacles (see fig. 20.3). The **mouth** is located in the center of the hypostome but is difficult to see.

Obtain some *Daphnia,* a freshwater crustacean, that *Hydra* will feed on. Small pieces of aquatic annelids such as California black worms also elicit a strong feeding response. Transfer some food organisms immediately next to your *Hydra* with a Pasteur pipette and watch closely through a dissecting microscope. *Hydra,* like other cnidarians, is a carnivore. It traps food by stinging and paralyzing other animals with specialized epidermal cells called **cnidocytes** located on its tentacles. Each cnidocyte has a special organelle, the nematocyst, which contains a coiled tube. When stimulated, a nematocyst shoots out its thread, which either entangles or pierces and poisons the prey. Seeing the *Hydra* capture food will take patience, but the time is well spent.

Record below how *Hydra* feeds. Include drawings showing how prey are transferred from the tentacles to mouth.

Once food is in *Hydra*'s gastrovascular cavity, digestion begins. Enzymes secreted by certain cells lining the cavity's begin **extracellular digestion.** Partially digested bits of food material are later taken up by phagocytic cells in the cavity lining, and further digestion occurs inside the food vacuoles in these cells. This is **intracellular digestion.** Food absorbed by the cells lining the gastrovascular cavity supplies all cells of the body.

2▷ Study the longitudinal section of *Hydra* in figure 20.3. Note that the body wall consists of the **epidermis,** and the inner **gastrodermis,** which lines the **gastrovascular cavity.** This cavity extends from the body cavity into each of the tentacles. There are several specialized cell types in each of the body layers. Some of these specialized cells are the cnidocytes on the tentacles and enzyme-secreting cells and flagellated cells in the gastrodermis. How many cell layers thick is the body wall?_____. Estimate how many cell layers thick the body wall of a mammal is?_____.

When animals increase in size and complexity, diffusion from the digestive cavity to the cells, as occurs in *Hydra,* is no longer adequate to supply the tissues' demands. Increased complexity means special systems for the transport of material from cell to cell. The relationship of the vertebrate digestive and circulatory systems represents the height of this development from physiological and evolutionary perspectives.

Mammalian Digestive System

3▷ If you have not already done so, get a fetal pig and dissection tray. Read the general instructions on page 377.

Anatomy of the Mouth

Located in the upper neck region beneath the skin are three pairs of **salivary glands: parotid, submaxillary,** and the **sublingual,** which is difficult to find. They produce saliva containing the enzyme salivary amylase that hydrolyzes starch during chewing.

To view the salivary glands, remove the skin and muscle layer from one side of the face and neck, as in figure 28.1 to reveal the rather dark, triangular-shaped parotid gland. Note the difference in appearance between muscle tissue and glandular tissue. If you dissected carefully, you should find the duct that drains the gland into the mouth near the upper premolar teeth. Try to trace the duct. The other salivary glands lie beneath and below the parotid gland.

4▷ With heavy scissors, a razor blade, or scalpel, cut through the corners of the mouth and extend the cut to a point below and caudal to the eye.

Open the mouth, as in figure 28.2 and observe the **hard palate,** composed of bone covered with mucous membrane, and the **soft palate,** which is a caudal continuation of the soft tissue covering the hard palate. The oral cavity ends and the **pharynx** starts at the base of the tongue.

Investigating Digestive and Gas Exchange Systems   **381**

Figure 28.1   Dissected fetal pig's head showing salivary glands.

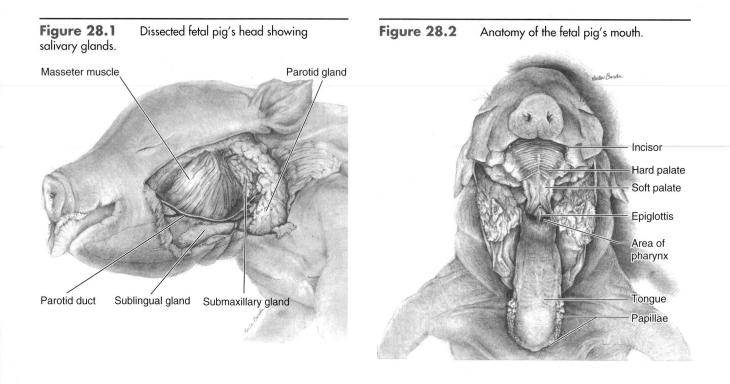

Masseter muscle

Parotid gland

Parotid duct   Sublingual gland   Submaxillary gland

Figure 28.2   Anatomy of the fetal pig's mouth.

Incisor

Hard palate

Soft palate

Epiglottis

Area of pharynx

Tongue

Papillae

Figure 28.3   Air and food passages across the pharynx.

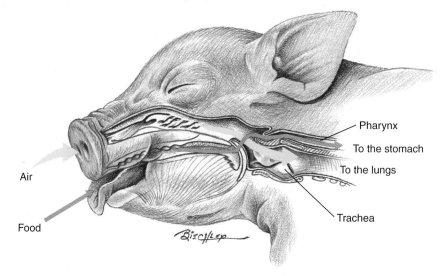

Pharynx

To the stomach

To the lungs

Trachea

Air

Food

The pharynx is a common passageway for the digestive and respiratory tracts, as seen in figure 28.3. The opening to the **esophagus** may be found by passing a blunt probe (not a needle) down along the back of the pharynx on the midline. This collapsible tube connects the pharynx with the stomach. The **glottis** is the opening into the **trachea** or windpipe and lies ventral to the esophagus. It is covered by a small white tab of cartilage, the **epiglottis.** The epiglottis may be hidden from view in the throat; if so, you will have to pull it forward with forceps or a probe to see it.

Alimentary Canal Anatomy

5 You will see the path of the esophagus to the stomach when you dissect the respiratory system. To view the rest of the alimentary tract and associated glands, use a scalpel or pair of scissors to make incisions into the **abdominal** cavity as indicated in fig. I.1, page 378. Cut carefully through only the skin and muscles to avoid damaging the internal organs.

The flap containing the umbilical cord will be held in place by blood vessels. Tie both ends of a 15-cm piece of thread to the blood vessel about 1 cm apart. Cut the vessels

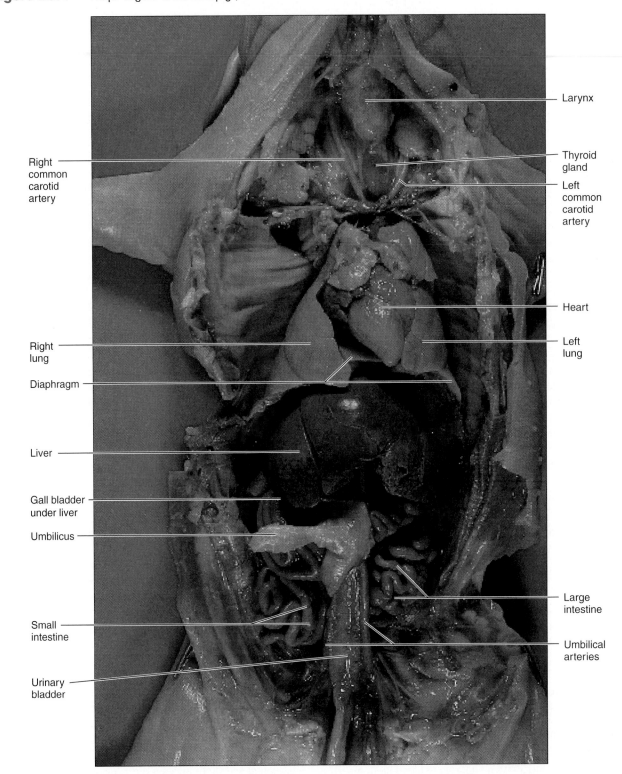

Larynx

Thyroid gland

Right common carotid artery

Left common carotid artery

Heart

Right lung

Left lung

Diaphragm

Liver

Gall bladder under liver

Umbilicus

Large intestine

Small intestine

Umbilical arteries

Urinary bladder

between the two knots and lay this tissue flap back. Leave the thread in place so you can later trace the circulatory system.

Find the thin, transparent membranes, the **mesenteries,** which suspend and support the internal organs in the body cavity. The dark brown, multilobed **liver** should be visible caudal to the **diaphragm** (fig. 28.4). If you trace the umbil-ical vein from the thread to the liver, you will see a green-colored sac, the **gallbladder,** located just below the en-trance of the vein into the liver. It stores bile produced in the liver. Bile travels from the gallbladder to the small in-testine via the bile duct but the duct is quite small. Bile is an emulsifying agent that aids in digestion of fats.

Under the liver is the **stomach.** Locate the point where the esophagus enters the **cardiac region** of the stomach. Gastric glands in the wall of the stomach secrete pepsinogen, hydrochloric acid, and rennin. Pepsinogen is activated by hydrochloric acid to become pepsin, which digests proteins. Rennin is an enzyme that hydrolyzes milk protein. Food leaves the stomach as a fluid suspension, **chyme.** It enters the **duodenum,** the first part of the small intestine.

Find the **pancreas,** a glandular mass lying in the angle between the curve of the stomach and duodenum. It secretes several enzymes into the duodenum that digest proteins, lipids, carbohydrates, and nucleic acids. Certain cells in the pancreas act as endocrine cells and secrete the hormones insulin and glucagon. In fact, insulin used in human diabetes therapy can be extracted from the pancreases of pigs collected at slaughterhouses.

Although it is not part of the digestive system, identify the **spleen** attached by mesenteries to the outer curvature of the stomach. It is made of lymphatic tissue and is important in development of immunity and the scavenging of iron from red blood cells when they break down.

6 ▶ Slit the stomach lengthwise, cutting through the cardiac and pyloric **sphincters,** muscles that regulate passage of material into and out of the stomach. The internal surface of the stomach is covered by gastric mucosal cells, which secrete mucus that prevents the stomach from digesting itself. When this protection fails, a peptic ulcer develops.

The small intestine is made up of three sequentially arranged regions: **duodenum, jejunum,** and **ileum.** These areas are difficult to differentiate from each other. Cut out a 2-cm section of the small intestine about 5 cm posterior from the stomach, slit it open, and place it under water in a dish. Use your dissecting microscope to observe the velvety internal lining made up of numerous fingerlike projections called **villi.** The villi are highly vascularized, containing capillaries and lymphatics that transport the products of digestion to other parts of the body, especially the liver.

The ileum opens into the large intestine, or **colon.** They join at an angle, forming a blind pouch, the **cecum,** which in primates and some other mammals often ends in a slender appendage, the **appendix.** In many herbivores, the cecum is very large and contains microorganisms that aid digestion by breaking down cellulose.

The **rectum** is the caudal part of the large intestine, where compacted, undigested food material is temporarily stored before being released through the **anus.** The colon of vertebrates contains large numbers of symbiotic bacteria, especially *Escherichia coli.* These bacteria produce vitamin K, which is absorbed and plays a vital role in blood clotting.

Histology of Small Intestine

7 ▶ Put your fetal pig aside for the moment and obtain a prepared slide of a cross section of a mammalian small intestine. Examine it under scanning power with the compound microscope. Compare what you see to figure 28.5.

The central opening is called the **lumen** and is the space through which food passes as chyme during digestion. Switch to low power and observe the small fingerlike projections of the intestine's inner surface. These are villi and are covered by a layer of cells called the **mucosa.** You should be able to distinguish two cell types in the intestinal mucosa: **goblet cells** and **columnar epithelial cells.** Examine them with the high-power objective. The goblet cells secrete mucus into the small intestine, serving as a lubricant for the passage of chyme. Epithelial cells are involved in absorption.

Return to the low-power objective and observe the **submucosa,** a layer of connective tissue that underlies the mucosa. Look for the blood vessels and lymphatic vessels that ramify through this layer. Sugars, amino acids, glycerides, and other components of digested food must move through the mucosal cells into the submucosa before they can enter the circulatory system and be distributed throughout the body.

To the outside of the submucosa are two smooth muscle layers: an **inner circular layer** and **outer longitudinal layer.** The inner circular muscles change the diameter of the intestine, and the outer muscles alter its length. These muscles contract in a wavelike motion called **peristalsis,** which pushes chyme through the digestive tract. The small intestine is covered by a layer of peritoneal cells, that together with underlying connective tissue, is called the **serosa.**

Figure 28.5 shows scanning electron micrographs of the three-dimensional arrangement of the small intestine. Note how the villi and microvilli increase the surface area. **?** What important process following digestion is facilitated by this increased surface area?

Invertebrate Respiratory Systems

Insect Tracheal System

The mammalian respiratory system represents only one method of gas exchange in terrestrial animals. Insects and other terrestrial arthropods make use of another, a rather remarkable **tracheal system,** which allows the exchange of gases to occur independent of the circulatory system.

In a tracheal system, air enters through several small, lateral openings called **spiracles** and passes into a branching system of tubules called **tracheae** and **tracheoles.** The tracheoles continue branching into a system of fluid-filled tubes where gas exchange occurs (fig. 28.6). The tracheoles branch so extensively that they are never very far from any cell in the body. Gas exchange occurs by diffusion from the cell to the tracheole with little participation of the circulatory system. There are apparently no special muscles that circulate air in the tracheal system, but contraction of the abdominal and flight muscles aid the exchange of air. Large insects often have **air sacs.** When compressed by movement of surrounding organs, they force large volumes of air through the system.

8 ▶ If your instructor has not prepared a demonstration dissection of a tracheal system, obtain a live, large insect (cricket, grasshopper, or roach) and a petri dish filled with wax to be used as a dissecting pan. Anesthetize the insect

Lumen Mucosa layer Submucosal layer Columnar epithelium Goblet cell Muscle layer

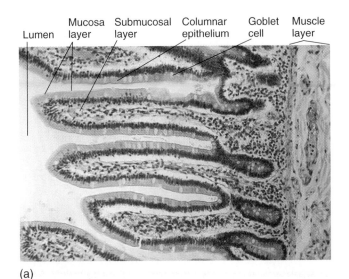

(a)

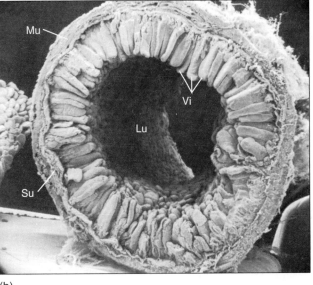

(b)

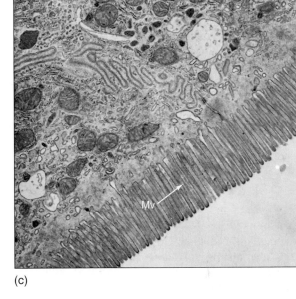

(c)

Figure 28.5 Microstructure of intestine. (*a*) Photo taken through a light microscope of cross section of small intestine. (*b*) Scanning electron micrograph of cross section of small intestine showing villi (Vi), lumen (Lu), submucosa layer (Su), and muscle layers (Mu). The epithelial cells on the surface of the villi have highly folded membranes, microvilli (Mv), which greatly increase the absorptive surface area of the cell layer, as seen in (*c*) a transmission electron micrograph showing highly folded cell membrane. (*b*) From R. G. Kessel and R. H. Kardon. *Tissues and Organs: A Text-Atlas of Scanning Electron Microscopy.* 1979. W. H. Freeman and Company.

with carbon dioxide or by placing it in a freezer. Remove the wings and legs and locate the spiracles along the side. With fine scissors, cut through the dorsal surface close to the lateral margin but do not cut through the spiracles. (Keep the points of your scissors against the inside of the exoskeleton and take small snips so that you do not damage any interior organs.) Run the cuts along the full length of the insect and join them together at the anterior and posterior ends. Remove the dorsal strip of cuticle and pin the insect in the dish, dorsal side up. Next flood the insect with saline solution to prevent drying and to float tissues for observation. Place the dish under a dissecting microscope and observe.

Gently remove the glistening yellow fat bodies. Vary the lighting angle while looking for a silver-colored tube originating from a spiracle. Remove this trachea with a forceps, mount it in a drop of saline on a microscope slide, and add a coverslip. Observe the trachea with your compound microscope and sketch it in the circle to the right. Compare your observations to figure 28.6*c*.

Investigating Digestive and Gas Exchange Systems

Figure 28.6 Tracheal systems. (*a*) Tracheal system of a flea penetrates all parts of its body, allowing direct gas exchange. (*b*) Large insects pump air through their tracheal systems by contracting muscles that press against air sacs. (*c*) Photomicrograph of tracheae shows reinforcing rings of chitin that prevent the tubules from collapsing.

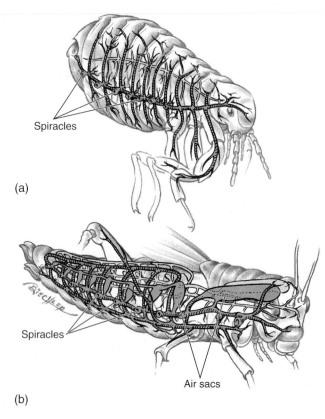

(a)

Spiracles

(b)

Spiracles

Air sacs

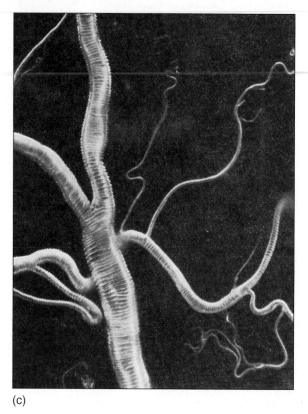

(c)

Gills

In aquatic organisms, gills function in respiratory gas exchange and ion regulation. Gills are found in many different kinds of animals, including larval amphibians, fish, arthropods, and molluscs. Except in molluscs, gills originate in the embryo as featherlike pocketings of the body wall that are amply supplied with blood. In most cases, a special mechanism exists to ventilate the gills, that is, to move water over them. In crayfish, this involves a specialized mouthpart, the **gill bailer,** which moves water forward over the gills in a sculling motion (fig. 28.7).

In fish, gill covers called **opercula** are raised, to draw water in through the open mouth. The mouth is then closed and depression of the opercula forces water over the gills and out through the external gill openings (fig. 28.7*b*).

In some fish, the pumping mechanism is not sufficient to satisfy the metabolic demands for oxygen, so that such fish must constantly swim with its mouth open to improve the rate of water flow over the gills. In clams as well as in many other animals, the gills function in gas exchange and also in filtering food particles from the water stream that passes over and through the gills.

9▶ In your lab, you may have some living crayfish in an aquarium. Place one in a bowl of water and give it time to resume normal gill ventilation. Take a Pasteur pipette and place a drop of india ink next to the side of the crayfish just

Figure 28.7 Direction of water movement over the gills of two animals: (*a*) crayfish; (*b*) fish.

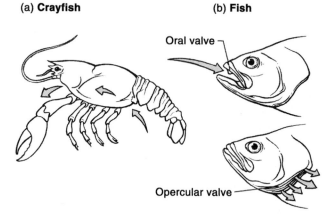

(a) Crayfish

(b) Fish

Oral valve

Opercular valve

behind the thorax. Watch the direction of water flow using the ink as a tracer. Record your observations below.

Mammalian Respiratory System

Respiratory Tree Anatomy

10▶ Return to your fetal pig and examine the external openings (external nares) on the snout. Cut across the snout with a scalpel about 1 to 2 cm from the end and remove the tip. The nasal passages are separated from each other by the nasal septum. The curved turbinate bones in the sinus area increase the surface area of the passageways, creating eddy currents that, along with hairs, cilia, and mucus, help remove dust in the inhaled air and humidify it. The floor of the nasal passages is made up of the hard palate and the soft palate (posterior to hard palate).

11▶ Look into the pig's mouth. Behind and above the soft palate is the **nasopharynx.** It may be necessary to slit the soft palate to observe this. Air enters the nasopharynx from the posterior end of the nasal passages, then passes into the **pharynx,** through the **glottis,** and into the **larynx** and ultimately the **trachea.** If food accidentally enters the glottis, choking results. The Heimlich maneuver can often "save" a person who has food wedged in the glottis. Your instructor will demonstrate this maneuver for you.

In the nasopharyngeal area, look for the openings of the **eustachian tubes.** They are very difficult to find. These tubes allow air pressure to equilibrate between the middle ear chamber and the atmosphere. (This is why changes in altitude cause the ears to "pop.") Throat infections often spread to the ears through the eustachian tubes.

12▶ Run your fingers over the pig's throat and locate the hard, round **larynx.** Make a medial incision in the skin of the throat and extend the end cuts laterally, folding back the skin flaps. Repeat this procedure for the muscle layers.

Note the large mass of glandular material, the **thymus,** in this area. In the young pig, the thymus produces **lymphocytes,** an important component of the immune system. Later in life, the thymus atrophies and is of little consequence.

As you approach the larynx and trachea in your dissection, use a blunt probe to separate the muscles and expose these structures. Ventral to the trachea, observe the brownish-colored **thyroid gland.** Note how both the trachea and larynx are supported by rings of cartilage. The hyoid apparatus is anterior to the larynx and is a supporting frame for the tongue extensor muscles.

The **larynx,** or voice box, contains folds of elastic tissue, which are stretched across the cavity. These vocal cords vibrate when air passes over them, producing sound, and attached muscles vary the cord tension, allowing variations in pitch. Slit the larynx longitudinally and observe the vocal cords. Continue the slit posteriorly into the trachea and observe its lining. The esophagus is located behind the trachea. Pass a blunt probe into the esophagus from the mouth to help identify it.

13▶ If the **thorax** of your animal is not already opened, make a longitudinal cut with heavy scissors through the ribs just to the animal's right of the **sternum,** or breastbone. Always keep the lower scissor tip pointed upward against the inside of the sternum to avoid catching and cut-

ting internal structures. Be careful! Several major blood vessels are under the sternum and should not be damaged.

The **diaphragm** is a sheet of muscle that separates the **abdominal cavity** from the **thoracic cavity.** The thoracic cavity is divided into three areas by membranes: the right and left **pleural cavities,** which surround the lungs, and the **pericardial cavity** where the heart is located.

If the pleural membranes are removed, the **lung** structure can be seen. The trachea, when it enters the thorax, divides into two **bronchi,** which are hidden from direct view beneath the heart and blood vessels. These bronchi, in turn, divide into progressively smaller **bronchioles,** which finally end in clusters of microscopic air sacs called **alveoli.** Alveoli have walls only a single cell layer thick and they are covered by capillaries (fig. 28.8). In these air sacs, oxygen and carbon dioxide are exchanged between the blood and the inhaled air. Have you ever known someone with bronchitis? Where do you think the infection was?

Bronchioles have smooth muscles in their walls. Asthma results from the spasmodic contraction of these muscles, preventing air flow to the alveoli.

Microscopic Examination of Mammalian Lung

14▶ Remove a piece of the lung and put it in a small bowl of water. Observe it with a dissecting microscope and find the alveoli and bronchioles. Also look at the demonstration slide of a section across several alveoli. How many cell thicknesses separate the air in a mammalian lung from the red blood cells in the capillaries?_____. The inset in figure 28.8 gives you the distances in micrometers between an erythrocyte in a capillary and the air in an alveolus. What is the diffusion distance for a molecule of oxygen in the alveolus to an erythrocyte and its hemoglobin?_____.

Figure 28.9 contains a scanning electron micrograph of an alveolus and surrounding capillaries. The alveoli form an interconnecting system of chambers, and macrophages in the alveoli scavenge for microorganisms and particulate material.

Lung Ventilation Mechanism

Air enters the lungs as a result of the combined effects of the contraction of the diaphragm and several sets of muscles. When the **external intercostal** (between the ribs) muscles contract, the volume of the thoracic cavity increases as the rib cage elevates and the diaphragm depresses, causing the air pressure in the cavity to decrease. Air rushes in through the respiratory passageways and expands the alveoli. This causes the pressure between the atmosphere and the pleural cavities to equilibrate. When the diaphragm and intercostal muscles relax, the rib cage drops and the diaphragm rises decreasing the volume of the thoracic cavity.

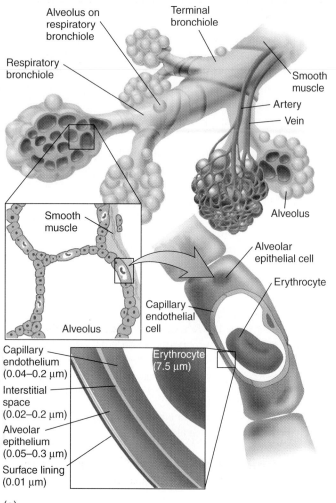

Alveolus on respiratory bronchiole

Terminal bronchiole

Respiratory bronchiole

Smooth muscle

Artery

Vein

Alveolus

Smooth muscle

Alveolar epithelial cell

Erythrocyte

Capillary endothelial cell

Alveolus

Capillary endothelium (0.04–0.2 μm)

Erythrocyte (7.5 μm)

Interstitial space (0.02–0.2 μm)

Alveolar epithelium (0.05–0.3 μm)

Surface lining (0.01 μm)

(a)

Figure 28.8 (a) Microscopic structure of lung and close relationship of alveoli and capillaries; (b) Inspiration and expiration mechanics. Contraction of intercostal muscles lifts rib cage and contraction of diaphragm muscles flattens diaphragm: both enlarge chest cavity and cause its pressure to drop. During resting expiration, elastic recoil of lung tissues, together with dropping of the rib cage and recoil of diaphragm because of organ pressure, decrease size of chest cavity.

Intercostal muscles

Diaphragm down

Intercostal muscles

Diaphragm up

Inhalation

External intercostal muscles contract, pull rib cage upward.

Exhalation

Internal intercostal muscles contract, help pull rib cage downward.

(b)

Consequently, the pressure in the pleural cavity increases and collapses the alveoli, driving air out of the "respiratory tree." Figure 28.8*b* illustrates these mechanics in humans.

15▷ Note the convex shape of the diaphragm in your pig and imagine how contraction of the diaphragm increases the volume of the thoracic cavity. Push the sternum up to mimic the contraction of the involved muscles and note the expansion of the chest cavity.

You are now finished with the fetal pig for this lab. Return it to the storage area according to the directions given by your lab instructor.

Measuring Human Respiratory Volumes

Respiratory volumes vary with a person's size, age, gender, and physical condition. Normal breathing results in approximately 500 ml of air moving into and out of the lungs with each breath. This is called the **tidal volume.** Each of us can draw much more into our lungs because we have an **inspiratory reserve volume** as well as force more from our lungs, our **expiratory reserve volume.** However, regardless of how much we exhale, there is always a **residual volume** of air in the lungs because the alveoli do not completely collapse. These concepts are summarized in figure 28.10.

Figure 28.9 Scanning electron micrograph of an alveolus, showing thickness of alveolar wall.

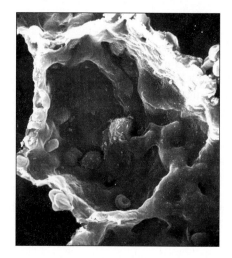

Figure 28.10   Approximate capacities of human lungs.

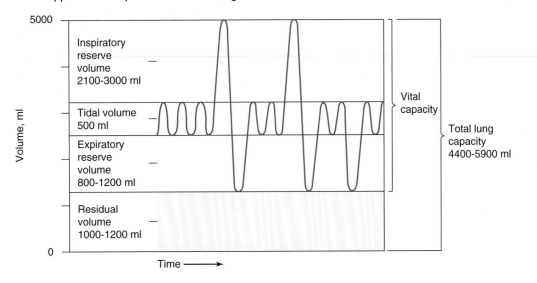

16   In this section of the exercise, you will measure these volumes with a **spirometer.** Your instructor will explain how the instrument works. You may work in pairs in this part of the exercise, or your instructor may choose to do it as a demonstration.

     Set the spirometer to zero and install a new disposable mouthpiece, or sterilize the old one by dipping it in 70% alcohol.

     To measure tidal volume, sit in a chair for three minutes. Then, breathe in and out at a normal rate five or six times. On the last breath, exhale as you normally would through the spirometer mouthpiece. Repeat this measurement three times and calculate an average volume. What is your tidal volume?

17   To measure inspiratory reserve volume, again breathe normally five or six times, then make the deepest possible inspiration. Now breathe out through the respirometer, exhaling only to the normal expiratory end point. Do not force the expiration beyond where you would normally stop exhaling. Repeat the measurement three times and calculate an average. To obtain the actual inspiratory reserve volume, subtract the tidal volume from the average of the just measured volumes. What is your inspiratory reserve volume?

18   Expiratory reserve volume is the amount of air that can be forcibly exhaled after a normal expiration. Devise a procedure to measure this and repeat the measurements three times. Calculate an average. What is your expiratory reserve volume?

     Your **vital capacity** is the sum of the three values you just measured. What is your vital capacity?

     Total lung capacity involves a fourth volume that cannot be easily measured. (See figure 28.10.) What is it?

Comparative Summary

To gain an understanding of comparative anatomy, fill in tables 28.1 to 28.4. Write a brief description of each structure and its function. Include a summary statement of the organ system for each animal.

TABLE 28.1   *Hydra*—digestive system summary

Cnidocytes

Mouth/anus

Gastrovascular cavity

Mesoglea

Description of digestive process

TABLE 28.2   Fetal pig—mammal—digestive system summary

Mouth

Salivary glands

Esophagus

Stomach

Small intestine

Liver

Pancreas

Large intestine

Rectum

Anus

Description of digestive process

TABLE 28.3   Insects—gas exchange system summary

Spiracles

Trachea

Tracheoles

Tissues

Air sacs

Description of gas exchange

TABLE 28.4   Fetal pig—mammal—gas exchange system summary

External nares

Nasopharynx

Larynx

Trachea

Bronchi

Bronchioles

Alveoli

Blood

Diaphragm

Rib cage muscles

Description of gas exchange

Learning Biology by Writing

Your instructor may ask you to answer the Lab Summary and Critical Thinking Questions that follow.

Internet Sources

The World Wide Web has become an important resource for those who work in the medical field. You can explore this resource by using GOOGLE to search for information on "pulmonary pathology." When the list of citations is returned, look for those that have images of black lung disease, silicosis, and asbestos accumulation. From your observations of the photos, write a short paragraph describing how these diseases affect the lung. Be sure to include the URL for future reference.

Lab Summary Questions

1. Create a flowchart that traces the pathway of food through the mammalian digestive system, and indicate to the right of the chart the functions of each organ.

2. The digestive system of *Hydra* is saclike, whereas that of mammals and many other animals is tubelike. What are the advantages of tubular digestive systems?
3. Outline the pathway of air from the nostrils to the alveoli in the fetal pig.
4. Describe how the diaphragm and rib cage function in moving air into and out of a mammal's lungs.
5. Contrast how cells in a mammal receive oxygen with how cells in an insect receive oxygen.
6. Explain what is meant by term vital capacity of the lungs. How is it different from total capacity?

Critical Thinking Questions

1. Explain why the small intestine is the longest part of the mammalian digestive system.
2. Consider air and water as respiratory media. Which contains a greater concentration of oxygen? Which takes more energy to move over the respiratory surface? Which is potentially more damaging to the respiratory surface?
3. What affect does the giraffe's long neck have on breathing? (Consider volume of trachea.)

LAB TOPIC 29

Investigating Circulatory Systems

Supplies

Preparator's guide available on WWW at
http://www.mhhe.com/dolphin

Equipment

Dissecting microscopes
Compound microscopes
Stethoscope
Sphygnomanometer

Materials

Fetal pig
Demonstration dissection of beef heart
Dissection pans and instruments
Microscope slides
    Cross section artery
    Cross section vein
    Wright-stained human blood smear
Live guppies, tadpoles, or frogs
Live crayfish
Small petri dishes
Coverslips
Absorbent cotton

Solutions

Ringers invertebrate saline

Student Prelab Preparation

Before doing this lab, you should read the introduction
and sections of the lab topic that have been scheduled
by the instructor.

You should use your textbook to review the
definitions of the following terms:

aorta	leucocyte
artery	lymphatic system
atrium	lymphocyte
capillary	pulmonary artery
carotid artery	pulmonary vein
erythrocyte	vein
hemolymph	vena cava
jugular vein	ventricle

You should be able to describe in your own words
the following concepts:

Open and closed circulatory systems
The pumping action of a four-chambered heart

The role of capillaries in exchange
General patterns of mammalian circulation to the
    lungs, brain, gut, and body muscles and return
    path to heart

As a result of this review, you most likely have
questions about terms, concepts, or how you will do
the experiments included in this lab. Write these
questions in the space below or in the margins of the
pages of this lab topic. The lab experiments should
help you answer these questions, or you can ask your
instructor during the lab.

Objectives

1. To dissect the heart and major vessels of the
   mammalian circulatory system
2. To be able to distinguish between a cross section
   of an artery and vein and between a red blood
   cell and a white blood cell
3. To observe blood flow in capillary beds of a fish
4. To measure human blood pressure, using a
   sphygmomanometer.
5. To contrast the pumping mechanism of a
   mammalian heart with that of an arthropod

Background

Invertebrates, such as cnidarians, flatworms, and roundworms, lack circulatory systems. Simple diffusion is sufficient for the necessary exchanges of respiratory gases, waste
products, and nutrients. Larger, more complex animals require a circulatory system to supply the needs of their tissues.

The circulatory system consists of a special internal
body fluid called blood or hemolymph, a pumping system,
and a vascular system consisting of vessels for moving the
blood rapidly from one location to another within an animal. The circulating fluid often contains a respiratory pigment, a protein that aids in transporting oxygen and carbon

dioxide between the tissues and the respiratory surface. Hemocyanin and hemoglobin (which in vertebrates occurs in red blood cells) are common pigments. Blood also contains cells or proteins that protect against invasion by microorganisms and proteins that are involved in clotting, the sealing of leaks. The blood-vessel system often has anatomical provisions so that the blood is brought into close contact with three other physiological systems: lung or gill, where gas exchange occurs; excretory, where salt, water, and waste exchange occur; and digestive, where nutrients are absorbed.

Circulatory systems may be either open or closed. In **open circulatory systems,** found in molluscs and arthropods, the arterial system is not connected to the venous system through capillary beds. Instead, the small arteries simply terminate, emptying their contents into the tissue spaces, and the blood (properly called hemolymph) directly bathes the tissues, eventually finding its way back to the heart.

In **closed circulatory systems,** found in some invertebrates and all vertebrates, the flow of blood is always within blood vessels. The arterial system is connected to the venous system by means of capillaries which have very thin walls only one cell thick. Blood entering the capillaries is under relatively high pressure, and part of the fluid portion is filtered through the capillary walls, entering the tissue spaces. On the venous side of the capillary bed, most of this fluid flows back into the capillaries due to osmosis. Gaseous, waste, and nutrient exchanges between the blood and tissues occur by way of this fluid exchange as well as by diffusion. Blood flow in each capillary bed is regulated by the opening and closing of the **precapillary sphincter,** as seen in figure 29.1.

The capillary bed, and only the capillary bed, is the functional site of the closed circulatory system where all exchanges take place. The **lymphatic system** consists of small open-ended lymphatic capillaries that conduct fluid into larger lymphatic ducts. Fluid that does not return to the blood capillaries enters the lymphatic capillaries. This fluid is collected in lymphatic ducts and returns to the venous system near the heart.

Vertebrate circulation is summarized in figure 29.2. Consider the changes that occur in the blood as it passes through the various circuits. It is more important to understand the purpose of the circulatory system than to know a long list of names of vessels.

LAB INSTRUCTIONS

You will observe the gross and microscopic features of the mammalian circulatory system and circulation in the capillary beds of a living vertebrate. You will learn how to measure the blood pressure of a human. Finally, you will observe a living arthropod heart in order to compare open and closed circulatory systems.

Figure 29.1 Anatomy of a capillary bed. Fluid leaves the capillaries because of the pumping force of the heart, raising the osmotic pressure of the blood. Pumping pressure falls across the capillary bed due to drag and volume loss. On the venous side, fluids are drawn into the capillary by osmosis. Excess fluids enter the lymphatic capillaries.

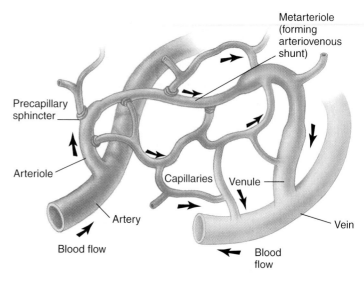

Figure 29.2 Schematic of major mammalian blood vessels and their relationship to one another.

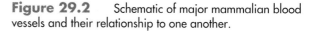
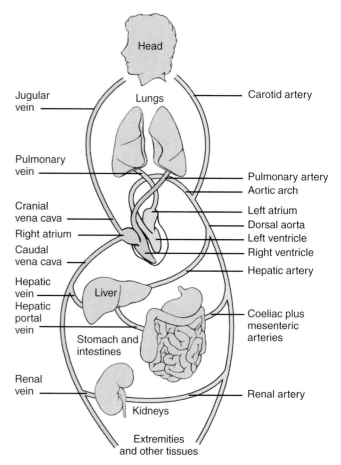

Invertebrate Circulatory System

Most molluscs, arthropods, and many other invertebrates (but not all) have open circulatory systems. Of these, the crayfish's is most easily observed. See figure 22.7 for a diagram showing the crayfish's circulatory system.

Optional Demonstration of Open Circulation

1▶ To observe the heart of a crayfish, first anesthetize an animal by packing it in crushed ice in a glass finger bowl. Cover the abdomen with wet cotton. The dorsal part of the carapace, the exoskeleton covering the thorax, should be removed by inserting scissors under its posterior edge 1 cm to the left of the midline and cutting forward to the region of the eye. This procedure should be repeated on the right side, and the strip of exoskeleton should be carefully lifted and removed, so that none of the underlying membranes are torn.

The heart can now be seen beating in the **pericardial sinus** covered by the epidermal and pericardium membranes. These membranes can be removed to expose the heart, which should be bathed in Ringer's solution to keep it from drying.

In an open circulatory system such as this one, **hemolymph,** the circulating fluid, leaves the heart in arteries but returns in open **sinuses** instead of veins. Under a dissecting microscope, you will be able to see the fluid surrounding the heart enter it through three pairs of slitlike openings called **ostia.** These ostia open when the heart relaxes and allow hemolymph to flow in. When the heart contracts, flaps of tissue inside the heart close the ostia, and hemolymph is forced out of the heart through the arteries.

You should be able to see the **dorsal abdominal artery,** which carries hemolymph to the tail, and the **ophthalmic** and **antennary arteries,** which carry hemolymph to the head region. Other arteries lie beneath the heart. Make a diagram of how the crayfish heart works.

Mammalian Circulatory System

2▶ If your fetal pig has not previously been opened, make a series of cuts as diagramed in the figure I.1, page 378. If you have followed the lab sequence in this manual, complete the opening as follows:

1. Make a longitudinal cut 1 cm to the left of the sternum from the lower ribs to the region of the forelimbs and parallel to the previous cut. Sever all ribs.

2. Lift up the center section, labeled (A) in figure I.1 (page 378), and cut any tissues adhering underneath. A transverse cut at the anterior end will detach this center piece, which should be discarded.

3. The heart and lungs will be easier to observe if the diaphragm is cut away from the rib cage on the animal's left side only. Cut close to the ribs.

4. In the region of the throat, remove the thymus glands, thyroid, and muscle bands, but do not cut or tear any major blood vessels.

The Heart and Its Vessels

3▶ Find the heart encased in the **pericardial sac.** Remove the sac and identify the four heart chambers. The paired **atria,** thin-walled, distensible sacs, collect blood as it returns to the heart. The two **ventricles** are the large, muscular pumping chambers of the heart.

Blood returning from the systemic circulation enters the right atrium from the cranial and caudal **vena cavae.** After passing into the right ventricle, it is pumped to the lungs through the **pulmonary trunk,** which divides into the left and right **pulmonary arteries.** The trunk is visible passing from the heart's lower right to upper left and passing between the two atria (fig. 29.3). In the mammalian fetus, the pulmonary trunk and aorta are connected by a short, shunting vessel, the **ductus arteriosus.** Find this vessel in your animal. During the intrauterine life, when the lungs are not functional, most blood entering the pulmonary circuit does not pass to the lungs. Instead, it is shunted to the aorta. At birth, the shunting vessel constricts so that blood enters the lungs. The constricted vessel fills with connective tissue to become a solid cord seen in adults as the arterial ligament.

Trace the pulmonary arteries to the lungs. Following gas exchange in the capillaries of the lungs, blood collects in the **pulmonary veins** and flows to the left atrium. These veins enter on the dorsal side of the heart and will be difficult to find. If you remove some of the lung tissue from the left side, you may be able to locate these vessels.

From the left ventricle, blood is pumped at high pressure through the **aorta** to the systemic circulation. Find the aorta. It will be partially covered by the pulmonary trunk but can be identified as the major vessel that curves 180° to the pig's left, forming the **aortic arch** (fig. 29.3).

Investigating Circulatory Systems   **395**

Figure 29.3 External ventral and dorsal views of the fetal pig's heart.

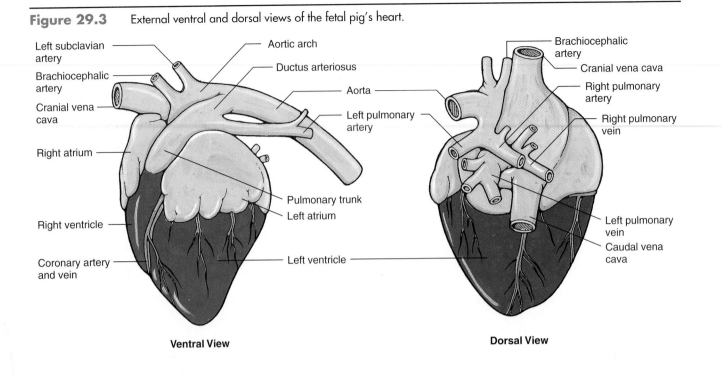

Ventral View

Dorsal View

Left subclavian artery
Aortic arch
Ductus arteriosus
Brachiocephalic artery
Aorta
Cranial vena cava
Left pulmonary artery
Right atrium
Pulmonary trunk
Right ventricle
Left atrium
Coronary artery and vein
Left ventricle

Brachiocephalic artery
Cranial vena cava
Right pulmonary artery
Right pulmonary vein
Left pulmonary vein
Caudal vena cava

Vessels Cranial to the Heart

Veins

4 ▷ Because the venous system is generally ventral to the arterial system, it will be studied first in the congested region of the heart. Refer to figure 29.4 and place a check next to each vein identified.

Trace the **cranial vena cava** forward from the heart to where it is formed by the union of the two very short **brachiocephalic veins.** Each of these in turn is formed by the union of the five major veins: the **internal** and **external jugular veins,** which drain the head and neck; the **cephalic vein,** which lies beneath the skin anterior to the upper forelimb and typically enters at the base of the external jugular; the **subscapular vein** from the dorsal aspect of the shoulder; and the **subclavian vein** from the shoulder and forelimb. As the latter passes into the forelimb, it is known as the **axillary vein** in the armpit and the **brachial vein** in the upper forelimb. Caudal to the union of the brachiocephalic veins, is a pair of **internal thoracic veins** that drain the chest wall. These veins were most likely cut when you opened the animal.

Arteries

5 ▷ Find the aortic arch and trace it back to the heart. Note the several arteries that branch off to supply the anterior region of the animal. Refer to figures 29.4 and 29.5 to identify the vessels. Check off the arteries as they are identified.

Find the small **coronary arteries** that arise from the base of the aorta behind the pulmonary trunk. They supply the muscles of the heart. The first major artery to branch

from the aorta is the **brachiocephalic artery.** It gives rise to the two **carotid arteries,** which pass anteriorly to supply the head, and the **right subclavian artery,** which passes to the right forelimb. Just to the left of the brachiocephalic artery, find the **left subclavian artery** arising as a separate branch from the aortic arch. Blood in this vessel goes to which region of the body?

Once the aorta runs posteriorward along the dorsal wall of the thorax, it gives rise to intercostal arteries, which supply the walls of the chest. The aorta then passes through the diaphragm to become the **abdominal aorta.**

Return to the left subclavian artery and trace it into the forelimb, removing skin and separating muscles as necessary. In the armpit it is known as the **axillary artery,** and in the upper forelimb as the **brachial artery.** The subscapular artery branches from the axillary artery and supplies the shoulder muscles. The brachial artery divides in the lower forelimb to give rise to the **radial** and **ulnar arteries.**

6 ▷ Find another student in the lab who is at the same stage in the dissection as you are. Quiz one another about the path blood takes as it flows from the forelimb through the heart to the head.

Figure 29.4 Ventral view of the fetal pig's major arteries (a) and veins (v).

Common carotid a
Thyrocervical a
Axillary a
Brachial a
Radial a
Ulnar a
Right atrium
Aortic arch
Ductus arteriosus
Pulmonary trunk
L. pulmonary a
Coronary a
Right ventricle
Hepatic a
Allantoic duct
Umbilical a
Renal a
Femoral a
External iliac a
Gonadal a
Internal iliac a
Internal jugular v
External jugular v
Brachiocephalic v
Cephalic v
Internal thoracic v
Cranial vena cava
Subscapular v
Radial v
Ulnar v
Axillary v
Brachial v
Subclavian v
Hepatic v
Caudal vena cava
Diaphragm
Coeliac a
Umbilical v
Abdominal aorta
Kidney
Renal v
Caudal mesenteric a
External iliac a
Femoral v
Internal iliac v
Median sacral a
Median sacral v

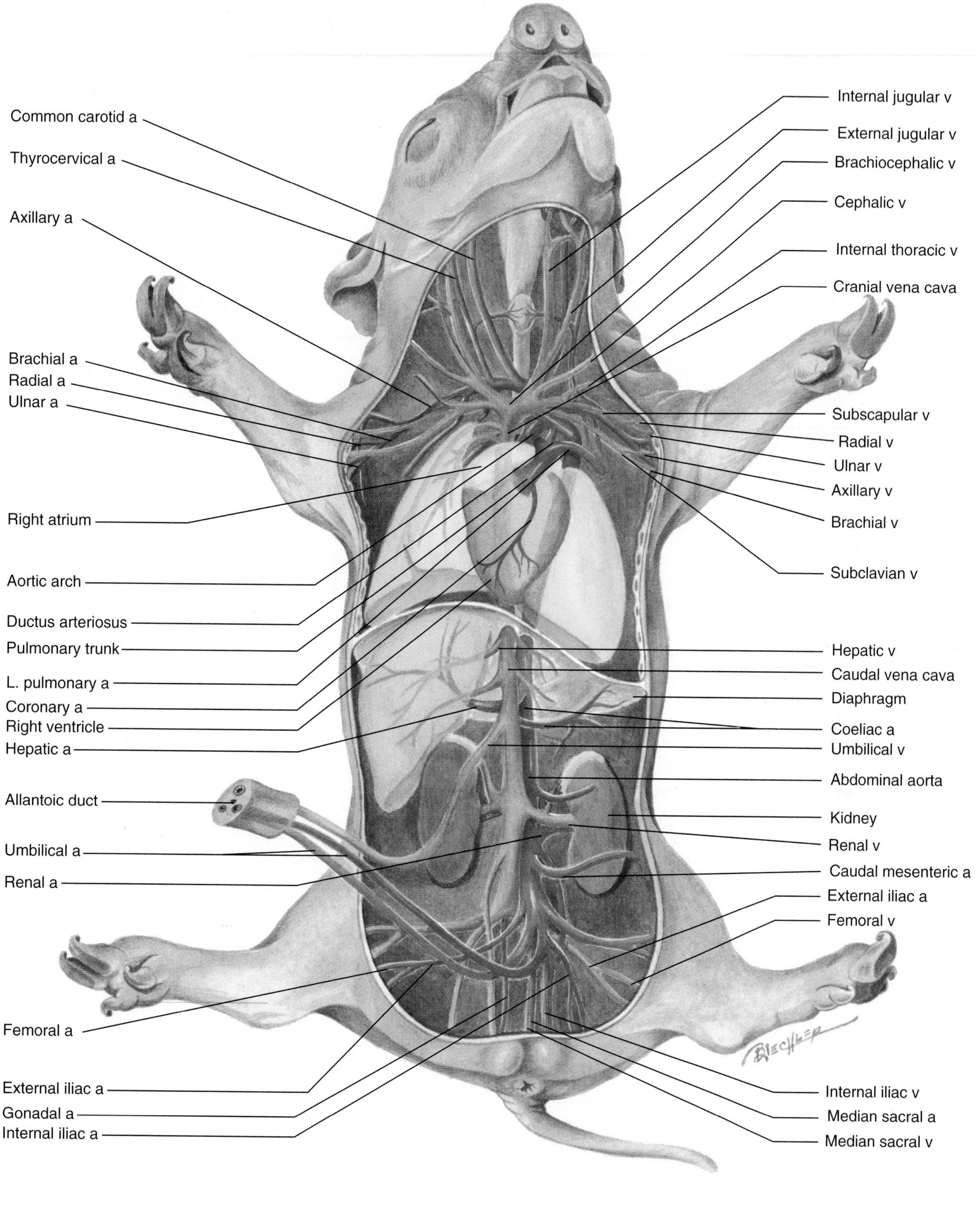

Figure 29.5   Major arteries in the region of the fetal pig's heart.

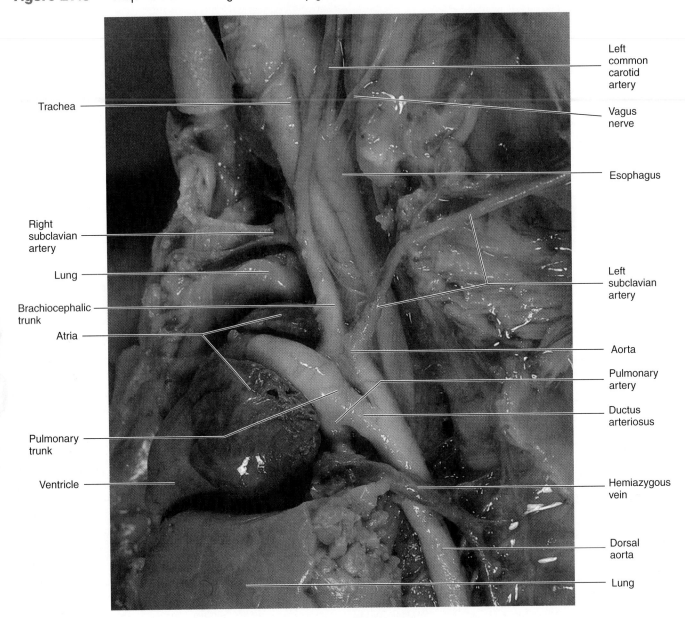

Left
common
carotid
artery

Trachea

Vagus
nerve

Esophagus

Right
subclavian
artery

Lung

Left
subclavian
artery

Brachiocephalic
trunk

Atria

Aorta

Pulmonary
artery

Ductus
arteriosus

Pulmonary
trunk

Ventricle

Hemiazygous
vein

Dorsal
aorta

Lung

Vessels Caudal to the Heart

Veins

7 ▷ If the heart is lifted and tilted forward, the **caudal vena cava** can be viewed at the point where it enters the right atrium. As this vein is traced caudally, several veins will be found flowing into it. After it passes through the diaphragm, the paired **hepatic veins** and single **umbilical vein** enter first. The umbilical vein carries oxygenated, nutrient-laden blood from the placenta. This vein passes through the liver where it is known as the ductus venosus.

The hepatic **portal vein** runs next to the common bile duct under the lobes of the liver. It will be difficult to find. Nutrient-laden blood flows through this vein to the liver where exchanges occur between the blood and the liver across the walls of the portal system capillaries. The blood then flows into the **hepatic veins**, which enter the caudal vena cava.

Follow the vena cava caudally to where the **renal veins** enter from the kidneys. In the male, the **spermatic veins,** and in the female, the **ovarian veins,** enter next. On the left side, these veins may enter the renal vein first. Below the kidneys, the vena cava splits into the **internal** and **external iliac veins** and the **median sacral vein,** a small vein that comes from the tail. The external iliac veins collect blood from the **femoral veins** in the hind legs, whereas the internal iliacs collect blood from the pelvic area.

Figure 29.6 Lateral view of circulatory system in fetal pig (arteries—red, veins—blue). Label the major arteries and veins.

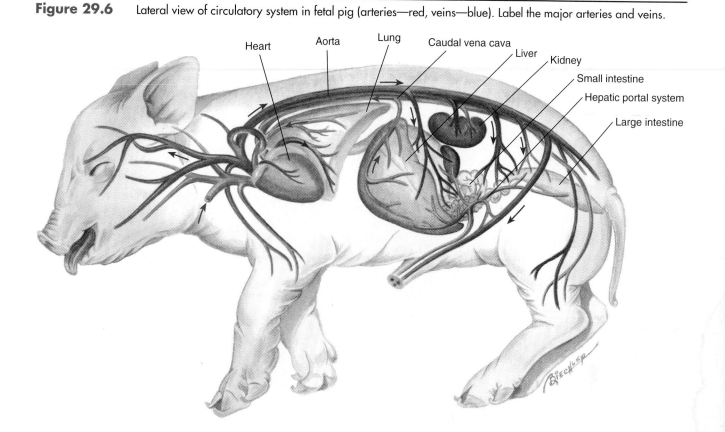

Heart   Aorta   Lung   Caudal vena cava   Liver   Kidney   Small intestine   Hepatic portal system   Large intestine

Arteries

8▶ After the aorta enters the abdominal cavity, a large, single **coeliac artery** arises from it at the cranial end of the kidneys. You will have to remove some connective tissues to obtain a full view of this artery. The coeliac artery eventually divides into three arteries supplying the stomach, spleen, and liver. The **mesenteric artery** next arises from the aorta and supplies the pancreas, small intestine, and large intestine. The **renal arteries** are short, paired arteries supplying the kidneys. The next large arteries arising from the aorta are the **external iliacs,** which supply the hind legs with a branch to the lower back. In the fetus, the **umbilical arteries** branch from the caudal end of the abdominal aorta and pass out through the umbilical cord. They form a capillary bed in the placenta for nutrient, gas, and waste exchange with the maternal circulatory system.

9▶ Find another student in the lab who is at the same stage in the dissection as you are. Quiz one another about the circulation paths to the major organs and hind limbs.

10▶ Look at the diagram in figure 29.6 and add labels to the major arteries and veins that you identified in your dissection.

Internal Heart Structure

11▶ Study the orientation of the heart so that you can later identify it in isolation. Now, free the heart from the body by cutting through all the vessels holding it in place. Be careful to

leave enough of each vessel so that they can be identified in the isolated heart. Alternative to removing the fetal pig's heart, your instructor may have a demonstration dissection of a beef heart or heart models for you to study. Whichever specimen you are using, orient yourself by identifying the **aorta, pulmonary artery, pulmonary vein,** and **vena cava.**

Place the heart in your dissecting pan, ventral side up. Make a razor cut along the pulmonary trunk down through the right ventricle. Spread the tissue open, pin it down, and remove the latex. You may wish to use a dissecting microscope to observe the open ventricle.

Identify the **semilunar valves** at the junction of the artery and ventricle. Consider how these valves work. The open flaps face into the pulmonary trunk, and any backflow in the pulmonary trunk fills the valve flaps with blood and closes the valve. You may have to cover the heart with water to float the valve flaps so that you can see them (fig. 29.7).

Now cut through the right atrium and remove the latex and coagulated blood. The **tricuspid valves** are between the atrium and ventricle. These valves also work on the backflow principle, allowing blood to flow only one way from the atrium into the ventricle. In the ventricle, fine fibers called **chordae tendinae** are attached to the valve flaps. These cords prevent the flaps from "blowing back" from high pressures developed when the ventricle contracts.

Cut into the left atrium and ventricle as you did on the right side. Identify the **bicuspid** or **mitral valve** between

Investigating Circulatory Systems   **399**

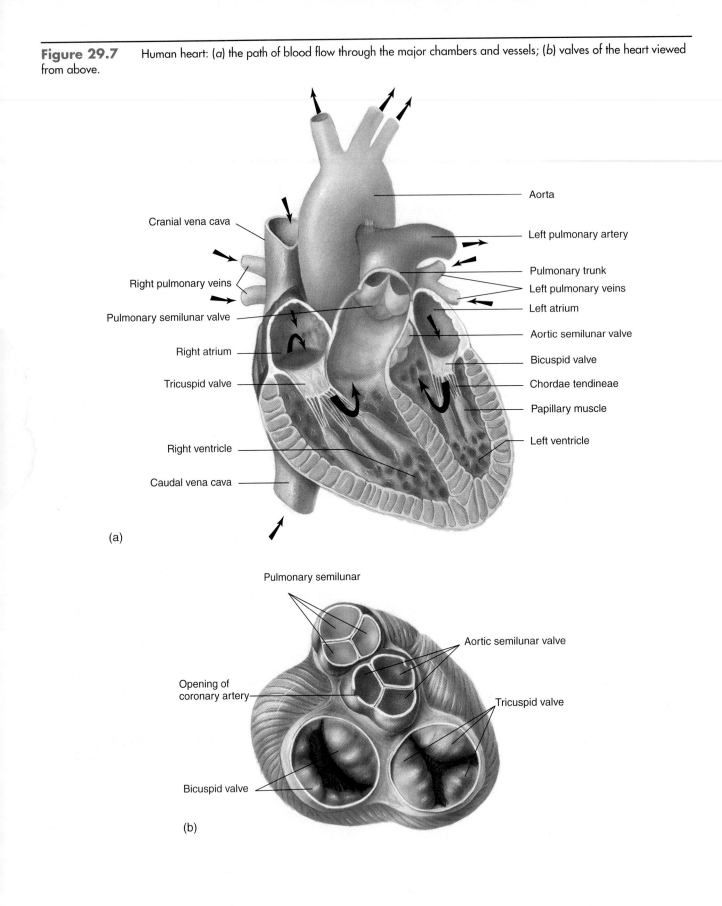

Aorta

Cranial vena cava

Left pulmonary artery

Right pulmonary veins

Pulmonary trunk

Left pulmonary veins

Pulmonary semilunar valve

Left atrium

Aortic semilunar valve

Right atrium

Bicuspid valve

Tricuspid valve

Chordae tendineae

Papillary muscle

Right ventricle

Left ventricle

Caudal vena cava

(a)

Pulmonary semilunar

Aortic semilunar valve

Opening of coronary artery

Tricuspid valve

Bicuspid valve

(b)

Figure 29.8 Histology of arteries and veins: (*a*) tissues in the wall of an artery; (*b*) tissues in the wall of a vein; valves in veins prevent backflow of blood in the venous sytem.

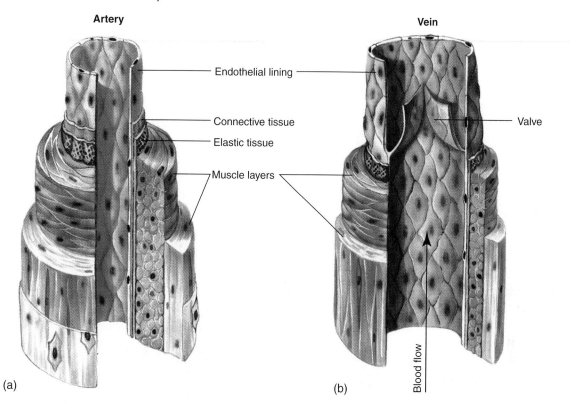

Artery

Vein

Endothelial lining

Connective tissue

Elastic tissue

Muscle layers

Valve

Blood flow

(a)

(b)

the atrium and ventricle with its associated chordae tendinae. After cutting into the ventricle and cleaning it, find the **aortic semilunar valve.** Blood flow through the human heart is shown in figure 29.7.

12 ▸ Pair off with another student and describe how blood returning to the heart from the foreleg travels to the lungs and then to the hindleg. Describe the operation of the heart valves.

Clean your dissecting instruments and tray and return your fetal pig to the storage area.

Histology of Vessels

13 ▸ Obtain prepared slides of cross sections of arteries and veins and observe them under low power with a compound microscope. Note that arteries have thicker walls than veins of the same size. Most of the difference in thickness is due to the increased amounts of muscle and connective tissue in the artery. Since arteries carry blood from the heart, they operate under relatively high pressure (average 120 mm of mercury equivalent). Veins experience only one-twentieth as much pressure.

Blood flows through veins because skeletal muscles press on them and move the blood along. **Valves** in the veins prevent backflow and make the passage of blood unidirectional (fig. 29.8).

Observe the tissues of a blood vessel under 10×. **Endothelial cells** are epithelial cells that line both arteries and veins; capillary walls consist of only endothelial cells. When the muscle layers are contracted in the smallest arter-

Figure 29.9 Scanning electron micrograph of a broken blood vessel, showing the smooth endothelial lining and several red blood cells (RBC).

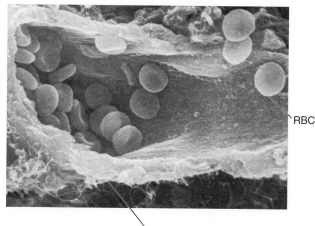

RBC

Vessel wall

ies, the total volume of the vascular system is reduced and the blood pressure rises.

Figure 29.9 shows the nature of the endothelial lining of blood vessels and red blood cells in an arteriole. The complexity of the microvasculature is evident in scanning electron micrographs of casts of the circulatory system. Note the capillaries and their relationship to arterioles in figure 29.10.

Investigating Circulatory Systems **401**

Figure 29.10 Scanning electron micrograph of a plastic cast of a capillary bed from skeletal muscle in which individual arterioles can be traced to capillaries. From R. G. Kessel and R. H. Kardon. *Tissues and Organs: A Text-Atlas of Scanning Electron Microscopy.* 1979. W. H. Freeman and Company.

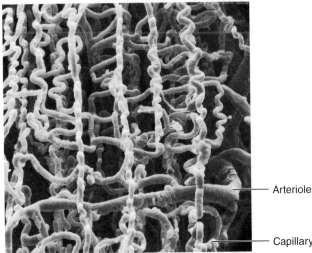
— Arteriole

— Capillary

Figure 29.11 Human blood stained with Wright's stain shows red blood cells and different types of white blood cells; (*a*) neutrophil; (*b*) lymphocyte.

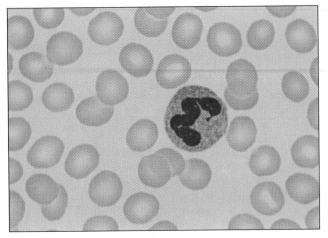
(a)

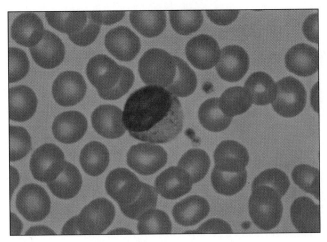
(b)

Blood

Human blood consists of 55% plasma and 45% cells by volume. Plasma is the fluid portion of the blood containing dissolved proteins, salts, nutrients, and waste products. Several different types of cells and cell fragments are contained in blood. By far the most common (about 95% of the cells) are **erythrocytes** (red blood cells) which are red because they contain hemoglobin. The other 5% are collectively called **leukocytes** (white blood cells) and **platelets** that are important in blood clotting. There are several types of leukocytes. The most common, neutrophils and lymphocytes, representing 95% of the white blood cells, are shown in figure 29.11. The remaining types of cells, basophils and monocytes, represent only 5% of the white blood cells.

14 Get a prepared slide of a Wright-stained human blood smear from the supply area and look at it with your compound microscope under medium power. Note the large number of red blood cells. Can you see a nucleus in these cells?_____. Why do you think that the red blood cells are lighter in color in the center and darker at the periphery?

As you look carefully at the slide, you will see occasional cells that look different. These are leukocytes and because of the staining they have a blue/purple color. Center a leukocyte in the field of view and observe it with high power. What structure in the cell is stained?_____. Return to medium power and scan the slide to locate other leukocytes. Try to find examples of each of the types shown in figure 29.11.

Neutrophils leave the blood early in the inflammation process and become phagocytic cells consuming cell debris and bacteria. **Lymphocytes** are important in the immune response. Some are involved in cellular immunity and others secrete antibodies that neutralize foreign proteins and other macromolecules.

Return your slide to the supply area.

Figure 29.12   Method for observing microcirculation in a fish tail.

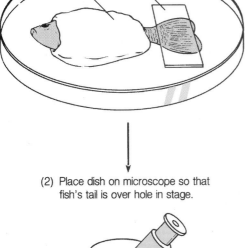

(1) Wrap the fish (except for the head and tail) with dripping wet cotton. Place fish in half of a petri dish. Place coverslip over tail.

Cotton   Coverslip

(2) Place dish on microscope so that fish's tail is over hole in stage.

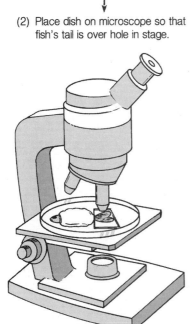

(3) Examine with low- and medium-power objectives of your microscope.

Circulation in Capillaries

15▷ To observe circulation in capillaries, net a small fish or tadpole from an aquarium and wrap it in dripping wet cotton, as shown in figure 29.12, being careful not to cover the head or the tail. Lay the wrapped fish in an open petri dish. About every five minutes return the fish to water. Place a few drops of water on the tail and add a coverslip over the tail.

Place the dish on a compound microscope stage and observe the tail under scanning power. Sketch your observations, answering the following questions:

1. Can you identify capillaries, venules, and arterioles?

2. Is blood flow faster in certain vessels compared to others?

3. Is blood flow continuous in all vessels? What might control this?

Dilute solutions of nicotine, caffeine, and adrenalin are available in the lab in dropper bottles. Devise an experiment to determine the effects of these chemicals on capillary circulation.

Measuring Blood Pressure

Blood pressure is the pressure that the blood exerts on the walls of blood vessels and is usually measured in arteries. Because of the contraction cycle of the heart, pressure varies from a high (**systolic**) to a low (**diastolic**) pressure during a cycle. Pressures are reported in mm of mercury (Hg) with the systolic pressure appearing first. Thus a blood pressure reading of 120/60 means that the pressure just following maximum contraction is 120 and just following maximum relaxation is 60 mm of Hg. In this section, you will learn how to take a blood pressure reading by the indirect method.

Method   An instrument called a **sphygmomanometer** and a **stethoscope** are used in this method. An inflatable cuff is placed around a person's arm (fig. 29.13) and inflated until the pressure exerted is greater than the systolic

Figure 29.13   Method to be used in measuring blood pressure.

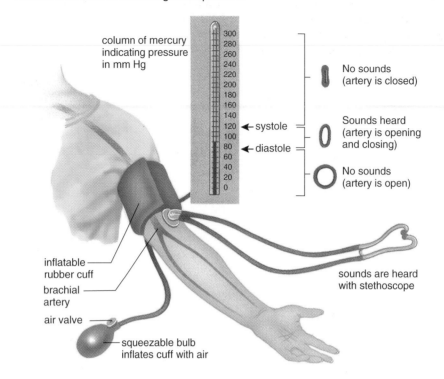

column of mercury indicating pressure in mm Hg

300
280
260
240
220
200
180
160
140
120 ← systole
100
80 ← diastole
60
40
20
0

No sounds (artery is closed)

Sounds heard (artery is opening and closing)

No sounds (artery is open)

inflatable rubber cuff

brachial artery

air valve

squeezable bulb inflates cuff with air

sounds are heard with stethoscope

pressure, thus collapsing the arteries in the upper arm. The pressure in the cuff is gradually reduced while the operator listens for resumption of blood flow in the arteries with a stethoscope (fig. 29.13). The pressure at which faint tapping sounds are heard is taken as the systolic pressure. As pressure in the cuff is further reduced, louder sounds are heard because of turbulent blood flow through the partially compressed artery. There is a pressure at which these sounds stop when blood flow becomes laminar. This pressure is considered the diastolic pressure.

16   Work with a partner to measure these pressures. The subject should sit with the arm extended and resting on a table. Your instructor may explain in more detail how to do the following procedures.

1. Wrap the cuff around the upper arm in such a way that the inflatable area is over the inner surface of the arm where the brachial artery is located.

2. Don the stethoscope and place the end over the pulse point located on the inside of the elbow joint (fig. 29.13).

3. Inflate the cuff to about 160 mm of pressure. DO NOT OVER INFLATE!

4. Slowly release the pressure by opening the valve on the squeeze bulb as you listen for the first thudding sounds of blood flow. Note the pressure when you hear them. Continue to reduce pressure until the sounds are no longer heard. Note the pressure reading when this happens.

Results   Record your results below:

First trial
    Systolic pressure
    Diastolic pressure

Second trial
    Systolic pressure
    Diastolic pressure

What range of pressure values were observed in the class? Were there any differences between genders? Were there any differences between athletes and non-athletes?

17 ▶ *Blood Pressures under Experimental Conditions*

Assuming that you now can measure blood pressures rather quickly, try doing so under these experimental conditions:

Have a fully cuffed subject recline on a table for two or three minutes and then quickly stand as you inflate the cuff. Take your measurements and record below:

First trial
    Systolic pressure
    Diastolic pressure

Second trial
    Systolic pressure
    Diastolic pressure

Does this measurement have any relationship to light-headedness that is sometimes felt when getting up quickly? Explain?

Learning Biology by Writing

Your instructor may ask you to answer the Lab Summary and Critical Thinking questions that follow.

Internet Sources

Many medical schools have extensive collections of pictures of pathological conditions. These collections are available over the WWW. Use your browser program and a search engine to locate pictures of a blood vessel with arteriosclerosis. Compare this picture to your observations of a normal artery in the lab. Describe the differences.

Lab Summary Questions

1. Create a flowchart showing the major vessels and heart chambers that a drop of blood passes through as it returns from the arm and passes to the back leg.
2. Create a flowchart showing the major vessels and heart chambers that a drop of blood passes through as it returns from the small intestine and passes to the brain.
3. List the valves of the heart and describe how they operate.

4. What are the structural differences between an artery and a vein? How do capillaries differ?
5. How does the open circulatory system of a crayfish differ from the closed system of a mammal? Describe how blood returns to and enters the heart of both types of animals.
6. Trace a molecule of oxygen from when it enters the pig's nostril to when it enters the tissues of the upper hind leg. Then trace a carbon dioxide molecule as it passes from the leg back to the nostril.
7. Explain what blood pressure is and what it means when a person says their blood pressure is 125/70.

Critical Thinking Questions

1. What are the roles of the lymphatic system and the venous system in returning fluid filtered through the microcirculation?
2. Since a giraffe's head is 15 feet above the ground, what circulation adaptations are necessary to allow adequate blood supply to the head?
3. If arthropods such as insects have open circulatory systems that lack veins, how does hemolymph return from the tissues to the heart?
4. Based on your lab observations, give an opinion on the following situation: A person has a blood pressure of 150/90.

LAB TOPIC 30

Investigating the Urogenital System

Supplies

Preparator's guide available on WWW at
http://www.mhhe.com/dolphin

Equipment

Compound microscopes
Dissecting microscopes

Materials

Fetal pig
Dissecting pan and instruments
Prepared slides
    Mammalian kidney cortex, demonstration
    Mammalian ovary with Graafian follicles
    Mammalian seminiferous tubules, cross section
Microscope slides and coverslips
Live earthworms

Solutions

0.5% methylene blue
10% ethanol

Student Prelab Preparation

Before doing this lab, you should read the introduction and sections of the lab topic that have been scheduled by the instructor.

You should use your textbook to review the definitions of the following terms:

bladder	oviduct
Bowman's capsule	seminiferous tubule
collecting duct	testis
epididymis	ureter
glomerulus	urethra
nephron	uterus
ovary	vas deferens

You should be able to describe in your own words the following concepts:

Glomerular filtration in kidney
Tubule reabsorption in kidney
Pathway sperm travel from formation to ejaculation
Pathway eggs travel from formation to uterus

As a result of this review, you most likely have questions about terms, concepts, or how you will do the experiments included in this lab. Write these questions in the space below or in the margins of the pages of this lab topic. The lab experiments should help you answer these questions, or you can ask your instructor during the lab.

Objectives

1. To observe nephridia from live earthworms
2. To observe the gross and microscopic anatomy of the mammalian excretory and reproductive systems in a fetal pig by dissection
3. To be able to trace a drop of water leaving the blood in the kidney until it shows up in urine being voided
4. To observe the microsopic structure of the seminiferous tubule and ovary

Background

The excretory systems of animals maintain a constant chemical state in the internal body fluids. The name *excretory system*, with its implicit emphasis on waste removal, does not suggest the three other important functions of these systems: (1) controlling water volume, (2) regulating salt concentrations, and (3) eliminating nonmetabolizable compounds absorbed from food.

Excretion involves the elimination of the waste products from the metabolism of nitrogen-containing compounds. **Ammonia,** a toxic compound at high concentrations, is produced by all animals when they metabolize amino acids and nucleotides. Depending on whether an animal inhabits an aquatic or terrestrial environment, this toxic product is handled in different ways. Aquatic organisms generally excrete the NH_3 directly into the surrounding water, often through their gills. Terrestrial animals convert it into less toxic compounds, such as **urea** in mammals or **uric acid** in insects, reptiles, and birds. These products are collected by excretory organs and periodically voided.

Marine, freshwater, and terrestrial environments present quite different challenges to animals in terms of internal water volume regulation and regulation of salt concentrations in body fluids. The following paragraphs outline these challenges.

Many marine invertebrates lack well-developed excretory systems. They have body fluids that resemble seawater in terms of osmotic concentration and ionic composition. Consequently, diffusion and active transport are sufficient to maintain the minor differences that exist. Furthermore, such animals usually have little tissue volume relative to total surface area. Ammonia diffuses from the body into the virtually infinite reservoir of the ocean at rates that are sufficient to prevent internal toxic levels.

In more advanced marine invertebrates and fish, the mass of tissues dictates that there be mechanisms other than diffusion to get rid of nitrogenous wastes. Body fluids circulating through gills often lose wastes and salts by diffusion while gas exchange is also taking place. This mechanism is augmented by development of excretory organs called kidneys.

In terrestrial animals two different mechanisms seem to have evolved. Insects, reptiles, and birds convert waste ammonia into uric acid, which is practically insoluble and requires little accompanying water when it is excreted. Mammals, on the other hand, have well-developed kidneys that can recover both water and salt from the urine after it is produced, while allowing urea to be excreted.

Mammalian kidneys function by filtering wastes and then selectively reabsorbing materials from the filtrate. The mammalian kidney consists of millions of capillary tufts, the **glomeruli.** They filter blood so that the fluid portion (that is, the portion with no blood cells or proteins) enters the tubules of the urinary system. The concentrations of water, salts, urea, sugars, amino acids, and so on, in the filtrate are very similar to those in the blood. As the filtrate passes through the tubular network of the **nephrons** toward the **collecting ducts** of the urinary system, many of these materials are reabsorbed by active and passive transport, leaving wastes, excess salts, and water. By the time the filtrate reaches the collecting ducts, it has been changed by the reabsorption of water and salts, but not waste products.

In the adult human, 180 liters of filtrate are produced each day, but the daily urine volume is only 0.6 to 2.0 liters. If we assume that the plasma volume is 3 liters, the production of 180 liters of filtrate means that all of the blood plasma must be filtered through the glomeruli 60 times in 24 hours. A volume of fluid equal to 178 to 179 liters is reabsorbed along with many salts. The reabsorption process is obviously of some magnitude.

In mammals the organs of the excretory and reproductive systems are closely associated especially in males, so the term urogenital system is often used although the systems are distinctly different in their functions. The gonads produce haploid gametes by meiosis. In oogenesis, the formation of eggs in the ovary, the first stages of meiosis have occurred while the animal was an embryo. The eggs will remain in an arrested stage until the female reaches sexual maturity. Eggs are then released by the ovary into the body cavity where they are picked up by ducts of the reproductive system. As eggs descend the reproductive tract, they will finish meiosis II. The elapsed time between meisosi I and II can be years. If fertilization occurs, then the embryo implants in the uterus and completes development. In spermatogenesis, the production of sperm in the testis, meiosis does not start until sexual maturity is reached. It can then be more or less continuous throughout the lifespan, producing millions upon millions of cells that develop into sperm.

LAB INSTRUCTIONS

You will observe the excretory system of the earthworm and mammal. Keep in mind that these systems are involved in water, salt, and waste balance. Because of the close anatomical location of the excretory and reproductive systems, you will also study the anatomy of the mammalian reproductive system.

Invertebrate Excretory Systems

Nephridial Systems

Invertebrates in several phyla have excretory organs called **nephridia.** The earthworm is one example of an animal with a nephridial system (refer to fig. 21.9). In each segment, except for a few anterior ones, is a pair of nephridia, which open independently to the outside. Figure 30.1 diagrams the anatomy of a nephridium. Body fluids enter the nephridium via a ciliated, funnel-like structure, the **nephrostome,** and pass through a convoluted **tubule** to a distensible **bladder.**

Figure 30.1 The anatomy of a nephridium in an earthworm segment.

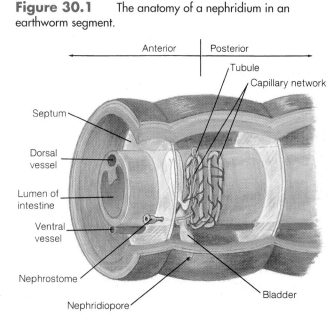

As fluids pass along the tubule, materials such as water and salts exchange between the tubule and the capillary network surrounding it. The contents of the bladder are voided to the outside through the **nephridiopore.**

1 ▷ If live earthworms are available in class, they can be used to view the nephridial system directly. Anesthetize a worm by placing it in 10% ethanol until it does not respond. Remove the worm and place it in a clean dissecting dish with a thick layer of wax on the bottom. Make a careful longitudinal cut into the posterior half of the worm and pin back the tissue flaps. Flood the body cavity with water. Surgically remove part of a septum containing a nephridium. Place it in a drop of 0.5% methylene blue dye on a microscope slide and spread the septum so that it is not folded. Add a coverslip and look at it with your compound microscope.

Observe the relationship between the capillaries and the tubules. Parasitic roundworms are often found in the bladder. Fill in the information required in table 30.1.

Mammalian Excretory System

Anatomy of Fetal Pig Excretory System

2 ▷ Get your fetal pig from the supply area. If you have a male pig, use figure 30.2 as a guide. If you have a female, refer to figure 30.3. Be sure to look at a pig of the opposite sex before you leave lab.

The kidneys are located on the dorsal wall of the abdominal cavity. Remove the **peritoneum,** the connective tissue membrane that holds them in place. Note how close the kidneys are to the **descending aorta.** The blood pressure drops very little as blood passes from the aorta into the kidneys via the **renal arteries.** A high blood pressure is essential for proper kidney function. **Renal veins** drain the kidney, returning blood to the caudal vena cava.

Find the **ureter,** which originates from the medial face of the kidney. Trace it caudally to where it empties into the **urinary bladder** where urine is stored. In the fetal pig, part of the urinary bladder extends between the two **umbilical arteries** and continues into the umbilical cord, where it is called the **allantoic duct.** After birth, the duct atrophies, becoming nonfunctional.

The **urethra** proceeds caudally from the bladder to its opening; its location depends on the sex of your pig. It is difficult to observe and trace. A sphincter muscle, under voluntary control, regulates release of urine from the bladder into the urethra. Smooth muscles in the bladder wall contract to force urine out of the bladder.

Though they are not involved in excretion, note the **adrenal glands,** which are located cranially and medially to the kidneys in the abdominal cavity. Adrenaline and noradrenaline, hormones produced by these glands, regulate a number of body functions, including heart rate, blood sugar levels, and arteriole constriction.

Remove one kidney. Make a longitudinal cut through the kidney with a sharp razor blade in a plane parallel with the front and back of the kidney. Study the cut surface with the dissecting microscope (fig. 30.4). Locate where the ureter entered the kidney and note how it subdivides into

TABLE 30.1	Excretory system summary		

Fill in the names of structures in the earthworm or the fetal pig that are involved in the named process.

	Earthworm	**Fetal Pig**
Filtration		
Absorption		
Secretion		
Voiding		

Investigating the Urogenital System

Figure 30.2   Ventral view of a male fetal pig's urogenital system.

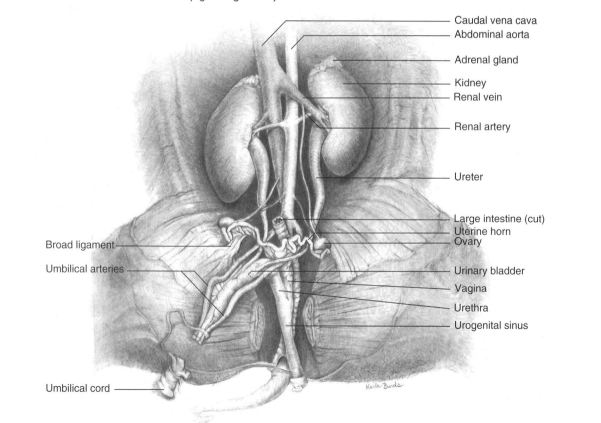

Caudal vena cava
Abdominal aorta
Adrenal gland
Kidney
Renal vein
Renal artery
Ureter
Testicular artery and vein
Large intestine (cut)
Cremasteric pouch
Urinary bladder
Testis
Inguinal canal
Epididymis
Ductus deferens
Umbilical arteries
Urethra
Penis
Bulbourethral glands

Figure 30.3   Ventral view of a female pig's urogenital system.

Caudal vena cava
Abdominal aorta
Adrenal gland
Kidney
Renal vein
Renal artery
Ureter
Large intestine (cut)
Uterine horn
Ovary
Broad ligament
Umbilical arteries
Urinary bladder
Vagina
Urethra
Urogenital sinus
Umbilical cord

Figure 30.4 Longitudinal section of a mammal's kidney.

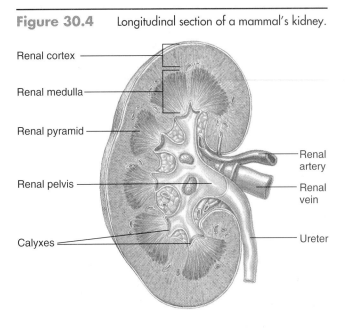

Renal cortex

Renal medulla

Renal pyramid

Renal pelvis

Calyxes

Renal artery

Renal vein

Ureter

Figure 30.5 Structure of nephron and relationship to circulatory system.

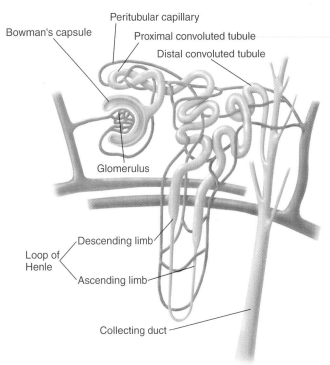

Bowman's capsule

Peritubular capillary

Proximal convoluted tubule

Distal convoluted tubule

Glomerulus

Loop of Henle

Descending limb

Ascending limb

Collecting duct

funnel-shaped **calyxes.** Urine is produced by the thousands of filtration-reabsorption units called **nephrons.** It flows through collecting ducts to the calyxes and passes by way of the ureter to the bladder for storage.

Nephron Structure

The functional unit of the kidney is the nephron. In a mature pig's kidney, there may be a million of these units. Its structure is shown in figure 30.5 and will be briefly discussed here. Blood from the renal artery finds its way to tufts of capillaries called glomeruli. There the high blood pressure forces the fluid portion of the blood, minus any blood cells or large proteins, out of the capillaries. The filtrate is captured by microscopic funnel-shaped structures called Bowman's capsules. From there, it flows through a U-shaped tubular system (fig. 30.5) where the filtrate is modified by selective absorption and secretion. Things like amino acids and sugars that passed through the filter are reabsorbed. Salts may be reabsorbed or secreted into the developing urine. Hormones such as antidiuretic hormone and aldosterone, regulate the filtration, absorption, and secretion processes. Eventually, the fluids flows into the collecting ducts. Collecting ducts from the million nephrons join together to form the ureter that leaves the kidney.

A whole nephron cannot be seen in a section of a kidney. The tubules and blood vessels pass in many different planes in the kidney and a slice virtually never contains a whole nephron. Your instructor may have a demonstration slide of a glomerulus with its associated Bowman's capsule for you to view.

3 ▷ If models of the kidney are available in the lab, you and a partner should trace how a drop of water can pass from the blood to the ureter.

To gain an understanding of comparative anatomy, fill in table 30.1. Name each structure involved in filtration, reabsorption, excretion, storage, and voiding.

Mammalian Reproductive System

Though not part of the excretory system, the organs of the reproductive system are located so close to the excretory system that they warrant a brief discussion here. Follow the directions according to the sex of your pig. After you have finished your dissection, explain the anatomy to someone who has a pig of the opposite sex. When finished, have them explain their dissection to you.

Male System

During embryonic development, the **testes** originate in the coelom caudal to the kidneys, and then descend to lie in an external pouch, the **scrotum,** in adult pigs.

4 ▷ To find the testes in a fetal pig, locate the testicular artery and vein on one side of the animal and trace them to the **inguinal canal** (see fig. 30.2). The testis passed through this canal during its descent guided by the gubernaculum, a mass of smooth muscle connecting the testis to the scrotal sac, which shortens during growth. Pass a blunt probe through the inguinal canal.

The probe should help you locate a thin-walled, elongated sac extending across the ventral surface of the thighs just ventral to the tail. Cut through the skin and reveal the **cremasteric pouch.** It is an outpocketing of the connective and muscle tissues of the abdominal wall. Cut it open to view the testis.

Sperm are produced in the microscopic seminiferous tubules inside the testis. They in turn pass into the **epididymis,** visible as a tightly coiled mass of tubules along one

Figure 30.6 Male fetal pig's urogenital system in sagittal section view.

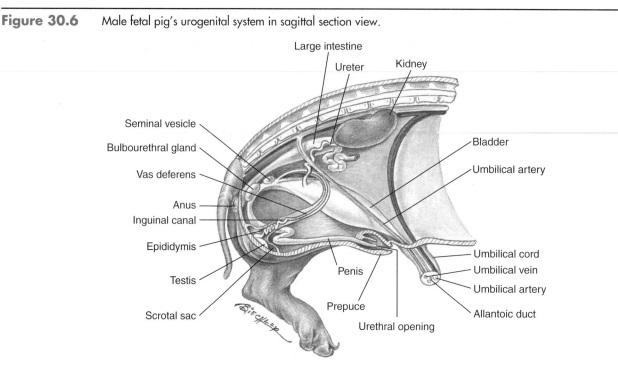

Large intestine
Ureter
Kidney
Seminal vesicle
Bulbourethral gland
Vas deferens
Anus
Inguinal canal
Epididymis
Testis
Scrotal sac
Penis
Prepuce
Urethral opening
Bladder
Umbilical artery
Umbilical cord
Umbilical vein
Umbilical artery
Allantoic duct

side of the testis. The epididymis flows into the **vas deferens,** which passes from the scrotal area through the abdominal wall and into the body cavity via the inguinal canal. After entering the body cavity, the vas deferens on each side loops over the ureter and enters the **urethra.** The paired **seminal vesicles** and single, bilobed **prostate gland** are also located at this juncture. The large **bulbourethral glands** lie on either side of the junction of the urethra with the **penis.** These three glands secrete additional fluids that carry the sperm during an ejaculation. The penis of a fetal pig is retracted and lies deep in the muscle layers of the groin area. The location of these structures is shown in sagittal section in figure 30.6.

Microanatomy of Seminiferous Tubules

5▶ Obtain a slide of a cross section of a mammalian testis containing **seminiferous tubules.** Using the compound microscope, look at it first with low power. Center a tubule in the field of view and then switch to high power. Compare the specimen to figure 30.7.

Note that the cells grow smaller as you scan from the outer tubule wall to the inside, where mature sperm may be present in the **lumen.** The largest outer cells are special stem cells that divide by mitosis to produce other cells. Half of these cells undergo meiosis and become sperm, while the other half function as stem cells to again divide by mitosis and keep the process going. Since a single ejaculation contains hundreds of millions of sperm, rejuvenation of the stem cell population is important.

Primary spermatocytes are cells destined to enter meiosis. When these cells finish meiosis I, two **secondary spermatocytes** are produced. They rapidly undergo meiosis II to

Figure 30.7 Cross section of a seminiferous tubule, showing stages in sperm development as they pass from outer wall to the lumen of tubule.

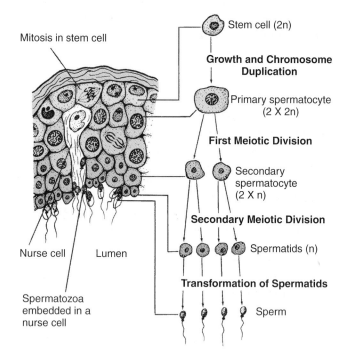

Mitosis in stem cell

Nurse cell   Lumen

Spermatozoa embedded in a nurse cell

Stem cell (2n)

Growth and Chromosome Duplication

Primary spermatocyte (2 X 2n)

First Meiotic Division

Secondary spermatocyte (2 X n)

Secondary Meiotic Division

Spermatids (n)

Transformation of Spermatids

Sperm

Figure 30.8   Ventral view of female urogenital organs. Intestines have been removed.

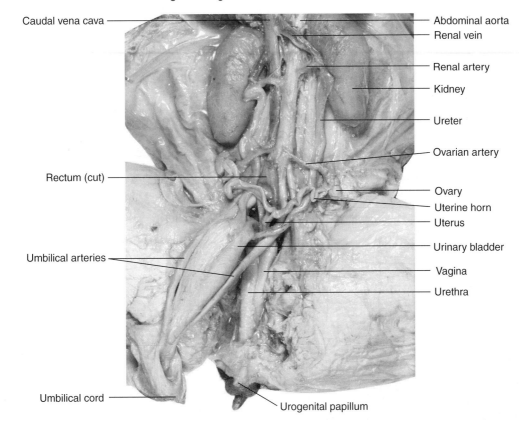

Caudal vena cava

Abdominal aorta
Renal vein

Renal artery
Kidney

Ureter

Ovarian artery

Rectum (cut)

Ovary
Uterine horn
Uterus

Umbilical arteries

Urinary bladder

Vagina
Urethra

Umbilical cord

Urogenital papillum

produce spermatids that mature into functional sperm. **Nurse cells** (or **Sertoli cells**) located in the seminiferous tubule walls aid in this process. Between the seminiferous tubules, **interstitial cells** secrete the male hormone testosterone.

Female System

6▷ The **ovaries** are small, bean-shaped organs on the dorsal wall near the caudal end of the kidneys (fig. 30.7). The ovaries are suspended by two sheets of connective tissue, the **broad ligament** (see fig. 30.3 and 30.8). On the dorsal side of each ovary is an **oviduct** (called Fallopian tubes in humans). Eggs bursting from the ovary enter the openings of the oviducts. Cilia on cells lining the oviduct move the egg toward the uterus. Fertilization usually occurs in the oviducts. The pig uterus has two convoluted uterine horns where multiple embryos can imbed and develop. In humans, the embryo develops in the uterus. The uterine horns unite to form the **uterus,** which opens to the outside through a muscular tube, the **vagina.** The vagina and the urethra open into a common area.

You are now finished with the fetal pig for this laboratory. Return it to the storage area.

Microanatomy of the Ovary

7▷ Obtain a slide of a section of a mammalian ovary and look at it under low power with the compound microscope. Large, clear areas will be visible in the section, as in figure 30.9. These are **Graafian follicles,** structures in which egg maturation occurs.

There are several hundred thousand follicles in each ovary of a female at the time of her birth. The ova in these follicles are in an arrested prophase of meiosis I. At sexual maturity, hormones from the pituitary gland cause the follicles to secrete the female hormones estrogen and progesterone.

If you are lucky, your slide will show such an enlarged follicle containing an ovum and surrounded by follicle cells. During ovulation, the follicle swells until it finally bursts, releasing the egg in a rush of fluid. Fingerlike projections of the oviduct surround the ovary and collect the egg. The egg moves down the oviduct by the action of cilia on the tube surface.

Sperm usually fertilize the egg as it passes down the oviduct. A second meiotic division will take place in the egg only if a sperm fertilizes it.

The collapsed follicle left in the ovary following ovulation assumes a star-shaped appearance and fills with a yellow fluid. At this stage, it is called a **corpus luteum** and secretes progesterone. If fertilization occurs and the ovum implants in the uterus, a hormone produced by the developing placenta will sustain the corpus luteum and progesterone production. If implantation does not occur, the corpus luteum will degenerate after several days and progesterone production will decrease. Falling levels of progesterone trigger the pituitary to produce follicle-stimulating hormone (FSH), which spurs another Graafian follicle toward ovulation. The total time course for these events in a human is approximately 28 days.

Figure 30.9 Photomicrograph of a section from a rhesus monkey's ovary.

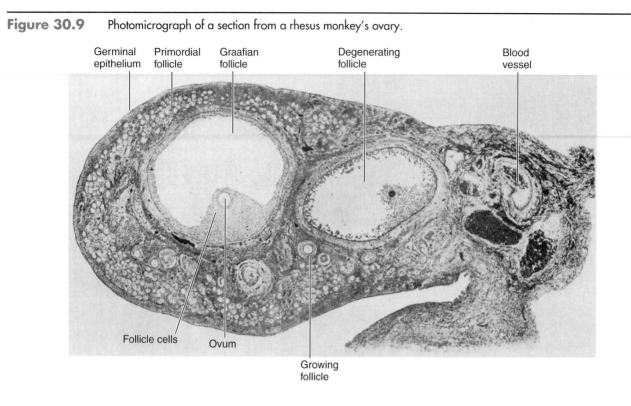

Germinal epithelium | Primordial follicle | Graafian follicle | Degenerating follicle | Blood vessel

Follicle cells | Ovum

Growing follicle

Learning Biology by Writing

The nephron is the basic functional unit of the kidney. Briefly describe the parts of the nephron and their functions, describing how water and small dissolved molecules pass from the blood and become urine. Describe the pathway that urine follows as it flows from the nephron to the time it is voided.

As an alternative assignment, your instructor may ask you to answer the Lab Summary and Critical Thinking Questions that follow.

Internet Sources

Search the World Wide Web for information on kidney stones. What are they and how are they formed? Be sure to record **URLs** for future reference.

Lab Summary Questions

1. How does the nephridial system of an earthworm work?
2. Create a flowchart showing how urine is formed in a mammal from fluids in the blood and is eventually voided. To the right of the structures named, briefly describe their function.
3. Discuss why filtration and reabsorption are both important processes in the kidney.

4. Name the structures that a sperm passes through from the time it is formed until it is ejaculated.
5. Name the structures an egg passes through from the time it is formed until it reaches the uterus.

Critical Thinking Questions

1. During embryonic development, the testes develop within the abdominal cavity. Shortly before birth, or shortly thereafter, the testes descend through the inguinal canal to lie outside the abdominal cavity within the scrotal sacs. Why is this important?
2. Some mammals, such as bears, badgers, and raccoons, enter a period of prolonged sleep during the winter months when temperatures are extreme and food is scarce. This is not considered "hibernation" because there is little if any drop in body temperature. Also, the heart rate in bears may drop only from 40 to 10 beats per minute. On the other hand, some mammals, such as woodchucks and ground squirrels, enter a period of true hibernation. Their body temperature cools to within a degree or two of the ambient temperature. Their respiratory rate decreases from 200 to around 5 breaths per minute, and heart rate drops from 150 to 5 beats per minute.

   What excretory problems do bears face during their long sleep? Would woodchucks have the same challenge? Suggest possible adaptations that these mammals might have evolved to deal with nitrogenous wastes during winter.

3. Is internal fertilization superior to external fertilization? Why do you say so?
4. Researchers interested in the physiology of the kidney made the measurements shown in table 30.2. Based on your understanding of how the kidney works, explain why some substances are found at the same concentration in the plasma and glomerular filtrate and why others are not. Also explain why most substances are found at different concentrations when the glomerular filtrate is compared to the urine.
5. Create a flowchart that traces a drop of water from the time it enters the mouth until it is voided in the urine of a mammal. You will have to list structures from the digestive, circulatory, and excretory systems.

TABLE 30.2 Concentrations of selected substances in mg per 100 ml of fluid

Substance	Plasma	Glomerular Filtrate	Urine
Albumin	4500	0	0
Glucose	100	0	0
Urea	26	26g	1820
Uric acid	4	4	53
Na	330	330	297
K	16	16	192
Cl	350	350	455

Assume that a normal adult has 31 liters of blood plasma; glomerular filtration rate is 180 liters per day and normal urine output is about 1.5 liters per day.

LAB TOPIC 31

Investigating the Properties of Muscle and Skeletal Systems

Supplies

Preparator's guide available on WWW at
http://www.mhhe.com/dolphin

Equipment

Compound microscopes
Intellitool Setups (Batavia, IL)
    Physiogrip and software
    Isolated, square wave stimulator
Computer

Materials

Fetal pig
Dissection pans and instruments
Millipedes, spiders, crabs, or insects
Bird skeleton
Human skeleton
Frog skeleton
Fresh beef long bones split longitudinally
Dry long bones
Prepared slides
    Skeletal muscle, longitudinal
    Smooth muscle section
    Cardiac muscle, longitudinal
    Bone
    Hyaline cartilage
    Tendon
    Earthworm, cross section
    Neuromuscular junction (demonstration)

Student Prelab Preparation

Before doing this lab, you should read the introduction and sections of the lab topic that have been scheduled by the instructor.
    You should use your textbook to review the definitions of the following terms:

antagonistic muscles	pelvic and shoulder girdles
appendicular skeleton	sarcomere
axial skeleton	smooth, cardiac, and
endoskeleton	skeletal muscle
exoskeleton	summation
extensor	tetany
flexor	threshold effect
hydrostatic skeleton	twitch

You should be able to describe in your own words the following concepts:

    Sliding filament theory of muscle contraction
    How a nerve impulse causes a muscle to contract
    Basic structure of bone

    As a result of this review, you most likely have questions about terms, concepts, or how you will do the experiments included in this lab. Write these questions in the space below or in the margins of the pages of this lab topic. The lab experiments should help you answer these questions, or you can ask your instructor during the lab.

Objectives

1. To compare the microanatomy of skeletal muscle to that of smooth and cardiac muscle
2. To investigate the surface muscles in the fetal pig
3. To investigate the physiological concepts of threshold, recruitment, and temporal summation during muscle contraction
4. To view examples of hydrostatic skeletons and exoskeletons
5. To observe the microanatomy of bone
6. To compare the endoskeletons of several vertebrates

Background

Simple small organisms, such as bacteria, algae, protozoa, and sponges, are capable of movement without muscle systems. They use cilia or flagella to move through their aqueous environments. All higher animals, whether aquatic or terrestrial, depend on muscle systems for movement and on nervous systems to control that movement. These muscle systems are also intimately associated with the skeletal system, which converts muscular contraction into effective movement and locomotion.

Figure 31.1 Antagonistic muscle arrangements in human limb. Contraction of human biceps flexes the forearm, while contraction of the triceps extends the forearm.

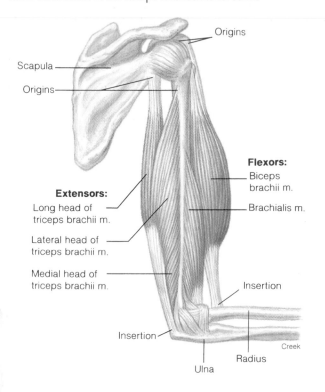

Origins

Scapula

Origins

Extensors:
Long head of
triceps brachii m.

Lateral head of
triceps brachii m.

Medial head of
triceps brachii m.

Insertion

Ulna

Radius

Creek

Flexors:
Biceps
brachii m.

Brachialis m.

Insertion

Muscles are nearly always found arranged as **antagonistic** pairs. As a result, contraction of one member of the pair causes an action, while contraction of the second member restores the body to its original position. In earthworms, the longitudinal and circular muscles in the body wall are antagonists. Contraction of the circular muscles elongates the segments by forcing the fluid of the hydrostatic skeleton forward and back in each segment, but contraction of the longitudinal muscles shortens the segments, leading to an increase in girth. In antagonistic pairs, neuronal activity is such that the contraction of one antagonist is usually accompanied by the relaxation of the second.

Figure 31.1 shows the antagonistic arrangement of human muscles associated with an internal skeleton. Muscles that increase the angle between two bones at a joint are called **extensors. Flexor** muscles are antagonistic to extensors and decrease the angle between two bones. Straightening your elbow is extension and bending it is flexion.

About 80% of the mammalian body mass is muscle. Most of this is **skeletal muscle** (sometimes called **striated** or **voluntary muscle**). Skeletal muscle is characterized by a high degree of cellular organization. Muscle proteins, consisting mostly of actin and myosin, occupy most of the cytoplasm of the cell. These proteins are arranged in a series of parallel filaments. Many cylindrical muscle cells make up a muscle, and the contractile proteins in these fibers are arranged in such a way that it gives a striated microscopic pattern to the muscle.

During contraction, actin and myosin filaments interact with each other, shortening the muscle in much the same manner as a deck of cards is aligned after it is shuffled (see fig. 31.4). The force for movement comes from the interaction of the filaments themselves rather than from a shortening of the filaments.

Along with skeletal muscle, two other types of muscle tissue are found in animals. They function in movements within the body, rather than in movement of the body from one location to another. **Smooth muscle tissue,** also called **involuntary muscle,** is found in the gut, blood vessels, pupil, and reproductive system. **Cardiac muscle** is found in the heart.

An integral part of any animal's locomotory system is its skeleton. While the word *skeleton* usually brings to mind the bony **endoskeleton** of vertebrates, other types of skeletons occur in the animal kingdom. **Hydrostatic skeletons** are found in earthworms and other soft-bodied invertebrates in which the pressure of internal fluids gives shape and allows the organism to move. **Exoskeletons** are found among such diverse animals as nematodes, arthropods, molluscs, and corals. This tough outer covering protects the organism and supports it in an upright position while allowing for muscle attachment and movement.

Skeletons reflect adaptations to an organism's environment and lifestyle. Frogs and birds not only have limbs modified to fit their environments, but also have bones that show either thickening or shortening to fit their forms of locomotion. Consider the skeletal adaptations of a human, horse, bat, and seal to their respective forms of locomotion. Careful study of the skeleton by the trained observer can reveal much about an animal, because skeletons are excellent examples of the biological principle that form reflects function. Skeletons serve many other functions besides locomotion. They protect soft tissue, serve as a reservoir for calcium and phosphate, and are sites of blood cell formation.

Skeletons are often the only parts of animals that survive from the past as fossils. Because the muscles attach to the skeleton, it is possible to see which muscles might have been highly developed and to speculate about how an extinct animal lived on a day-to-day basis.

LAB INSTRUCTIONS

You will study the anatomy and properties of muscle and compare several types of skeletal systems.

Muscular System

Microscopic Anatomy of Muscle

1. Obtain prepared slides of smooth muscle, cardiac muscle, and a longitudinal section of skeletal muscle. Compare the tissues on your slide to figure 31.2.

Figure 31.2 Three Types of Muscle. (*a*) Smooth muscle has spindle-shaped cells, each with one nucleus (*b*) Skeletal muscle cells are cylindrical and have many nuclei. (*c*) Cardiac muscle is unique to the heart. Its striated cells join at connections called intercalated disks. The cylindrical cells branch. Cardiac and skeletal muscle cells are striated due to the orderly organization of contractile proteins.

Smooth muscle (a)

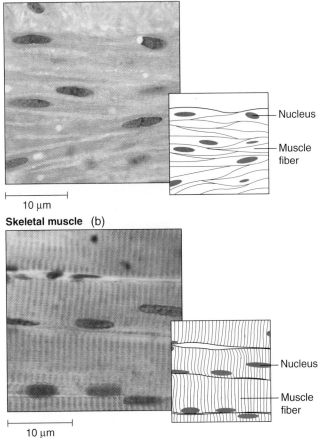

10 µm

Skeletal muscle (b)

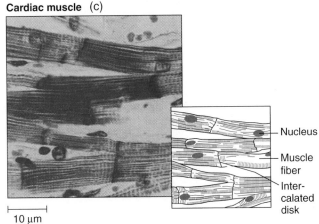

10 µm

Cardiac muscle (c)

10 µm

Smooth Muscle Observe the smooth muscle slide first. Note the shape of individual cells, the presence of nuclei, and the absence of filament organization in the cytoplasm. These muscle cells are spindle shaped, and their actin and myosin are not organized in parallel arrays. Smooth muscle is innervated by the autonomic nervous system and is not under voluntary control. It responds to various hormones. Its response is characterized by slow, rhythmic contractions. In some cases, contractions arise spontaneously in smooth muscles and are propagated along the length of the organ. Sketch two or three examples of smooth muscle cells.

2 *Skeletal Muscle* Now examine the slide of skeletal muscle under the medium-power objective and compare its organization to that of smooth muscle. Note the substantial differences. Identify the individual cylindrical fibers that run the length of the tissue specimen on the slide. Find a very thin area of the section that is only one fiber thick and examine it under high power. On the periphery of each fiber, you will see several nuclei. Each skeletal muscle fiber is **multinucleated** because it is formed by the fusion of numerous smaller, uninucleated cells during embryonic development. Thus a muscle fiber is really a composite of several smaller cells. Though probably not visible, several mitochondria are also present in this peripheral cytoplasmic area. The central area of the cell consists of many parallel fibers called **myofilaments** running lengthwise in the cell. These fibers give a cross-banding appearance to the cytoplasm. Draw two or three of these cells.

To understand how your drawing relates to the structure of a whole muscle, such as the biceps in your upper arm, look at figure 31.4. What most people call a muscle contains thousands of muscle fibers arranged parallel to one another and surrounded by a sheath of connective tissue. Within a single muscle cell are many **myofibrils,** which consist of contractile units called **sarcomeres** joined end to end (fig. 31.4). Each sarcomere contains two types of filamentous proteins. **Actin** is the protein in thin filaments and **myosin** is the protein in thick filaments. The actin filaments at each end are anchored in the Z discs, which mark the ends of the sarcomere. The myosin filaments are suspended between and surrounded by actin filaments. It is the interdigitation of these filaments and the areas of overlap that create the alternating light and dark banding patterns that you saw on your slide of skeletal muscle.

During contraction, the thin filaments slide toward the middle over the thick filaments, pulling the Z discs inward and shortening the overall length of the sarcomere. Because this process is repeated simultaneously in each sarcomere, muscle cells can dramatically shorten.

In recent years, research has revealed the molecular mechanisms involved in muscle contractions. When a nerve impulse arrives at the **neuromuscular junction,** it causes the release of a chemical called acetylcholine (fig. 31.3). It depolarizes the muscle cell membrane and triggers an action potential in the muscle cell. As the action potential moves along the muscle cell, it penetrates to the interior of the cell via the T tubule system (fig. 31.4). Membrane depolarization at the Z disks causes the sarcoplasmic reticulum to release calcium ions into the sarcomere. The calcium ions allow the actin to react chemically with the myosin filaments. This pulls the actin filament over the surface of the myosin filament. The coordinated effect of the interaction between actin and myosin is the shortening of the muscle.

3▷ Look at the demonstration slide of a neuromuscular junction that is available in the lab. What do you find interesting about this slide?

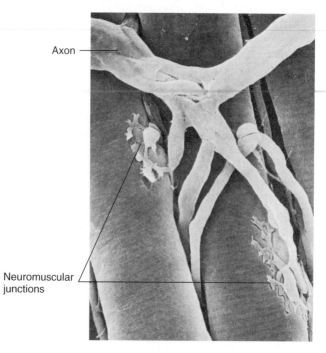

Figure 31.3 Scanning electron micrograph showing axon ending at the neuromuscular junction on a muscle cell.

Axon

Neuromuscular junctions

Look at a slide of cardiac muscle through your compound microscope. Would you say that it is similar to smooth muscle or to striated muscle? Why?

Use your 43× objective and look closely at one muscle fiber. Move the slide so that you can follow a single fiber for some distance. You should find two things in cardiac muscle that are different from what you saw in striated muscle. What are these differences?

Cardiac Muscle

4▷ As the name implies, cardiac muscle is found in the heart, where it makes up the mass of the heart wall along with connective and nervous tissue. It is capable of rhythmic contraction which can be modulated by the pacemaker of the heart. The contraction of heart muscle is not under voluntary control.

The actin and myosin components in cardiac muscle are found in interdigitating linear arrays, as in skeletal muscle, but the fibers are capable of spontaneous contraction, as in smooth muscle. The muscle cells are shorter than in

Figure 31.4 Skeletal muscle structure and function. A muscle fiber or cell contains many myofibrils, consisting of sarcomeres. When the myofibrils of a muscle fiber contract, the sarcomere shortens as the actin filaments move toward the center of the sarcomere and Z discs move closer.

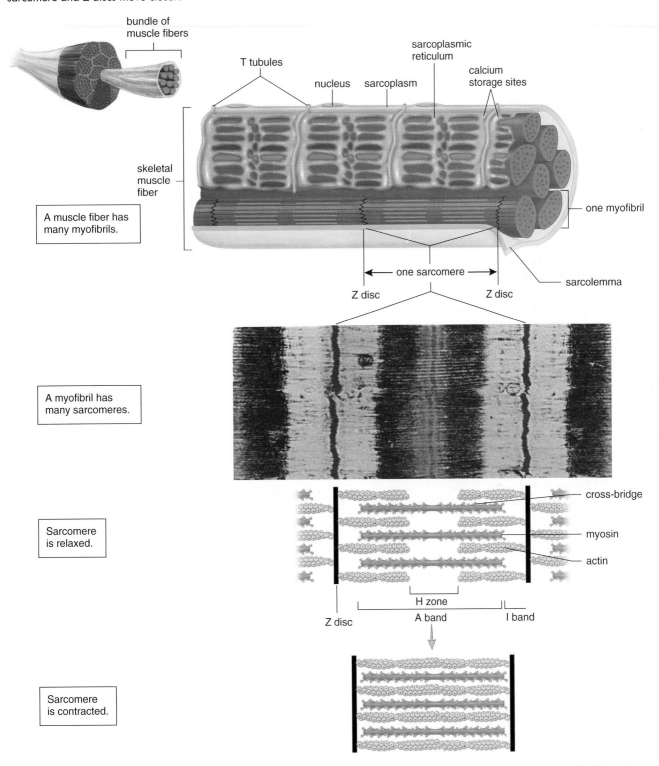

skeletal muscle, often branch, and are joined end to end by tight junctions, called **intercalated disks,** which electrically couple the cells and allow contraction to spread from cell to cell independent of the nervous system.

Fetal Pig Superficial Muscles

This dissection may be done by students or by the instructor as a demonstration, depending on the time available and the emphasis of the course on gross anatomy.

Three terms are used to describe the anatomical orientations of muscles: origin, belly, and insertion. The **origin** is the end attached to the less mobile portion of the skeleton. The **belly** is the central part of the muscle, and the **insertion** is the end of the muscle attached to the freely moving part of the skeleton. **Tendons** are fibrous connective tissues that connect muscles to the skeleton.

5 ▷ If your fetal pig is not already skinned, skin it—being careful not to tear away muscle as you remove the skin. Use a probe or finger to separate the two. Under the skin, there may be adipose (fat) tissue, which should be removed to reveal the muscles. When the skin is removed, dry the carcass with paper towels to improve viewing. In young fetal pigs, the muscles are not fully developed and are often tightly connected to the skin so that they are torn during skinning. Identification will be difficult.

6 ▷ Starting at the head, identify the major muscles indicated in figure 31.5.

1. The **latissimus dorsi** is a broad muscle running obliquely around the sides of the thoracic region. It originates on the vertebrae and inserts on the proximal end of the humerus. The latissimus dorsi is involved in moving the foreleg.

2. The **trapezius** originates on the occipital bone of the skull and from the first ten vertebrae and inserts on the scapula or shoulder blade. This broad muscle draws the scapula medially. When this muscle contracts, how does the leg move?

Figure 31.5 The major muscles of the fetal pig.

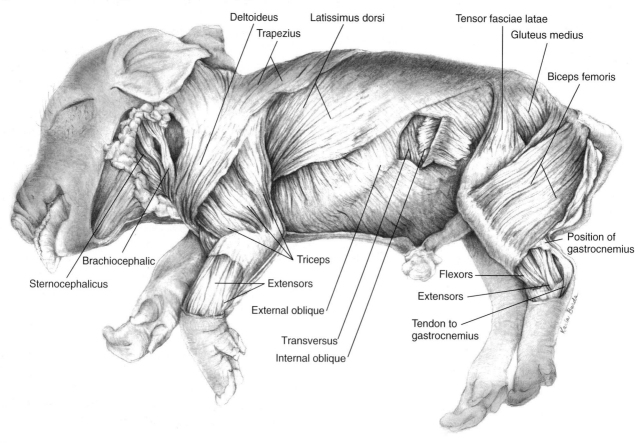

3. The **brachiocephalic** muscle extends obliquely as a flat belt from the back of the skull (the occipital bone) to the foreleg (distal end of the humerus). It is involved in moving the leg anteriorly.

4. The **sternocephalic** is a long muscle below the brachiocephalic muscle. It controls the flexing of the head. It originates on the sternum and inserts on the mastoid process of the skull by means of a long tendon. Remove the parotid salivary gland to see the tendon.

5. The **brachialis** is a small muscle located in the angle formed by the flexed foreleg. It originates on the humerus and inserts on the ulna. What is its function?

6. The **deltoid** is a broad shoulder muscle originating on the scapula and inserting on the humerus. It aids in flexing the humerus.

7. The **extensors** are several muscles in the lower foreleg that extend and rotate the wrist and digits.

8. The **triceps** (brachii) is a large muscle making up practically the entire outer surface of the forelimb. It originates on the humerus and inserts on the proximal end of the ulna. When it contracts, the forelimb extends.

9. The **external oblique** muscles make up the outer wall of the abdomen and run obliquely from the ribs to a ventral **longitudinal ligament** along the ventral midline. Contraction of these muscles constricts the abdomen.

10. The **internal oblique** fibers lie just under the external oblique muscle and run almost at right angles to them. Contraction of these fibers also results in abdominal constriction.

11. The **tensor fasciae latae** is the most cranial of the thigh muscles. It originates on the pelvis and continues ventrally as a thin, triangular muscle until it becomes a sheet of connective tissue called the **fasciae latae,** which inserts on the kneecap and extends the leg.

12. The **gluteus medius** is a thick muscle covered by the tensor fasciae latae in the hip region. If the overlying muscle is removed, its origin on the hip and insertion on the femur can be seen. What action does it cause?

13. The **biceps femoris** is a large muscle making up most of the back half of the thigh. It originates on the pelvis and inserts on the lower femur and upper part of the tibia.

14. The **gastrocnemius** is the large muscle of the calf, originating on the lower end of the femur and attaching to the heel by means of the Achilles tendon. Its action extends the foot. The **soleus** muscle lies close to the gastrocnemius and has a similar function.

15. The **digital flexors** and **extensors** originate on the tibia and fibula and insert on the metatarsals. What are their functions?

16. The **peroneus** muscles have origins and insertions similar to digital muscles. They are involved in moving the whole foot.

Physiology of Muscle

This part of the lab may be performed by the instructor as a demonstration or by students working in groups of four or more, depending on the time and equipment available.

Overview of Experiment

In this section of the lab, you will investigate four properties of human muscles: threshold stimulus, recruitment, temporal summation, and tetany. You will use special equipment and software that allows a computer to collect and analyze data. The experimental subject(s) will be a volunteer member(s) of your class. Anyone who has heart problems or does not wish to participate directly in this experiment should not be a subject. They can operate the computer or record notes during the experiment.

To do this experiment, you need to make adjustments to four major pieces of equipment—a stimulator, a transducer, an interface unit, and a computer running special software. A **stimulator** is an electronic device that produces a pulse of electricity that can be varied by volts (strength), duration of the pulse (length), and frequency (number of pulses per second). The **transducer** is a device that converts the movement of a trigger in a pistol handle into an electrical resistance that can be recorded. The **interface** is a box that connects various cables together into a single lead that is connected to a port of your computer. The **software** allows your computer to record raw data from the transducer and process the data, converting it into graphical displays. Before you start investigating the biology of muscle, this equipment must be calibrated.

Equipment Calibration

Stimulator Look at the stimulator and note the dials. Make sure that the power switch is turned off. The various dials and switches allow you to adjust the characteristics of the stimulus. Set the duration of the pulse to 10 msec and the voltage to 30 volts. Note that there may be a voltage dial and a multiplier dial. If the voltage dial is set at 3.0 and the multiplier at 10, then the pulse will be 30 volts. There should also be a switch that allows you to select between two modes of operation: single stimulus or continuous (multiple stimuli). In multiple mode a dial should allow you to set the number of stimuli per second. To start, set the switch to single mode. Connect a cable from the stimulator to the interface box according to the manufacturer's directions. This will send a signal to the computer each time a stimulus is administered. Remember that red connectors should always be connected to positive terminals and black to negative ground terminals. Now connect wires from the stimulus electrodes to the stimulator output terminals according to the manufacturer's directions. One electrode will have a flat plate ground electrode (–) and the other will be a pencil-like probe electrode (+). Note the button that must be depressed to administer a stimulus. This is a safety feature that prevents shocks from happening. Do not turn on the stimulator power until you are ready to do the experiment.

Transducer Find the Physiogrip transducer. This device allows you to measure the amount of muscle contraction. When you squeeze the trigger, the electrical resistance of a wire passing through the device changes. The Physiogrip is not a device to measure the strength of your grip, and it should be treated gently. It is a sensitive device that measures how rapidly and forcefully a single muscle can contract.

Computer Calibration Connect the Physiogrip to the interface box according to the manufacturer's directions. Connect the interface box to the computer. The software should have been previously installed on the computer. Run the Physiogrip software according to directions from your instructor. Click on NEW under the FILE menu and a dialog box will appear to lead you through transducer calibration. Follow the directions that appear on the computer screen.

Trial Use of Software After calibrating the computer and Physiogrip, take a moment to familiarize yourself with the software. In the Acquiring Data mode, you can record the output from that Physiogrip. Do this by clicking on START on the Acquiring Data screen and then squeeze the trigger of the Physiogrip in and out. You should see a movement of the trace on the computer screen and a record of the contractions. You can stop the trace by clicking AUTOSTOP or by letting it go for the preset time of 50 seconds. The data just collected can be printed by selecting PRINT SCREEN from the FILE menu or it can be saved by choosing SAVE from the FILE menu.

You are now ready to use this equipment in your investigation of muscle physiology.

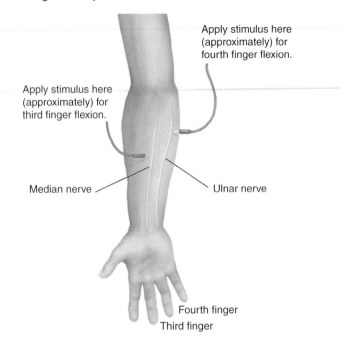

Figure 31.6 Approximate location of nerve innervating flexor digitorum superficialis.

Apply stimulus here (approximately) for fourth finger flexion.

Apply stimulus here (approximately) for third finger flexion.

Median nerve

Ulnar nerve

Fourth finger

Third finger

Determining a Motor Point

Motor points are locations on the body surface where nerves innervating certain skeletal muscles pass close to the skin. If a stimulating electrical current is given at this point, the nerve passing to the muscle is stimulated and the muscle contracts. In this part of the lab, you will determine the motor point for the nerve innervating the **flexor digitorum superficialis,** a muscle involved in the rapid flexing of the third and fourth digits. This nerve is located in the belly area on the underneath surface of the forearm (fig. 31.6).

7▶ To start the experiment, one person should volunteer to be the subject. Thin males have less subcutaneous fat than other students and will be the best subjects. Be sure that the stimulator is turned off. Attach the negative plate electrode to the back of the hand **on the arm that will be stimulated,** using a liberal amount of electrode gel between the plate and the skin (fig. 31.7). When handling the electrodes, use caution.

CAUTION

Guard against any situation in which one electrode is held in one hand and the second electrode is held in the other hand. In such a situation, when the stimulator is turned on an electric current could flow across the chest and cause an arrhythmia in the heartbeat.

The three remaining people in the group should each take a different job. One should handle the stimulating electrode, a second should work at the computer, and the third should work the controls on the stimulator and depress the stimulator button to activate the stimulator probe.

Figure 31.7   Position of electrode attachment.

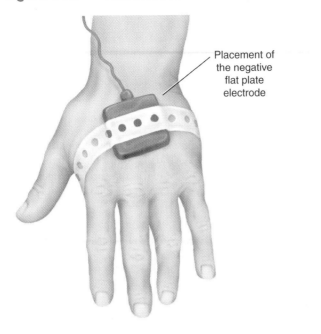

Placement of
the negative
flat plate
electrode

When the first electrode has been attached to the back of the hand, have the person rest their arm on the table with the palm up and muscles relaxed. Check that the stimulator is set at 10 msec duration, 30 volts at a frequency of 1 per second. Turn it on. Put a dab of electrode gel on the end of the probe stimulator and press it on a spot near the center of the belly of the forearm muscles (fig. 31.6). Try several different spots in this general vicinity. You may have to press rather hard to get the best results. When you find the motor point, the third and fourth digits will contract involuntarily and almost all discomfort from the stimulus disappears. If you cannot find the spot, increase the strength of the stimulus to 40 volts and try again. When a motor point is identified, mark it with marker or a ballpoint pen.

8▷ After the motor point is identified, have the subject grip the Physiogrip and start the software. Set the controls so that:

Metronome and Autostop are off;

Samples per second are 100;

Seconds per data set are 50.

Apply the stimulus to the previously identified motor point and work out a procedure that allows you to record twitches on the screen. Once you can cause involuntary contractions by stimulating the motor point and record data, you are ready to do the experiment. (*Note:* Changing the position of the subject's arm may require that you locate the motor point again.)

Determining Threshold Stimulus Intensity

Nerve cells are excitable cells. When subjected to an electrical current, ions move across the cell membrane causing a change in the membrane potential that is proportional to

the current. If the change in potential exceeds a **threshold value,** an action potential is triggered in the excitable cell. You will determine this threshold for the nerve innervating the flexor digitorum superficialis. What do you hypothesize will happen when the threshold potential is reached?

9▷ The person operating the computer should start a new data file by selecting NEW from the FILE menu, with the computer screen controls set as before. The subject should grasp the Physiogrip so that a very slight pressure is put on the trigger, raising the baseline a few millimeters on the computer screen. The pressure should be steady. This takes the slack out of the system and ensures that small twitches will be recorded.

The person applying the stimulus should turn the voltage down to 20, but keep the duration at 10 msec and the frequency at 1 per sec. Turn on the stimulator. Place the stimulating probe at the previously determined motor point and see if there is any response. If not, raise the voltage by 5 volts and repeat the stimulation. Continue doing so until a stimulus intensity is reached that elicits a twitch. Because the threshold could be up to 4 millivolts less than the stimulus just used, decrease the intensity by 4 volts and see if a twitch results. If not, increase the intensity by 2 volts and apply the stimulus. Repeat until the threshold is determined. What is the voltage of this threshold?_____ Why is it called a threshold?

Recruitment of Motor Units

Muscles are composed of **motor units,** groups of muscle cells that are separately innervated. Different motor units have different thresholds. Consequently, if the strength of the stimulus is increased, more muscle cells contract, a phenomenon known as **recruitment.** What should happen to the deflection of the trace on the computer screen as recruitment occurs?

10 Use the same setup as before. Be sure to use NEW from the FILE menu to select a new data file.

Start with the previously determined threshold voltage and record a twitch. Slowly increase the voltage and continue recording twitches. Continue this procedure until a maximum height of contraction is reached and further increments of stimulus intensity do not evoke stronger contractions. What is the voltage at which no further increase in strength of contraction occurs?_____ Turn off the stimulator.

If excitable cells exhibit an all-or-none response when exposed to a threshold stimulus, explain why the height of contraction increased with stimulus strength above the threshold? (*Hint:* Consider the phenomenon of recruitment.)

Temporal Summation and Tetany

11 So far in your investigations of muscle you have used a low frequency of stimulation. In this section, you will investigate the effects of increasing frequency of contraction on the strength of muscle contraction. With increasing frequency of contraction, there is insufficient time for the muscle to relax before the next stimulating impulse arrives. Consequently, the contractions begin to summate with a new contraction starting in a partially contracted muscle. This phenomenon is called **temporal summation,** and when the muscle is fully contracted and cannot relax between stimuli, **tetany** occurs. What would you hypothesize is the effect on the strength of muscle contraction? State your hypothesis.

Test your hypothesis, using the following procedure. Choose NEW from the FILE menu to create a new data file. Turn on the stimulator. Have the subject grasp the Physiogrip and put a slight pressure on the trigger to raise the baseline a few millimeters. Set the stimulus parameters to 10 msec duration, frequency to 1 per sec in continuous mode, and a voltage that gives a single contraction.

Once this base is established, gradually increase the frequency of stimulation over a 10-second interval. Can you see individual contractions at low frequencies?_____ At higher frequencies?_____ What happens to the strength of contraction as the frequency of stimulation increases?

If the strength of contraction is so strong that the trigger reaches the end of its travel, stop the experiment and change the spring in the Physiogrip to a stronger one. Erase the previous trace from memory in the computer and repeat the experiment.

When the strength of contraction no longer increases with increasing frequency of stimulation, stop the experiment and turn off the stimulator.

Analyze the data that you have collected in the following way. First, print a copy of the screen by choosing PRINT from the FILE menu. This will provide a record for any report that you may have to write.

From the FILE menu, select ANALYZE and then DISPLACEMENT. You should see a record of the contractions on the screen and at the top should be a series of hash marks indicating how often a stimulus was given. Using the ANALYZE cursor, determine the frequency of stimulation just before temporal summation started.

What is the value in number per second?_____ Now repeat this measurement at the point where tetanic contraction was reached. What is the frequency of stimulation that results in tetany?_____

Measure the height of an individual twitch at the beginning of this experiment. What is it?_____

Now measure the height of contraction reached in tetany. What is it?_____

What is the percent change due to tetany?_____

Skeletal Systems

Skeletal systems support and protect an animal and make locomotion possible. In some animals, especially those with endoskeletons, the skeleton is a storage site for calcium and phosphate ions needed in metabolism. Three types of skeletal systems are found in the animal kingdom: hydrostatic (sometimes called hydraulic), exoskeletons, and endoskeletons.

Types of Skeletons

Hydrostatic Skeletons

Many soft-bodied invertebrates, such as roundworms, cnidarians, and annelids, do not appear to have specialized, differentiated skeletal systems. However, close examination reveals that the body wall muscles act on the incompressible fluids in the body cavity and intracellular spaces to facilitate very effective movement and support. Such skeletal systems are called **hydrostatic skeletons.**

12 Obtain a microscope slide of a cross section of an earthworm. Look at it first under the low-power objective and note the general anatomy of the animal by comparing the slide to figure 21.9.

The external surface of the earthworm is covered by a highly flexible **cuticle** secreted by the underlying cells of the dermis. The fibrous proteins of the cuticle protect the worm and help maintain its form. Some cells in the dermis secrete mucus, which lubricates the passage of the earthworm through the soil and also maintains a moist surface through which respiratory gases are exchanged.

Note that the body wall is made up of two layers of muscles: an outer circular set and a featherlike inner longitudinal set. Since the longitudinal set runs parallel to the long axis of the body, it will appear in cross section in the slide. Also note the well-developed body cavity, which is filled with fluid in live animals.

What happens to the worm's shape when the circular muscles contract? (Consider the effects of both the muscles and internal fluids.)

What happens to the worm's shape when the longitudinal muscles contract? (Consider the effects of both the muscles and internal fluids.)

The body cavity of the earthworm is divided into compartments, corresponding to each segment, by cross walls called **septa.** These prevent fluids from "sloshing" from one end of the worm to the other and allow earthworms to make a variety of movements with fine gradations.

Although they are not part of the hydrostatic skeleton, you should note the spinelike **setae** in the body wall. Muscles attached to each seta extend or retract them. When extended they project into the soil and anchor the worm. Describe how you think an earthworm could crawl forward using its setae, circular muscles, and longitudinal muscles.

Exoskeletons

Exoskeletons are characteristic of several animal phyla. In addition to supporting the animal, exoskeletons prevent water loss, protect the organs, and serve as points of attachment for muscles. Animals in the phylum **Arthropoda,** which includes crayfish, insects, spiders, and millipedes, have a well-developed exoskeleton composed of a complex polysaccharide called **chitin** combined with proteins to give flexibility. Several examples of these animals should be available in the laboratory.

12 Look at an insect through your dissecting microscope and note the segmentation of the exoskeleton. The **head** and **thorax** are composed of several fused body segments. In insects, the thorax bears the walking legs and wings. The segments should be clearly visible in the **abdomen.**

An insect's exoskeleton is made up of hardened plates called **sclerites.** In the head and thorax, the sclerites are rigidly joined to one another along **suture** lines. In the abdomen, the dorsal and ventral sclerites are enlarged and joined to one another by **pleural membranes,** which correspond to thin, reduced vestiges of the lateral sclerites. Similar membranous areas are found between adjacent dorsal or ventral sclerites. These thin, flexible areas allow the animal to expand, contract, and curl its abdomen. Flexible membranous areas are also found between the joints of the various appendages.

Endoskeletons

Defined as an internal supporting system of hardened material, endoskeletons are highly characteristic of vertebrates. Cartilage and bone make up the vertebrate endoskeleton. You will start your study of endoskeletons by examining the structure of bone.

Bone Structure

13 Examine a fresh beef femur that has been cut longitudinally in half. The central shaft of a long bone such as this one is called the bone's **diaphysis,** while the enlarged ends are the bone's **epiphyses** (fig. 31.8). Though difficult to observe in

Figure 31.8 Structure of a bone: (*a*) a bone in partial longitudinal section; (*b*) scanning electron micrograph of cancellous (spongy) bone; (*c*) scanning electron micrograph of Haversian canal (HC) systems in compact bone. Canals allow blood vessels to pass into bone. Bone cells are located in the lacunae (La). (*c*) From R.G. Kessel and R.H. Kardon. *Tissues and Organs: A Text-Atlas of Scanning Electron Microscopy.* 1979. W.H. Freeman and Co.

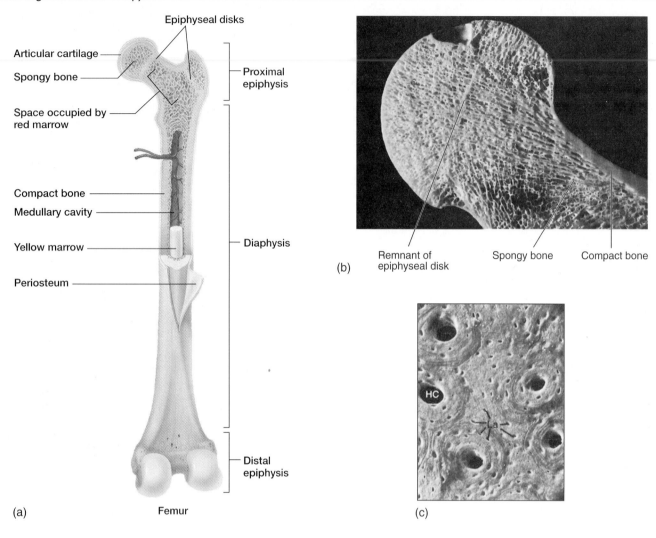

older bones, a narrow zone of cartilage, the **epiphyseal disk** (growth center) of the bone, separates the epiphysis from the diaphysis.

The rounded projection from the upper end of the bone (fig. 31.8a) is called the **head,** and it articulates with the **acetabulum,** a depression where all three pelvic bones intersect. **Condyles** (rounded, articular prominences) are located on the lateral and medial sides of the distal end of the femur. With which bone(s) do they articulate?

Find the **periosteum,** the tough connective tissue layer surrounding the diaphysis. The surfaces of the condyles are covered by a glassy smooth **articular cartilage,** which, along with synovial fluid, reduces frictional erosion between articulating bones at the joints. Residues from the ligament that held the joint together may also be visible.

On the outer surface of the bone, find the narrow ridges called crests and the small, round, elevated tubercles, where muscles attach by tendons to the bones. The surface of the bone is perforated by openings called **foramina** (sing. **foramen**) that allow blood vessels and nerves to penetrate the bone.

The center of the bone is a hollow chamber called the **medullary cavity.** It is filled with **marrow.** Red marrow is where new red blood cells are formed, but in older animals much of the marrow is fatty yellow matter, which no longer produces blood cells.

Compare the structure of the bone in the diaphysis with that in epiphysis. Compact bone is very dense, whereas spongy bone has many interior cross braces. Spongy bone is less dense than compact bone.

Microanatomy of Bone

14▶ Obtain a prepared slide of ground bone and examine it under low power. Using figure 31.8*c* as a guide, find a **Haversian canal.** These canals allow nerves, blood vessels, and lymphatics to penetrate bone. Note the chambers, called **lacunae,** arranged in concentric circles around the canal. **Osteocytes,** mature bone cells, line in these spaces and are responsible for depositing and removing calcium salts. Their activity is regulated by hormones. Note the **canaliculi** that radiate from the lacunae. They are transportation networks and allow osteocytes to extend their cytoplasm and form intimate contact with the bone matrix.

Comparative Vertebrate Endoskeletons

The vertebrate skeleton has two components: the **axial skeleton,** consisting of the **skull** and **vertebral column;** and the **appendicular skeleton,** which includes the **pelvic** and **pectoral girdles** as well as the bones of the appendages.

Skeletal Comparisons

15▶ In the laboratory, you will observe the skeletons of three vertebrates: a frog, a bird, and a human (figs. 31.9–31.11). The task before you is to determine how these skeletons are similar and how they are different. For the differences, you should offer explanations of how these differences correlate with locomotion or life style. For each skeleton you should be able to quickly identify the axial and appendicular skeletons. Then you should be able to find and identify the bones listed below.

Axial skeleton

    Skull

    cranium (fused bones encasing brain and sense
        organs) consisting of the following major bones:
        frontal
        parietal
        temporal
        occipital

    facial bones consisting of the following major bones:
        zygomatic
        maxilla
        mandible (lower jaw), the only movable bone in
          skull

    Vertebral column, which supports and protects the
        spinal cord. Cartilaginous discs are found between
        the vertebrae in living animals. The vertebral
        column is divided into five regions:
        cervical vertebrae (neck)
        thoracic vertebrae (on which the ribs articulate)
        lumbar vertebrae (abdominal region)

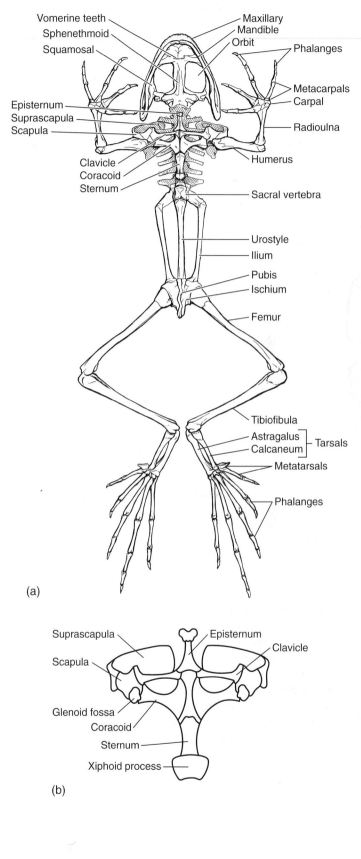

Figure 31.9 (*a*) Skeleton of frog (ventral view), including (*b*) ventral view of pectoral girdle.

Figure 31.10   Skeleton of a bird.

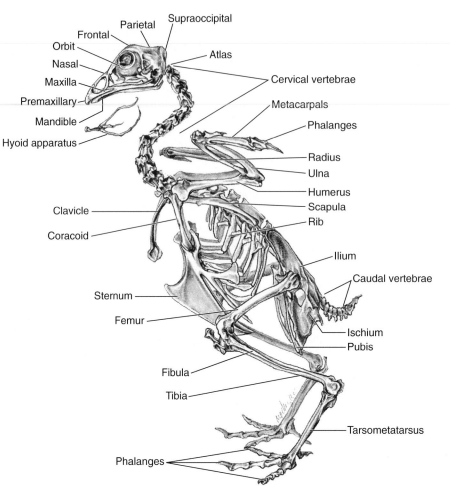

Supraoccipital
Parietal
Frontal
Orbit
Nasal
Maxilla
Premaxillary
Mandible
Hyoid apparatus
Atlas
Cervical vertebrae
Metacarpals
Phalanges
Radius
Ulna
Humerus
Scapula
Rib
Clavicle
Coracoid
Ilium
Caudal vertebrae
Sternum
Femur
Ischium
Pubis
Fibula
Tibia
Tarsometatarsus
Phalanges

sacral vertebrae (enclosed by pelvic girdle)
caudal vertebrae or tail (coccyx in human)

Ribs (attached and floating)
  Sternum
  Xyphoid process

Appendicular skeleton

Pectoral girdle consisting of:
  clavicle (collarbone)
  scapula (shoulder blade)
  coracoid

Forelimbs consisting of:
  humerus (long bone of lower limb)
  radius (long bone of lower limb, which forms a
      pivot joint with the ulna and is also part of the
      hinge joint of the elbow)
  ulna (other long bone of lower arm, forming hinge
      joint at elbow)

carpals (small bones of wrist)
metacarpals (bones of palm)
phalanges (finger bones)

Pelvic girdle consisting of:
  ilium
  ischium
  pubis

Hind limbs consisting of:
  femur (long bone of thigh)
  patella (kneecap)
  tibia (larger of two long bones of lower limb)
  fibula (smaller long bone of lower limb)
  tarsals (bones of ankle)
  calcaneus (heel bone)
  metatarsals (slender foot bones)
  phalanges (toe bones)

Figure 31.11   Human skeleton.

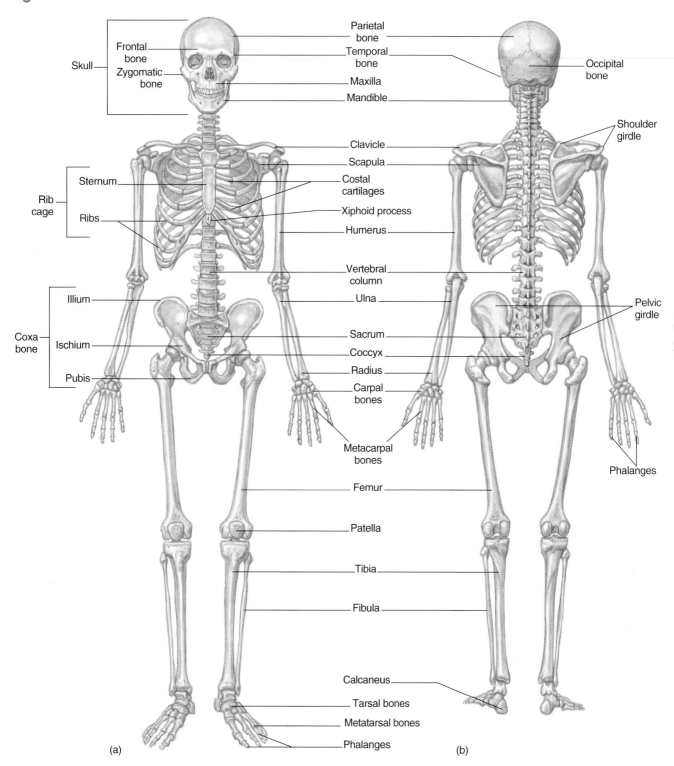

(a)

(b)

Learning Biology by Writing

Write a lab report that describes your experiments with human muscles. It should have three sections in the results:

Determining threshold
Demonstrating motor unit recruitment
Demonstrating temporal summation and tetany

The discussion section should describe how recruitment and summation are important in graded responses such as are used in sports or playing a musical instrument.

Your instructor may ask you to answer the Lab Summary and Critical Thinking Questions that follow.

Internet Sources

The disease rickets affects the human skeletal system. Use GOOGLE to search the WWW to learn what this disease is. Where is it common? Is it found in the U.S.? How can it be treated?

Lab Summary Questions

1. What evidence do you have that smooth, cardiac, and skeletal muscle cell are different? Describe. Give examples of where you would find each in your body.
2. Describe the all-or-none response in muscle cells. How is it possible for us to have graded responses in our muscles if muscle cells respond in an all-or-none manner to stimuli?

3. Both recruitment and temporal summation lead to greater force of muscle contraction. Explain what is happening in both of these phenomena.
4. Describe the structure of a bone, such as the humerus, at the macroscopic and microscopic levels.
5. Describe the major common features of skeletons found in all vertebrates.
6. "Form reflects function" is a statement that applies to the skeletal system. How does the form of the vertebrate skeleton relate to the means of locomotion in a frog compared to a human? In a bird compared to a human?

Critical Thinking Questions

1. About 99% of the body's calcium is found as calcium phosphate salts in bone tissue. The calcium in plasma is in an ionic form, Ca^{++}. Although the level of Ca^{++} in the blood and tissues is closely regulated by parathyroid hormone and calcitonin, imbalances in Ca^{++} can occur. What effect would an elevated Ca^{++} concentration have on muscle activity? What effect would a Ca^{++} deficiency have on muscle activity?
2. As can be seen in figure 31.10, the skeleton of a bird is modified for flight. Aside from the size and arrangement of the bones in the skeleton, what other feature of its bones might affect the ability of a bird to fly?
3. How do scientists deduce the appearance of hominids or dinosaurs when all they have to study are a few bone fragments?

LAB TOPIC 32

Investigating Nervous and Sensory Systems

Supplies

Preparator's guide available on WWW at
    http://www.mhhe.com/dolphin

Equipment

Compound microscopes
Dissecting microscopes

Materials

Fetal pig
Pig brain, sagittal section
Whole pig brains
Prepared slides
    Mammalian nerve cord with dorsal and ventral
        roots, cross section
    Neurons or multipolar neurons
    Muscle with nerve end plate (demonstration slide)
    Mammalian retina (demonstration slide)
    Mammalian cochlea, cross section
Models of human eye and ear
Overhead transparencies made from black paper with
    one inch squares cut in center and covered with
    red, blue, or green cellophane so that a small
    square is projected. Alternatively, photographic or
    PowerPoint® slides can be made to project one of
    the colors with a black background.

Student Prelab Preparation

Before doing this lab, you should read the introduction
and sections of the lab topic that have been scheduled
by the instructor.

You should use your textbook to review the
definitions of the following terms:

axon	meninges
cerebellum	myelin sheath
cerebrum	organ of Corti
cochlea	pupil
cornea	reflex arc
cranial nerves	retina
dendrite	semicircular canals
gray matter	spinal cord
iris	spinal nerves
medulla oblongata	white matter

You should be able to describe in your own words
the following concepts:

How a nerve cell functions
Types of nervous systems found in animals
Major regions of the mammalian brain
Reflex arc
How the eye "sees" light
How the ear "hears" sound

As a result of this review, you most likely have
questions about terms, concepts, or how you will do
the experiments included in this lab. Write these
questions in the space below or in the margins of the
pages of this lab topic. The lab experiments should
help you answer these questions, or you can ask your
instructor during the lab.

Objectives

1. To observe the microscopic structures of nerves,
   nerve cells, and neuromuscular junctions
2. To learn the gross anatomy of the mammalian
   central nervous system, eye, and ear
3. To determine experimentally some of the properties
   of rods and cones

Background

Coordination of the several different types of specialized
tissues in multicellular animals is necessary for the organism to operate as an integrated whole. The nervous system
coordinates the body's relatively rapid responses to
changes in the environment. The endocrine system regulates longer term adaptive responses to changes in body
chemistry between meals, as the seasons change, or as developmental changes occur during maturation. Because the
functions of these two systems often complement one another, biologists often speak of the neuroendocrine system.

Figure 32.1   Nervous systems in invertebrates: (*a*) nerve net in *Hydra* lacks central tracts; (*b*) flatworms have longitudinal nerve cords with cross connectives; (*c*) ventral central nervous system of earthworm shows cephalization and has integrating ganglia in each segment.

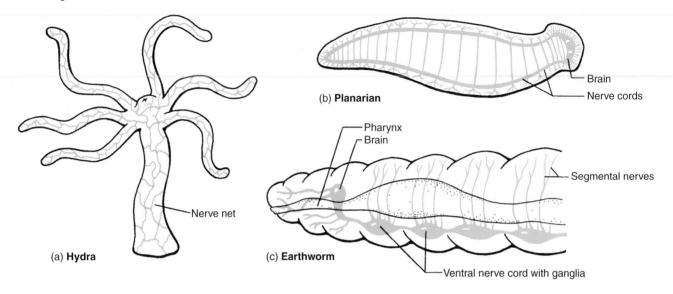

(a) **Hydra**

Nerve net

(b) **Planarian**

Brain

Nerve cords

Pharynx

Brain

Segmental nerves

(c) **Earthworm**

Ventral nerve cord with ganglia

   The nervous system has three functions: (1) to receive signals from the environment and from within the body through the sense organs, (2) to process the information received, which can involve integration, modulation, learning, and memory, and (3) to cause a response in appropriate muscles or glands. **Receptors** are usually specialized cells outside of the central nervous system that detect physical and chemical changes. There are separate receptor cells for heat, cold, light, and so on. The function of receptor cells is to convert an environmental signal into a change in the cellular membrane's ion permeability, leading to a voltage change across the cell membrane. If the voltage change is of sufficient magnitude, a nerve impulse will be created in an adjacent nerve cell, and travel from the sensing zone to the central nervous system.

   The central nervous system consists of neuronal networks, which are involved in decision making. A decision may involve the simple passage of nervous impulses from an **afferent sensory fiber** to an **efferent motor (effector) fiber,** assuring that perception results in a response, as in a **reflex arc.** In other cases, many **interneurons** in the nervous system may become involved, tempering the inputs with memory or reasoning. The response resulting from this interaction of nerve signals may differ according to conditions. For example, if a person is frightened by a loud sound, he or she will gasp, but if this stimulus is repeated when the person is under water, higher nerve centers inhibit the response.

   To gain insights into how nervous systems operate, many biologists study the simpler systems of invertebrates. These scientists have found that neurons are similar in form and function no matter what type of animal they are taken from. The **action potential,** the electrical potential changes comprising a nerve impulse, also seems to be universal in animals. The differences between simple and complex nervous systems involve the types of receptors present, the organization of the central nervous system, and the neurotransmitter chemicals that function in cell-to-cell impulse transmission.

   One of the simplest nervous systems is found in the radially symmetrical cnidarians, such as *Hydra.* The **nerve net** found in these organisms is not organized into tracts or centers (fig. 32.1*a*). Bilaterally symmetrical animals as simple as the Platyhelminthes have tracts running the length of the organism, with the anterior end of the tract serving as a coordinating center (fig. 32.1*b*). In higher phyla, the coordinating centers become more complex, creating an evolutionary trend referred to as **cephalization,** in which sense organs and the coordinating centers are concentrated at the animal's anterior.

   The central nervous systems of annelids and arthropods consist of two ventral nerve cords with interspersed **ganglia,** areas of the cord where the cell bodies of neurons are concentrated. The cords rapidly transmit signals, whereas the ganglia serve as processing centers (fig. 32.1*c*). In arthropods, the ganglia are clustered in the head and thorax regions, where most activity takes place.

   In the vertebrates, it is obvious that the brain dominates the nervous system. It is estimated that the brain contains between 10 and 100 billion neurons, which make over 100 trillion points of contact with each other. Well over 99% of these neurons are interneurons; only 2 to 3 million are motor neurons.

   The nervous system receives its information about the environment and the position of an animal in its environment from sensory cells. Sensory cells usually do not occur as isolated units but are part of a larger organized unit, the sensory organ. The mammalian eye consists not only of

Figure 32.2 Neurons in a smear from a bovine spinal cord.

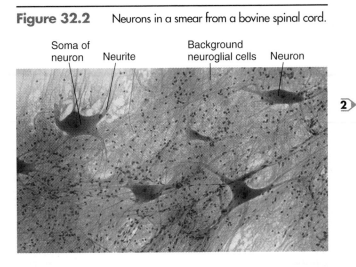

Soma of neuron Neurite Background neuroglial cells Neuron

light-sensitive rod and cone cells, but also of a focusable lens, a light-regulating iris, a tear gland, and a covering eyelid. The proper functioning of a sense depends not only on the sensory and neuronal cells, but also on the anatomical structures that collect, amplify, and direct environmental signals to the receptor cells.

Senses are "hard-wired" into the organism. You feel cold because specific cold receptors in the skin are stimulated, and impulses are carried by specific neuronal routes to specific areas of the brain where that information is integrated with other sensory inputs and memory. You see something because light has stimulated individual cells, and these cells each stimulate different neurons that carry nerve impulses to the brain. The brain integrates those signals with memory. The perception of a face, especially a familiar face, is actually a comparison of new neuronal inputs with a record of previous neuronal inputs. Sensory cells do not perceive, they receive. Perception is an integrative process in which nerve impulses from receptors, not environmental stimuli, are processed and may result in some action by the animal.

LAB INSTRUCTIONS

You will study the major components of the mammalian central nervous system and the anatomy of nerves, the eyes, and the ear.

Microanatomy of the Nervous System

1▷ Obtain a slide of neurons and observe it through a compound microscope, comparing what you see to figure 32.2. Small cells surrounding the neurons are neuroglial cells which supply nutrients for the neurons and produce chemicals that modulate the functions of neurons. Find a neuron's cell body called the **soma,** and the cytoplasmic extensions called **neurites. Axons** are neurites that conduct

impulses away from the soma. **Dendrites** are neurites conducting impulses toward the soma. You cannot morphologically distinguish between axons and dendrites on a slide like this. Why?_____ Neurites often extend long distances. In a 7-foot basketball player, axons extend from the base of the spine to the toes, a distance of 3 to 4 feet!

2▷ Obtain a slide of a spinal cord cross section through the **dorsal root ganglion.** Observe it at first under low power with a compound microscope or with a dissecting microscope against a white background.

Note the general organization and compare with figure 32.3. Sensory nerves are dendrites and always enter the top (dorsal). Motor neurons are axons and always leave the bottom (ventral) of the cord. Note that the nerve cord is hollow, containing a **central canal,** a vestige from its embryonic development. The central canal is connected with the ventricle spaces of the brain.

Examine the peripheral white matter surrounding the central gray matter with a high-power objective. Note that the white matter consists of neurites that have been cut in cross section. They are myelinated nerves and are surrounded by layers of lipid which electrically insulates them, allowing rapid action potential transmission. They extend up and down the spinal cord. When the spinal cord is severed in an accident, it is impossible to rejoin these thousands of neurites and numbness and paralysis result.

Examine the central gray matter and find the cell bodies in this area. Interneurons here process information and send motor axons out the ventral root to muscles. They lack the insulating myelin.

3▷ Examine the demonstration slide your instructor set up, showing a neuromuscular junction where a motor nerve innervates muscle fibers. The axon of the motor nerve divides into fine branches which terminate as **motor end plates** on the muscle fiber surface (see fig. 31.3). Nerve impulses arriving at the end plates cause the release of acetylcholine, a neurotransmitter chemical that diffuses across the neuromuscular junction cleft and triggers an action potential in the muscle cell. The action potential leads to the muscle contracting. Sketch and label a junction below.

Investigating Nervous and Sensory Systems   **435**

Figure 32.3   Cross section of spinal cord showing reflex arc components.

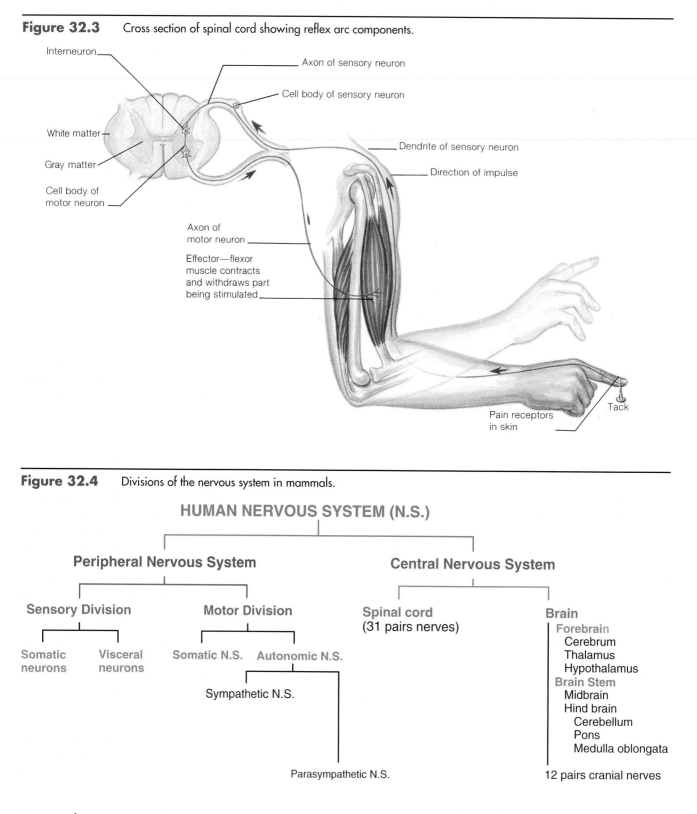

Figure 32.4   Divisions of the nervous system in mammals.

Mammalian Nervous System

Figure 32.4 provides an organizational overview of the mammalian nervous system. It consists of two major parts: the peripheral nervous system and the central nervous system, which in turn can be further subdivided into their component parts. As you work through the anatomy of the nervous system, refer to this chart to refresh your memory of the relationships.

4▷   If you have not already done so, remove the skin of your fetal pig. Lay the pig on its stomach. Expose the **spine** by removing the muscles that cover and extend on each side of the vertebrae. Since it is time consuming for each student to expose the entire spine, groups in the laboratory

Figure 32.5   The relationship of the spinal cord, spinal nerves, and sympathetic nervous system to the vertebrae and meninges.

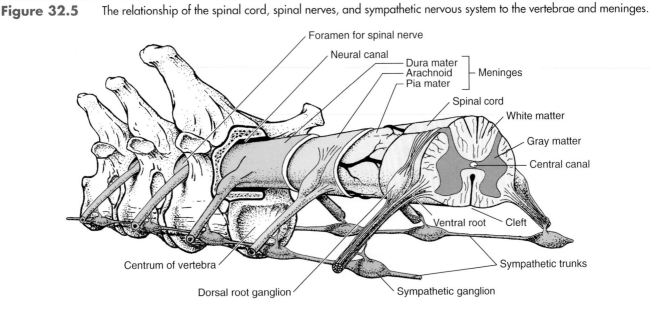

Foramen for spinal nerve
Neural canal
Dura mater ⎤
Arachnoid ⎬ Meninges
Pia mater ⎦
Spinal cord
White matter
Gray matter
Central canal
Ventral root   Cleft
Sympathetic trunks
Centrum of vertebra
Dorsal root ganglion
Sympathetic ganglion

should expose only short sections of about five to eight vertebrae each. Coordinate your efforts so that representative parts along the entire spinal cord are exposed.

Spinal Cord

5 ▷ Once the vertebrae are exposed, take a sharp scalpel and gradually remove the cartilaginous spines and neural arches of several vertebrae, exposing the spinal cord (fig. 32.5). The spinal cord will be surrounded by three membranes, or **meninges.** Slit the meninges to observe the spinal cord.

Note the **spinal nerves** branching from the cord. There are 33 pairs. Each nerve is composed of several small roots or fibers. As discussed earlier, sensory neurons carrying impulses into the spinal cord enter by the **dorsal root,** and motor neurons carrying impulses to effectors leave by the **ventral root.** These two roots come together to form the spinal nerves. You may be able to better see the nerves by turning your animal over and looking closely at the back wall of the body cavity. Each of these nerves consists of thousands of neurites extending from different neurons in the spinal cord. They are bundled by surrounding connective tissues.

The cranial end of the spinal cord gradually enlarges to become the **medulla oblongata.** This region of the brain contains the cardiac and respiratory centers as well as numerous sensory and motor nerve tracts, which transmit impulses to and from the higher brain centers.

Autonomic System

The **autonomic nervous system** consists of the **sympathetic** and **parasympathetic systems,** which are involved in the control of glands, smooth muscle, and viscera. These two systems have opposing effects to slow or speed muscle movements in the viscera. This system is part of the **peripheral nervous system** and is connected to the central nervous system by many nerve fibers and ganglia.

The **sympathetic trunks** can be viewed by turning the fetal pig over and looking at the dorsal wall of the body cavity on either side of the spinal column in the thoracic and abdominal cavities. They are thin strands of nerves with small enlargements, the **ganglia,** along their length, running parallel to the spine.

Brain

This dissection takes some time to perform; your instructor may provide a demonstration of the larger brain from an adult animal so that you do not have to do the dissection.

6 ▷ If you are to do the dissection, expose the skull by making a longitudinal cut from the base of the snout to the base of the skull. Make lateral cuts from the ends of the first cut to the angle of the jaws at the anterior end and to the level of the ears at the caudal end. Remove the muscle layers. Make a shallow longitudinal cut in the skull and then lateral cuts from this incision at 2 cm intervals. Break off chips of the skull and gradually expose the brain. At the base of the skull, the tough **occipital** bone will have to be dissected out separately. The spinal cord passes through this bone.

The brain is surrounded by three meninges as was the spinal cord. Remove the membranes and observe the gross features of the brain (fig. 32.6). A longitudinal fissure separates the right and left hemispheres of the **cerebrum.** Higher order functions, such as memory, intelligence, and perception, are associated with this part of the brain. Caudal to the cerebrum is the smaller **cerebellum,** which coordinates motor activity and equilibrium. The **medulla oblongata** lies ventral and caudal to the cerebellum and is continuous with the spinal cord.

The cerebrum has a convoluted surface. The ridges are called **gyri** and the valleys, **sulci.** The outer part of the cerebrum, the **cortex,** is made up of **gray matter,** which is composed of nerve cell bodies and supporting, but not conducting,

Figure 32.6   Dorsal view of the fetal pig's brain.

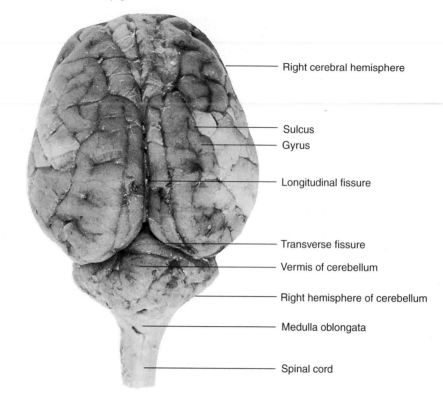

- Right cerebral hemisphere
- Sulcus
- Gyrus
- Longitudinal fissure
- Transverse fissure
- Vermis of cerebellum
- Right hemisphere of cerebellum
- Medulla oblongata
- Spinal cord

neuroglial cells. The inner part of the cerebrum is made up of **white matter,** which is composed of nerve-cell extensions or fibers encased in an insulating, fatty myelin sheath. In the nerve cord, the relative positions of white and gray matter are reversed, with myelinated nerve fibers on the outside and cell bodies on the inside.

The **transverse fissure** separates the cerebrum from the cerebellum. There are three parts to the cerebellum, the two **lateral hemispheres** and a central **vermis.** If the caudal edge of the cerebellum is raised, a thin vascular membrane called the posterior choroid plexus may be observed covering the medulla. The space beneath the choroid plexus is the **fourth ventricle** of four ventricles, or cavities, found in the brain. **Cerebrospinal fluid** created by filtration from the capillaries located in the ventricles passes into the spinal cord spaces. This fluid carries nutrients to the cells and protects the nervous system from mechanical shock. The ventricles are best seen in a sagittal section of the brain (fig. 32.7).

7▷ Remove the brain from the skull (or obtain demonstration material) and orient it with the ventral surface up (fig. 32.8). At the anterior end of the cerebrum, find the **olfactory bulbs.** In the midportion of the cerebrum, two large nerve trunks, the **optic tracts,** cross to form the **optic chiasma.** Just posterior to the crossover is the **pituitary,** but this small body is often broken off when the brain is removed from the cranium. Dorsal to the pituitary comprising the ventral surface of the brain is a conspicuous oval area, the **hypothalamus.** It integrates many autonomic functions, such as sleep, body temperature, appetite, and water bal-

ance. Posterior to this is a wide transverse group of fibers, the **pons,** which serves as a passageway for nerves running from the medulla to higher centers. Twelve pairs of **cranial nerves** directly enter the brain, bringing sensory information concerning sight, sound, taste, equilibrium, and touch. Motor fibers to the head region also exit via these nerves. The nerves are named in figure 32.8.

In table 32.1 write a brief description of each structure and a summary statement of its function.

Sensory Systems

Eye Anatomy

Dissection of beef eyes may be substituted for fetal pig dissection, if desired. If doing so, skip the next paragraph.

8▷ To remove an eye from the fetal pig, make an incision that extends from the external corner of the eye completely around the eye, removing the upper and lower lids. A thin mucous membrane, the **conjunctiva,** covers the eye and folds back to line the undersurface of the lids, preventing foreign material from entering the socket. Now remove the connective tissue, the muscle, and part of the bony orbit surrounding the eye. Push the eye to the side and up and down, noting the seven thin strips of the ocular muscles that originate on the skull and insert on the eye. Cut these muscles as far as possible from the eye. Draw the eye out of the orbit and snip the optic nerve. Once the eye is free, note the muscle mass that surrounds the eyeball and holds it in orbit.

Figure 32.7 Sagittal section of the fetal pig's brain.

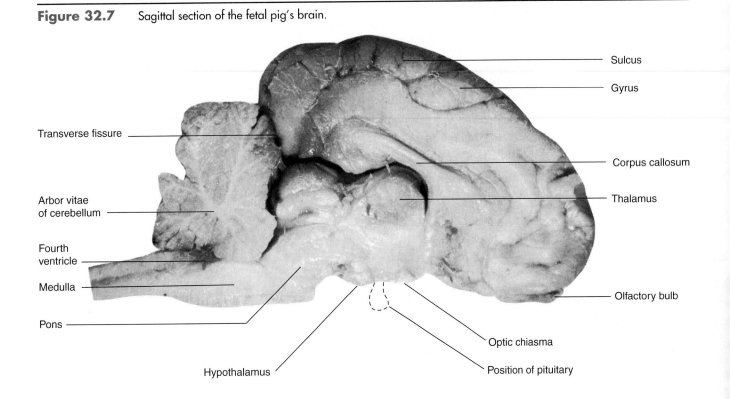

Sulcus

Gyrus

Transverse fissure

Corpus callosum

Thalamus

Arbor vitae of cerebellum

Fourth ventricle

Medulla

Pons

Olfactory bulb

Optic chiasma

Hypothalamus

Position of pituitary

Figure 32.8 Ventral view of the fetal pig's brain with cranial nerves indicated by standard numbering system.

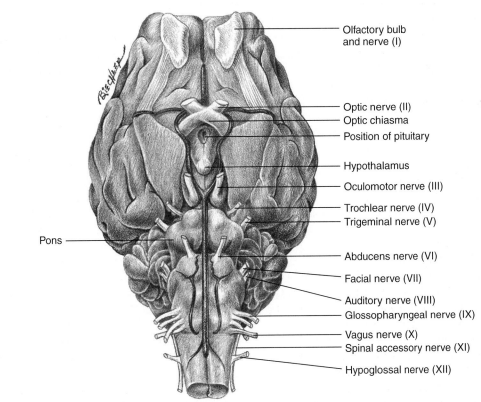

Olfactory bulb and nerve (I)

Optic nerve (II)
Optic chiasma
Position of pituitary

Hypothalamus
Oculomotor nerve (III)

Trochlear nerve (IV)
Trigeminal nerve (V)

Pons

Abducens nerve (VI)

Facial nerve (VII)

Auditory nerve (VIII)
Glossopharyngeal nerve (IX)

Vagus nerve (X)
Spinal accessory nerve (XI)

Hypoglossal nerve (XII)

TABLE 32.1	Summary description of nervous system components in fetal pig

Spinal nerves

Sensory nerve tracts

Motor nerve tracts

Medulla oblongata

Pons

Cerebellum

Cerebrum

Meninges

Cranial nerves

With a sharp razor blade, cut longitudinally through the eye to one side of the **optic nerve** and place the halves in a dish of water. Identify the three layers making up the wall of the eye: the white **sclera,** the dark **choroid,** and the **retina** (fig. 32.9). The front surface of the eye is covered by a transparent layer of connective tissue called the **cornea.** The large white structure within the eye at the front is the **lens.** (It has turned white from the preservative.) The colored material surrounding the lens is the **iris,** and the central opening in the iris is the **pupil.** Muscles in the iris regulate the opening of the pupil, depending on the light intensity in the environment.

The large posterior space of the eye is filled with a gelatinous fluid, the **vitreous humor,** whereas the small chamber anterior to the lens is filled with **aqueous humor.** Where the optic nerve enters the eye, the retina is devoid of rods and cones. This area is called the **blind spot** (optic disk) since light falling on this area does not trigger nerve impulses. The small, yellowish spot in the center of the retina is called the **macula lutea.** The depression (**fovea**) in the center of this spot has a very high density of cones, and is the area producing the greatest visual acuity. Figure 32.9 shows the structure of the eye.

Color Vision

In the human retina, there are two kinds of light-sensitive cells: **rods** and **cones.** These cells are located on the outside of the retina, away from the vitreous humor and against the choroid layer, and are covered by a layer of bipolar and

Figure 32.9   Section of the eye.

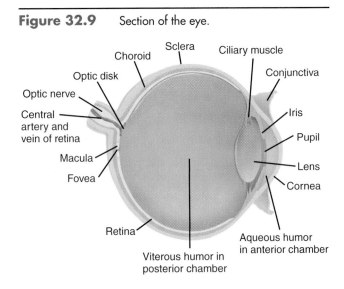

ganglion cells as well as neurons, as shown in figure 32.10. Light enters the posterior chamber through the lens and must pass through several cell layers before it is absorbed by the rods and cones. Look at the demonstration slide of the mammalian retina and identify rods and cones.

Cones are involved in color vision at daylight intensities. Figure 32.10 shows that each cone is associated with one bipolar cell. Generator potentials in a cone excite one neuron, giving high-level resolution vision. Cones will not cause neuronal firing at low light intensities and are, therefore, nonfunctional at night.

Figure 32.10 Diagram of the cells in the retina. Note the light direction. Light must pass through the layer of nerve cells before it reaches the rods and cones.

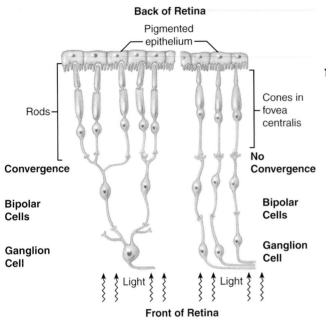

Back of Retina

Pigmented
— epithelium —

Rods—

Cones in
—fovea
centralis

Convergence

**No
Convergence**

**Bipolar
Cells**

**Bipolar
Cells**

**Ganglion
Cell**

**Ganglion
Cell**

Light

Light

Front of Retina

Rods provide black-and-white vision at low light intensities. A bipolar cell associates with several rods, so that generator potentials sum in their effect. This assures neuronal firing at low light intensities and provides a highly sensitive vision system but with less resolution. Why?

When light strikes a receptor cell, it triggers a photochemical reaction that leads to neuronal firing. If a rod or cone receives high-intensity light, the visual pigments will be bleached temporarily and will not absorb any more light. Within a matter of a few seconds, enzyme systems in the cells will restore the visual pigment to its light-absorbing form. This can be easily demonstrated. If, after viewing a bright light, the eye is quickly shut or turned to a dark wall, the bright image will persist as a positive **afterimage** because the high generator potential causes the continued firing of neurons. If the eye is cast on a lighter background after a few seconds, a negative **afterimage**, or a dark image of the object, will appear. This is because the still bleached cells are not receptive to the light coming from the light background.

According to the **Young-Helmholtz theory of color vision,** there are three types of cone cells, which respond respectively to red, green, and blue light. All other colors are perceived as the brain interprets impulses coming from a mix of these receptors. If an object of one color is viewed for a long period of time at high light intensity, the cones for that color will become bleached. The afterimage of the object will be "seen" in the complementary color.

10► Your instructor will demonstrate afterimages to you. A small square of red light will be projected on the screen in the dark lab room. After staring intensely at the square for 20 seconds or so without shifting your gaze, the slide will be changed to a soft white. Continue staring at the screen as the change is made. What do you "see"? What was the color of the afterimage?

Repeat this experiment using a blue square. Before doing the experiment, hypothesize what color the afterimage will be according to the Young-Helmholtz theory.

Hypothesis to be tested:

What did you see when the experiment was conducted?

? Must you accept or reject your hypothesis?

? Explain the colors seen in these afterimages using the Young-Helmholtz theory.

Ear Anatomy

The mammalian inner ear is difficult to dissect because it is part of the bony skull. However, the anatomy is not difficult to visualize. Study figure 32.11 and models of the ear provided in the laboratory.

11 ▷

Sound waves are collected by the external ear and pass down the auditory canal, causing the eardrum (**tympanic membrane**) to vibrate. The movement of the eardrum is transmitted by the three **auditory ossicles** of the middle ear to the flexible membrane covering the **oval window** of the **cochlea** or inner ear. Sound energy is amplified in this process for two reasons: the area of the tympanic membrane is about 30 times larger than the oval window membrane; and the mechanical leverage systems of the ossicles, the **malleus, incus,** and **stapes,** greatly amplify any movement. The back-and-forth movement of the oval window causes the fluid in the cochlea to move. The cochlea is divided lengthwise into three canals by soft-tissue walls.

Pressure waves generated at the oval window travel up the **vestibular canal** to the apex of the cochlea and down the **tympanic canal.** The **round window** at the base of the tympanic canal serves as a pressure-release valve, allowing the fluid of the cochlea to move back and forth. Thus, these gross anatomical structures of the ear convert the movement of air molecules, or sound, into the movement of the fluid (perilymph) in the inner ear.

The pressure waves cause the thin, longitudinal **basilar membrane** to vibrate. **Sensory hair cells** which are found

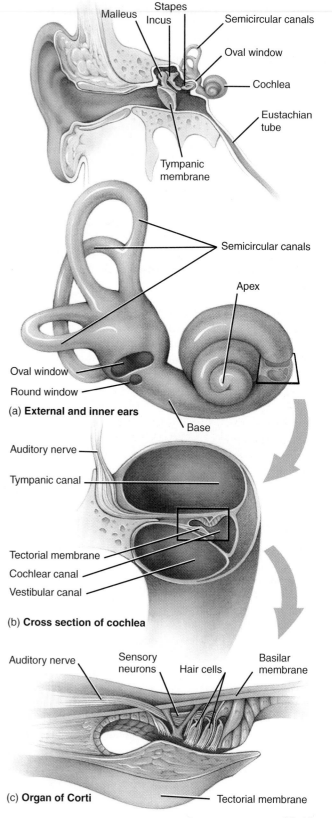

Figure 32.11   Anatomy of the human ear: (*a*) the relationship between the external and inner ears, and a cross section of the cochlea, showing the canal and membrane organization; (*b*) magnified view of the cochlear canal showing organ of Corti and the relationship of hair cells to the tectorial membrane; (*c*) the organ of Corti in magnified view.

Stapes
Malleus
Incus
Semicircular canals
Oval window
Cochlea
Eustachian tube
Tympanic membrane

Semicircular canals
Apex
Oval window
Round window

(a) **External and inner ears**
Base

Auditory nerve
Tympanic canal
Tectorial membrane
Cochlear canal
Vestibular canal

(b) **Cross section of cochlea**

Auditory nerve
Sensory neurons
Hair cells
Basilar membrane

(c) **Organ of Corti**
Tectorial membrane

on this membrane in an area known as the **organ of Corti** (fig. 32.11*c*), are displaced upward and brush against a stiff, overhanging structure called the **tectorial membrane.** Distortion of the hair cells causes generator potentials, which in turn cause the firing of the cochlear neurons that travel along the auditory nerve to the brain.

Loudness is encoded in the neurons by how frequently nerve impulses are generated in the optic nerve. Pitch is detected in a different way. High-frequency sounds stimulate cells near the base of the cochlea and low-frequency sounds stimulate cells near the apex. Separate neurons lead from hair cells at different points along the cochlea to the brain. Location of receptors thus determines pitch discrimination.

Obtain a slide of a cross section of the cochlea. First look at the slide under the 10× objective. Identify the structures shown in figure 32.11*b* and *c*. Switch to the high-power objective and observe the hair cells and their relationship to the tectorial membrane in the organ of Corti.

Pair off with another student and quiz one another about how the ear functions. Use the model and explain to one another how sound passes into the inner ear and is transduced into a nerve impulse.

Learning Biology by Writing

Write a 150-word essay on how a sensory nerve impulse travels from a muscle spindle in the calf to the central nervous system and back to a muscle cell in the leg via a motor neuron. Discuss the ways in which action potentials, synapses, and conduction velocities are involved.

As an alternative assignment, your instructor may ask you to answer the Lab Summary and Critical Thinking Questions that follow.

Internet Sources

Neuromuscular junctions are fascinating anatomical sites where the nervous system meets the muscular system. These are still the topics of much research. Use a search engine to search the World Wide Web to discover the type of research that is being conducted now. Jump to at least three of the sites returned in the list and summarize the research that is being done. Include the URLs for future reference.

Lab Summary Questions

1. Describe the organizational structure of nervous systems in the animal kingdom as you proceed from Cnidaria to Vertebrata.
2. How does the structure of a nerve cell reflect its function?
3. What is the difference between white matter and gray matter in the mammalian central nervous system?
4. Name the major anatomical features of the brain as viewed whole from the dorsal perspective.
5. Describe how light entering the eye is regulated and focused on the retina. Describe the response of rods and cones to different colors of light.
6. Explain the nature of afterimages that are seen after looking at a bright light. Why are afterimages a different color than the actual images?
7. Trace the pathway of sound energy as it enters the auditory canal and passes into the inner ear. How are pitch and loudness encoded by the nervous system?

Critical Thinking Questions

1. Hearing impairment can result from mechanical failure of parts of the ear, or from sensory cell damage, or from auditory nerve damage. Which of these might be easily corrected by a hearing aid? Why?
2. If you examined the retinas of various animals with a microscope, what differences would you expect to find between animals that are active at night versus those active during the day?
3. As humans age, the lens of the eye often hardens and the ciliary muscles will not bend it as readily as when we are young. How does this affect what can be seen?
4. A person suffers a spinal injury. He can move his legs but has no sensation from the legs. What has been damaged and what is prognosis for recovery?

LAB TOPIC 33

Statistically Analyzing Simple Behaviors

Supplies

Preparator's guide available on WWW at
   http://www.mhhe.com/dolphin

Equipment

Lamp with 100-watt bulb

Materials

Ring stand, two clamps and one ring clamp
Ruler
Paper towel tube. At one end punch two holes
   opposite each other 1/4" in from edge. Close that
   end with a #8 or #9 rubber stopper.
Plastic vials that fit into paper towel tube when taped
   together end to end
Tape (clear)
Thermometer
400-ml and 800-ml beakers filled with water
Wooden stick 1/8" diam × 6"
Fruit flies *(Drosophila melanogaster)*
Marking pencil
Black cloth 24" × 24"

Student Prelab Preparation

Before doing this lab, you should read the introduction
and sections of the lab topic that have been scheduled
by the instructor.

   You should use your textbook to review the
definitions of the following terms:

   innate
   instinct
   kinesis
   taxis

   You should be able to describe in your own words
the following concepts:

   Nature versus nurture debate
   Chi-square test
   Page reference
   Read appendix C, page 478

   As a result of this review, you most likely have
questions about terms, concepts, or how you will do
the experiments included in this lab. Write these
questions in the space below or in the margins of the
pages of this lab topic. The lab experiments should

help you answer these questions, or you can ask your
instructor during the lab.

Objectives

1. To test hypotheses regarding phototactic and
   geotactic behavior, using fruit flies
2. To use statistical tests to reject or accept
   experimental hypotheses

Background

Behavior has been studied from two philosophical perspectives. The older view involves observing animals and trying
to determine which behaviors are due to **instinct** or **innate
behavior patterns** and which to **learning.** This is the **nature-
versus-nurture** perspective. Behaviorists, such as Konrad
Lorenz and Niko Tinbergen, developed another approach,
called **ethology,** or the study of the evolution of behavioral
characteristics of a species.

   Although ethology embraces the nature-versus-nurture
viewpoint, it also asks *how* the behavior developed and *why*
it is an advantage to the species. This viewpoint should not
be confused with an anthropomorphic, or human-oriented,
view of animal behavior. For example, an anthropomorphic
explanation of why a bird sings on a beautiful spring morn-
ing might be because it is happy. An ethological explanation for the same event might be that the bird is singing to
communicate two things to other members of the population: availability for mating and territorial occupation.

   Like physiological processes, behavior must be considered in terms of survival value. Like other biological phenomena, complex behavior ultimately depends on cells:
sensory cells, neurons, and effector cells acting in coordination with one another. The functioning of these cells is ultimately related to the genes that dictate the behavioral reper-
toire. Experiments with fruit flies have confirmed this idea
by showing that some behaviors, such as circadian rhythms
and mating behavior, are inherited in accordance with
Mendelian genetic mechanisms.

   However, biologists are a long way from knowing the
molecular mechanisms involved in the development of these
behaviors. Neurobiologists are just now beginning to map
the neuronal circuits involved in many complex behaviors.

Some day research may reveal how innate behavior and memory are stored in the nervous system. When scientists understand the mechanisms of storage and the way in which incoming stimuli are processed, they may be able to understand exactly why an animal behaves as it does in certain situations.

Until that time occurs, the study of behavior is a "black box" approach. This scientific approach to studying a process where the mechanisms are unknown involves observing an animal first under normal conditions and then under experimental conditions with the changes in response noted. In this way, the observer obtains some information about the capabilities of the underlying mechanism and may also gain information that proves useful in managing the system or organism. The black box approach has led to great advances in our understanding of animal communications, orientation, and sociobiology.

The simplest behaviors are **kineses** and **taxes.** Kineses involve a change in the rate of random movement in response to an environmental stimulus. For example, pill bugs, usually found in moist places, get there because of kineses. When in dry conditions, they move quickly in random directions, but in moist conditions, they move slowly or not at all. Thus, they tend to congregate where there is moisture but they do not seek it. In contrast to kineses, taxes are directed movements. An animal that senses and moves toward a stimulus is said to show positive taxis; one that moves away shows negative taxis. Both taxes and kineses are innate behaviors.

LAB INSTRUCTIONS

You will investigate taxes in fruit flies, specifically, their response to light and gravity.

Hypothesis to Be Tested

A statistical test always tests a **null hypothesis (H_o)**, which is based on a scientific model of what the outcomes from an experiment should be. The null hypothesis proposes that there is no significant difference between the results actually obtained and the results predicted by a model for an experiment.

A null hypothesis is always paired with an **alternative hypothesis (H_a)**, which proposes an alternative explanation of the results, not based on the same scientific model as was the null hypothesis. If the statistical test indicates that you cannot accept H_o, then H_a is true by inference: the variation in your data compared to that predicted by the model is greater than that expected by chance, and some other factor is responsible for the variation between observed and predicted results. A more thorough discussion of scientific hypotheses may be found in appendix C.

For these experiments, we will propose a simple null hypothesis. (Since the response of fruit flies cannot be predicted, the model used will be a random distribution model.)

H_o: Fruit flies randomly distribute themselves in the container regardless of the direction of light and/or gravity.

H_a: Fruit flies orient and move either positively or negatively in response to light or gravity.

Therefore, to show taxis is to reject H_o and infer H_a.

1 ▷ To collect data to test your hypotheses, obtain one small plastic vial, some transparent tape, and 15 fruit flies in a second plastic vial from the supply table. Tap the vial containing the flies on the table, so that all the flies fall to the bottom. Quickly remove the stopper and tape the open end of the empty vial to the open end of the vial containing the flies, so that you have a double vial experimental chamber. The tape should pass completely around the seam between vials (fig. 33.1).

Use a marker to place lines on the double vial, dividing it into thirds. Take a 5-inch piece of tape and turn it back on itself with a 2-inch overlap. Attach this as a pull tab on one end of the double vial. Do the same for the other end. This chamber and the flies it contains will now be used to test hypotheses about behavior.

Phototaxis

You should work in groups of three for all experiments in this lab topic.

2 ▷ To determine if fruit flies respond to light, slide the chamber into a cardboard tube and put stoppers in the ends of the tube so no light enters. Place the tube horizontally on a table. Place a lamp (off) 12 inches from one end of the tube with a 800-ml beaker of water in between the tube and the lamp to serve as a heat filter. Wait for at least five minutes and measure the temperature at the end of the tube nearest the lamp. Record the temperature in table 33.1. Leave the thermometer in front of the tube but elevated off the table. Why is it necessary to measure the temperature before and after the experiment? If temperature rises, how would it affect your analysis of the data?

Now remove the stopper nearest the lamp and turn the light on for five minutes. Record the temperature again and slip the tube off while holding the experimental chamber by its tab to prevent movement. Immediately count the number of flies in each third of the chamber and record your results in table 33.1.

Figure 33.1 Experimental setup for studying phototaxis.

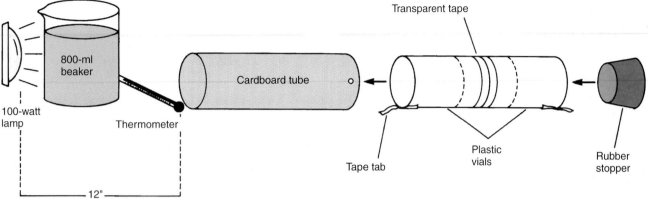

TABLE 33.1 Results from your phototaxis experiments

Temperature (°C)

	Before	After
Trial 1		
Trial 2		
Trial 3		

Fruit Fly Distribution

	Nearest ⅓	Middle ⅓	Farthest ⅓
Sum across trials			

Repeat the experiment twice more, but each time turn the experimental chamber end-over-end before slipping it back into its tube. Why?

On the blackboard, record your sums across trials from table 33.1. Copy down the class data in table 33.4. The methods for analyzing this data will be discussed later. Turn off the light used in this part of the experiment.

Geotaxis

4▷ To determine if the fruit flies will react to gravity, slip the chamber into the cardboard tube, stopper both ends, and lay the tube horizontally on the table for three minutes.

Now remove one stopper and stand the tube and the chamber upright on the table with the remaining stopper on the top. After five minutes, quickly slip off the tube and count the flies in each of the three sections. Record your results in table 33.2.

Reverse the vial and repeat the procedure for measuring the response to gravity twice more. Be sure to precede each trial with a three-minute horizontal equilibration period.

5▷ Put your sum across trials data for geotactic response on the board and transfer the class data into table 33.5.

Why should you reverse the vials for each new trial?

TABLE 33.2 Results from your geotaxis experiments

Fruit Fly Distribution

	Top ⅓	Middle ⅓	Bottom ⅓
Trial 1			
Trial 2			
Trial 3			
Sum across trials			

Statistically Analyzing Simple Behaviors

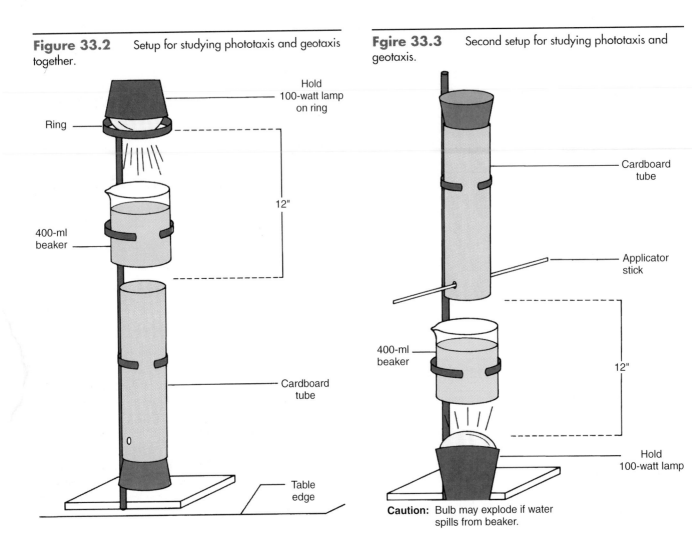

Figure 33.2 Setup for studying phototaxis and geotaxis together.

- Hold 100-watt lamp on ring
- Ring
- 12"
- 400-ml beaker
- Cardboard tube
- 0
- Table edge

Fgire 33.3 Second setup for studying phototaxis and geotaxis.

- Cardboard tube
- Applicator stick
- 400-ml beaker
- 12"
- Hold 100-watt lamp

Caution: Bulb may explode if water spills from beaker.

Combined Test of Phototaxis and Geotaxis

6 You will now test the relative strength of the responses to light and gravity, simultaneously. Place the container of flies in the cardboard tube and stopper both ends. Clamp the tube and container horizontally on a ring stand for a three-minute equilibration period. Clamp a 400-ml beaker full of water above the tube to serve as a heat filter. Measure the temperature at the cardboard tube. Record your results in table 33.3.

Remove the stopper from one end of the tube and rotate the clamp holding the tube so that the open end is upward (fig. 33.2). Turn on the light and place it 12 inches above the open end of the tube for five minutes. Have someone hold a thermometer next to the open end of the tube during this time. Read the temperature after five minutes. Remove the stopper from the bottom of the tube and carefully slip the container out without jarring it. Count the number of flies in each section of the container and record the results in table 33.3.

Repeat this procedure twice more and record the results in the top section of table 33.3.

The procedure will now be reversed, so that both light and gravity come from the bottom of the tube (fig. 33.3). Put

a stopper in both ends and clamp the tube in a horizontal position on the ring stand. Adjust the position of the ring stand on the table, so that the tube hangs over the edge. Clamp a 400-ml beaker of water below the tube. Measure the temperature at the tube. Record the temperature in table 33.5. After three minutes, remove the stopper from the end with the two holes and insert a wooden applicator stick through the holes in the wall of the cardboard tube. Rotate the clamp so the end with the stick is downward.

Turn on the lamp and hold it 12 inches below the open end of the tube for five minutes.

C A U T I O N

If water spills on the bulb, it will explode.

Have someone check the temperature and record it in table 33.3. Now remove the stopper and carefully slip the vial upward, using the tab handle. Note the number of flies in each container section and record it in the lower section of table 33.5. Repeat the procedure twice more.

Light from top	Temperature (°C)			Fruit Fly Distribution		
	Before	After		Top ⅓	Middle ⅓	Bottom ⅓
Trial 1						
Trial 2						
Trial 3						
Sum across trials						

Light from bottom	Before	After		Top ⅓	Middle ⅓	Bottom ⅓
Trial 1						
Trial 2						
Trial 3						
Sum across trials						

7▷ Place your sum across trials data on the blackboard and copy the class data into tables 33.6 and 33.7.

Statistical Analysis of Results

In these four experiments, you collected quantitative data that describe the taxis response of fruit flies. If all fruit flies went to one end of the tube, either toward or away from the stimulus, there is no doubt that they exhibit a taxis. However, if only 6 out of 15 flies were at one end of the tube, does this represent a significant response or has the experiment failed to show a taxis?

Statistical tests can help us make up our minds in situations when the results are not clear-cut, when we need to make an unbiased judgment. Statistical tests are based on two premises: that you have a hypothesis to test and that you are willing to reject an answer that could be true.

Statistical Testing of Hypotheses

The **chi-square (χ^2) goodness of fit test** can be used to test the null hypothesis (H_o), that is, whether the flies are equally distributed throughout the chamber within an acceptable level of deviation or whether they are clumped toward or away from the stimuli tested. The χ^2 (chi-square) test is discussed further in appendix C and you should read it before continuing.

8▷ The class data tables (tables 33.4 through 33.7) for each experiment have been set up with a χ^2 work table at the bottom. To determine the **observed** results, sum the entries in the columns. To determine the **expected** results, take the total number of flies (sum across observed results) and divide by 3. If flies randomly distribute in the chambers during the experiment, you would expect that one-third of the flies would be in each area. Enter only whole numbers as expected results.

The next three lines of the work tables are self-explanatory in terms of the arithmetic that needs to be done. What you are doing is developing a single number that is a measure of the amount of variation seen in the experiment compared to the expected random distribution proposed in the null hypothesis. If you sum across the last line of the table, you will have an estimation of the total variability in the results of the experiment. This is the calculated χ^2 for the experiment. Thus,

$$\chi^2 = \sum \frac{(\text{obs-exp})^2}{\text{exp}}$$

When the observed and expected results are the same, χ^2 will equal zero: the greater the deviation of the observed results from those expected, the larger will be the value of χ^2.

How do you judge whether the χ^2 value you obtained is significant? You must refer to a table of critical χ^2 values, such as table C.4 on page 476. To use an χ^2 table, you must know the degrees of freedom and the acceptable confidence level.

The **degrees of freedom (d.f.)** is a number equal to the number of categories into which your data fall minus one. In your data, you categorized flies according to which one-third of the chamber they were found in. How many degrees of freedom should you use?_____

Students trained in statistics will realize that an assumption is being made here. To keep the analysis simple, we will assume that there is no interaction occurring between replication of the experiment by different laboratory groups and the location of the flies in a vial. If some groups consistently bump an upright vial knocking flies to the bottom, this assumption will be false and a contingency table analysis would have to be done. Students wanting to do such an analysis should consult a statistics textbook.

Statistically Analyzing Simple Behaviors   **449**

TABLE 33.4 Class data from phototaxis experiments

Fruit Fly Distribution

Group	ΔTemp (°C)	Top ⅓	Middle ⅓	Bottom ⅓
1				
2				
3				
4				
5				
6				
7				
8				
Sum across groups (observed)				
Expected by chance				
(obs-exp)				
(obs-exp)²				
$\dfrac{(obs-exp)^2}{exp}$				

$\chi^2 = \dfrac{\Sigma(obs-exp)^2}{exp}$ across chamber areas = _____

Critical value for χ^2 from table C.4, p. 476 _____

d.f. = _____: Probability level _____

State null hypothesis for this experiment _____

_____

_____

_____

Total _____

State the alternative hypothesis _____

_____

_____

Accept or reject null hypothesis _____

TABLE 33.5 Results from class geotaxis experiments

Fruit Fly Distribution

Group	Top ⅓	Middle ⅓	Bottom ⅓
1			
2			
3			
4			
5			
6			
7			
8			
Sum across groups (observed)			
Expected by chance			
(obs-exp)			
(obs-exp)²			
$\dfrac{(obs-exp)^2}{exp}$			

$\chi^2 = \dfrac{\Sigma(obs-exp)^2}{exp}$ across chamber areas = _____

Critical value for χ^2 from table C.4, p. 476 _____

d.f. = _____: Probability level _____

State null hypothesis for this experiment _____

_____

_____

_____

Total _____

State the alternative hypothesis _____

_____

_____

Accept or reject null hypothesis _____

TABLE 33.6 Class data on combined phototaxis and geotaxis: light from the top

Fruit Fly Distribution

Group	ΔTemp (°C)	Top ⅓	Middle ⅓	Bottom ⅓
1				
2				
3				
4				
5				
6				
7				
8				
Sum across groups (observed)				
Expected by chance				
(obs-exp)				
(obs-exp)²				
$\frac{(obs-exp)^2}{exp}$				

$\chi^2 = \dfrac{\Sigma(obs-exp)^2}{exp}$ across chamber areas = _____

Critical value for χ^2 from table C.4, p. 476 _____

d.f. = _____: Probability level _____

State null hypothesis for this experiment _____

_____

_____

_____

Total _____

State the alternative hypothesis _____

_____

_____

Accept or reject null hypothesis _____

TABLE 33.7 Class data on combined phototaxis and geotaxis: light from the bottom

Fruit Fly Distribution

Group	ΔTemp (°C)	Top ⅓	Middle ⅓	Bottom ⅓
1				
2				
3				
4				
5				
6				
7				
8				
Sum across groups (observed)				
Expected by chance				
(obs-exp)				
(obs-exp)²				
$\frac{(obs-exp)^2}{exp}$				

$\chi^2 = \dfrac{\Sigma(obs-exp)^2}{exp}$ across chamber areas = _____

Critical value for χ^2 from table C.4, p. 476 _____

d.f. = _____: Probability level _____

State null hypothesis for this experiment _____

_____

_____

_____

Total _____

State an alternative hypothesis _____

_____

_____

Accept or reject null hypothesis _____

Each column in table C.4 contains **critical values** for χ^2 at several degrees of freedom and confidence levels. Scientists usually use the 95% confidence level, which means that 95% of the chance differences between observed and expected values will give an χ^2 value equal to or less than the value indicated for a specified degree of freedom. In other words, only 5% of the χ^2 values will be greater than this number due to chance.

If your calculated χ^2 value at your specified degrees of freedom is smaller than or equal to the value in the table, you cannot reject the null hypothesis because the agreement between the observed and the expected values is exceptionally good: fruit flies are randomly distributed in the vial. On the other hand, if your calculated χ^2 value is greater than the value in the table, you must reject the null hypothesis and infer the alternative: fruit flies respond to the variables being tested. You must look at the data to determine if they are positive or negative in their taxes.

9 ▷ For each experiment, compare your χ^2 value to that in the table. Must you accept or reject the null hypothesis in each experiment? What does that mean in terms of the model you proposed? Record your decisions below.

Experiment 1: Phototaxis

Experiment 2: Geotaxis

Experiment 3: Geotaxis with light above

Experiment 4: Geotaxis with light below

Learning Biology by Writing

This lab is a well-defined experiment. The objectives and the hypotheses being tested are clear, and a lab report should be easy to write. In the introduction to your lab report, describe the purposes and hypotheses. In the methods section, describe briefly how you made your observations. Finally, in the data section, present your summary tables and anecdotal observations. The discussion should answer the questions in the introduction and generally discuss the significance of your conclusions. Be sure to include a discussion of sources of error. Finally, you might want to adopt an ethological approach to behavior and speculate on the survival value to the fruit fly of these taxes or lack of taxes.

Internet Sources

In this lab topic, you studied, among other things, the response of fruit flies to gravity. Using GOOGLE search the World Wide Web for information about geotaxis in fruit flies. Several responses will be for a database called Flybase, a database for the genetics of fruit flies. Read the reports that are cited and determine why geotaxis was mentioned in the report.

As an alternative assignment, your instructor may ask you to answer the Lab Summary Questions.

Lab Summary Questions

1. What evidence do you have from your experiments that fruit flies are positively or negatively phototactic? Geotactic? Is the influence of phototaxis or geotaxis stronger?
2. In your own words, describe what a χ^2 (chi-square) goodness of fit test does? What does it mean? Why do we use it?
3. What is a null hypothesis? What is an alternative hypothesis? Can both be true?
4. If your χ^2 value leads you to reject a null hypothesis, is the alternative hypothesis automatically true? Why?

Critical Thinking Questions

1. In the phototaxis experiments, how could you determine whether the flies are responding to light or to heat?
2. Speculate on the evolutionary advantage of a fruit fly being positively phototactic and negatively geotactic.
3. What information would you need to have to decide whether the behavior you have observed is really a taxis or could be a kinesis?
4. How could you test whether the behaviors observed were genetically determined?

LAB TOPIC 34

Estimating Population Size and Growth

Supplies

Preparator's guide available on WWW at
http://www.mhhe.com/dolphin

Equipment

Balances, with 0.01 g sensitivity

Materials

Wooden pegs or applicator sticks
Small plastic bags
4-m string
40-cm string
About 1,000 beads per student lab group (all one
    color, in coffee cans)
Beads of another color
Markers
Means of generating pairs of random numbers
If lab done outside
    Maps of quadrat sampling area
    Resource books on identification of common lawn
        weeds in area
    Identified samples of lawn plants in lab for
        reference
If lab done inside
    Table top covered by 1-m^2 paper ruled in 10-cm grid
    Beads to scatter on table
    Boards 1 m × 5 cm to edge grid
Microcomputer and population growth software

Student Prelab Preparation

Before doing this lab, you should read the introduction
and sections of the lab topic that have been scheduled
by the instructor.

You should use your textbook to review the
definitions of the following terms:

dispersion	population
exponential	population density
logarithm	quadrat
mark and recapture	

You should be able to describe in your own words
the following concepts:

Population growth curves
Exponential growth
Logistic growth
Carrying capacity

As a result of this review, you most likely have
questions about terms, concepts, or how you will do
the experiments included in this lab. Write these
questions in the space below or in the margins of the
pages of this lab topic. The lab experiments should
help you answer these questions, or you can ask your
instructor during the lab.

Objectives

1. To use quadrat sampling techniques to estimate the
   size of nonmobile populations
2. To simulate mark-and-recapture methods for
   estimating the size of large, mobile populations
3. To plot logistic growth curves, using prepared data
4. To use semilog graph paper for plotting population
   growth
5. To investigate factors influencing population
   growth, using computer simulations

Background

Organisms do not usually exist as isolated individuals in nature but are parts of larger biological units called **populations.** These reproductive and evolutionary units consist of
the members of a single species residing in a defined geo-
graphical area, for example, in a small park, in a mountain
range, or on a continent. A population of clover in a lawn
may be easy to recognize because of its compactness,
whereas a widespread population of moose in northern Minnesota may be less apparent. Organisms within each population are interlinked by reproductive gene flow. Natural selection operates on individuals within these populations to
shape a population's adaptation to an environment over several generations. Populations are the fundamental units of
species on which the mechanisms of evolution operate.

Biologists study populations for several reasons: to understand gene flow, selection, adaptation mechanisms and to
manage the populations. Management addresses the problem

of the relationship between humankind and the environment. In some cases, management involves controlling the size of populations of noxious organisms, such as rabid skunks, mosquitoes, or parasites of humans. In other cases, the goal may be to increase populations of beneficial organisms, such as lumber-producing trees, edible fish, or game species. To manage populations, it is necessary to know the basic biology of the organisms involved as well as to understand the physical and biological environment.

Studies of populations start with such basic questions as: How many individuals are in the population? How is the population distributed in the study area? Is the population increasing or decreasing in size? To answer these questions, biologists must use sampling procedures that allow them to estimate the number of individuals in the population distributed in a certain space, that is, the **population density.** Rarely is it practical to count the entire population. In this lab, you will use two often-employed techniques of estimation: **quadrat sampling** for nonmobile animals and plants and **mark-and-recapture techniques** for mobile animals. With the passage of the National Environmental Protection Act in 1970, these techniques have been and are, frequently used by biologists who are trying to determine the environmental impact of changes in ecosystems.

Both techniques are based on random sampling statistical procedures; care must be taken to assure that randomness occurs when these techniques are applied to an actual biological problem. Sampling plots must be randomly located and representative of the study areas, and animals must be randomly captured and released so that they can freely mix with the total population. If randomness is not realized, the sampling procedure will lead to erroneous estimates. Sampling procedures repeated over time allow the investigator to determine whether a population is growing or declining.

Estimation of population size alone, however, is not sufficient for making management decisions. Additional information is required, such as what resources are needed by and available to the population, how other organisms influence the population, and what the population age structure is like. The question of resources involves the physical environment: for example, water quality, soil type, temperature, or available nesting sites. The influence of other organisms on a population affects food availability, parasitism, and predation. Population structure refers to age distributions in the population. A stand of white pine trees could consist of all young, all old, or a mixture of young and old trees. Harvest practices and replanting programs would depend on the population structure in a particular timber tract. Similar problems exist in managing deer herds and coho salmon and in assessing the effects of agricultural pesticides on harmful insects.

Many biologists feel that population ecology is the most demanding and comprehensive field of biology because the investigator must understand and deal with environmental variables, metabolic efficiency, physiological reactions, hereditary mechanisms, and the interactions of organisms.

Quadrat Sampling of Vegetation

The structure and composition of nonmobile organisms, such as terrestrial plant communities, can be sampled by a number of techniques. The most common procedure involves using randomly located plots of standard size called **quadrats.** Quadrat sampling varies in terms of the size, number, shape, and arrangement of the sample plots, all of which depend on the information sought and the nature of the populations or communities being studied.

Technique

The size of a plot is determined by the size and density of the plants being sampled. It should be large enough to include a number of individuals but small enough to allow easy separation, counting, and measuring of those individuals. Obviously, plots of different sizes would be used to estimate the number of trees in a forest in contrast to the number of cattails in a swamp.

For the sampling procedures to be statistically valid, quadrats must be chosen randomly within the study site. To do this, a baseline is laid out along one side of the study area. A pair of random numbers is chosen that corresponds (1) to distance along the baseline and (2) to perpendicular distance from the baseline to the center of the sampling quadrat. The units of length will depend on the size of the study area and may be in feet, meters, paces, or any other appropriate linear measure.

Because weather might not allow you to perform quadrat sampling in the field, the technique can be simulated in the laboratory. If the indoor simulation is to be used, your instructor will have 1 m^2 of paper taped to a table top. It will be ruled into a 10-cm grid. How many 10-cm squares are in 1 m^2?_____

Inside Quadrats

One edge of the paper will have the 10-cm squares numbered 1 through 10. An adjacent edge will have them lettered A through J. The location of each 10-cm square is given by a letter-number combination. For example, H8 is found at the intersection of column 8 with row H.

Randomly scattered on the grid will be several hundred beads, some of one color and some of another. The colors represent juveniles and adults of the same species. Some squares will have boxes taped to them to exclude beads. These areas are uninhabitable by the species and should be subtracted from the total area available. Imagine that the bead distribution represents trees in a forest and that you are looking down on it or that you have taken an aerial photo on which a grid was superimposed.

Your task is to estimate the total population size in the study area and the immature population size by quadrat sampling.

Student pairs will each be assigned three random coordinates, such as A5, D2, and F7. Count the number of beads of each color in each assigned square and calculate an average per square. Record below.

2 Place these estimates on the blackboard and calculate a grand average across all lab groups. Record below.

3 Determine the total number of inhabitable squares in the study area and multiply your individual average estimates per square by this number to get your estimation of total population size and number of immatures. Record below.

4 Repeat these calculations, using the grand averages from the combined lab data.

5 Now return to the study area and help others count the total number of beads of each type present. This will give you the actual population size. How many individuals were actually in the population? _____

What percentage were immatures? _____

Why would you not simply count all of the individuals in a natural population?

Calculate the percent error for the estimations based on your data and those based on the class data, using the relationship:

$$\% \text{ error} = \frac{\text{estimate} - \text{real} \times 100}{\text{real}}$$

Why are the % errors different for the estimations based on your data and the class data?

Use the actual numbers of beads in the population to answer this question. If all adults were to die and only the juveniles produce offspring during the next year, is the population expanding or contracting? (Assume each pair of juveniles will produce six offspring and that the number of males equals the number of females.)

Outside Quadrats (Optional)

6 If you are to do this lab outside, your instructor will provide you with a map of a lawn on your campus that will serve as your study area. You will also be given a set of random coordinates, a piece of string 40 cm long and one 4 m long each tied into a loop, and five pegs. Study the map so that you understand where the study area is and the orientation of the baseline.

TABLE 34.1   Number of nongrass plants in quadrat

Species	Number	Comments

Take your lab manual, a pencil, and your strings to the study area. Using your coordinates, pace out the distances along the baseline and perpendicular from it to locate the center of your sampling quadrat. Place a peg at this point. Tie the ends of the 4-m string together to form a loop. Using the other four pegs, stretch the string as a square around the center peg (the area inside the string will be 1 m^2).

7 Count the number of nongrass plants contained within your quadrat. Keep separate counts for each species. If you cannot identify what kind of plants they are, make up a temporary name and then check resource books available in the laboratory. (Specimens can be collected and placed in plastic bags to prevent drying before later identification.) Record your counts in table 34.1.

If, by chance, you have a quadrat with thousands of clovers in it, you will not be able to accurately count them in the time provided. If this is the case, switch to the 40-cm string and lay out a smaller square around the center peg. Count the clover within this area and record the count separately.

8 To determine the density of grass in the lawn, use the 40-cm string to lay out the quadrat around the center peg or the upper-right corner. Count and record the number of grass plants in this quadrat. Consider any aboveground stem to be a grass "plant" but realize they will be connected to other plants by underground stems. Convert the count per small quadrat to the count per large quadrat by multiplying by 100 (= 1m$^2 \div$ [0.1 m $\times$ 0.1 m]).

Note any unusual features in your study area. For example:

Are there any ant hills?

Is the quadrat located on a path? Under a tree? In the open?

Is there drainage into the area?

Are plants uniformly distributed or in clumps?

9 Remove one of each type of broadleaf plant from the lawn by cutting it off at the soil level. Collect several dozen grass plants in the same way.

Analysis

After you have completed your counts, notes, and collection, return to the lab. Put your data on the blackboard, including brief summarizations of your notes. After all lab groups have reported, the data should be discussed to see if any should be rejected because of atypical localized situations, such as chemical spills, trench excavation, or other isolated interferences. Combine the acceptable data and record them as a class summary in table 34.2.

TABLE 34.2 Class results from quadrat analysis

Group	Area Size	No. of Grass Plants	No. of Other Plants (Specify)
1			
2			
3			
4			
5			
6			
7			
8			
9			
10			

10 Calculate the mean and standard deviation for the number of each type of plant per square meter. (Refer to lab topic 5 or appendix C for a description of how to calculate standard deviations.)

A **hectare** is a unit in the metric system equivalent to 2.47 acres or 10,000 m². This would be equivalent to an area of 100 meters by 100 meters. How many square meters were contained in the total study area? Remember that length times width equals area. How many hectares are in the study area?

Using your measurements of grass and broadleaf weed density, how many of each type of plant would be found in a hectare? In an acre?

How many grass plants and other species are contained in the entire study area as defined by your instructor?

11 Weigh each type of plant you collected. Calculate the average weight of a grass plant. Since you know the number of plants in a hectare and the weight of each type of plant, calculate the aboveground wet biomass in a hectare of lawn.

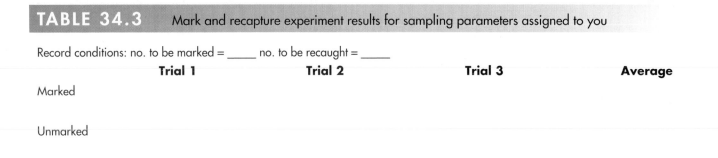

TABLE 34.3   Mark and recapture experiment results for sampling parameters assigned to you

Record conditions: no. to be marked = _____ no. to be recaught = _____

	Trial 1	Trial 2	Trial 3	Average
Marked				
Unmarked				

Mark and Recapture

In the field, it is difficult to measure the size of a population of randomly dispersed, mobile animals, but an estimation can be obtained by mark-and-recapture methods. In these methods, living individuals are trapped or collected, marked, and released at the site of capture. A record is kept of the number marked. At a later time that allows for dispersal, animals are again collected from the population. Some of these will be marked and some will be unmarked due to immigrations and emigrations of individuals to and from the collection sites.

Assuming the animals are randomly dispersed, the frequency of marked and unmarked animals in the second collection will allow you to estimate the total population size, using the following proportion (Lincoln-Peterson method): *The number of animals marked is to the total number of animals in the population as the number of marked animals recaptured is to the number of animals recaptured.* Or,

$$\frac{M}{N} = \frac{R}{C}$$

solving for N

$$N = \frac{MC}{R}$$

where

N = total number of individuals in population

M = number of animals marked and released (50, 100, or 150 depending on your assignment by the instructor)

C = total number of animals caught in second sample (60 or 120 depending on your assignment)

R = number of marked animals caught in second sample (recaptured)

When using the mark-and-recapture technique in natural situations, several assumptions are made. These are:

1. Capturing the animals the first time does not influence whether they will be caught a second time;

2. Marking does not influence behavior and chances of a second capture;

3. All animals in the population are equally catchable;

4. Animals are not migrating into or out of the population;

5. Traps are randomly spread through the population's territory.

These assumptions are not always met. For example, some animals prefer to feed on bait in traps rather than on natural food while others do not. Or, traps might provide shelter for animals that they prefer compared to natural shelters. Thus, one needs to carefully analyze any situation where traps are used.

Technique

In the lab are several cans containing 700 to 1,000 beads of one color. These beads are to simulate animal populations. Each lab group will be given a can and will be asked to estimate the population size, using mark-and-recapture methods. Each group will use a slightly different sampling regime. The results of the different groups will be compared.

12▶ Two groups should each mark 50 beads in their population, while two others should mark 100 and two others, 150. Do this by removing the assigned number of beads from the can and replacing them by adding the same number of different color beads to the can. By comparing results with other groups, you will determine if the size of the marked sample influences accuracy. One group from each of the above pairs should recapture 60 beads from its population, while the other group will recapture 120. This will test the effect of the recapture sample size on accuracy.

Once you are sure of your group's assignment for the number of beads to be marked and the number to be recaptured, start the experiment. Add "marked" individuals to the population in the can. Close the can and shake well to ensure random mixing. Without looking, withdraw a bead and tally its color on a piece of scrap paper. Stir the beads, and withdraw another, adding to the tally. Continue sampling, one bead at a time, until you withdraw the number of beads corresponding to your assigned recapture sample size. After every five draws put the top back on the can and shake it to ensure random mixing.

Count how many beads in the recapture sample are marked and how many unmarked. Record the counts in table 34.3. Return the beads to the can and repeat the recapture sampling twice more. Record the results. Calculate an average value for your samples.

TABLE 34.4 — Effects of sampling regime on % error

	Recaptured Sample Size	
	60	**120**
Size of marked sample		
50		
100		
150		

? What fraction of the total population would this number be?

? Describe the relationship between % error, marked sample size, and recapture sample size.

? If you were optimizing effort and accuracy, what sampling strategy would be best? Why?

Analysis

13▶ Using the average values and the Lincoln-Peterson formula, estimate the population size in your can.

Now count all the beads (both colors) in your can and compare the real population size to the estimated population size.

Estimated number in population _____
Actual number in population _____
Calculate the % error in your estimation:

$$\% \text{ error} = \frac{\text{estimate} - \text{real} \times 100}{\text{real}}$$

Lab groups should now share their results by placing their % error values on the board in a facsimile of table 34.4. Record class values in table 34.4.

14▶ Take these class data and plot them on graph paper with the absolute value (no plus or minus sign) of % error (y-axis) as a function of marked sample size (x-axis). Two lines will be drawn on the graph: one for the recapture sample of 60 and the other for 120. If both of these lines are extrapolated to zero percent error, how many individuals must be marked before you will get correct estimates?

A Growth Curve Problem

Populations of organisms, whether they be oak trees, frogs, amoebas, or bacteria, have the inherent reproductive potential to increase exponentially in number over generations. Actual size of the population is limited by the available resources, such as food, water, and habitat.

Table 34.5 lists data about the growth of laboratory populations of the protozoan *Acanthamoeba,* a soil amoeba commonly used in research laboratories because it is easily grown in cultures. This amoeba can be grown on a peptone medium much like bacteria are in flasks. At low population levels, the amoebas will grow and reproduce by mitosis, an asexual process. The data in the table come from 13 different cultures in the author's research lab and have been combined to give an average value plus or minus one standard deviation.

You will plot these data to see what a population growth curve looks like. The same data will be plotted twice, once on arithmetic paper and once on semilog graph paper. You may not be familiar with semilog paper. Look at the sheet on page 465. Note that the spacing of vertical lines is equal, but the horizontal lines have unequal spacings numbered 1 through 9, and the cycle repeats four times on the page.

The spacing of these lines corresponds to the logarithm (base 10) of the number indicated on the line. Each cycle on the paper corresponds to a power of 10. Thus, if we designate the lowermost 1 as being 1,000, the 1 in the next cycle above is 10,000 and the 1's in the remaining cycles above are 100,000 and 1,000,000, respectively.

The numbers designated in each cycle take on the place value of the 1 below them. Thus 4 in the second cycle from the bottom in our example would be 40,000 because that is the place value of that cycle. The number 1,500,000 would be located on the heavy line between 1 and 2 in the uppermost cycle. Semilog paper is a convenient way of plotting the logarithm of a number without looking for the logarithm in a table.

TABLE 34.5 Cell counts from
Acanthamoeba cultures

Age (hrs)	Average Cell Density (Cells/ml)	Standard Deviation
0	1,000	± 50
10	2,100	± 120
30	10,000	± 450
50	44,000	± 3,000
70	160,000	± 10,000
90	330,000	± 15,000
110	600,000	± 40,000
150	1,050,000	± 40,000
210	1,400,000	± 30,000
250	1,430,000	± 40,000

15 ▶ At the end of the exercise, on both types of graph paper, plot the cell densities from table 34.5 as a function of time.

Look at the two plots and note the differences. One is a straight line that plateaus at the end, and the other sigmoid. Biologists recognize three general areas on these curves: the exponential growth phase, the declining growth phase, and the stationary growth phase.

During **exponential growth,** every cell is dividing to give two cells, and they divide to give four, and then eight, and so on in an exponential progression. A measure of exponential growth is the **doubling time** of the population, the time that it takes the population to double. Doubling time represents the balance between birth and death rates in a population. The doubling time can be read directly from the graph you have drawn. What is the exponential doubling time on the following:

arithmetic paper _____
semilog paper _____

? Do the doubling times significantly differ from each other in the two types of plots? How can they be different if you used the same data?

The problem is that you probably read the curves in different regions. Note that on semilog paper, the population

has a constant doubling time until it reaches 100,000 cells/ml and then the growth rate slows. On arithmetic paper, the slope of the line constantly changes, making it nearly impossible to define the period of exponential growth. This is one of the reasons semilog paper is used to plot growth curve data. The end of exponential growth is clearly shown as the place where linearity is lost. Exponential growth can be described by the exponential equation: $N_T = N_0 e^{rt}$

where

N_T = number of cells at some time (t)
N_0 = number of cells to start
e = base of natural logarithms
t = elapsed time
r = a constant

This equation can be written as

$$\log N_T = \log N_0 + 2.3\, rt$$

which is the equation of a straight line. The reason that the growth curve is a straight line on semilog paper is that you have really plotted $\log N_T$ as a function of time (t). That is, you have plotted the above equation, where N_0 is the intercept and r is the slope of the line.

? If it took approximately 52 hours for the cultures to reach a population density of 50,000 amoebas per ml, how many more hours will it take to reach 10^5 amoebas per ml? (Hint: look at graph you made.)

Now look at the growth curve and find the cell concentration at which the culture is in stationary growth, neither increasing nor decreasing in number. This is called the **carrying capacity** of the environment. In the culture, amoebas are dying at the same rate as they are dividing, so that the number stays constant.

Name at least three factors that would determine the carrying capacity of an environment.

Computer-Generated Growth Curves

Because the underlying mathematics of population growth curves are well understood, a number of computer programs are available to simulate population growth. Your lab instructor will describe the programs available in your lab and how to use the computers.

Most of the available programs allow you to vary the starting number of individuals in the population, the doubling time (generation time), the number of offspring produced per mating, and the carrying capacity of the environment. Use the program to understand the following situations.

16 Compare two populations in which population 1 starts with twice as many individuals as population 2, but population 1 produces twice as many offspring per mating as population 2. Test the following null hypothesis: No differences in the shape of the growth curves between these two populations are expected. What would be the alternative hypothesis?

Run the simulation program under the above conditions. Must you accept or reject the null hypothesis? Why?

17 Explore the other variables that can be changed in your program. Describe below what you have discovered.

Learning Biology by Writing

Your instructor may ask you to turn in answers to the Critical Thinking or Lab Summary Questions that follow.

Internet Sources

Mark and recapture is a standard technique for estimating the population size of mobile animals. Use a search engine such as **Google** (at www.google.com) to locate information on the World Wide Web for this technique. Connect to three of the URLs and read what they have to say about this technique. Summarize in a paragraph what you read, and record the URLs for future reference.

Lab Summary Questions

1. Under what conditions should quadrat sampling be used instead of mark-and-recapture methods?
2. What are the underlying assumptions of the mark-and-recapture technique and what suggestions would you make to improve the accuracy?

3. Why is random sampling important in the techniques used to estimate the sizes of populations?
4. Distinguish between arithmetic and exponential increases. Explain why it is better to use semilog paper than arithmetic paper to plot exponential increases.
5. Tear out the population growth curves you drew for *Acanthamoeba* and turn them in with this summary.

Critical Thinking Questions

1. List several sources of error in estimating population sizes by the quadrat sampling technique and indicate whether these would lead to an overestimation or an underestimation of population size.
2. Population growth results from a balance between the rate of birth and the rate of death in the population. For a hypothetical species of tree in a national forest, make a long list of natural factors that would influence the "birth rate." Do the same for those factors that influence the death rate.
3. The world human population during the twentieth century increased exponentially. Speculate on the logical outcome if this rate of growth continues. How can the rate of growth be reduced? If the practices that you suggest were implemented, would there be any ethical dilemmas?

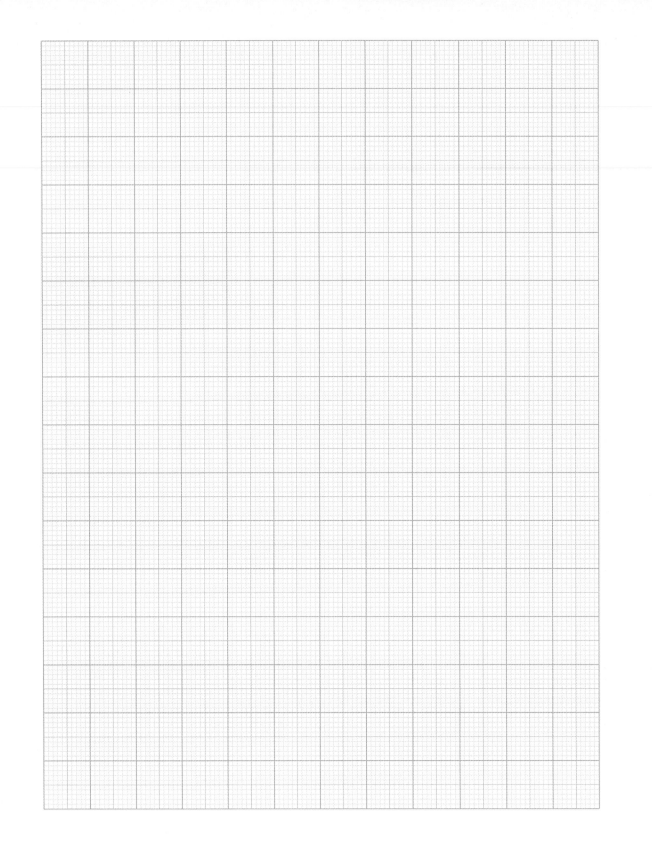

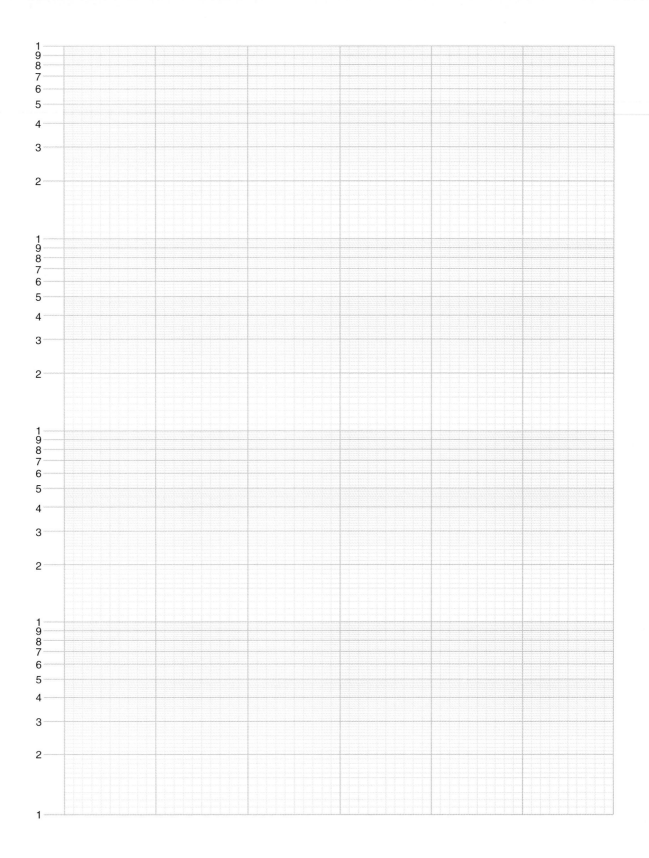

APPENDIX A

Significant Figures and Rounding

In the laboratory, students often ask how precise they should be in recording measurements during experiments. The question of precision also arises when doing calculations to solve laboratory problems. A few simple explanations and rules are given here to guide you in these quantitative aspects of laboratory biology.

What Are Significant Figures?

Significant figures are defined as *the necessary number of figures required to express the result of a measurement so that only the last digit in the number is in doubt.* For example, if you have a ruler that is calibrated only in centimeters and find that a pine needle is between 9 and 10 cm long, how do you record the length?

The definition of significant figures tells you that you should estimate the additional fraction of a centimeter in tenths of a centimeter and add it to 9 cm, thus indicating that the last digit is only an estimate. You would never write this additional fraction as hundredths or thousandths of a centimeter because it would imply a precision that did not exist in your measuring instrument.

However, suppose you have a ruler calibrated in millimeters and measure the same pine needle, finding that the needle is between 93 and 94 mm long. You should then estimate the additional fraction in tenths of a millimeter and add it to 93 mm which would be 9.3 cm plus the estimate.

Memorize and use this rule throughout the course: *When recording measurements, include all of the digits you are sure of plus an estimate to the nearest tenth of the next smaller digit.*

Doing Arithmetic with Significant Figures

Other rules apply when doing calculations in the laboratory. Several situations you will encounter are discussed in the following paragraphs.

When converting measurements from one set of units to another in the metric system, be sure not to introduce greater precision than exists in the original number. For example, if you have estimated that something is 4.3 cm long and wish to convert it to millimeters, the correct answer is 43mm, not 43.0 mm because the number of centimeters was known only with precision to a tenth of a centimeter and not a hundredth of a centimeter as 43.0 mm implies.

When performing multiplication or division involving numbers with different levels of significant figures, recognize that the answer should be expressed only with the precision of the number in the calculation that shows the least number of significant figures. For example, if you wish to calculate the weight of 10.1 ml of water and you are told the density of water is 0.9976 g/ml, you would multiply the density times the volume to obtain the weight. However, the correct answer would be 10.1 g, not 10.07576 g. Because the water volume measurement is known only to three significant figures, the latter number conveys a precision that is not justified given the uncertainty of the water volume measurement.

When performing additions or subtractions, the answer should contain no more decimal places than the number with the least number of digits following the decimal place. Thus, 7.2° C subtracted from 7.663° C yields a correct answer of 0.5° C not 0.463° C. If the first number had been known with a precision of 7.200° C, then the latter answer would have been correct.

What Is Rounding?

The last example introduces the concept of rounding to the appropriate number of significant figures. The rules governing this are straightforward. You should not change the value of the last significant digit if the digit following it is less than five. Therefore, 3.449 would round off to 3.4 if two significant figures were required. If the value of the following number is greater than five, increase the last significant digit by one. Therefore, 88.643 would round off to 89 if two significant figures were required.

There is some disagreement among scientists and statisticians as to what to do when the following number is exactly five, as in 724.5, and three significant numbers are required. Some will always round the last significant figure up (in this case 725), but others claim that this will introduce a significant bias to the work. To eliminate this problem, they would flip a coin (or use another random event generator) every time exactly five is encountered, rounding up when heads was obtained and leaving the last significant digit unchanged when tails was obtained. Recognize, however, that if the number were 724.51 or greater, the last significant digit would always be rounded up to 725.

Examples of Rounding

49.5149 rounded to 5 significant figures is 49.515
$$(= 4.9515 \times 10)$$
49.5149 rounded to 4 significant figures is 49.51
$$(= 4.951 \times 10)$$
49.5149 rounded to 3 significant figures is 49.5
$$(= 4.95 \times 10)$$
49.5149 rounded to 2 significant figures is 50
$$(= 5.0 \times 10)$$
49.5149 rounded to 1 significant figure is 50
$$(= 5 \times 10)$$

APPENDIX B

Making Graphs

Graphs are used to summarize data—to show the relationship between two variables. Graphs are easier to remember than are numbers in a table and are used extensively in science. You should get in the habit of making graphs of experimental data, and you should be able to interpret graphs quickly to grasp a scientific principle.

In using this lab manual, you will be asked to make two kinds of graphs—line graphs and histograms. **Line graphs** show the relationship between two variables, such as amount of oxygen consumed by a tadpole over an extended period of time (fig. B.1). **Histograms** are bar graphs and are usually used to represent frequency data, that is, data in which measurements are repeated and the counts are recorded, such as the values obtained when an object is weighed several times (fig. B.2).

Line Graphs

When you make line graphs, always follow these rules.

1. Decide which variable is the dependent variable and which is the independent variable. The **dependent variable** is the variable you know as a result of making experimental measurements. The **independent variable** is the information you know before you start the experiment. It does not change as a result of the dependent variable but changes independently of the other variable. In figure B.1, time does not change as a result of oxygen consumption. Therefore, time is the independent variable and the amount of oxygen consumed (which is dependent on time) is the dependent variable.

2. Always place the independent variable on the x-axis (the horizontal one), and the dependent variable on the y-axis (the vertical one).

3. *Always label* the axes with a few words describing the variable, and *always put the units* of the variable in parentheses after the variable description (fig. B.1).

4. Choose an appropriate *scale* for the dependent and independent variables so that the highest value of each will fit on the graph paper.

5. Plot the data set (the values of y for particular values of x). Make the plotted points dark enough to be seen. Always use pencil not pen in case you need to erase. If two or more data sets are to be plotted on the same coordinates, use different plotting symbols for each data set (·, ×, ⊙, ⊗, ▫, *etc.*).

6. Draw *smooth curves* or *straight lines* to fit the values plotted for any one data set. Do not connect the points

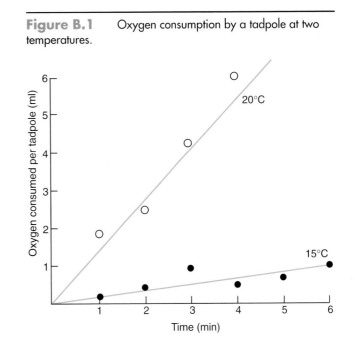

Figure B.1   Oxygen consumption by a tadpole at two temperatures.

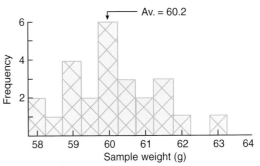

Figure B.2   Histogram of a hypothetical series of sample weights obtained by weighing the same sample several times. An average was calculated and is indicated by an arrow.

with short lines. A smooth curve through a set of points is a visual way of averaging out chance variability in data. Do not extrapolate beyond a data set unless you are using it as a prediction technique, because you do not know from your experiment whether the relationship holds beyond the range tested.

7. Every graph should have a legend, a sentence, explaining what the graph is about.

Histograms

In making histograms, the count data are always on the y-axis. The categories in which the data fall are on the x-axis. For example, the data from which figure B.2 was drawn are as follows:

Class results from a series of weighings of the same sample (in grams)				
61.0	60.0	60.0	59.8	58.0
61.5	61.5	61.0	60.9	58.0
59.0	60.0	60.2	61.7	60.6
59.7	59.0	60.3	63.0	60.4
62.0	59.0	60.7	58.5	59.0

To make the histogram, the x-axis was laid out with a range of 58 to 64 so that all values would be included. The values were then marked on the graph as lightly penciled Xs, one in each square of the graph paper for each observation. After all data were plotted, bars were then drawn to show the frequencies of measurements. On bar graphs, it is a good idea to show the average value across all measurements with an arrow. In calculating an average, remember the significant figure rule (appendix A).

A final word should be said about how to draft graphs. Be neat! Remember most people do not trust sloppy work. Always print labels and use a sharp pencil, not a pen. Use a ruler to draw straight lines and a drafting template called a French curve to draw curved lines.

APPENDIX C

Simple Statistics

Quantitative data may be expressed in two forms, **count data** or **measurement data.** Count data are discontinuous variables that always consist of whole numbers. They are derived by counting how the results from an experiment fall into certain categories; for example, in a genetic cross between two heterozygotes, you expect to obtain a genotypic ratio of 1:2:1. Measurement data are continuous variables obtained by using some measuring instrument. The precision of the measurement depends on the fineness of the scale on the instrument; for example, a ruler calibrated in centimeters is not as precise as one calibrated in millimeters. Accuracy differs from precision in that it depends not on the scale used but on the calibration against a standard and the proper reading of the scale.

Whenever measurements are made, there are potential sources of error. Instruments and the humans who read them make random errors that affect accuracy. If the instrument is properly calibrated and if the person who is reading it is careful, the percent error is small and will be randomly distributed around the true measurement. If several readings are taken, the true value will be closer to the mean than to any single measurement.

In some cases, another source of error may be introduced. Bias occurs when an instrument is improperly calibrated or when the operator makes a consistent error in reading or sampling. For example, if a watch that is five minutes slow is used to measure the time of sunset for several days, the data will reflect a consistent bias, showing sunset as occurring five minutes earlier than it really did. Similarly, if a balance, spectrophotometer, or pipette is improperly calibrated, it will consistently yield a biased estimation either over or under the true value; and if the operator of the device misreads the instrument, additional bias enters. For example, if one looks at a car speedometer from the passenger's seat, the speed of the car appears to be lower than it actually is because of the viewing angle, called parallax. No statistical procedure can correct for bias; there is no substitute for proper calibration and care in making measurements.

Assuming that all sources of bias have been ruled out, there is still the problem of dealing with random fluctuations in measurement and in the properties (size, weight, color, and so on) of samples. In biological research, this is especially important since variation is the rule rather than the exception. For example, white pine needles are approximately 4 inches long at maturity, but in nature, the length of the needles varies due to genetic and environmental differences. A biologist must constantly be aware that biological variability and bias influence the results of experiments and any analysis should include procedures to minimize the effects.

Dealing with Measurement Data

When several measurements are made by an individual or by a class, most would agree that it is best to use the **mean** or **average** of those measurements to estimate the true value. However, determining the average alone may mean that important information concerning variability is ignored. Look at table C.1, which contains two sets of data: the average temperature in degrees centigrade at 7:00 P.M. each day in September for two different geographic locations.

The average temperature for each location can be calculated by adding the readings for that location and dividing by the total number of readings:

$$\text{average temperature} = \frac{\Sigma \text{ Readings}}{N}$$

where Σ equals "sum of"

For both data sets, the averages are the same (15°C). Based on the averages alone, one might conclude that the two locations have similar climates. However, by simply scanning the table, one can see that location A has a more variable temperature than location B. Such temperature variations, especially those below 0°C, may have a tremendous effect on organisms; many plants and small insects may die at subzero temperatures. Therefore, reporting only the average temperature from this set of data does not convey crucial information on variability.

The **range** of values can convey some of this information. Location A had a mean temperature of 15°C with a range from –5 to 35, while location B's mean temperature was 15°C with a range of 10 to 20. Unfortunately, the range of values has a limited usefulness because it does not indicate how often a given temperature occurs. For example, if another location (location C) had 15 days at –5°C and 15 days at 35°C, it would have the same mean and range as location A, but the climate would be harsher.

To overcome some of the limitations of range, the data could be plotted in frequency histograms, with the x-axis showing the daily temperature and the y-axis showing the frequency of that temperature, or how often it occurs. The data for all three locations are plotted in figure C.1. It is now obvious, looking at the plotted data, that these three locations have quite different climates in September even though the mean temperatures are the same and the ranges overlap for two of the three. However, this method of reporting is cumbersome.

An efficient way to report information about variability and its frequency is to calculate the **standard deviation**

TABLE C.1 September temperatures in °C at 7:00 P.M. for two locations

Day	Location A	Location B
1	0	15
2	0	20
3	5	20
4	10	20
5	15	20
6	15	15
7	20	10
8	25	10
9	30	10
10	35	10
11	30	10
12	35	15
13	30	20
14	30	13
15	25	17
16	25	18
17	20	19
18	20	20
19	19	10
20	15	11
21	15	12
22	11	13
23	10	17
24	10	15
25	5	10
26	5	20
27	0	10
28	0	10
29	–5	20
30	–5	20

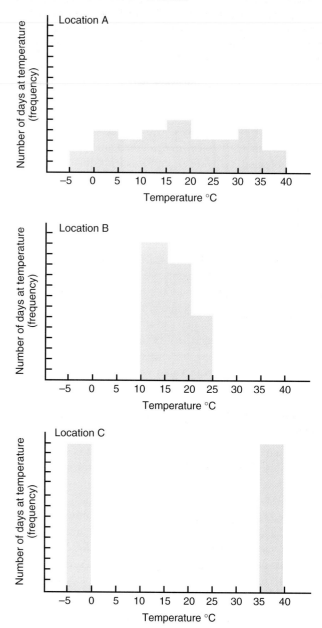

Figure C.1   Frequency histograms of temperatures at three locations each having the same mean temperature of 15° C. Note differences in variation.

from the mean. This value is calculated by determining the difference between each observation and the average. If a group of measurements is distributed randomly and symmetrically about the mean, then the standard deviation defines a range of measurements in which 68% of the observed values fall. This is demonstrated in figure C.2.

Figure C.3 shows that distributions can differ in three ways: the means may be different, the standard deviations may be different, or both the means and the standard deviations may be different. Obviously, when the standard deviation is large, the variability is great, and when the variability is small, the standard deviation is small.

Figure C.2 A frequency distribution representing a normal curve. On normal curves, the mean ($\bar{x}$) coincides with the mode, or the most frequently occurring value, and also with the median (the value at which 50% of all values are higher and 50% are lower). The shaded area corresponds to plus or minus one standard deviation about the mean, and will always include 68% of all the observations.

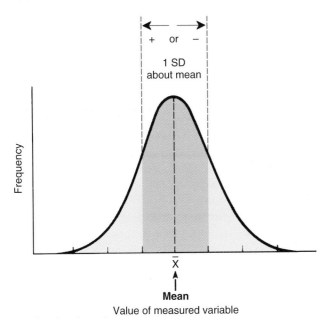

The following formula is used to calculate standard deviation:

$$\text{standard deviation } (\sigma) = \pm\sqrt{\frac{\Sigma\,(x_o - \bar{x})^2}{n - 1}}$$

where

$\sqrt{}$ = square root

Σ = sum of

x_o = an observed value

$\bar{x}$ = average of all observed values

n = number of observed values

The computation of standard deviation is best performed by setting up a calculation sheet as in table C.2, which contains the data from location A in the earlier example. The value for x_o is the temperature observation on each day, and $\bar{x}$ is the average temperature for the month, 15° C.

The mean and the standard deviation for this set of data are 15 ± 12° C. This means that the average temperature of 15° C plus or minus 12° C (3° C to 27° C) represents 68% of the temperatures you would expect to measure in this location in September. If the same figures are

Figure C.3 Three comparisons of normal curves that differ in their means and standard deviations. The dotted line represents the mean, and the shaded area represents plus or minus one standard deviation. In (a), the curves differ in their means but the standard deviations are the same. In (b), the means are the same but the standard deviations are different. In (c) both the means and standard deviations differ.

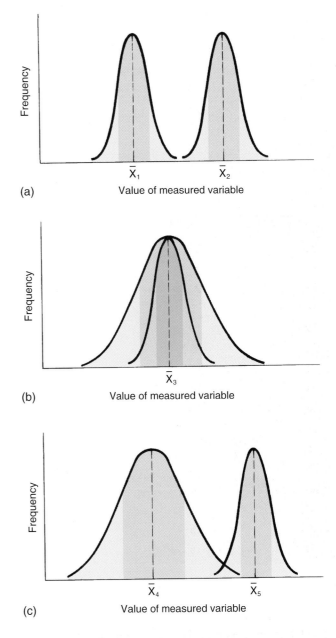

calculated for the data set from location B, the results are 15 ± 4° C (not 4.2 because of significant figure rule; see appendix A). The mean and the standard deviation thus help convey an accurate impression of the similarities and differences between the two sets of data.

n	x_o	$(x_o - \bar{x})$	$(x_o - \bar{x})^2$
1	0	−15	225
2	0	−15	225
3	5	−10	100
4	10	−5	25
5	15	0	0
6	15	0	0
7	20	5	25
8	25	10	100
9	30	15	225
10	35	20	400
11	30	15	225
12	35	20	400
13	30	15	225
14	30	15	225
15	25	10	100
16	25	10	100
17	20	5	25
18	20	5	25
19	19	4	16
20	15	0	0
21	15	0	0
22	11	−4	16
23	10	−5	25
24	10	−5	25
25	5	−10	100
26	5	−10	100
27	0	−15	225
28	0	−15	225
29	−5	−20	400
30	−5	−20	400
	$\Sigma = 450$		$\Sigma = 4182$

$$\bar{X} = \frac{\Sigma X_0}{n} = \frac{450}{30} = 15$$

$$SD = \pm \sqrt{\frac{4182}{30-1}} = \pm 12$$

In teaching laboratories, each student usually makes only one or two measurements of a biological phenomenon. The concept of standard deviation is not appropriate for so few readings; a range and average are more suitable. However, if the class combines its readings so that there are 20 or 30 measurements, it is better to calculate the average and the standard deviation. Form the habit of reporting averages with some estimate of the variability when presenting data.

Comparing Count Data: Dealing with Variability

Often in experiments, theories are used to predict the results of experiments and data are recorded as counts or frequencies in categories. For example, Mendelian genetics can be used to predict the progeny from a cross between two heterozygotes. In cases where there is clear dominance and recessiveness, we would use our understanding of genetics to predict that offspring would occur in a ratio of 3:1. What happens when the actual results from a genetic cross of this type do not exactly fit the predictions? First, we would review the data for any obvious errors of technique or arithmetic and eliminate the data based on identified errors. If the data still do not conform exactly to our predictions, we must decide whether the variations are within the range expected by chance.

When variations from the expected are small, we usually just round off and say that there is a good fit between the data and the expected. However, this is somewhat arbitrary and can become a problem as deviations from the expected become large. When are deviations no longer small and insignificant in comparison to the expected results (i.e., the theory is not a good predictor)?

Statisticians confront this subjectivity by using a statistical test called the **chi-square (χ^2) goodness of fit test.** This test creates a number, the χ^2, which summarizes the differences between the data and what was expected. If that number is large, then the theory may be wrong. If it is small, the theory is a reasonable predictor of the results. The steps in using this test are outlined below.

Scientific Hypothesis

The χ^2 test is always used to test a hypothesis known as the **null hypothesis (H_o).** It is based on the scientific theory known as the **model,** which is the basis for predicting the results. A null hypothesis proposes that there is no difference between the results **actually obtained** and those **predicted** from a model within an acceptable range of variation.

A null hypothesis is paired with an **alternative hypothesis (H_a),** which proposes an alternative explanation of the results not based on the same scientific model as was the null hypothesis. The acronym "HoHa" emphasizes the coupling of hypotheses in statistical testing.

If a statistical test indicates that the results of an experiment do not fit the expected, then the null hypothesis must be rejected and the alternative hypothesis is true by inference: the data variation is greater than that expected by chance, and a factor other than that tested is influencing the results, implying that the model is not correct.

Rejection of H_o does not prove H_a! This is a common mistake in the use of statistics. Rejection of H_o only rejects the model on which H_o is based.

TABLE C.3 — Frequency table of observed and expected results from hypothetical fruit fly cross

	Dom-Dom	Dom-Rec	Rec-Dom	Rec-Rec	Total
Observed (O)	150	60	67	23	300
Expected (E)	169	56	56	19	300

Let us suppose that you are conducting breeding experiments. You are looking at a dihybrid cross involving dominant and recessive traits located on autosomes. Mendel's principle of independent assortment predicts offspring in the phenotypic ratio of 9:3:3:1. When the experiments were finished, the following results were obtained:

150 phenotypes dominant for both traits

60 phenotypes dominant for the first trait and recessive for the second

67 phenotypes recessive for the first trait and dominant for the second

23 phenotypes recessive for both traits

How would you determine whether these results were an acceptable variation from the expected 9:3:3:1 ratio? The actual ratio in this case is close to 7:3:3:1. Has the experiment failed to show Mendelian inheritance, or are the results simply within the limits of chance variability?

First, a null hypothesis should be formulated and then an alternative, mutually exclusive, hypothesis proposed. For the above experiment, these would be:

H_o There is no significant difference between the results obtained and those predicted by the Mendelian principle of independent assortment.

H_a Independent assortment does not predict the outcomes of this experiment.

Note how H_o and H_a are mutually exclusive and both cannot be true.

Testing the Null Hypothesis

Because we are dealing with count (frequency) data, which should conform to those predicted by a model, the chi-square goodness of fit test can be used to compare the actual and predicted results.

First, a frequency table is created (table C.3). The entries on the first line are the actual (observed) results from an experimental cross in the lab. Those on the second line are the expected results obtained by multiplying the total by the fractions expected in each category (9/16 × 300, 3/16 × 300, etc.).

The summary chi-square statistic is calculated using the following formula:

$$\chi^2 = \sum^n \frac{(O_i - E_i)^2}{E_i}$$

where

Σ indicates "sum of"
O_i is the observed frequency in class i
E_i is the expected frequency in class i
n is the number of experimental classes (in this case 4)
χ^2 is the Greek letter *chi,* squared

It should be clear from inspecting this formula that the value of χ^2 will be 0 when there is perfect agreement between the observed and expected results, whereas the χ^2 value will be large when the difference between observed and expected results is large.

To calculate χ^2 for the data in table C.3, the following steps are required:

$$\chi^2 = \frac{(150 - 169)^2}{169} + \frac{(60 - 56)^2}{56} +$$

$$\frac{(67 - 56)^2}{56} + \frac{(23 - 19)^2}{19}$$

$$= \frac{(-19)^2}{169} + \frac{(+4)^2}{56} + \frac{(+11)^2}{56} + \frac{(+4)^2}{19}$$

$$= 2.14 + 0.29 + 2.16 + 0.84$$

$$\chi^2 = 5.43$$

Therefore, for this experiment, the variability from the expected result is now summarized as a single number, $\chi^2 = 5.43$.

Making a Decision about the Null Hypothesis

To determine if a value of 5.43 is large and indicates poor agreement between the actual and predicted results, the calculated χ^2 value must be compared to a **critical χ^2 value** obtained from a table of standard critical values (table C.4). These values represent acceptable levels of variability obtained in random experiments of similar design.

To use table C.4, you must know a parameter called the **degrees of freedom (d.f.)** for the experiment; it is always numerically equal to one less than the number of classes in the outcomes from the experiment. In our example, the experiment yielded data in four classes. Therefore, the degrees of freedom would be 3.

In addition to knowing the degrees of freedom for an experiment, you must also decide on a **confidence level,** a percentage between 0 and 99.99, which indicates the confidence that you wish to have in making a decision to reject

TABLE C.4 Critical values for χ^2 at different confidence levels and degrees of freedom (d.f.)

d.f	Confidence Levels $\chi^2.90$	$\chi^2.95$	$\chi^2.98$	$\chi^2.99$	$\chi^2.999$
1	2.7	3.8	5.4	6.6	10.8
2	4.6	6.0	7.8	9.2	13.8
3	6.3	7.8	9.8	11.3	16.3
4	7.8	9.5	11.7	13.3	18.5
5	9.2	11.1	13.4	15.1	20.5
6	10.6	12.6	15.0	16.8	22.5
7	12.0	14.1	16.6	18.5	24.3
8	13.4	15.5	18.2	20.1	26.1
9	14.7	16.9	19.7	21.7	27.9
10	16.0	18.3	21.2	23.2	29.6

the null hypothesis. Most scientists use a 95% confidence level. Any other confidence level can be used and several (but not all) are given in table C.4.

If you now read table C.4 by looking across the row corresponding to 3 degrees of freedom to the column for 95% confidence, you see the value 7.8. This critical value means that 95% of the χ^2 values for experiments with 3 degrees of freedom in which the H_o is true fall below 7.8 due to chance variation and that only 5% will be above that value due to chance.

The next step is to compare the calculated χ^2 value to the critical value for χ^2. When this is done, two outcomes are possible:

1. The calculated value is less than or equal to the critical value. This means that the variation in the results is of the type expected by chance in 95% of the experiments of a similar design; the null hypothesis cannot be rejected.

2. The calculated χ^2 value is greater than the critical value. This means that results would occur only 5% of the time by chance. Stated another way, these results should not be accepted as fitting the model; the null hypothesis should be rejected.

When these decision-making rules are applied to our calculated χ^2 value, 5.43 is obviously less than 7.8. Therefore, although our ratio was not 9:3:3:1, it is within an acceptable limit of variation and we cannot reject the H_o. A model based on the Mendelian principle of independent assortment predicts the results of the experiment. Note, however, that you have not proven your H_o; you simply failed to reject it with 95% confidence in your decision. This is more than a subtle difference because it indicates that scientific knowledge is probabilistic and not absolute.

A Hypothetical Alternative

For illustration, let us suppose that another group got different results. When they calculated their χ^2 value, it was 9.4. If they went through the same decision making steps that we just did, they would reject the null hypothesis: Mendelian genetics did not have predictive power for their experiment. In rejecting the null hypothesis, they can have 95% confidence that they are not making a mistake by rejecting what is actually a true null hypothesis.

Those who have studied table C.4 might suggest a change in strategy here. If this second group changed its confidence level to 97.5%, the critical value becomes 9.8, which is greater than the calculated value of 9.4, thus keeping the group from rejecting the null hypothesis. This illustrates that by choosing a higher value for a confidence level at a constant number of degrees of freedom, the critical value will be larger, allowing one to accept almost any null hypothesis. Is this not arbitrary, the very situation we sought to avoid by invoking this statistical test?

The solution to this dilemma is found in what is really tested by the chi-square goodness of fit test. The comparison of calculated and critical χ^2 values is done to attempt to falsify the H_o, to reject it at a predetermined confidence level. If H_o cannot be rejected, then by default it is accepted. *You have not proven H_o, you have simply failed to disprove it.* The confidence level that is stated before each test represents the confidence you have in rejecting the null hypothesis, not the confidence you have in accepting it. As you increase the confidence level, you decrease the likelihood of rejecting H_o.

Statisticians often speak of type I and type II errors in statistical testing. A **type I error** is the probability that you will reject a true null hypothesis. A **type II error** is the probability that you will accept a false null hypothesis. As you increase the confidence level, you reduce the probability that you will reject a true null hypothesis (type I error), but you increase the probability that you will accept a false one (type II error). The confidence level of 95% is used by convention because it represents a compromise between the probabilities of making type I versus type II errors. The basis of this conservatism in science is the recognition that it is better to reject a true hypothesis than it is to accept a false one. Once a confidence level is stated for an experiment, it should not be changed according to the whim of the experimenter who wants a model to have predictive power. Those who change the confidence level run the risk of accepting fiction as fact.

APPENDIX D

Writing Lab Reports and Scientific Papers

Verbal communication is temporal and easily forgotten, but written reports exist for long periods and yield long-term benefits for the author and others. Gregor Mendel's work is a perfect example. When he finished his research he gave a verbal presentation to a scientific meeting, but few understood it. That was in 1872. Fortunately, he also wrote a paper and published it. About 30 years later that paper was read by scientists who understood it and Mendel was given credit for founding the modern study of genetics.

Scientific research is a group activity. Individual scientists perform experiments to test hypotheses about biological phenomena. After experiments are completed and duplicated, researchers attempt to persuade others to accept or reject their hypotheses by presenting the data and their interpretations. The lab report or the scientific paper is the vehicle of persuasion; when it is published, it is available to other scientists for review. If the results stand up to criticism, they become part of the accepted body of scientific knowledge unless later disproved.

In some cases, a report may not be persuasive in nature but instead is an archival record for future generations. For example, data on the distribution and frequency of rabid skunks in a certain year may be of use to future epidemiologists in deciding whether the incidence of rabies is increasing. Regardless of whether a report is persuasive or archival, the following guidelines apply.

Format

A scientific report usually consists of the following:

1. Title
2. Abstract
3. Introduction
4. Materials and methods
5. Results
6. Discussion
7. Literature cited

There is general agreement among scientists that each section of the report should contain specific types of information.

Title

The title should be less than ten words and should reflect the factual content of the paper. Scientific titles are not designed to catch the reader's fancy. A good title is straightforward and uses keywords that researchers in a particular field will recognize.

Abstract

The purpose of an abstract is to allow the reader to judge whether it would serve his or her purposes to read the entire report. A good abstract is a concise (about 100 words) summary of the purpose of the report (hypotheses tested), the data obtained, and the author's major conclusions.

Introduction

The introduction defines the subject of the report. It must outline the scientific purpose(s) and hypotheses tested, giving the reader sufficient background to understand the rest of the report. Care should be taken to limit the background to whatever is pertinent to the experiment. A good introduction will answer several questions, including the following:

Why was this study performed?

Answers to this question may be derived from observations of nature or from the literature.

What knowledge already exists about this subject?

The answer to this question must review what is known about the topics, showing the historical development of an idea and including the conflicts and gaps in existing knowledge.

What is the purpose of the study?

The specific hypotheses being tested should be stated and the experimental design described.

Materials and Methods

As the name implies, the materials and methods used in the experiments should be reported in this section. The difficulty in writing this section is to provide enough detail for the reader to understand the experiment without overwhelming him or her. When procedures from a lab book or another report are followed exactly, simply cite the work, noting that details can be found there. However, it is still necessary to describe special pieces of equipment and the general theory of the assays used. This can usually be done in a short paragraph, possibly along with a drawing of the experimental apparatus. Generally, this section attempts to answer the following questions:

What materials were used?

How were they used?

Where and when was the work done? (This question is most important in field studies.)

Results

The results section should summarize the data from the experiments without discussing their implications. The data should be organized into tables, figures, graphs, photographs, and so on. But data included in a table should not be duplicated in a figure or graph.

All figures and tables should have descriptive titles and should include a legend explaining any symbols, abbreviations, or special methods used. Figures and tables should be numbered separately and should be referred to in the text by number, for example:

1. Figure 1 shows that the activity decreased after five minutes.
2. The activity decreased after five minutes (fig. 1).

Figures and tables should be self-explanatory; that is, the reader should be able to understand them without referring to the text. All columns and rows in tables and axes in figures should be labeled. See appendix B for graphing instructions.

This section of your report should concentrate on general trends and differences and not on trivial details. Many authors organize and write the results section before the rest of the report.

Discussion

In writing this section, you should explain the logic that allows you to accept or reject your original hypotheses. You should not just restate your results, but should emphasize interpretation of the data, relating them to existing theory and knowledge. Speculation is appropriate, if it is so identified. Suggestions for the improvement of techniques or experimental design may also be included here. You should also be able to suggest future experiments that might clarify areas of doubt in your results.

Literature Cited

This section lists all articles or books cited in your report. It is not the same as a bibliography, which simply lists references regardless of whether they were cited in the paper. The listing should be alphabetized by the last names of the authors. Different journals require different formats for citing literature. The format that includes the most information is given in the following examples:

For articles:

Fox, J. W. 2000. Nest-building behavior of the catbird, *Dumetella carolinensis. Journal of Ecology* 47: 113–17.

For books:

Bird, W. Z. 1990. *Ecological aspects of fox reproduction.* Berlin: Guttenberg Press.

For chapters in books:

Smith, C. J. 2003. Basal cell carcinomas. In *Histological aspects of cancer,* ed. C. D. Wilfred, pp. 278–91. Boston: Boston Medical Press.

When citing references in the text, do not use footnotes; instead, refer to articles by the author's name and the date the paper was published. For example:

1. Fox in 2000 investigated the effects of hormones on the nest-building behavior of catbirds.
2. Hormones are known to influence the nest-building behavior of catbirds (Fox, 2000).

When citing papers that have two authors, both names must be listed. When three or more authors are involved, the Latin *et al. (et alia)* meaning "and others" may be used. A paper by Smith, Lynch, Merrill, and Beam published in 2004 would be cited in the text as:

Smith et al. (2004) have shown that . . .

This short form is for text use only. In the Literature Cited, all names would be listed, usually last name preceding initials.

There are a number of style manuals that provide detailed directions for writing scientific papers. Some are listed in further readings at the end of this section.

More and more students turn to the Word Wide Web (WWW) and search engines to locate background information for reports. This is certainly acceptable but you should be aware that mere publication on the WWW does not assure that something is true. Because electronic sources may disappear overnight or a source may be changed (updated) in the time between a first and second access, such sources must be cited differently from printed materials. Citation of an electronic source should include:

Author's last and first names with middle initial; Date of publication on the Internet, including revision dates; Title of the electronic document; The URL contained within angle brackets; and the date on which you accessed it. An example is:

Darwin, C. 1845 and 1997. *The voyage of the Beagle.* Project Guttenberg. <ftp://ibiblio.org/pub/docs/books/guttenberg/etext97/vbgle10.zip> Accessed 2001.

General Comments on Style

1. All scientific names (genus and species) must be italicized. (Underlining indicates italics if fonts are not available.)
2. Use the metric system of measurements. Abbreviations of units are used without a following period.
3. Be aware that the word *data* is plural while *datum* is singular. This affects the choice of a correct verb. The

word *species* is used both as a singular and as a plural. Never use the word *specie;* it refers to a coin not an organism.

4. Numbers should be written as numerals when they are greater than ten or when they are associated with measurements; for example, 6 mm or 2 g but *two* explanations or *six* factors. When one list includes numbers over and under ten, all numbers in the list may be expressed as numerals; for example, 17 sunfish, 13 bass, and 2 trout. Never start a sentence with numerals. Spell all numbers beginning sentences.

5. Be sure to divide paragraphs correctly and to use starting and ending sentences that indicate the purpose of the paragraph. A report or a section of a report should not be one long paragraph.

6. Every sentence must have a subject and a verb.

7. Avoid using the first person, I or we, in writing. Keep your writing impersonal, in the third person. Instead of saying, "We weighed the frogs and put them in a glass jar," write, "The frogs were weighed and put in a glass jar."

8. Avoid the use of slang and the overuse of contractions.

9. Be consistent in the use of tense throughout a paragraph—do not switch between past and present. It is best to use past tense.

10. Be sure that pronouns refer to antecedents. For example, in the statement, "Sometimes cecropia caterpillars are in cherry trees but they are hard to find," does "they" refer to caterpillars or trees?

After writing a report, read it over, watching especially for lack of precision and for ambiguity. Each sentence should present a clear message. The following examples illustrate lack of precision:

1. "The sample was incubated in mixture A minus B plus C." Does the mixture lack both B and C or lack B and contain C?

2. The title "Protection against Carcinogenesis by Antioxidants" leaves the reader wondering whether antioxidants protect from or cause cancer.

The only way to prevent such errors is to read and think about what you write. Learn to reread and edit your work.

Further Readings

CBE Style Manual Committee. 1994. *Scientific style and format: CBE style manual for authors, editors, and publishers.* 6th ed. New York: Cambridge University Press.

McMillan, V. E. 1997. *Writing papers in the biological sciences.* 2d ed. New York: St. Martin's Press, Inc.

Pechenik, J. A. 1997. *A short guide to writing about biology.* 3d ed. White Plains: Longman Publishing Group.

CREDITS

Illustrations

Chapter 3

3.9b: From John W. Hole, Jr., *Human Anatomy and Physiology*, 5th ed. Copyright © 1990 The McGraw-Hill Companies, Inc. Reprinted by permission. All Rights Reserved.

Chapter 4

4.1: From Ricki Lewis, et al., *Life*, 5th ed. Copyright © 2004 The McGraw-Hill Companies, Inc. Reprinted by permission. All Rights Reserved.

Chapter 6

6.1: From Burton S. Guttman, *Biology*, Copyright © 1999 The McGraw-Hill Companies, Inc. Reprinted by permission. All Rights Reserved. **6.2:** From George B. Johnson, *The Living World*, 3rd ed. Copyright © 2003 The McGraw-Hill Companies, Inc. Reprinted by permission. All Rights Reserved. **6.3:** From Ricki Lewis, et al., *Life*, 5th ed. Copyright © 2004 The McGraw-Hill Companies, Inc. Reprinted by permission. All Rights Reserved. **6.4:** From George B. Johnson, *The Living World*, 3rd ed. Copyright © 2003 The McGraw-Hill Companies, Inc. Reprinted by permission. All Rights Reserved. **6.5:** From Peter H. Raven and George B. Johnson, *Biology*, 6th ed. Copyright © 2002 The McGraw-Hill Companies, Inc. Reprinted by permission. All Rights Reserved. **6.6:** From Peter H. Raven and George B. Johnson, *Biology*, 6th ed. Copyright © 2002 The McGraw-Hill Companies, Inc. Reprinted by permission. All Rights Reserved. **6.7:** From Randy Moore, et al., *Botany*, 2nd ed. Copyright © 1998 The McGraw-Hill Companies, Inc. Reprinted by permission. All Rights Reserved. **6.8:** From Sylvia S. Mader, *Biology*, 8th ed. Copyright © 2004 The McGraw-Hill Companies, Inc. Reprinted by permission. All Rights Reserved. **6.9:** From Randy Moore, et al., *Botany*, 2nd ed. Copyright © 1998 The McGraw-Hill Companies, Inc. Reprinted by permission. All Rights Reserved. **6.10:** From George B. Johnson, *The Living World*, 3rd ed. Copyright © 2003 The McGraw-Hill Companies, Inc. Reprinted by permission. All Rights Reserved. **6.11:** From Peter H. Raven and George B. Johnson, *Biology*, 6th ed. Copyright © 2002 The McGraw-Hill Companies, Inc. Reprinted by permission. All Rights Reserved. **6.12:** From Randy Moore, et al., *Botany*, 2nd ed. Copyright © 1998 The McGraw-Hill Companies, Inc. Reprinted by permission. All Rights Reserved. **6.13:** From Randy Moore, et al., *Botany*, 2nd ed. Copyright © 1998 The McGraw-Hill Companies, Inc. Reprinted by permission. All Rights Reserved. **6.14:** From Peter H. Raven and George B. Johnson, *Biology*, 6th ed. Copyright © 2002 The McGraw-Hill Companies, Inc. Reprinted by permission. All Rights Reserved. **6.15:** From Randy Moore, et al., *Botany*, 2nd ed. Copyright © 1998 The McGraw-Hill Companies, Inc. Reprinted by permission. All Rights Reserved.

Chapter 9

9.1: From George B. Johnson, *The Living World*, 3rd ed. Copyright © 2003 The McGraw-Hill Companies, Inc. Reprinted by permission. All Rights Reserved.

Chapter 14

14.1: From Peter H. Raven and George B. Johnson, *Biology*, 6th ed. Copyright © 2002 The McGraw-Hill Companies, Inc. Reprinted by permission. All Rights Reserved.

Chapter 17

17.9: From Burton S. Guttman, *Biology*, Copyright © 1999 The McGraw-Hill Companies, Inc. Reprinted by permission. All Rights Reserved.

Chapter 18

18.3a: From Ricki Lewis, et al., *Life*, 5th ed. Copyright © 2004 The McGraw-Hill Companies, Inc. Reprinted by permission. All Rights Reserved. **18.5a:** From George B. Johnson, *The Living World*, 3rd ed. Copyright © 2003 The McGraw-Hill Companies, Inc. Reprinted by permission. All Rights Reserved. **18.8d:** From Ricki Lewis, et al., *Life*, 5th ed. Copyright © 2004 The McGraw-Hill Companies, Inc. Reprinted by permission. All Rights Reserved.

Chapter 19

19.6: From Darrell S. Vodopich and Randy Moore, Biology *Laboratory Manual*, 6th ed. Copyright © 2002 The McGraw-Hill Companies, Inc. Reprinted by permission. All Rights Reserved.

Chapter 21

21.4b: From Peter H. Raven and George B. Johnson, *Biology*, 6th ed. Copyright © 2002 The McGraw-Hill Companies, Inc. Reprinted by permission. All Rights Reserved. **21.10:** From Peter H. Raven and George B. Johnson, *Biology*, 6th ed. Copyright © 2002 The McGraw-Hill Companies, Inc. Reprinted by permission. All Rights Reserved.

Chapter 22

22.7: From Charles F. Lytle, *General Zoology Laboratory Guide*, 12th ed. Copyright © 1996 The McGraw-Hill Companies, Inc. Reprinted by permission. All Rights Reserved.

Chapter 24

24.4: From Sylvia S. Mader, *Inquiry Into Life*, 9th ed. Copyright © 2002 The McGraw-Hill Companies, Inc. Reprinted by permission. All Rights Reserved. **24.5:** From Randy Moore, et al., *Botany*, 2nd ed. Copyright © 1998 The McGraw-Hill Companies, Inc. Reprinted by permission. All Rights Reserved. **24.9:** From Ricki Lewis, et al., *Life*, 5th ed. Copyright © 2004 The McGraw-Hill Companies, Inc. Reprinted by permission. All Rights Reserved. **24.10c:** From Peter H. Raven and George B. Johnson, *Biology*, 5th ed. Copyright © 1999 The McGraw-Hill Companies, Inc. Reprinted by permission. All Rights Reserved.

Chapter 25

25.1: From Burton S. Guttman, *Biology*, Copyright © 1999 The McGraw-Hill Companies, Inc. Reprinted by permission. All Rights Reserved.

Chapter 26

26.4: From Kingsley R. Stern, *Introductory Plant Biology*, 6th ed. Copyright © 1994 The McGraw-Hill Companies, Inc. Reprinted by permission. All Rights Reserved.

Chapter 28

28.8: From Burton S. Guttman, *Biology*, Copyright © 1999 The McGraw-Hill Companies, Inc. Reprinted by permission. All Rights Reserved.

Chapter 29

29.1: From Kent M. Van De Graaff and Stuart Ira Fox, *Concepts of Human Anatomy and Physiology*, 4th ed. Copyright © 1995 The McGraw-Hill Companies, Inc. Reprinted by permission. All Rights Reserved. **29.7a:** From Kent M. Van De Graaff, *Synopsis of Human Anatomy and Physiology*. Copyright © 1997 The McGraw-Hill Companies, Inc. Reprinted by permission. All Rights Reserved. **29.7b:** From David Shier, Jackie Butler, and Ricki Lewis, *Hole's Human Anatomy and Physiology*, 7th ed. Copyright © 1996 The McGraw-Hill Companies, Inc. Reprinted by permission. All Rights Reserved. **29.8:** From Kent M. Van De Graaff, *Human Anatomy*, 4th ed. Copyright © 1995 The McGraw-Hill Companies, Inc. Reprinted by permission. All Rights Reserved. **29.13:** From Sylvia S. Mader, *Human Biology*, 8th ed. Copyright © 2004 The McGraw-Hill Companies, Inc. Reprinted by permission. All Rights Reserved.

Chapter 30

30.4: From Kent M. Van De Graaff and Stuart Ira Fox, *Concepts of Human Anatomy and Physiology*, 4th ed. Copyright © 1995 The McGraw-Hill Companies, Inc. Reprinted by permission. All Rights Reserved. **30.5:** From Burton S. Guttman, *Biology*, Copyright © 1999 The McGraw-Hill Companies, Inc. Reprinted by permission. All Rights Reserved.

Chapter 31

31.1: From Kent M. Van De Graaff and Stuart Ira Fox, *Concepts of Human Anatomy and Physiology*, 4th ed. Copyright © 1995 The McGraw-Hill Companies, Inc. Reprinted by permission. All Rights Reserved. **31.2:** From Ricki Lewis, et al., *Life*, 5th ed. Copyright © 2004 The McGraw-Hill Companies, Inc. Reprinted by permission. All Rights Reserved. **31.3:** From Sylvia S. Mader, *Biology*, 8th ed. Copyright © 2004 The McGraw-Hill Companies, Inc. Reprinted by permission. All Rights Reserved. **31.8a:** From David Shier, Jackie Butler, and Ricki Lewis, *Hole's Human Anatomy and Physiology*, 7th ed. Copyright © 1996 The McGraw-Hill Companies, Inc. Reprinted by permission. All Rights Reserved. **31.11:** From Kent M. Van De Graaff, *Human Anatomy*, 4th ed. Copyright © 1995 The McGraw-Hill Companies, Inc. Reprinted by permission. All Rights Reserved.

Chapter 32

32.3: From David Shier, Jackie Butler, and Ricki Lewis, *Hole's Human Anatomy and Physiology*, 7th ed. Copyright © 1996 The McGraw-Hill Companies, Inc. Reprinted by permission. All Rights Reserved. **32.10:** From Kent M. Van De Graaff and Stuart Ira Fox, *Concepts of Human Anatomy and Physiology*, 4th ed. Copyright © 1995 The McGraw-Hill Companies, Inc. Reprinted by permission. All Rights Reserved.

Illustrators

Dean Biechler: 1.1, 1.2, 3.6, 3.7B, 4.3, 4.4, 8.1, 8.3, 11.1, 15.12, 16.10, 17.10, 17.11, 18.2, 18.3A, 18.4, 18.5A, 18.7A, 19.2, 19.7, 20.4, 20.5, 20.6, 20.7, 20.9, 20.10A, 20.11, 21.7, 22.1, 22.2, 22.5, 22.8, 22.9, 23.3B, 23.4B&C, 26.1B, 26.3, 26.6, I-1, 29.2, 29.3, 29.4, 29.6, 31.6, 31.7, 32.4, 32.9, 32.11A&B;

Karla Burds: 28.1, 28.2, 30.2, 30.3, 31.5; **Chris Peterson:** 16.5, 16.9, 23.2, 27.3; **Meredith McLain:** 12.4, 12.5, 13.2, 23.3C, 25.8; **Ann Mackey-Weiss:** 27.1, 27.2, 31.10; **Margaret Hunter:** 3.5, 20.3; **Shawn Gould:** 18.9, 32.5; **Mike Gipple:** 4.2.

Photos

Chapter 2

2.2: Leica, Inc.; **2.7:** © Cabisco/Phototake; **2.8:** © Inga Spence/Visuals Unlimited; **2.9:** Courtesy Joseph Viles, Iowa State University; **2.10:** © E.H. Newcomb & W.P. Wergin/Biological Photo Service.

Chapter 3

3.1a-c: © David M. Phillips/Visuals Unlimited; **3.2:** © John J. Cardamone, Jr./Biological Photo Service; **3.7a:** © Ed Reschke; **3.8:** © Dennis Strete, photographer/The McGraw-Hill Companies, Inc.; **3.9a, 3.10:** © Ed Reschke.

Chapter 4

4.6: © M. Abbey/Visuals Unlimited.

Chapter 5

5.2: Courtesy Brinkman Instruments, Inc.; **5.5a,c:** Courtesy of Bausch & Lomb; **5.5b:** Warren Dolphin.

Chapter 6

6.8b: © CNRI/SPL/Photo Researchers, Inc.

Chapter 9

9.2a(1-4), 9.2b: © Kingsley Stern, photographer/The McGraw-Hill Companies, Inc.

Chapter 10

10.2, 10.4a-g: © Cabisco/Visuals Unlimited; **10.5b:** © Fred Hostler/Visuals Unlimited; **10.6:** © Cabisco/Phototake.

Chapter 12

12.1: © K.G. Marti/Visuals Unlimited; **12.2:** © David Dresser/Huntington Potter/Tiepin/Getty.

Chapter 15

15.1: © E. Whitney/Visuals Unlimited; **15.2:** © David M. Phillips/Photo Researchers, Inc.; **15.4, 15.6a:** © Carolina Biological/Visuals Unlimited; **15.6b:** Courtesy of Dr. Yuuji Tsukii, Hosei University (Protist Information Server, *http://protist.i.hosei.ac.jp/Protist_menuE.html*); **15.6c:** © Carolina Biological/Visuals Unlimited; **15.6d:** © Nalco Chemical Co./Fundamental Photographs; **15.6e:** © Roland Birke/Peter Arnold Inc; **15.6f:** © Bruce Russell/BioMedia Associates; **15.7:** © John D. Cunningham/Visuals Unlimited; **15.9:** © Douglas P. Wilson; Frank Lane Picture Agency/CORBIS; **15.10a-d, 15.14:** © Cabisco/Visuals Unlimited; **15.15a:** © Will Troyer/Visuals Unlimited; **15.16a:** © Phil A. Harrington/Peter Arnold Inc.; **15.16b:** © Eric V. Grave/Photo Researchers, Inc.; **15.16c:** © Wim Van Egmond; **15.16d-e:** Courtesy of Dr. Yuuji Tsukii, Hosei University (Protist Information Server, http://protist.i.hosei.ac.jp/Protist_menuE.html).

Chapter 16

16.3a: © Heather Angel; **16.3b:** © Cabisco/Phototake; **16.4:** © Runk/Schoenberger/Grant Heilman; **16.7a:** © David S. Addison/Visuals Unlimited; **16.7b:** © Runk/Schoenberger/Grant Heilman; **16.7c:** © Ed Reschke; **16.8a-b:** © John D. Cunningham/Visuals Unlimited; **16.8c:** © S. Elms/Visuals Unlimited.

Chapter 17

17.2a: © George Loun/Visuals Unlimited; **17.2b:** © Runk/Schoenberger/Grant Heilman; **17.2c:** © Science/Visuals Unlimited; **17.3:** © John D. Cunningham/Visuals Unlimited; **17.4:** © Jack M. Bostrack/Visuals Unlimited; **17.5a:** © Doug Sokell/Visuals Unlimited; **17.5b:** © Dr. John Cunningham/Visuals Unlimited; **17.5c:** © George J. Wilder/Visuals Unlimited; **17.7a-b:** © Cabisco/Visuals Unlimited.

Chapter 18

18.1: © Biophoto Associates/Photo Researchers, Inc.; **18.3b:** © Runk/Schoenberger/Grant Heilman Photography; **18.3c:** © Ed Reschke; **18.3d:** © Bruce Iverson/Visuals Unlimited; **18.5b:** © James Richardson/Visuals Unlimited; **18.5c:** © Ed Degginger/Color-Pic, Inc.; **18.5d:** © Doug Sherman/Geofile; **18.6a:** © William Ormerod/Visuals Unlimited; **18.6b:** © Richard Thom/Visuals Unlimited; **18.6c:** © Bill Keogh/Visuals Unlimited; **18.7b:** © Biophoto Associates; **18.8a:** © Stephen Kraseman/Peter Arnold, Inc.; **18.8b:** © Bob Ross/RARE Photographer; **18.8c:** © John Shaw/Tom Stack & Associates; **18.8e:** © V. Ahmadjian/Visuals Unlimited.

Chapter 19

19.1a-b: © R. Calentine/Visuals Unlimited; **19.1c-h, 19.3a-d, 19.8:** © Cabisco/Visuals Unlimited; **19.9:** © Peter Parks/Oxford Scientific Films.

Chapter 20

20.10c: Courtesy Dr. Fred Whittaker; **20.10d:** © Ed Reschke; **20.12a-b:** © Stan W. Elms/Visuals Unlimited.

Chapter 21

21.2a: Courtesy of David Barnes; **21.2b:** © Dr. John Cunningham/Visuals Unlimited; **21.3a:** © Runk/Schoenberg/Grant Heilman; **21.3b:** © Kjell Sandved/Visuals Unlimited; **21.3c:** © Robert & Linda Mitchell; **21.4a:** Courtesy of University of Saskatchewan Archives; **21.6a-b:** Courtesy of Dr. Clyde Herreid, University at Buffalo, SUNY; **21.11a:** © Ray Coleman/Photo Researchers, Inc.; **21.11b:** © Daniel W. Gotshall/Visuals Unlimited; **21.11c:** © Mindy E. Klarman/Photo Researchers, Inc.; **21.14:** © John D. Cunningham/Visuals Unlimited; **21.15a-b:** Courtesy of Dr. Clyde Herreid, University at Buffalo, SUNY.

Chapter 22

22.6: © Ken Taylor.

Chapter 23

23.1a: © Michael DiSpezio; **23.1b:** © Daniel Gotshall/Visuals Unlimited; **23.1c:** © Carl Toessler/Tom Stack & Associates; **23.1d:** © Daniel Gotshall/Visuals Unlimited; **23.1e:** © Bill Ober/Visuals Unlimited; **23.1f:** © Museum of New Zealand TePapa Tongarewa; **23.3b:** © Michael DiSpezio; **23.4b-c:** © John D. Cunningham/Visuals Unlimited.

Chapter 24

24.1: © Harry Horner, Iowa State University; **24.2:** © Biophoto Associates/Photo Researchers; **24.3:** © Biophoto Associates/Photo Researchers; **24.5c:** Courtesy Eva Frei and R.D. Preston;

24.6: © John D. Cunningham/Visuals Unlimited; **24.7:** © Biophoto Associates/Photo Researchers, Inc.; **24.8c:** © Bruno P. Zahnder/Peter Arnold, Inc.; **24.8a:** © Biophoto Associates/Photo Researchers, Inc.; **24.9a:** © John D. Cunningham/Visuals Unlimited; **24.8b:** © Randy Moore/Visuals Unlimited; **24.9b:** © George J. Wilder/Visuals Unlimited; **24.8d:** © Kingsley R. Stern; **24.10a:** © Randy Moore/Visuals Unlimited; **24 10b:** © George J. Wilder/Visuals Unlimited.

Chapter 25

25.2: © Jeremy Burgess/SPL/Photo Researchers, Inc.; **25.4:** © BioPhot; **25.6:** © Cabisco/Phototake; **25.7:** © John D. Cunningham/Visuals Unlimited; **25.9:** © Jack M. Bostrack/Visuals Unlimited; **25.10a:** © Cabisco/Phototake; **25.11a:** Courtesy John Troughton and Lesley Donaldson; **25.11b:** © Cabisco/Phototake; **25.12:** © Ed Reschke; **25.14a:** © Cabisco/Phototake; **25.14b:** © Runk/Schoenberger/Grant Heilman Photography.

Chapter 26

26.1a: © W.P. Wergin & E.H. Newcomb/Biological Photo Service; **26.6:** © Nancy Vander Sluis.

Chapter 27

27.4: © Cabisco/Visuals Unlimited; **27.5:** © Ed Reschke; **27.6a, 27.8a-c:** © Cabisco/Visuals Unlimited; **27.11a-c:** © William H. Allen, Jr.

Chapter 28

28.4: © Ken Taylor; **28.5a:** © Lester V. Bergman/CORBIS; **28.5b:** From R. G. Kessel and R. H. Kardon, *Tissues and Organs: A Text-Atlas of Scanning Electron Microscopy*, 1979, W. H. Freeman and Company; **28.5c:** © Keith Porter; **28.6c:** © Thomas Eisner; **28.9:** Courtesy Gregory J. Highison, provided by Frank N. Low.

Chapter 29

29.5: © Ken Taylor; **29.9:** © Warren Rosenberg, Iona College/Biological Photo Service; **29.10:** From R. G. Kessel and R. H. Kardon, *Tissues and Organs: A Text-Atlas of Scanning Electron Microscopy*, 1979, W. H. Freeman and Company; **29.11a-b:** © Ed Reschke.

Chapter 30

30.8: Courtesy of William Radke; **30.9:** From W. Bloom and D.S. Fawcett, *A Textbook of Histology*, 10th edition, Fig. 33.2, p. 859. Copyright © 1975 Chapman & Hall.

Chapter 31

31.2a: © Ed Reschke; **31.2b:** © Manfred Kage/Peter Arnold, Inc.; **31.2c:** © Ed Reschke; **31.3:** © Janzo Desaki, Ehime University School of Medicine, Shigenobu, Japan; **31.4:** Courtesy of Hugh E. Huxley; **31.8b:** Courtesy of John W. Hole; **31.8c:** From R. G. Kessel and R. H. Kardon, *Tissues and Organs: A Text-Atlas of Scanning Electron Microscopy*, 1979, W. H. Freeman and Company.

Chapter 32

32.2: © Biodisc/Visuals Unlimited; **32.6:** Courtesy Theron O. Odlaug; **32.7:** Courtesy Theron O. Odlaug.